PART I

STUDENT'S SOLUTIONS MANUAL

CALCULUS AND ANALYTIC GEOMETRY 7TH

THOMAS/FINNEY

ALEXIA B. LATIMER
BENITA H. ALBERT
JUDITH BROADWIN

ADDISON-WESLEY PUBLISHING COMPANY, INC.
Reading, Massachusetts · Menlo Park, California · New York
Don Mills, Ontario · Wokingham, England · Amsterdam · Bonn
Sydney · Singapore · Tokyo · Madrid · Bogotá · Santiago · San Juan

CONTENTS

Reproduced by Addison-Wesley from camera-ready copy supplied by the authors.

ISBN 0-201-16325-X
BCDEFGHIJ-BA-898

CHAPTER 1

THE RATE OF CHANGE OF A FUNCTION

1.1 COORDINATES FOR THE PLANE

	P(x,y)	Q(x,-y)	R(-x,y)	S(-x,-y)	T(y,x)
1.	(1,-2)	(1,2)	(-1,-2)	(-1,2)	(-2,1)
3.	(-2,2)	(-2,-2)	(2,2)	(2,-2)	(2,-2)
5.	(0,1)	(0,-1)	(0,1)	(0,-1)	(1,0)
7.	(-2,0)	(-2,0)	(2,0)	(2,0)	(0,-2)
9.	(-1,-3)	(-1,3)	(1,-3)	(1,3)	(-3,-1)
11.	$(-\pi,-\pi)$	$(-\pi,\pi)$	$(\pi,-\pi)$	(π,π)	$(-\pi,-\pi)$
13.	(x,y)	(x,-y)	(-x,y)	(-x,-y)	(y,x)

15.

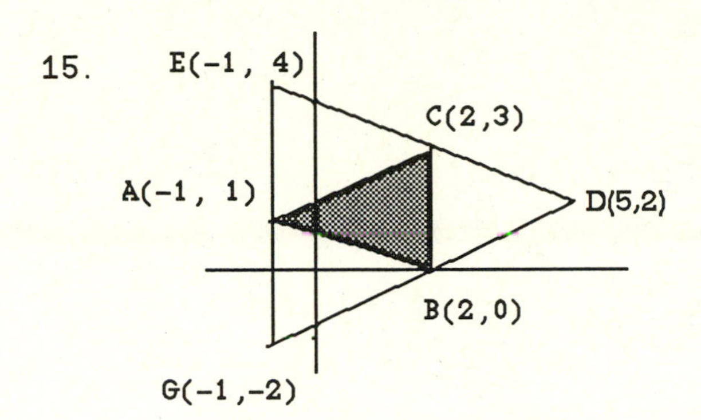

BC = 3 units.

In ▱ACBG, G (−1, 1 − 3) = G(−1, −2)

and in ▱ABCE, E(−1, 1 + 3) = E(−1, 4).

$m_{AD} = \frac{-1}{3}$.

In ▱ABDC, D(3 − 1, 2 + 3) = D(5, 2)

17. l = 3w and 2l + 2w = 56. Therefore, l = 21 and w = 7. The coordinates are A(-12,2), B(-12,-5) and C(9,-5).

19. b = 2

1.2 THE SLOPE OF A LINE

1. $\Delta x = 1 - (-1) = 2$ $\quad \Delta y = 2 - 1 = 1$

3. $\Delta x = -1 - (-3) = 2$ $\quad \Delta y = -2 - 2 = -4$

5. $\Delta x = -8 - (-3) = -5$ $\quad \Delta y = 1 - 1 = 0$

7. (a) $\Delta x = 57 - 0 = 57$, $\Delta y = 22 - 32 = -10$ (b) $\Delta x = 26 - 28 = -2$, $\Delta y = 6 - 18 = -12$ (c) $\Delta x = 40 - 39 = 1$, $\Delta y = 4 - 18 = -14$

9. right, above 11. left 13. right, below 15. below

17. $m = \frac{-2-1}{1-2} = 3$; $m_\perp = -\frac{1}{3}$

19. $m = \frac{-1-(-2)}{-2-1} = -\frac{1}{3}$; $m_\perp = 3$

21. $m = \frac{0-1}{1-0} = -1$; $m_\perp = 1$

23. $m = \frac{3-3}{2-(-1)} = 0$; $m_\perp$ is undefined

25. $m = \frac{-2-0}{0-(-2)} = -1$; $m_\perp = 1$

27. $m = \frac{-4-0}{-2-0} = 2$; $m_\perp = -\frac{1}{2}$

29. $m = \frac{y-0}{x-0} = \frac{y}{x}$; $m_\perp = \frac{x}{y}$ if $y \neq 0$, undefined if $y = 0$

31. $m =$ undefined; $m_\perp = 0$

33. $\tan\alpha = \frac{5}{16} \Rightarrow \alpha = 17.35^\circ$; $\tan\beta = \frac{9}{8} \Rightarrow \beta = 48.37^\circ$

35. $\tan 40^\circ = 0.8391$

37. Fiberglass is the best with -16°/in.

Gypsum is the poorest with -3°/in.

39. (a) A(1,0); B(0,1); C(2,1)

$m_{AB} = -1$; $m_{AC} = 1$ ∴ not collinear

(b) A(-2,1); B(0,5); C(-1,2)

$m_{AB} = \frac{4}{2} = 2$; $m_{AC} = \frac{1}{1} = 1$ ∴ not collinear

(c) A(-2,1); B(-1,1); C(1,5); D(2,7)

$m_{AB} = 0$; $m_{BC} = 2$; $m_{CD} = 2$; $m_{BD} = \frac{6}{3} = 2$ ∴ B,C,D collinear

(d) A(-2,3); B(0,2); C(2,0)

$m_{AB} = \frac{1}{-2}$; $m_{BC} = -1$ ∴ A,B,C not collinear

(e) A(-3,-2); B(-2,0); C(-1,2); D(1,6)

$m_{AB} = \frac{-2}{-1} = 2$; $m_{BC} = \frac{2}{1} = 2$; $m_{CD} = \frac{4}{2} = 2$ ∴ A,B,C,D collinear

41. A(-2,3): $x = -2 + \Delta x = -2 + 5 = 3$; $y = 3 - \Delta x = 3 - 6 = -3$
New position is P(3,-3)

43. Moves from A(x,y) to B(3,-3) with $\Delta x = 5$ and $\Delta y = 6$
Original $x = 3 - 5 = -2$; original $y = -3 - 6 = -9$

1.3 EQUATIONS FOR LINES

1. Vertical: $x = 2$ Horizontal: $y = 3$

3. Vertical: $x = 0$ Horizontal: $y = 0$

5. Vertical: $x = -4$ Horizontal: $y = 0$

7. Vertical: $x = 0$ Horizontal: $y = b$

9. $y - 1 = x - 2 \Rightarrow y = x$

11. $y - 1 = x + 1 \Rightarrow y = x + 2$

13. $y - b = 2x \Rightarrow y = 2x + b$

15. $m = \frac{3}{2}$; $y = \frac{3}{2}x$

17. $x = 1$

19. $x = -2$

21. $m = \frac{-F_o}{T}$ $y = \frac{-F_o}{T}(x - T) \Rightarrow Ty = -F_o x + TF_o$
$F_o x + Ty = TF_o$

23. $x = 0$

25. $m = \frac{.7}{-2.1} = \frac{-1}{3} \Rightarrow y - 1.5 = \frac{-1}{3}(x + 7) \Rightarrow 15y = -5x + 19$

27. $m = \frac{y_1 - y_0}{x_1 - x_0}$ $y - y_1 = \frac{y_1 - y_0}{x_1 - x_0}(x - x_1)$

29. $y = 3x - 2$

31. $y = x + \sqrt{2}$

33. $y = -5x + 2.5$

35. $y = 3x + 5$: $(-\frac{5}{3}, 0)$ $(0,5)$; $m = 3$

37. $x + y = 2$: $(2,0)$ $(0,2)$; $m = -1$

39. $x - 2y = 4$: $(4,0)$ $(0,-2)$; $m = \frac{1}{2}$

41. $4x - 3y = 12$: $(3,0)$ $(0,-4)$; $m = \frac{4}{3}$

43. $\frac{x}{3} + \frac{y}{4} = 1$: $(3,0)$ $(0,4)$; $m = \frac{-4}{3}$

45. $\frac{x}{2} - \frac{y}{3} = 1$; $(-2,0)$ $(0,3)$; $m = \frac{3}{2}$

47. $105x - 35y = 700 \Rightarrow 105x - 700 = 35y \Rightarrow 3x - 20 = y$

$(0,-20)$ $(\frac{20}{3},0)$ $m = \frac{20}{\frac{20}{3}} = 3$

49. $\frac{x}{a} + \frac{y}{b} = 1$ $(0,b)$ $(a,0)$ $m = \frac{-b}{a}$

51. $m = \frac{-\sqrt{3}}{3}$; $y - 1 = -\frac{\sqrt{3}}{3}x \Rightarrow 3y - 3 = -x\sqrt{3} \Rightarrow 3y - 3 = -x\sqrt{3}$

53. $P(2,1)$ L: $y = x + 2$

$\parallel$: $y - 1 = x - 2 \Rightarrow y = x - 1$

$\perp$: $y - 1 = -(x - 2) \Rightarrow y = -x + 3$

$d = \frac{|ax_1 + by_1 - c|}{\sqrt{a^2 + b^2}} = \frac{|2 - 1 + 2|}{\sqrt{2}} = \frac{3}{\sqrt{2}}$ (See Misc. Ex. 42)

55. $P(0,0)$, L: $y\sqrt{3} = -x + 3 \Rightarrow y = -\frac{1}{\sqrt{3}}x + \sqrt{3}$

$\parallel$: $y = \frac{-1}{\sqrt{3}}x$

$\perp$: $y = x\sqrt{3}$

$d = \frac{|-3|}{\sqrt{1 + 3}} = \frac{3}{2}$

57. $P(-2,2)$, L: $y = -2x + 4$

$\parallel$: $y - 2 = -2(x + 2) \Rightarrow y = -2x - 2$

$\perp$: $y - 2 = \frac{1}{2}(x + 2) \Rightarrow y = \frac{1}{2}x + 3$

$2y = x + 6$

$-6 = x - 2y$

$d = \frac{|2 - 4 - 4|}{\sqrt{1 + 4}} = \frac{6}{\sqrt{5}}$

59. $P(1,0)$, L: $2x - y = -2 \Rightarrow 2x + 2 = y$

$\parallel$: $y = 2(x - 1) \Rightarrow y = 2x - 2$

$\perp$: $y = \frac{-1}{2}(x - 1) \Rightarrow x + 2y = 1$

$d = \frac{|2 + 2|}{\sqrt{5}} = \frac{4}{\sqrt{5}}$

61. $P(3,2)$, L: $x = -5$

$\parallel$: $x = 3$ $\perp$: $y = 2$ $d = 8$

63. $P(a,b)$, L: $x = -1$

$\parallel$: $x = a$ $\perp$: $y = b$ $d = |a + 1|$

65. P(4,6). L: $3y = -4x + 12 \Rightarrow y = \frac{-4}{3}x + 4$

$\parallel$: $y - 6 = \frac{-4}{3}(x - 4) \Rightarrow y = \frac{-4}{3}x + \frac{34}{3}$

$\perp$: $y - 6 = \frac{3}{4}(x - 4) \Rightarrow y = \frac{3}{4}x + 3$

$d = \frac{|16 + 18 - 12|}{5} = \frac{22}{5}$

67. $y = x + 2 \qquad m = 1 \Rightarrow \phi = 45^\circ = \frac{\pi}{4}$ radians

69. $y\sqrt{3} = -x + 3$

$y = \frac{-1}{\sqrt{3}}x + \sqrt{3}, \quad m = \frac{-1}{\sqrt{3}} \Rightarrow \phi = 150^\circ = \frac{5\pi}{6}$ radians

71. $y = -2x + 4$; $m = -2 = \tan\phi \Rightarrow \phi = 180° - \tan^{-1} 2$ or 116.6^0

73. $3y = -4x + 12$; $m = \frac{-4}{3}$; $\alpha \approx 180^0 - \text{Arctan}\,\frac{4}{3}$ or 126.9^0

75. $m = \tan\phi = \sqrt{3} \Rightarrow y - 4 = \sqrt{3}(x - 1)$ or $y = x\sqrt{3} + (4 - \sqrt{3})$

77. $x = -2$

79. $p = kd + 1$

$d = 100 \Rightarrow p = 10.94$ so $10.94 = 100k + 1$ or $k = \frac{9.94}{100}$

$k = .0994$

$p = (.0994)d + 1$

$p = (.0994)(50) + 1 \approx 5.97$ atm

81. $m = \frac{35 - 35.16}{65 - 135} = \frac{-.16}{-70} = \frac{.16}{70}$

$s - 35 = \frac{.16}{70}(t - 65)$

$70s - 2450 = .16t - 10.4$

$70s = .16t + 2439.6$

$s = 0.0023t + 34.85$

83. $\sin 37.1 = \frac{14}{z} \Rightarrow z = \frac{14}{\sin 37.1} \approx 23$ ft

1.4 FUNCTIONS AND GRAPHS

1. $y = 2\sqrt{x}$ D: $x \geq 0$ R: $y \geq 0$

3. $y = -\sqrt{x}$ D: $x \geq 0$ R: $y \leq 0$

5. $y = \sqrt{x + 4}$ D: $x \geq -4$ R: $y \leq 0$

7. $y = \frac{1}{x - 2}$ D: $x \in R$, $x \neq 2$ R: $y \neq 0$

9. $y = 2\cos x$ D: $x \in R$ R: $-2 \leq y \leq 2$

11. $y = -3\sin x$ D: $x \in R$ R: $-3 \leq y \leq 3$

13. $y = x^2 + 1$
 (a) D: $x \varepsilon R$
 (b) R: $y \geq 1$

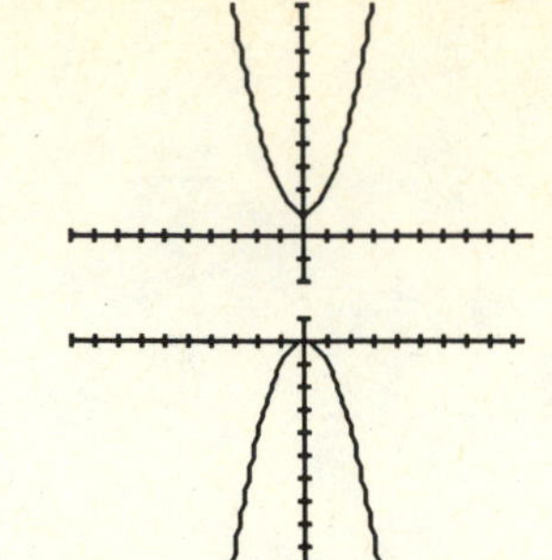

15. $y = -x^2$
 (a) D: $x \varepsilon R$
 (b) R: $y \leq 0$

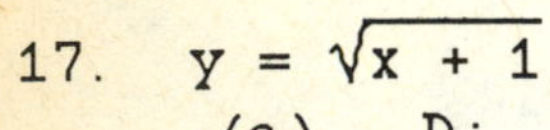

17. $y = \sqrt{x + 1}$
 (a) D: $x \geq -1$
 (b) R: $y \geq 0$

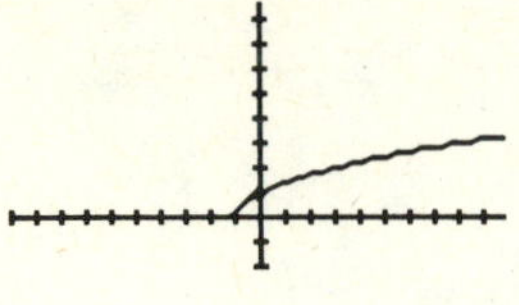

19. $y = 1 + \sqrt{x}$
 (a) D: $x \geq 0$
 (b) R: $y \geq 1$

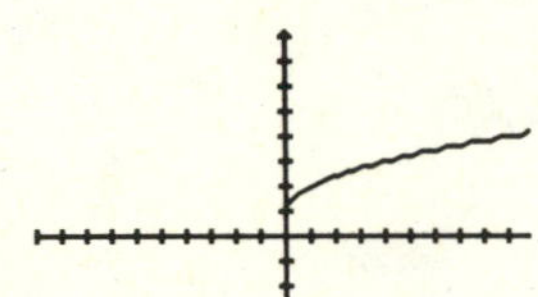

21. $y = (\sqrt{2x})^2$
 (a) D: $x \geq 0$
 (b) R: $y \geq 0$

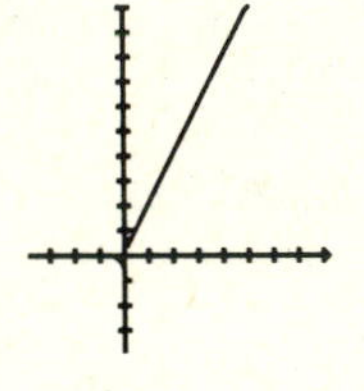

23. $y = \frac{-1}{x}$
 (a) D: $x \varepsilon R,\ x \neq 0$
 ((b) R: $y \varepsilon R,\ y \neq 0$

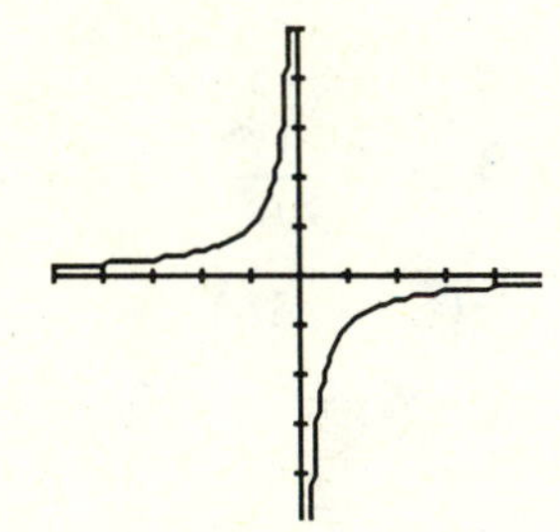

25. $y = \sin 2x$
 (a) D: $x \varepsilon R$
 (b) R: $-1 \leq y \leq 1$

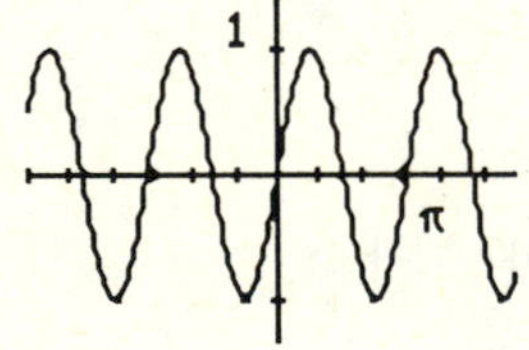

27. $y = \sin^2 x$
 (a) D: $x \varepsilon R$
 (b) R: $0 \leq y \leq 1$

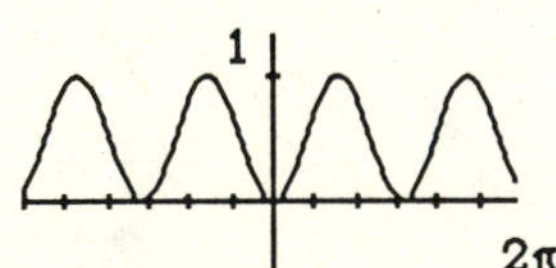

29. $y = 1 + \sin x$
 (a) D: $x \varepsilon R$
 (b) R: $0 \leq y \leq 2$

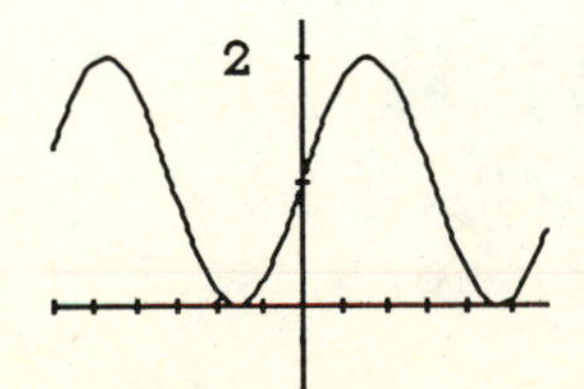

31. $y = \frac{1}{\sqrt{x}}$ (a) no, $x > 0$ (b) no, $x \neq 0$ (c) $x > 0$

33. $f(x) = \sqrt{\frac{1}{x} - 1}$

$\frac{1}{x} - 1 \geq 0 \Leftrightarrow \frac{1 - x}{x} \geq 0$

(a) no (b) no (c) no (d) $0 < x \leq 1$

35. $y = \tan \frac{x}{2}$

(a) $\frac{x}{2} \neq \pm \frac{\pi}{2}, \pm \frac{3\pi}{2}, \ldots, \pm \frac{(2n-1)\pi}{2}$, n a positive integer

(b) $x \neq \pm\pi, \pm 3\pi, \pm 5\pi, \ldots, \pm (2n - 1)\pi$, n a positive integer

(c) $-\infty < y < \infty$

(d) D: $x \neq \pm\pi, \pm 3\pi, \ldots, \pm(2n - 1)\pi$ R: $-\infty < y < \infty$

37. $y = 4x^2$. Since $y \geq 0$ it is not (i); $f(0) = 0$ so cannot be (iii) or (iv)

39.

x	f(x)
0	0
1	1
2	0

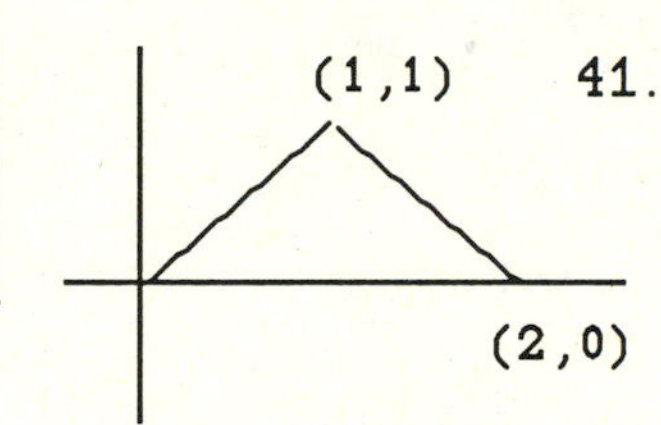

41. $y = \begin{cases} \frac{1}{x} & x < 0 \\ x & x > 0 \end{cases}$

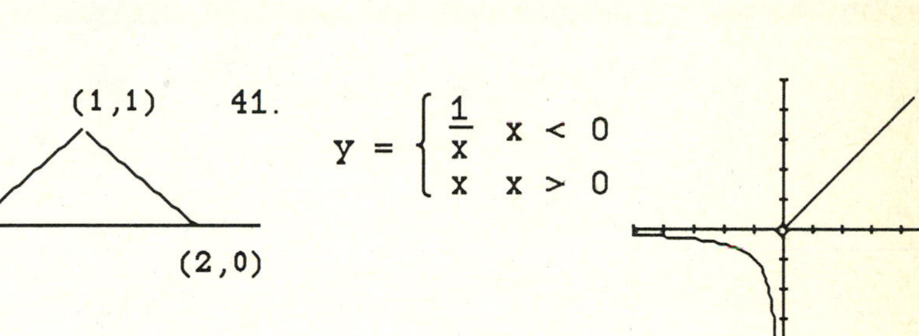

43. $y = \begin{cases} 1 & x < 0 \\ \sqrt{x} & x > 0 \end{cases}$

45. $f(x) = x$ $g(x) = \sqrt{x - 1}$ D_f: $x \in R$ D_g: $x \geq 1$

D_{f+g}: $x \geq 1$ $D_{f/g}$: $x > 1$

D_{f-g}: $x \geq 1$ D_{fg}: $x \geq 1$

$D_{g/f}$: $x \geq 1$

47. $f(x) = \sqrt{x}$ $g(x) = \sqrt{x + 1}$ D_f: $x \geq 0$ D_g: $x \geq -1$

$D_{f+g} = D_{f-g} = D_{fg}$: $x \geq 0$

$D_{f/g}$: $x \geq 0$ $D_{g/f}$: $x > 0$

49. $h = 1 + \frac{5}{x}$

(a) $h(-1) = -4$ (b) $h(\frac{1}{2}) = 11$ (c) $h(5) = 2$

(d) $h(5x) = 1 + \frac{1}{x}$ (e) $h(10x) = 1 + \frac{1}{2x}$

(f) $h(\frac{1}{x}) = 1 + 5x,\ x \neq 0$

51. $f(x) = \frac{x - 1}{x}$ $f(1 - x) = \frac{-x}{1 - x}$

$f(x)\cdot f(1 - x) = \frac{x - 1}{x} \cdot \frac{-x}{1 - x} = 1$

53. $F(t) = 4t - 3$

$\frac{F(t + h) - F(t)}{h} = \frac{4(t + h) - 3 - 4t + 3}{h} = \frac{4h}{h} = 4$

1.5 ABSOLUTE VALUES

1. $|-3| = 3$

3. $|-2 + 7| = 5$

5. $|x| = 2 \Rightarrow x = \pm 2$

7. $|2x + 5| = 4 \Rightarrow 2x + 5 = \pm 4 \Rightarrow 2x = -1$ or $2x = -9 \Rightarrow x = -\frac{1}{2}$ or $-\frac{9}{2}$

9. $|8 - 3x| = 9 \Rightarrow 8 - 3x = \pm 9 \Rightarrow 17 = 3x$ or $-1 = 3x \Rightarrow x = \frac{17}{3}$ or $\frac{-1}{3}$

11. $|x| < 4 \Rightarrow -4 < x < 4$, f

13. $|x - 5| < 2 \Rightarrow -2 < x - 5 < 2 \Rightarrow 3 < x < 7$, c

15. $|1 - x| < 2 \Rightarrow -2 < 1 - x < 2 \Rightarrow -3 < -x < 1 \Rightarrow -1 < x < 3$, b

17. $|2x + 4| < 1 \Rightarrow -1 < 2x + 4 < 1 \Rightarrow -5 < 2x < -3 \Rightarrow \frac{-5}{2} < x < \frac{-3}{2}$, d

19. $|\frac{2x + 1}{3}| < 1 \Rightarrow -3 < 2x + 1 < 3 \Rightarrow -4 < 2x < 2 \Rightarrow -2 < x < 1$, a

21. $|x| < 2 \Rightarrow -2 < x < 2$

23. $|x - 1| \leq 2 \Rightarrow -2 \leq x - 1 \leq 2 \Rightarrow -1 \leq x \leq 3$

25. $|x + 1| < 3 \Rightarrow -3 < x + 1 < 3 \Rightarrow -4 < x < 2$

27. $|2x + 2| < 1 \Rightarrow -1 < 2x + 2 < 1 \Rightarrow -3 < 2x < -1 \Rightarrow \frac{-3}{2} < x < \frac{-1}{2}$

29. $|1 - 2x| \leq 1 \Rightarrow -1 \leq 1 - 2x \leq 1 \Rightarrow -2 \leq -2x \leq 0 \Rightarrow 0 \leq x \leq 1$

31. $\frac{1}{|x|} \leq 1 \Rightarrow 1 \leq |x| \Rightarrow x \geq 1$ or $x \leq -1$

33. $|x| < 8$

35. $-5 < x < 1$; the midpoint of the segment is -2; the distance to either endpoint is 3; the interval is open. $\therefore |x + 2| < 3$

37. $|y| < a$

39. $|y - L| < \varepsilon$

41. $|x - x_0| < 5$

43. $|1 - x| = 1 - x$ when $x \le 1$; $|1 - x| = x - 1$ when $x \ge 1$

45. $f(x) = -|x|$

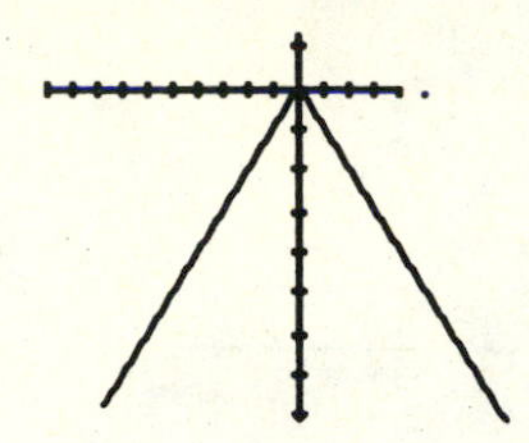

47. $f(x) = \frac{|x| - x}{2} = \begin{cases} 0 & x \ge 0 \\ -x & x < 0 \end{cases}$

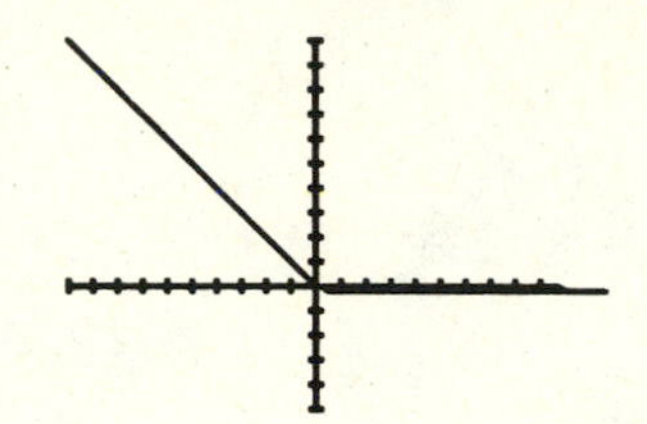

49. $y = \sqrt{x^2} = |x|$ has domain D: $x \in R$ and range R: $y \ge 0$
$y = (\sqrt{x})^2 = x,\ x \ge 0$ has domain D: $x \ge 0$ and range R: $y \ge 0$

51. $f(x) = x^2 + 2x + 1 = (x + 1)^2$ and $g(f(x)) = |x + 1|$ $\therefore$ $g(x) = \sqrt{x}$

53. (a) This function may be expressed as:

$$f(x) = \begin{cases} 3 & \text{if } x = 3 \\ x - 2 & 2 \le x < 3 \\ x - 1 & 1 \le x < 2 \\ x & 0 \le x < 1 \\ x + 1 & -1 \le x < 0 \\ x + 2 & -2 \le x < -1 \\ x + 3 & -3 \le x < -2 \end{cases}$$

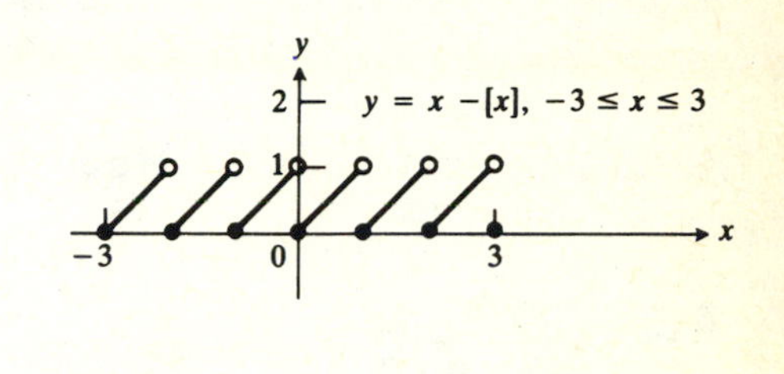

(b) This function may be expressed as:

$$f(x) = \begin{cases} 1 & 2 \le x \le 3 \\ 0 & 0 \le x < 2 \\ -1 & -2 \le x < 0 \\ -2 & -3 \le x < 2 \end{cases}$$

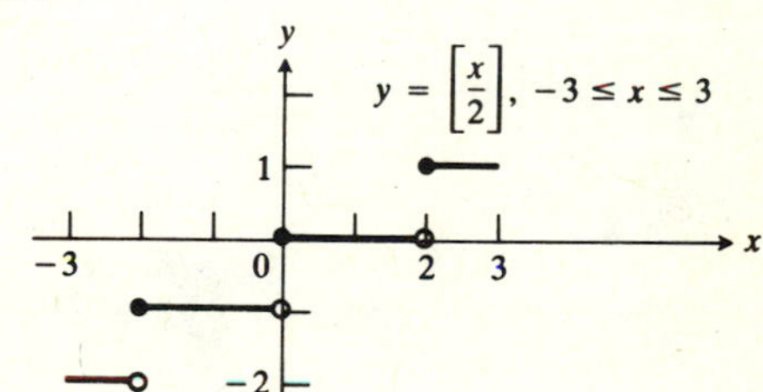

(c) $y = [2x] - 2[x]$

x	y
-2.75	0
-2.25	1
-1.75	0
-1.18	1
.5	1
0.35	0
0.55	1
1.25	0

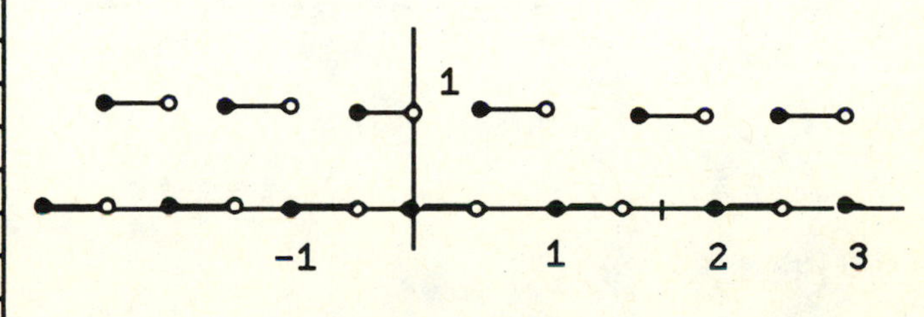

(d) This function may be expressed as:

$$f(x) = \begin{cases} 3 & \text{if} \quad x = 3 \\ \frac{1}{2}(x+2) & 2 \le x < 3 \\ \frac{1}{2}(x+1) & 1 \le x < 2 \\ \frac{1}{2}x & 0 \le x < 1 \\ \frac{1}{2}(x-1) & -1 \le x < 0 \\ \frac{1}{2}(x-2) & -2 \le x < -1 \\ \frac{1}{2}(x-3) & -3 \le x < -2 \end{cases}$$

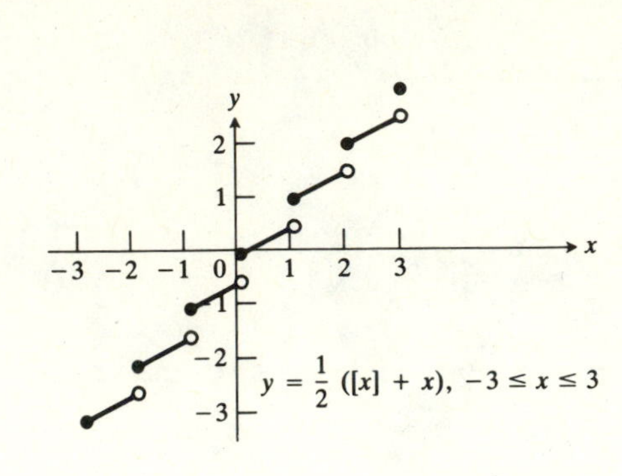

55. If $x < 0$ is not an integer, then $[x]$ is one less that the integer part of the decimal representation of x.

1.6 TANGENT LINES, SLOPES OF QUADRATIC AND CUBIC CURVES

1. $\frac{\Delta F}{\Delta t} = \frac{140}{40} = 3.5$ flies/day

3. $y = x^3 - 3x + 3 \qquad Q(h,\ h^3 - 3h + 3)$

(a) $m_{sec} = \frac{f(h) - f(0)}{h - 0} = \frac{h^3 - 3h + 3 - 3}{h} = \frac{h(h^2 - 3)}{h} = h^2 - 3$

(b) $\lim_{h \to 0} (h^2 - 3) = -3$

(c) $y - 3 = -3x$ or $y + 3x = 3$

5. $y = x^2 + 1$, $P(2, 5)$

(a) $m_{sec} = \frac{[(x + \Delta x)^2 + 1] - (x^2 + 1)]}{\Delta x} = \frac{2x\Delta x + (\Delta x)^2}{\Delta x} = 2x + \Delta x$

$m_{tan} = 2x$

(b) $x = 2 \Rightarrow m = 4$. $\therefore y - 5 = 4(x - 2)$ or $y = 4x - 3$

(c) $2x = 0 \Leftrightarrow x = 0$

7. $y = 4 - x^2$, $P(-1,3)$

(a) $m_{sec} = \dfrac{[4-(x+\Delta x)^2]-[4-x^2]}{\Delta x} = \dfrac{-2x\,\Delta x-(\Delta x)^2}{\Delta x} = -2x - \Delta x$

$m_{tan} = -2x$

(b) $x = -1 \Rightarrow m = 2$. $\therefore y - 3 = 2(x+1)$ or $y = 2x + 5$

(c) $-2x = 0 \Leftrightarrow x = 0$ $(0, 4)$

9. $y = x^2 + 3x + 2$, $P(-1,0)$

(a) $m_{sec} = \dfrac{[(x+\Delta x)^2 + 3(x+\Delta x) + 2] - [x^2 + 3x + 2]}{\Delta x}$

$= \dfrac{2x\,\Delta x + (\Delta x)^2 + 3\Delta x}{\Delta x} = 2x + \Delta x + 3$; $m_{tan} = 2x + 3$

(b) $x = -1 \Rightarrow m = 1$ $\therefore y = x + 1$

(c) $2x + 3 = 0 \Leftrightarrow x = -\dfrac{3}{2}$. $\left(-\dfrac{3}{2}, -\dfrac{1}{4}\right)$

11. $y = x^2 + 4x + 4$, $P(-2,0)$

(a) $m_{sec} = \dfrac{[(x+\Delta x)^2 + 4(x+\Delta x) + 4] - [x^2 + 4x + 4]}{\Delta x}$

$= \dfrac{2x\,\Delta x + (\Delta x)^2 + 4\Delta x}{\Delta x} = 2x + \Delta x + 4$; $m_{tan} = 2x + 4$

(b) $x = -2 \Rightarrow m = 0$ $\therefore y = 0$

(c) $2x + 4 = 0 \Leftrightarrow x = -2$ $(-2, 0)$

13. $y = x^2 - 4x + 4$, $P(1,1)$

(a) $m_{sec} = \dfrac{[(x+\Delta x)^2 - 4(x+\Delta x) + 4] - [x^2 - 4x + 4]}{\Delta x}$

$= \dfrac{2x\,\Delta x + (\Delta x)^2 - 4\Delta x}{\Delta x} = 2x + \Delta x - 4$; $m_{tan} = 2x - 4$

(b) $x = 1 \Rightarrow m = -2$ $\therefore$ $y - 1 = -2(x-1)$ or $y = -2x + 3$

(c) $2x - 4 = 0 \Leftrightarrow x = 2$ $(2, 0)$

15. $y = x^3$, $P(1,1)$

(a) $$m_{sec} = \frac{(x+\Delta x)^3 - x^3}{\Delta x} = \frac{x^3 + 3x^2\Delta x + 3x(\Delta x)^2 + (\Delta x)^3 - x^3}{\Delta x}$$

$$= 3x^2 + 3x\Delta x + (\Delta x)^2 \ ; \quad m_{tan} = \ = 3x^2$$

(b) $x = 1 \Rightarrow m = 3 \ \therefore \ y - 1 = 3(x-1)$ or $y = 3x - 2$

(c) $3x^2 = 0 \Leftrightarrow x = 0 \quad (0,0)$

17. $y = x^3 - 3x$, $P(-1,2)$

(a) $$m_{sec} = \frac{(x+\Delta x)^3 - 3(x+\Delta x) \ -x^3 + 3x}{\Delta x}$$

$$= \frac{x^3 + 3x^2\Delta x + 3x(\Delta x)^2 + (\Delta x)^3 - 3x - 3\Delta x - x^3 + 3x}{\Delta x}$$

$$= 3x^2 + 3x\Delta x + (\Delta x)^2 - 3; \quad m_{tan} = \ 3x^2 - 3$$

(b) $x = -1 \Rightarrow m = 0 \ \therefore \ y = 2$

(c) $3x^2 - 3 = 0 \Leftrightarrow x = \pm 1$. $(1,-2)$ and $(-1,2)$

19. $y = x^3 - 3x^2 + 4$, $P(1,2)$

(a) $$m_{sec} = \frac{(x+\Delta x)^3 - 3(x+\Delta x)^2 + 4 - x^3 + 3x^2 - 4}{\Delta x}$$

$$= \frac{x^3 + 3x^2\Delta x + 3x(\Delta x)^2 + (\Delta x)^3 - 3x^2 - 6x\Delta x - 3(\Delta x)^2 - x^3 + 3x^2 - 4}{\Delta x}$$

$$= 3x^2 + 3x\Delta x + (\Delta x)^2 - 6x - 3\Delta x; \quad m_{tan} = \ = 3x^2 - 6x$$

(b) $x = 1 \Rightarrow m = -3 \ \therefore \ y - 2 = -3(x-1)$ or $y = -3x + 5$

(c) $3x^2 - 6x = 0 \Leftrightarrow x = 0$ or 2 $(0,4)$ and $(2,0)$

1.7 THE SLOPE OF THE CURVE y = f(x). DERIVATIVES

1. $f(x) = x^2$

$$f'(x) = \lim_{h\to 0} \frac{f(x+h) - f(x)}{h} = \lim_{h\to 0} \frac{(x+h)^2 - x^2}{h}$$

$$= \lim_{h\to 0} \frac{2xh + h^2}{h} = \lim_{h\to 0} 2x + h = 2x$$

$f'(3) = 6$; $f(3) = 9$; $y - 9 = 6x - 18$ or $y = 6x - 9$

3. $f(x) = 2x + 3$

$$f'(x) = \lim_{h\to 0} \frac{f(x+h) - f(x)}{h} = \lim_{h\to 0} \frac{2(x+h) + 3 - 2x - 3}{h}$$

$$= \lim_{h\to 0} \frac{2h}{h} = 2$$

$f(3) = 9$; $y - 9 = 2(x - 3)$ or $y = 2x + 3$

5. $f(x) = 1 + \sqrt{x}$

$$f'(x) = \lim_{h\to 0} \frac{f(x+h) - f(x)}{h} = \lim_{h\to 0} \frac{1 + \sqrt{x+h} - (1 + \sqrt{x})}{h} =$$

$$\lim_{h\to 0} \frac{\sqrt{x+h} - \sqrt{x}}{h} \cdot \frac{\sqrt{x+h} + \sqrt{x}}{\sqrt{x+h} + \sqrt{x}} = \lim_{h\to 0} \frac{1}{\sqrt{x+h} + \sqrt{x}} = \frac{1}{2\sqrt{x}}$$

$f'(3) = \frac{1}{2\sqrt{3}}$; $f(3) = 1 + \sqrt{3}$; $y - (1 + \sqrt{3}) = \frac{1}{2\sqrt{3}}(x - 3)$ or $x - 2\sqrt{3}y = -(3 + 2\sqrt{3})$

7. $f(x) = \frac{1}{2x + 1}$

$$f(x+h) - f(x) = \frac{1}{2x + 2h + 1} - \frac{1}{2x + 1}$$

$$= \frac{-2h}{(2x + 2h + 1)(2x + 1)}$$

$$\frac{f(x+h) - f(x)}{h} = \frac{-2}{(2x + 2h + 1)(2x + 1)}$$

$$f'(x) = \lim_{h\to 0} \frac{-2}{(2x + 2h + 1)(2x + 1)} = \frac{-2}{(2x + 1)^2}\Bigg|_{x=3} = \frac{-2}{49}$$

$f(3) = \frac{1}{7}$; $y - \frac{1}{7} = \frac{-2}{49}(x - 3)$ or $49y + 2x = 13$

9. $f(x) = 2x^2 - x + 5$

$$f'(x) = \lim_{h\to 0} \frac{f(x+h) - f(x)}{h} = \lim_{h\to 0} \frac{2(x+h)^2 - (x+h) + 5 - 2x^2 + x - 5}{h}$$

$$= \lim_{h\to 0} \frac{4xh + 2h^2 - h}{h} = \lim_{h\to 0} (4x + 2h - 1) = 4x - 1$$

$f'(3) = 11$; $f(3) = 20$

$y - 20 = 11(x - 3)$ or $y = 11x - 13$

11. $f(x) = x^4$; $f'(x) = \lim_{h\to 0} \frac{f(x+h) - f(x)}{h} = \lim_{h\to 0} \frac{x^4 + 4x^3h + 6x^2h^2 + 4xh^3 + h^4 - x^4}{h}$

$$= \lim_{h\to 0} (4x^3 + 6x^2h + 4xh^2 + h^3) = 4x^3\Big]_{x=3} = 108$$

$f(3) = 81;\ y - 81 = 108(x - 3)$ or $y = 108x - 243$

13. $f(x) = x - \frac{1}{x}$

$$f(x + h) = x + h - \frac{1}{x + h}$$

$$f(x + h) - f(x) = h - \frac{1}{x + h} + \frac{1}{x} = h + \frac{-x + x + h}{x(x + h)}$$

$$f'(x) = \lim_{h\to 0} \frac{f(x + h) - f(x)}{h} = \lim_{h\to 0} 1 + \frac{1}{x(x + h)} = 1 + \frac{1}{x^2}\bigg|_{x=3} = \frac{10}{9}$$

$f(3) = \frac{8}{3};\ y - \frac{8}{3} = \frac{10}{9}(x - 3)$ or $10x - 6 = 9y$

15. $f(x) = \sqrt{2x}$

$$f'(x) = \lim_{h\to 0} \frac{f(x + h) - f(x)}{h} = \lim_{h\to 0} \frac{\sqrt{2x + 2h} - \sqrt{2x}}{h}$$

$$= \lim_{h\to 0} \frac{\sqrt{2x + 2h} - \sqrt{2x}}{h} \cdot \frac{\sqrt{2x + 2h} + \sqrt{2x}}{\sqrt{2x + 2h} + \sqrt{2x}}$$

$$= \lim_{h\to 0} \frac{2}{\sqrt{2x + 2h} + \sqrt{2x}} = \frac{1}{\sqrt{2x}}$$

$f'(3) = \frac{1}{\sqrt{6}};\quad f(3) = \sqrt{6}$

$$y - \sqrt{6} = \frac{1}{\sqrt{6}}(x - 3)$$

$$y = \frac{1}{\sqrt{6}}x - \frac{3}{\sqrt{6}}\frac{\sqrt{6}}{\sqrt{6}} + \sqrt{6} = \frac{1}{\sqrt{6}}x + \frac{1}{2}\sqrt{6}$$

$$y\sqrt{6} = x + 3$$

17. $f(x) = \sqrt{2x + 3}$

$$f'(x) = \lim_{h\to 0} \frac{f(x + h) - f(x)}{h}$$

$$= \lim_{h\to 0} \frac{\sqrt{2x + 2h + 3} - \sqrt{2x + 3}}{h} \cdot \frac{\sqrt{2x + 2h + 3} + \sqrt{2x + 3}}{\sqrt{2x + 2h + 3} + \sqrt{2x + 3}}$$

$$= \lim_{h\to 0} \frac{2}{\sqrt{2x + 2h + 3} + \sqrt{2x + 3}} = \frac{2}{2\sqrt{2x + 3}} = \frac{1}{\sqrt{2x + 3}}$$

$f'(3) = \frac{1}{3};\ f(3) = 3$

$y - 3 = \frac{1}{3}(x - 3)$ or $y = \frac{1}{3}x + 2$

19. $f(x) = \dfrac{1}{\sqrt{2x + 3}}$

$$f'(x) = \lim_{h\to 0} \frac{f(x + h) - f(x)}{h} = \lim_{h\to 0} \frac{\dfrac{1}{\sqrt{2x + 2h + 3}} - \dfrac{1}{\sqrt{2x + 3}}}{h}$$

$$= \lim_{h\to 0} \frac{\sqrt{2x + 3} - \sqrt{2x + 2h + 3}}{h\sqrt{(2x + 3)(2x + 2h + 3)}} \cdot \frac{\sqrt{2x + 3} + \sqrt{2x + 2h + 3}}{\sqrt{2x + 3} + \sqrt{2x + 2h + 3}}$$

$$= \lim_{h\to 0} \frac{-2}{h\sqrt{(2x + 3)(2x + 2h + 3)}\ \sqrt{2x + 3} + \sqrt{2x + 2h + 3}}$$

$$= \frac{-2}{(2x + 3)(2\sqrt{2x + 3})} = \frac{-1}{(2x + 3)^{3/2}}\bigg|_{x=3} = \frac{-1}{27}$$

$f(3) = \frac{1}{3}$; $y - \frac{1}{3} = \frac{-1}{27}(x - 3)$ or $y = \frac{-1}{27}x + \frac{4}{9}$

21. The function $f(x) = \dfrac{|x|}{x}$ can be written as $f(x) = \begin{cases} x & x < 0 \\ -x & x < 0 \end{cases}$, and has the same derivative at that shown in Example 4 for $f(x) = |x|$.
The derivative for all $x \neq 0$ is

$$f'(x) = \begin{cases} 1 & x > 0 \\ -1 & x < 0 \end{cases}$$ Therefore, $f'(|x|) = \dfrac{|x|}{x}$ if $x \neq 0$.

For emphasis:

$$\lim_{h\to 0+} \frac{|0 + h| - |h|}{h} = \lim_{h\to 0+} \frac{|h|}{h} = +1 \text{ but } \lim_{h\to 0^-} \frac{|h|}{h} = -1$$

23. (a)

x	f'
(-3,0)	$\frac{2}{3}$
(0,2)	$\frac{-5}{2}$
(2,4)	0
(4,5)	2
(5,7)	$\frac{1}{2}$

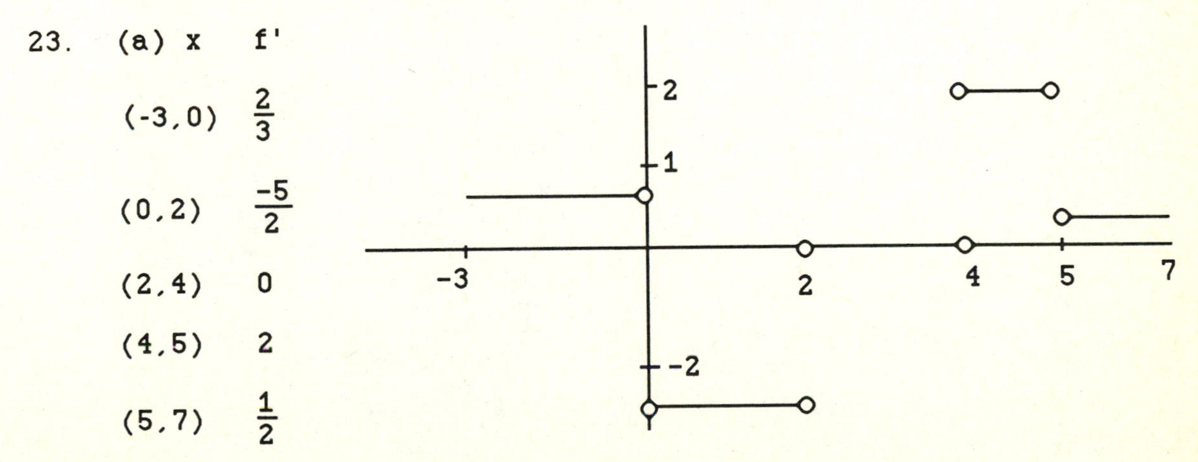

(b) $f'(x)$ does not exist at $x = 0, 2, 4,$ and 5

25. When the number of rabbits is the largest ($t = 40$ days), the derivative is zero. When the number of rabbits is the smallest ($t \geq 135$ days), the derivative is zero.

27. foxes/day

1.8 VELOCITY AND OTHER RATES OF CHANGE

1. $F(t + \Delta t) = a(t + \Delta t)^2 + b(t + \Delta t) + c$

$$\frac{f(t + \Delta t) - f(t)}{\Delta t} = \frac{a(2t\Delta t + \Delta t^2) + b\Delta t}{\Delta t} = 2at + a\Delta t + b$$

$$\lim_{\Delta t \to 0} (2at + a\Delta t + b) = 2at + b$$

3. $s = .8t^2$

(a) $s(0) = 0 \qquad s(2) = 3.2$

$\Delta s = s(2) - s(0) = 3.2, \ \frac{\Delta s}{\Delta t} = \frac{3.2}{2} = 1.6$

(b) $\frac{ds}{dt} = 2at + b = 1.6t$

(c) $\frac{ds}{dt}\Big|_{t=2} = 3.2$

5. $s = 2t^2 + 5t - 3$

(a) $s(0) = -3, \ s(2) = 15, \ \Delta s = 18, \ \frac{\Delta s}{\Delta t} = \frac{18}{2} = 9$

(b) $\frac{ds}{dt} = 4t + 5$

(c) $\frac{ds}{dt}\Big|_{t=2} = 13$

7. $s = 4 - 2t - t^2$

(a) $s(0) = 4, \ s(2) = -4, \ \Delta s = -8, \ \frac{\Delta s}{\Delta t} = -4$

(b) $\frac{ds}{dt} = -2t - 2$

(c) $\frac{ds}{dt}\Big|_{t=2} = -6$

9. $s = 4t + 3$

(a) $s(0) = 3, \ s(2) = 11, \ \Delta s = 8, \ \frac{\Delta s}{\Delta t} = 4$

(b) $\frac{ds}{dt} = 4$

(c) $\frac{ds}{dt}\Big|_{t=2} = 4$

11. (a) $s = 490t^2, \ v = 980t$

(b) $160 = 490t^2 \Rightarrow \frac{16}{49} = t^2 \Rightarrow t = \frac{4}{7}$ sec

$$V_{av} = \frac{s(\frac{4}{7}) - s(0)}{\frac{4}{7}} = \frac{160}{\frac{4}{7}} = 280 \text{ cm/sec}$$

(c) $\dfrac{\frac{4}{7}}{17} = \dfrac{4}{119} \approx 0.034$ sec/flash

13.

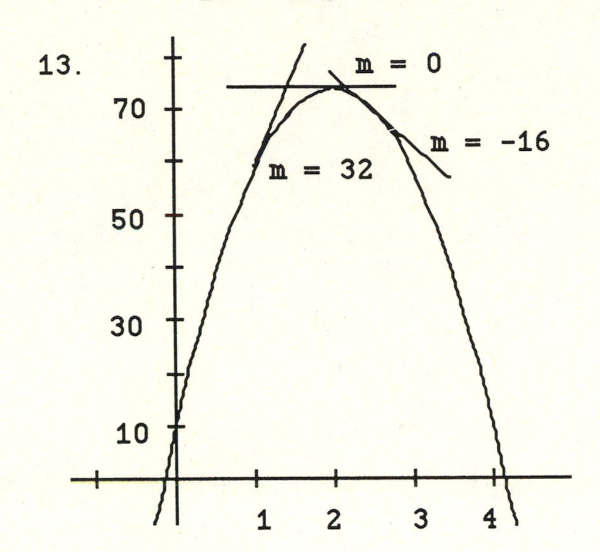

15. 190 ft/sec

17. Rocket was at maximum height at t = 8 sec. (v > 0 means rocket is rising). The velocity was equal to 0 then.

19. 2.8 sec

21. $Q = 200(30 - t)^2$

(a) $\dfrac{dQ}{dt} = -400(30 - t)\Big|_{t=10} = -400(20) = -8000$ gal/min

(b) $Q(10) = 200(400) = 80000$; $Q(0) = 200(900) = 180000$

$$\frac{Q(10) - Q(0)}{10} = \frac{-100000}{10} = -10000 \text{ gal/min}$$

23. $f(x) = 2000 + 100x - 0.1x^2$

(a) $f(100) = 11{,}000$; $f(0) = 2000$

$$\frac{f(100) - f(0)}{100} = \frac{9000}{100} = \$90\text{/machine}$$

(b) $f' = 100 - .2(x)\Big|_{x=100}$

$f' = 100 - 20 = \$80$/machine, marginal cost

(c) $f(101) = 2000 + (100)(101) - .1(101)^2 = 11{,}080$

$f(101) - f(100) = \$80$.

1.9 LIMITS

1. $\lim_{x\to 2} 2x = 4$

3. $\lim_{x\to 4} 4 = 4$

5. $\lim_{x\to 1}(3x - 1) = 2$

7. $\lim_{x\to 5} x^2 = 25$

9. $\lim_{x\to 0} (x^2 - 2x + 1) = 1$

11. $\lim_{\Delta x\to 0} (2x + \Delta x) = 2x$

13. $\lim_{x\to 1} |x - 1| = 0$

15. $\lim_{x\to 0} 5(2x - 1) = -5$

17. $\lim_{x\to 2} 3x(2x - 1) = 6(3) = 18$

19. $\lim_{x\to 2} 3x^2(2x - 1) = (12)(3) = 36$

21. $\lim_{x\to -1} (x + 3) = 2$

23. $\lim_{x\to -1} (x^2 + 6x + 9) = 1 - 6 + 9 = 4$

25. $\lim_{x\to -4} (x + 3)^{1984} = 1$

27. If $\lim_{x\to c} f(x) = 5$ and $\lim_{x\to c} g(x) = -2$, then:

(a) $\lim_{x\to c} f(x)\cdot g(x) = 5(-2) = -10$

(b) $\lim_{x\to c} 2\, f(x)\cdot g(x) = 2(5)(-2) = -20$

29. If $\lim_{x\to b} f(x) = 7$ and $\lim_{x\to b} g(x) = -3$, then:

(a) $\lim_{x\to b} (f(x) + g(x)) = 7 - 3 = 4$

(b) $\lim_{x\to b} f(x)\cdot g(x) = (7)(-3) = -21$

(c) $\lim_{x\to b} 4g(x) = 4(-3) = -12$

(d) $\lim_{x\to b} \frac{f(x)}{g(x)} = \frac{7}{-3} = -\frac{7}{3}$

31. $\lim_{t\to 2} \frac{t + 3}{t + 2} = \frac{5}{4}$

33. $\lim_{x\to 5} \frac{x^2 - 25}{x + 5} = \frac{0}{10} = 0$

35. $\lim_{x \to 5} \frac{x + 5}{x^2 - 25} = \lim_{x \to 5} \frac{1}{x - 5}$ does not exist.

37. $\lim_{x \to 0} \frac{x^2(5x + 8)}{x^2(3x^2 - 16)} = -\frac{1}{2}$

39. $\lim_{x \to 2} \frac{(y - 2)(y - 3)}{y - 2} = \lim_{x \to 2} (y - 3) = -1$

41. $\lim_{x \to 4} \frac{x - 4}{(x - 4)(x - 1)} = \lim_{x \to 4} \frac{1}{x - 1} = \frac{1}{3}$

43. $\lim_{t \to 1} \frac{(t - 2)(t - 1)}{(t - 1)(t + 1)} = \lim_{t \to 1} \frac{t - 2}{t + 1} = \frac{-1}{2}$

45. $\lim_{x \to -3} \frac{x^2 + 4x + 3}{x - 3} = \frac{0}{-6} = 0$

47. $\lim_{x \to -2} \frac{(x + 2)(x - 1)}{(x + 2)(x - 2)} = \lim_{x \to -2} \frac{x - 1}{x - 2} = \frac{-3}{-4} = \frac{3}{4}$

49. $\lim_{x \to 2} \frac{x^2 - 7x + 10}{x - 2} = \lim_{x \to 2} \frac{(x - 2)(x - 5)}{x - 2} = \lim_{x \to 2} (x - 5) = -3$

51. $\lim_{x \to a} \frac{(x - a)(x^2 + ax + a^2)}{(x - a)(x + a)(x^2 + a^2)} = \frac{3a^2}{2a(2a^2)} = \frac{3}{4a}$

53. Let $f(x) = \begin{cases} \frac{1}{x} & x \neq 0 \\ 0 & x=0 \end{cases}$ and $g(x) = \begin{cases} \frac{-1}{x} & x \neq 0 \\ 0 & x=0 \end{cases}$

Then the limits of $f(x)$ and $g(x)$ do not exist as $x \to 0$ but

$f(x) + g(x) = \frac{1}{x} - \frac{1}{x} = 0$ and $\lim_{x \to 0} ((f(x) + g(x)) = \lim_{x \to 0} 0 = 0$

55. Let $f(x) = \frac{1}{x}$ and $g(x) = \frac{1}{x^2}$. Then neither the limit of $f(x)$ nor $g(x)$ exists as $x \to 0$ but

$\lim \frac{f(x)}{g(x)} = \lim \frac{\frac{1}{x}}{\frac{1}{x^2}} = \lim x = 0$ as $x \to 0$.

57. If $\lim_{h \to 0} \frac{|-1 + h| - |-1|}{h}$ then $f(x) = |x - 1|$

59. $f(x) = \begin{cases} \sqrt{1-x^2} & 0 \le x < 1 \\ 1 & 1 \le x < 2 \\ 2 & x = 2 \end{cases}$

(a) $0 < c < 1$
$1 < c < 2$
(b) $c = 2$
(c) $c = 0$

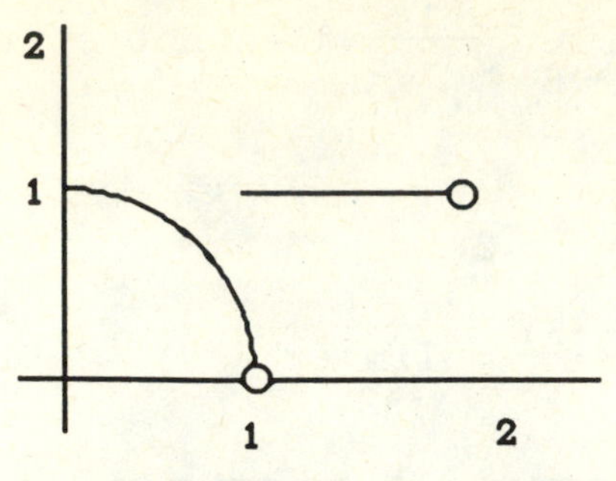

61. $\lim_{x\to 0} [x] = 0$

63. $\lim_{x\to .5} [x] = 0$

65. $\lim_{x\to 0^+} \frac{x}{|x|} = 1$

67. $\lim_{x\to 3^+} \frac{x^2 - 9}{|x - 3|} = \lim_{x\to 3^+} \frac{(x - 3)(x + 3)}{x - 3} = 6$

69. For all c not integers.

71. $\lim_{x\to 0} \frac{1 + \sin x}{\cos x} = 1$

73. $\lim_{t\to 0} \frac{t}{\sin t} = 1$

75. $\lim_{h\to 0} (\frac{\sin h}{h})^2 = 1$

77. $\lim_{x\to 0} \tan x = 0$

79. $\lim_{x\to 0^-} \sin x = 0$

81. $\lim_{x\to 0^-} \frac{\sin x}{-x} = -1$

83. $\lim_{x\to 0} \frac{2 \sin 2x}{2x} = 2\cdot 1 = 2$

85. $\lim_{y\to 0} \frac{\tan 2y}{3y} = \lim_{y\to 0} \frac{2}{3} \cdot \frac{\sin 2y}{2y} \cdot \frac{1}{\cos 2y} = \frac{2}{3}$

87. (a) $|\sin \frac{1}{x}| \le 1 \Rightarrow -1 \le \sin \frac{1}{x} \le 1$. If $x > 0$, then $-x \le x \sin \frac{1}{x} \le x$. If $x < 0$, then $x \le x \sin \frac{1}{x} \le -x$. Therefore, we must have $-|x| \le x \sin \frac{1}{x} \le |x|$.

(b) From the results in (a), $\lim_{x\to 0} -|x| = \lim_{x\to 0} x \sin \frac{1}{x} \le \lim_{x\to 0} |x|$

Since $\lim_{x\to 0} |x| = 0$, we must have, by the Sandwich Theorem, that $\lim_{x\to 0} x \sin \frac{1}{x} = 0$

89. (a) $F(t) = 2t + 3$, $c = 1$

$\lim_{t\to 1} (2t + 3) = 5$ if there exists $\delta > 0$ for which $|(2t + 3) - 5| < \varepsilon$ whenever $0 < |t - 1| < \delta$. But

$|2t + 3 - 5| < \varepsilon \Leftrightarrow 0 < |2t - 2| < \varepsilon \Leftrightarrow 0 < |t - 1| < \frac{\varepsilon}{2}$.

Therefore, choose $\delta \le \frac{\varepsilon}{2}$.

(b) $F(t) = 2t - 3$, $c = 1$

$\lim_{t \to 1} (2t - 3) = -1$ if there exists $\delta > 0$ for which $|(2t - 3) + 1| < \varepsilon$ whenever $0 < |x - 1| < \delta$. But

$|2t - 3) + 1| < \varepsilon \Leftrightarrow 2|t - 1| < \varepsilon \Leftrightarrow 0 < |t - 1| < \frac{\varepsilon}{2}$.

Therefore, choose choose $\delta \le \frac{\varepsilon}{2}$

(c) $F(t) = 5 - 3t$, $c = 2$

$\lim_{t \to 2} (5 - 3t) = -1$ if there exists $\delta > 0$ for which $|5 - 3t + 1| < \varepsilon$ whenever $0 < |x - 2| < \delta$. But

$|-3t + 6| < \varepsilon \Leftrightarrow 3|t - 2| < \varepsilon \Leftrightarrow 0 < |t - 2| < \frac{\varepsilon}{3}$.

Choose $\delta \le \frac{\varepsilon}{3}$.

(d) $F(t) = 7$, $c = -1$.

Since $|7 - 7| = 0 < \varepsilon$, choose any $\delta > 0$. Then $\lim_{t \to -1} 7 = 7$.

(e) $F(t) = \frac{t^2 - 4}{t - 2}$, $c = 2$

$\lim_{t \to 2} (t + 2) = 4$ if there exists $\delta > 0$ for which

$|t + 2 - 4| < \varepsilon$ whenever $0 < |t - 2| < \delta$. Since

$|t + 2 - 4| = |t - 2| < \varepsilon$, choose $\delta \le \varepsilon$.

(f) $F(t) = \frac{t^2 + 6t + 5}{t + 5}$, $c = -3$

$\lim_{t \to 5} \frac{t^2 + 6t + 5}{t + 5} = \frac{60}{10} = 6$ if there exists $\delta > 0$ for which

$\left| \frac{t^2 + 6t + 5}{t + 5} - 6 \right| < \varepsilon$ whenever $0 < |t + 3| < \delta$. Since

$\left| \frac{t^2 + 6t + 5}{t + 5} - 6 \right| = \left| \frac{(t + 5)(t + 1)}{t + 5} - 6 \right| = |t + 1 - 6| =$

$|t - 5| < \varepsilon$, choose $\delta \le \varepsilon$

(g) $F(t) = \dfrac{3t^2 + 8t - 3}{2t + 6}$, $c = -3$

$\lim_{t\to-3} \dfrac{(t + 3)(3t - 1)}{2(t + 3)} = \dfrac{-10}{2} = -5$ if there exists $\delta > 0$ for which

$\left|\dfrac{3t^2 + 8t - 3}{2t + 6} + 5\right| < \varepsilon$ whenever $0 < |t - 2| < \delta$. Since

$$\left|\frac{3t^2 + 8t - 3}{2t + 6} + 5\right| = \left|\frac{(t + 3)(3t - 1)}{2(t + 3)} + 5\right| = \left|\frac{3t - 1}{2} + 5\right| =$$

$$\left|\frac{3t - 1 + 10}{2}\right| = \left|\frac{3t + 9}{2}\right| = \frac{3}{2}|t + 3| < \varepsilon \text{ if } |t + 3| < \frac{2\varepsilon}{3},$$

choose $\delta \le \dfrac{2\varepsilon}{3}$.

(h) $F(t) = \dfrac{4}{t}$, $c = 2$

$\lim_{t\to2} \dfrac{4}{t} = 2$ if there exists $\delta > 0$ for which

$\left|\dfrac{4}{t} - 2\right| < \varepsilon$ whenever $0 < |t - 2| < \delta$. But $\left|\dfrac{4}{t} - 2\right| = \left|\dfrac{4 - 2t}{t}\right|$.

$= \dfrac{|4 - 2t|}{|t|}$. First we restrict the size of $|t|$. Choose $\delta = 1$

$|t - 2| < 1 \Rightarrow 1 < t < 3$ and $\dfrac{|4 - 2t|}{|t|} < \dfrac{|4 - 2t|}{1}$. Then

$|4 - 2t| < 3\varepsilon \Leftrightarrow 2|t - 2| < 3\varepsilon \Leftrightarrow |t - 2| < \dfrac{3\varepsilon}{2}$.

Choose $\delta \le \min\left(\dfrac{3\varepsilon}{2}, 1\right)$

(i) $F(t) = \dfrac{\frac{1}{t} - \frac{1}{3}}{t - 3}$, $c = 3$

$\lim_{t\to3} \dfrac{\frac{1}{t} - \frac{1}{3}}{t - 3} = \dfrac{3 - t}{3t} \cdot \dfrac{1}{t - 3} = \dfrac{-1}{9}$ if there exists $\delta > 0$ for which

$$\left|\frac{\frac{1}{t} - \frac{1}{3}}{t - 3} - \frac{-1}{9}\right| = \left|\frac{-1}{3t} + \frac{1}{9}\right| = \frac{1}{3}\left|\frac{1}{t} - \frac{1}{3}\right| = \frac{1}{3}\left|\frac{t - 3}{3t}\right|.$$

Choose $\delta = 1$ so that $|t - 3| < 1 \Rightarrow 2 < t < 4 \Rightarrow 6 < 3t < 12$.

Then $\dfrac{1}{3}\left|\dfrac{3 - t}{3t}\right| < \dfrac{1}{3}\left|\dfrac{3 - t}{6}\right| < \varepsilon \Rightarrow |t - 3| < 18\varepsilon$.

Choose $\delta \le \min(18\varepsilon, 1)$.

91. $f(t) = t^2 + t;\quad \lim_{t \to 3} (t^2 + t) = 12$

(a) $|t^2 + t - 12| < \frac{1}{10} \Leftrightarrow |(t + 4)(t - 3)| < \frac{1}{10}$.

Choose $\delta = 1$ so that $|t - 3| < 1 \Rightarrow 2 < t < 4$. Then

$|(t + 4)(t - 3)| < 8|t - 3| < \frac{1}{10}$ if $|t - 3| < \frac{1}{80}$.

Choose $\delta \le \frac{1}{80}$. (We know that $\frac{1}{80} < 1$)

(b) Choose $\delta \le \frac{1}{800}$.

(c) Choose $\delta \le \min\left(\frac{\varepsilon}{8}, 1\right)$

93. $f(x) = \sqrt{9 - x^2}$

$$\frac{f(\Delta x) - f(0)}{\Delta x} = \frac{\sqrt{9 - (\Delta x)^2} - 3}{\Delta x}$$

Δx	$\frac{\sqrt{9 - (\Delta x)^2} - 3}{\Delta x}$
0.1	-0.0166713
0.01	-0.0016667
0.001	-0.0001666...
-0.1	0.0166713
-0.01	0.001666...
-0.001	0.0001666...

1.10 INFINITY AS A LIMIT

1. $$\lim_{x \to \infty} \frac{2x + 3}{5x + 7} = \lim_{x \to \infty} \frac{2 + \frac{3}{x}}{5 + \frac{7}{x}} = \frac{2}{5}$$

3. $$\lim_{x \to \infty} \frac{x + 1}{x^2 + 3} = \lim_{x \to \infty} \frac{\frac{1}{x} + \frac{1}{x^2}}{1 + \frac{3}{x^2}} = 0$$

5. $$\lim_{y \to \infty} \frac{3y + 7}{y^2 - 2} = \lim_{y \to \infty} \frac{\frac{3}{y} + \frac{7}{y^2}}{1 - \frac{2}{y^2}} = 0.$$

7. $$\lim_{t \to \infty} \frac{t^2 - 2t + 3}{2t^2 + 5t - 3} = \lim_{t \to \infty} \frac{1 - \frac{2}{t} + \frac{3}{t^2}}{2 + \frac{5}{t} - \frac{3}{t^2}} = \frac{1}{2}$$

9. $\lim\limits_{x\to\infty} \dfrac{x}{x-1} = \lim\limits_{x\to\infty} \dfrac{1}{1-\dfrac{1}{x}} = 1$

11. $\lim\limits_{x\to-\infty} |x| = \infty$

13. $\lim\limits_{a\to\infty} \dfrac{|a|}{|a|+1} = \lim\limits_{a\to\infty} \dfrac{1}{1+\dfrac{1}{|a|}} = 1$

15. $\lim\limits_{x\to\infty} \dfrac{3x^3+5x^2-7}{10x^3-11x+5} = \lim\limits_{x\to\infty} \dfrac{3+\dfrac{5}{x}-\dfrac{7}{x^3}}{10-\dfrac{11}{x^2}+\dfrac{5}{x^3}} = \dfrac{3}{10}$

17. $\lim\limits_{s\to\infty} \left(\dfrac{s}{s+1}\right)\left(\dfrac{s^2}{5+s^2}\right) = \lim\limits_{s\to\infty} \left(\dfrac{1}{1+\dfrac{1}{s}}\right)\left(\dfrac{1}{\dfrac{5}{s^2}+1}\right) = 1$

19. $\lim\limits_{r\to-\infty} \dfrac{8r^2+7r}{4r^2} = \lim\limits_{r\to-\infty} \dfrac{8+\dfrac{7}{r}}{4} = 2$

21. $\lim\limits_{y\to\infty} \dfrac{y^4}{y^4-7y^3+7y^2+9} = \lim\limits_{y\to\infty} \dfrac{1}{1-\dfrac{7}{y}+\dfrac{7}{y^2}+\dfrac{9}{y^4}} = 1$

23. $\lim\limits_{x\to\infty} \dfrac{x-3}{x^2-5x+4} = \lim\limits_{x\to\infty} \dfrac{\dfrac{1}{x}-\dfrac{3}{x^2}}{1-\dfrac{5}{x}+\dfrac{4}{x^2}} = 0$

25. $\lim\limits_{x\to\infty} \dfrac{-2x^3-2x+3}{3x^3+3x^2-5x} = \lim\limits_{x\to\infty} \dfrac{-2-\dfrac{2}{x^2}+\dfrac{3}{x^3}}{3+\dfrac{3}{x}-\dfrac{5}{x^2}} = -\dfrac{2}{3}$

27. $\lim\limits_{x\to\infty} \dfrac{x+\sin x}{x+\cos x} = \lim\limits_{x\to\infty} \dfrac{1+\dfrac{\sin x}{x}}{1+\dfrac{\cos x}{x}} = 1$

29. $\lim\limits_{x\to\infty} \left(\dfrac{1}{x^4}+\dfrac{1}{x}\right) = 0$

31. $\lim\limits_{x\to\infty} \left(\cos\dfrac{1}{x}+1\right) = \lim\limits_{y\to 0} (\cos y + 1) = 2$

33. $\lim\limits_{x\to 0^+} \dfrac{1}{3x} = \infty$

35. $\lim\limits_{x\to 0^+} \dfrac{5}{2x} = \infty$

37. $\lim\limits_{t\to 2} \dfrac{t^2-4}{t-2} = \lim\limits_{t\to 2} (t+2) = 4$

39. $\lim\limits_{x\to 1^+} \dfrac{x}{x-1} = \infty$

41. $\lim\limits_{x\to 1^-} \dfrac{1}{x+1} = -\infty$

43. $\lim\limits_{x\to -2^+} \dfrac{1}{x+2} = \infty$

45. $\lim\limits_{x\to 3} \dfrac{x-3}{x^2} = \dfrac{0}{9} = 0$

47. $\lim\limits_{x\to 2^-} \dfrac{x^2+5}{x-2} = -\infty$

49. $\lim\limits_{x\to -5} \dfrac{x^2+3x-10}{x+5} = \lim\limits_{x\to -5} \dfrac{(x+5)(x-2)}{x+5} = -7$

51. (a) $\lim\limits_{x\to 0} \dfrac{x-1}{2x^2-7x+5} = -\dfrac{1}{5}$

(b) $\lim\limits_{x\to \infty} \dfrac{x-1}{2x^2-7x+5} = 0$

(c) $\lim\limits_{x\to 1} \dfrac{x-1}{2x^2-7x+5} = -\dfrac{1}{3}$

53. $f(x) = \dfrac{1}{x-2}$ has a vertical asymptote at $x = 2$ because $\lim\limits_{x\to 2^+} f(x) = +\infty$ and $\lim\limits_{x\to 2^-} f(x) = -\infty$.

f has a horizontal asymptote at $y = 0$ because $\lim\limits_{x\to \infty} f(x) = 0$.

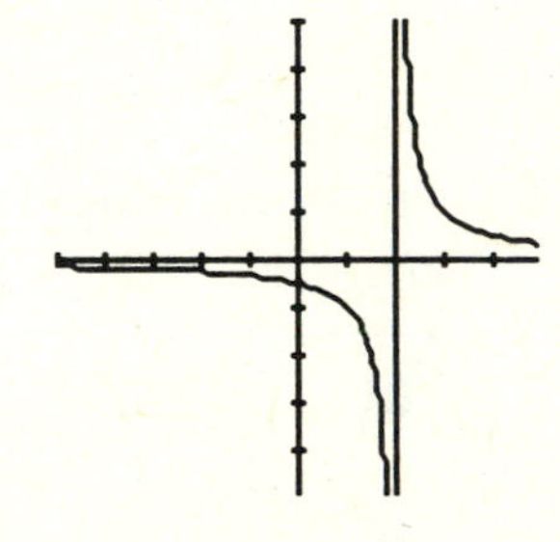

55. $f(x) = 1 + \dfrac{1}{x}$ has a vertical asymptote at $x = 0$ because $\lim\limits_{x\to 0^+} f(x) = +\infty$ and $\lim\limits_{x\to 0^-} f(x) = -\infty$.

f has a horizontal asymptote at $y = 1$ because $\lim\limits_{x\to \infty} f(x) = 1$.

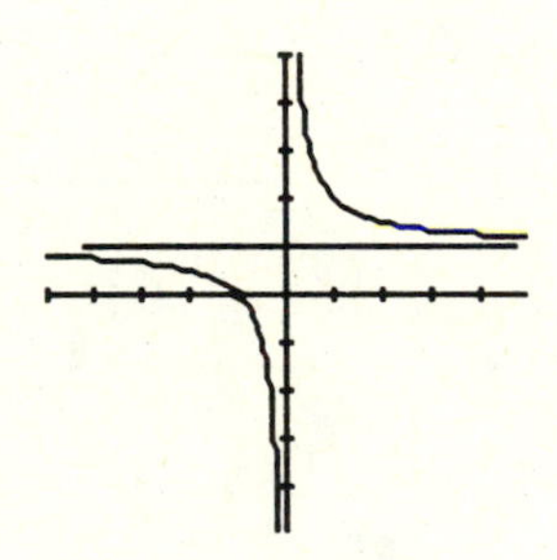

57. $\dfrac{2x-3}{x} < f(x) < \dfrac{2x^2+5x}{x^2} \Rightarrow \lim\limits_{x\to\infty} \dfrac{2x-3}{x} < \lim\limits_{x\to\infty} f(x) < \lim\limits_{x\to\infty} \dfrac{2x^2+5x}{x^2}$.

Since $\lim\limits_{x\to\infty} \dfrac{2x-3}{x} = 2$ and $\lim\limits_{x\to\infty} \dfrac{2x^2+5x}{x^2} = 2$,

by the Squeeze Theorem, $\lim\limits_{x\to\infty} f(x) = 2$.

1.11 CONTINUITY

1. (a) $f(-1) = 0$ yes
   (b) $\lim_{x\to -1^+} f(x) = 0$ yes
   (c) $\lim_{x\to -1^+} f(x) = f(-1)$ yes
   (d) f is continuous at $x = -1$

3. (a) f(2) does not exist no
   (b) f is NOT continuous at $x = 2$ no

5. (a) $\lim_{x\to 2} f(x) = 0$
   (b) $f(2) = 0$

7. $f(x) = \begin{cases} 0 & x < 0 \\ 1 & 0 \le x \le 1 \\ 0 & x > 1 \end{cases}$

   continuous for all x except $x = 0, 1$

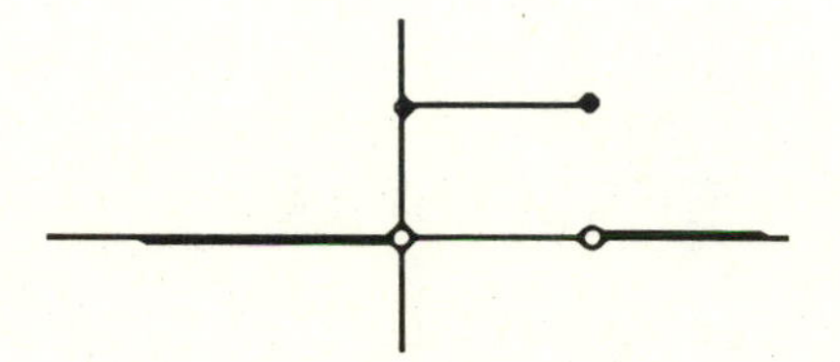

9. The function f is continuous when $-1 \le x < 1$ or $1 < x < 2$ or $2 < x \le 3$

11. The function f is continuous when $x < -1$ or $-1 < x < 0$ or $0 < x < 1$ or $x > 1$

13. The function $f(x) = \dfrac{1}{(x + 2)^2}$ is not continuous at $x = -2$.

15. The function $f(x) = \dfrac{(x + 1)}{(x - 1)(x - 3)}$ is discontinuous at $x = 1, 3$.

17. The function $f(x) = \dfrac{x + 3}{(x - 5)(x + 2)}$ is discontinuous at $x = -2, 5$

19. The function $f(x) = \dfrac{1}{x^2 + 1}$ is continuous for all x.

21. The function $f(x) = \dfrac{|x|}{x}$ is discontinuous at $x = 0$.

23. $\lim_{x\to 3} \dfrac{x^2 - 9}{x - 3} = \lim_{x\to 3} (x + 3) = 6$ so define $g(3) = 6$.

25. $f(x) = \dfrac{x^3 - 1}{x^2 - 1} = \dfrac{(x - 1)(x^2 + x + 1)}{(x - 1)(x + 1)} = \dfrac{x^2 + x + 1}{x + 1}$

    $\lim_{x\to 1} \dfrac{x^2 + x + 1}{x + 1} = \dfrac{3}{2}$ so define $f(1) = \dfrac{3}{2}$.

27. (a) $f(x) = \begin{cases} x & 0 \le x \le 1 \\ 2 - x & 1 < x \le 2 \end{cases}$

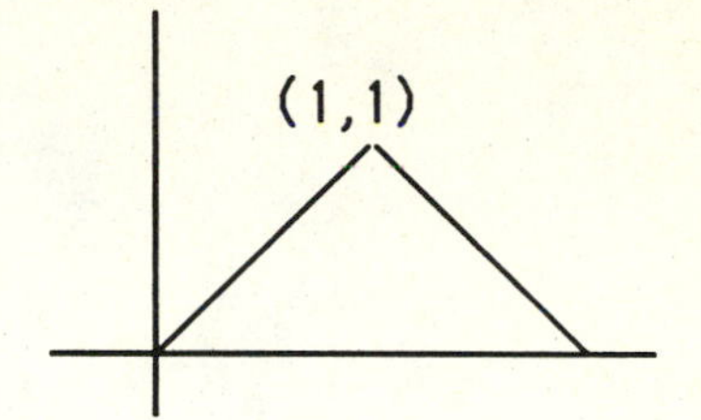

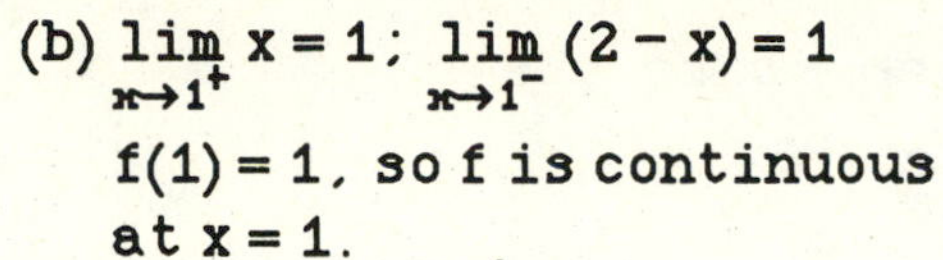

(b) $\lim_{x \to 1^+} x = 1$; $\lim_{x \to 1^-} (2 - x) = 1$

$f(1) = 1$, so f is continuous at $x = 1$.

(c) $f'(x) = \begin{cases} 1 & 0 \le x < 1 \\ -1 & 1 < x \le 2 \end{cases}$

Therefore, $f'(1)$ does not exist

29. $\lim_{x \to 3^+} (x^2 - 1) = 8$; $\lim_{x \to 3^-} 2ax = 8$. Therefore, $2a(3) = 6a = 8$ so $a = \frac{4}{3}$.

31. $\lim_{x \to 0} \frac{1 + \cos x}{2} = 1$

33. $\lim_{x \to 0} \tan x = 0$

35. $f(x) = |x| = \begin{cases} x & 0 \le x \le 1 \\ -x & -1 \le x < 0 \end{cases}$ has a maximum value of 1 and a minimum value of 0.

37. The function $f(x) = x^2$ has no maximum on $-1 < x < 1$ because the interval is open, but there is a minimum value of 0 at $x = 0$.

39. If $f(0) < 0$ and $f(1) > 0$, then there exists at least one value of x, say $x = c$, $0 < x < 1$, such that $f(c) = 0$ by the Intermediate Value Theorem.

41. $\lim_{x \to 0} f(x)$ does not exist. Therefore $f(x)$ is not continuous at $x = 0$ and does not have the Intermediate Value property.

Also, if $f(x)$ is to be the derivative of a function, $F(x)$, then

$F'(0) = \lim_{\Delta x \to 0} \frac{f(\Delta x) - f(0)}{\Delta x}$ must exist but $\lim_{\Delta x \to 0^+} \frac{1 - 1}{\Delta x} = 0$ and

$\lim_{\Delta x \to 0^-} \frac{-1 - 1}{\Delta x} = \infty$ so $F'(0)$ does not exist.

1.M MISCELLANEOUS PROBLEMS

1. If the particle ends at the point $B(u,v)$ after traveling $\Delta x = h$ and $\Delta y = k$ units, then it must have begun at the point $A(u - h, v - k)$.

3. On the curve $y = x^2$ between $A(1,1)$ and $B(a,a^2)$,
$$\frac{\Delta y}{\Delta x} = \frac{a^2 - 1}{a - 1} = a + 1, \quad a \neq 1.$$

5. (a) $2y = 3x + 4$ or $y = \frac{3}{2}x + 2$. Therefore the line through $P(1,-3)$ perpendicular to $L: 2y - 3x = 4$ would have slope $m_\perp = -\frac{2}{3}$. Thus $y + 3 = \frac{-2}{3}(x - 1)$ or

$$2x + 3y = -7 \text{ is the required line.}$$

(b) $$d = \frac{|-3 - 6 - 4|}{\sqrt{9 + 4}} = \frac{13}{\sqrt{13}} = \sqrt{13}$$

7. The equation of the circle is $(y - 1)^2 + (x - 2)^2 = 4$. Therefore the circle is tangent to the y-axis and one tangent line is $x = 0$. Let $P(x_1,y_1)$ be the other point of tangency. Then

(i) the slope between the center $(2,1)$ and P is $\frac{y_1 - 1}{x_1 - 2}$ and the slope between the origin and P is $\frac{y_1}{x_1}$;

(ii) these slopes are negative reciprocals, so

$$\frac{y_1 - 1}{x_1 - 2}\cdot\frac{y_1}{x_1} = -1 \Rightarrow y_1^2 - y_1 = -x_1^2 + 2x_1 \text{ or}$$

$$x_1^2 + y_1^2 - 2x_1 - y_1 = 0$$

(iii) The point P also satisfies the equation of the circle, so substituting and solving the two equations simultaneously,

$$x_1^2 + y_1^2 - 2x_1 - y_1 = 0$$
$$x_1^2 + y_1^2 - 4x_1 - 2y_1 = -1$$

$$2x_1 + y_1 = 1$$
$$y_1 = -2x_1 + 1$$
$$(-2x_1)^2 + x_1^2 - 4x_1 + 4 = 4$$
$$4x_1^2 + x_1^2 - 4x_1 = 0$$
$$x_1(5x_1 - 4) = 0 \quad \text{or} \quad x_1 = \frac{4}{5}$$
$$y_1 = -\frac{8}{5} + 1 = -\frac{3}{5}$$

(iv) The slope between the point $(\frac{4}{5}, \frac{-3}{5})$ and $(0,0)$ is

$$m = \frac{-\frac{3}{5}}{\frac{4}{5}} = -\frac{3}{4}$$

The required line is $y + \frac{3}{5} = -\frac{3}{4}(x - \frac{4}{5})$

or $y = -\frac{3}{4}x$

9. $Ax + By = C \Rightarrow By = -Ax + C$ or $y = \frac{-A}{B}x + \frac{C}{B}$

(a) $m = -\frac{A}{B}$

(b) y-intercept is $\frac{C}{B}$

(c) x-intercept is $\frac{C}{A}$

(d) $m_{\perp} = \frac{B}{A}$. Therefore $y = \frac{-A}{B}x + \frac{C}{B}$ or $Bx - Ay = 0$.

11. $x = p\cos\alpha$ and $y = p\sin\alpha$

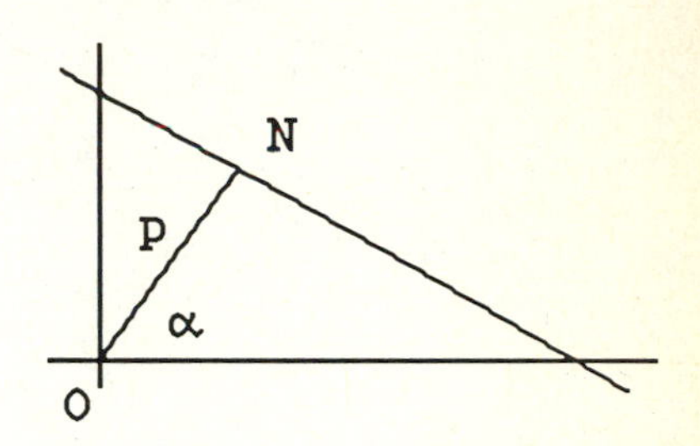

$m_{ON} = \frac{y}{x}$ and $m_1 = -\frac{x}{y} = \frac{-\cos\alpha}{\sin\alpha}$

$y - p\sin\alpha = -\frac{\cos\alpha}{\sin\alpha}(x - p\cos\alpha)$

$y\sin\alpha - p\sin^2\alpha = -x\cos\alpha + p\cos^2\alpha$

$x\cos\alpha + y\sin\alpha = p$

13. Let $P(x_1, y_1)$ be the center of one such circle. If the circle is to be tangent to the lines $x + y - 1 = 0$, $x - y + 1 = 0$, and $x - 3y - 1 = 0$ then P must be equidistant from each. That is,

$$\frac{|x_1+y_1-1|}{\sqrt{2}} = \frac{|x_1-y_1+1|}{\sqrt{2}} = \frac{|x_1-3y_1-1|}{\sqrt{10}}.$$ We separate these

into four cases and solve the resulting systems.

Case I: $x_1 + y_1 - 1 = +(x_1 - y_1 + 1) \Rightarrow 2y_1 = 2$ or $y_1 = 1$

$\sqrt{5}(x_1 - y_1 + 1) = +(x_1 - 3y_1 - 1)$ and $y_1 = 1 \Rightarrow$

$\sqrt{5}x_1 = x_1 - 3 - 1$ or $x_1(\sqrt{5} - 1) = -4$

$$x_1 = \frac{4}{1-\sqrt{5}}\,\frac{1+\sqrt{5}}{1+\sqrt{5}} = -1 - \sqrt{5}$$

The radius is $r = \frac{|-1-\sqrt{5}+1-1|}{\sqrt{2}} = \frac{1+\sqrt{5}}{\sqrt{2}}\,\frac{\sqrt{2}}{\sqrt{2}} = \frac{\sqrt{2}+\sqrt{10}}{2}$

Case II: $x_1 + y_1 - 1 = -(x_1 - y_1 + 1) = -x_1 + y_1 - 1$

$2x_1 = 0$ or $x_1 = 0$

$\sqrt{5}(x_1 + y_1 - 1) = -(x_1 - 3y_1 - 1)$
$= -x_1 + 3y_1 + 1$ and $-x_1 + 3y_1 + 1$

$$y_1\sqrt{5} - \sqrt{5} = 3y_1 + 1 \text{ or } -1 - \sqrt{5} = y_1(3 - \sqrt{5})$$

$$y_1 = \frac{1+\sqrt{5}}{\sqrt{5}-3} = \frac{1+\sqrt{5}}{\sqrt{5}-3}\cdot\frac{\sqrt{5}+3}{\sqrt{5}+3} = \sqrt{5} - 2$$

The radius is $r = \dfrac{|-2-\sqrt{5}-1|}{\sqrt{2}} = \dfrac{3 + \sqrt{5}}{\sqrt{2}}\cdot\dfrac{\sqrt{2}}{\sqrt{2}} = \dfrac{3\sqrt{2} + \sqrt{10}}{2}$

Case III: $\sqrt{5}(x_1 + y_1 - 1) = x_1 - 3y_1 - 1$

$$y_1\sqrt{5} - \sqrt{5} = -3y_1 - 1 \Rightarrow y_1(\sqrt{5} + 3) = \sqrt{5} - 1$$

$$y_1 = \frac{\sqrt{5}-1}{3+\sqrt{5}}\cdot\frac{3-\sqrt{5}}{3-\sqrt{5}}$$

$$y_1 = \frac{3\sqrt{5} - 3 - 5 + \sqrt{5}}{4} = \frac{4\sqrt{5} - 8}{4} = -2 + \sqrt{5}$$

The radius is $r = \dfrac{|-2 + \sqrt{5} - 1|}{\sqrt{2}} = \dfrac{3 - \sqrt{5}}{\sqrt{2}}\cdot\dfrac{\sqrt{2}}{\sqrt{2}} = \dfrac{3\sqrt{2} - \sqrt{10}}{2}$

Case IV: $\sqrt{5}(x_1 - y_1 + 1) = -x_1 + 3y_1 + 1$ and $y_1 = 1$

$$\sqrt{5}(x_1) = -x_1 + 4 \quad \Rightarrow \quad x_1(1 + \sqrt{5}) = 4$$

$$x_1 = \frac{4}{1+\sqrt{5}}\cdot\frac{1-\sqrt{5}}{1-\sqrt{5}} = \frac{4 - 4\sqrt{5}}{-4} = \sqrt{5} - 1$$

The radius is $r = \dfrac{|\sqrt{5}-1+1-1|}{\sqrt{2}} = \dfrac{|\sqrt{5}-1|}{\sqrt{2}} = \dfrac{\sqrt{10} - \sqrt{2}}{2}$

15. Let L_1: $a_1x + b_1y + c_1 = 0$ and L_2: $a_2x + b_2y + c_2 = 0$ be two lines. If L_1 not parallel to L_2 then

$$L_3: \quad a_1x + b_1y + c_1 + k(a_2x + b_2y + c_2) = 0$$

is the family of all lines through the point of intersection of L_1 and L_2. The reason is that if $P(x_1,y_1)$ is the point of intersection of L_1 and L_2 then the coordinates of P satisfy both equations (i.e. make them equal to zero) and $0 + k\cdot 0 = 0$. If $L_1 \parallel L_2$, then L_3 is the family of all lines $\parallel$ to L_1 and L_2.

17. The line $5x - y = 1$ has slope $m = 5$. Then $m_\perp = -\frac{1}{5}$.

There are two such lines:

$$\frac{b}{-a} = -\frac{1}{5} \text{ and } \frac{1}{2}ab = 5 \Rightarrow 5b = a \text{ and } \frac{1}{2}(5b)\cdot b = 5$$

Then $b^2 = 2$ or $b = \pm\sqrt{2}$.

L_1: $(5\sqrt{2}, \sqrt{2})$

$$y-\sqrt{2} = \frac{-1}{5}(x-5\sqrt{2})$$
$$5y-5\sqrt{2} = -x+5\sqrt{2}$$
$$x + 5y = 10\sqrt{2}$$

L_2: $(-5\sqrt{2}, \sqrt{2})$

$$y+\sqrt{2} = \frac{-1}{5}(x+5\sqrt{2})$$
$$5y+5\sqrt{2} = -x-5\sqrt{2}$$
$$x + 5y = -10\sqrt{2}$$

19. $A = \pi r^2 \Rightarrow \frac{A}{\pi} = r^2$ or $r = \sqrt{\frac{A}{\pi}}$; $C = 2\pi r \Rightarrow r = \frac{C}{2\pi}$

$$\sqrt{\frac{A}{\pi}} = \frac{C}{2\pi} \Rightarrow \frac{A}{\pi} = \frac{C^2}{4\pi^2} \text{ or } A = \frac{C^2}{4\pi}$$

21. $y = \dfrac{1}{1 + x}$ D: $x \neq -1$ R: $y \neq 0$

23. $y = \dfrac{1}{1 + \sqrt{x}}$ has domain D: $x \geq 0$. To find the range, we solve for x in terms of y:

$$1 + \sqrt{x} = \frac{1}{y} \Rightarrow \sqrt{x} = \frac{1}{y} - 1 = \frac{1 - y}{y}$$

$$\frac{1 - y}{y} \geq 0 \text{ if } 0 < y \leq 1.$$

Therefore the range is R: $0 < y \leq 1$.

25. If $f(x) = ax + b$ and $g(x) = cx + d$, then

$f(g(x)) = a(cx + d) + b$ and $g(f(x)) = c(ax + b) + d$.

Therefore, we must have

$$acx + ad + b = acx + bc + d$$

or

$$ad + b = bc + d.$$

27. (a) $f(\frac{1}{x}) = \dfrac{\frac{1}{x}}{\frac{1-x}{x}} = \dfrac{1}{1-x}$, $x \neq 0$

(b) $f(-x) = \dfrac{-x}{-x-1} = \dfrac{x}{x+1}, \quad x \neq -1$

(c) $f(f(x)) = \dfrac{\frac{x}{x-1}}{\frac{x}{x-1} - 1} = \dfrac{\frac{x}{x-1}}{\frac{x-x+1}{x-1}} = x, \quad x \neq 1$

(d) $f\left(\dfrac{1}{f(x)}\right) = \dfrac{\frac{x-1}{x}}{\frac{x-1}{x} - 1} = \dfrac{\frac{x-1}{x}}{\frac{x-1-x}{x}} = \dfrac{x-1}{-1} = 1 - x, \quad x \neq 0,1$

29. $|x| + |y| = 1$

In Quadrant I, $x \geq 0$, $y \geq 0$ and $x + y = 1$ or $y = 1 - x$
In Quadrant II, $x \leq 0$, $y \geq 0$ and $-x + y = 1$ or $y = x + 1$
In Quadrant III, $x \leq 0$, $y \leq 0$ and $-x - y = 1$ or $y = -x - 1$
In Quadrant IV, $x \geq 0$, $y \leq 0$ and $x - y = 1$ or $y = x - 1$.
The graph is the intersection of these four lines.

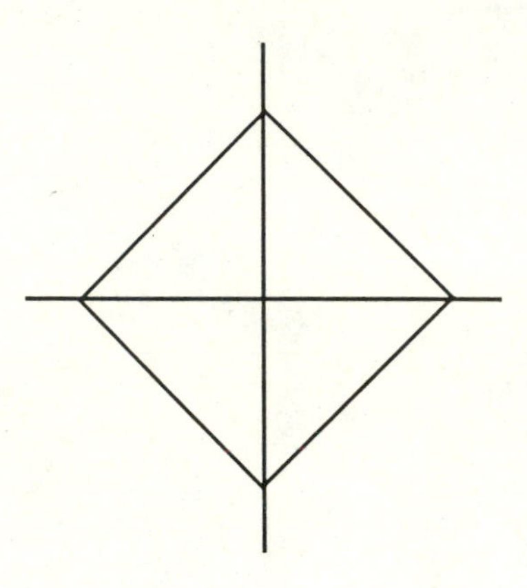

31. $M(a,b) = \dfrac{a+b}{2} + \dfrac{|a-b|}{2}$. If $a \geq b$,

$M = \dfrac{a+b}{2} + \dfrac{a-b}{2} = \dfrac{2a}{a} = a$; if $a \leq b$, $M = \dfrac{a+b}{2} + \dfrac{b-a}{2} = \dfrac{2b}{b} = b$.

The expression $M(a,b) = \dfrac{a+b}{2} - \dfrac{|a-b|}{2}$ gives the smaller, because when $a \leq b$ $m = \dfrac{a+b}{2} - \dfrac{b-a}{2} = a$ and when $b \geq a$ $m = \dfrac{a+b}{2} - \dfrac{a-b}{2} = b$

33. (a) $y = [x)$

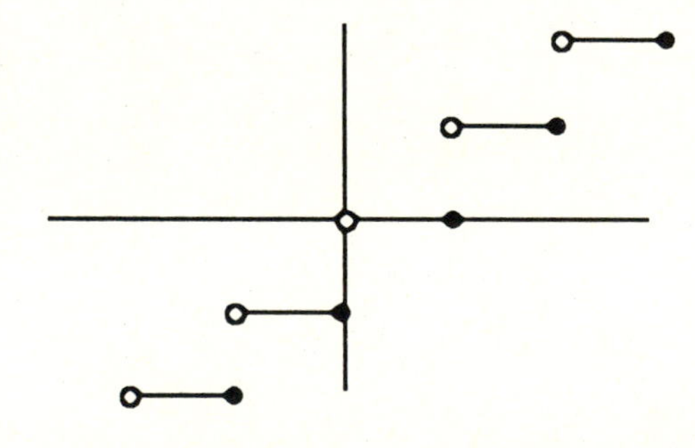

(b) $y = [x] - [x)$

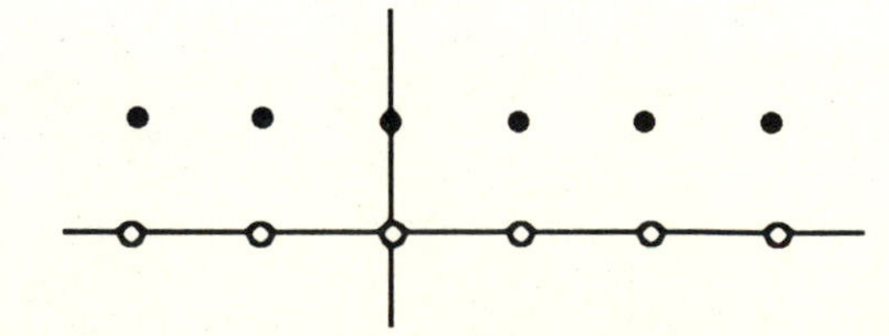

35. (a) $f(x) = \dfrac{x - 1}{x + 1}$

$$f'(x) = \lim_{h\to 0} \frac{\dfrac{x+h-1}{x+h+1} - \dfrac{x-1}{x+1}}{h} = \lim_{h\to 0} \frac{2h}{h(x+h+1)(x+1)}$$

$$= \lim_{h\to 0} \frac{2}{(x+h+1)(x+1)} = \frac{2}{(x + 1)^2}$$

(b) $f(x) = x^{3/2}$

$$\frac{f(x+h)-f(x)}{h} = \frac{(x+h)^{3/2}-x^{3/2}}{h}$$

$$= \frac{(x+h)^3-x^3}{h[(x+h)^{3/2} + x^{3/2}]}$$

$$= \frac{3x^2h + 3xh^2 + h^2}{h[(x+h)^{3/2} + x^{3/2}]}$$

$$f'(x) = \lim_{h\to 0} \frac{3x^2 + 3xh + h^2}{(x+h)^{3/2} + x^{3/2}} = \frac{3}{2}x^{1/2}$$

(c) $f(x) = x^{1/3}$

$$\frac{f(x+h)-f(x)}{h} = \frac{(x+h)^{1/3} - x^{1/3}}{h}$$

$$= \frac{1}{(x+h)^{2/3} + (x+h)^{1/3}x^{1/3} + x^{2/3}}$$

$$f'(x) = \lim_{h\to 0} \frac{1}{(x+h)^{2/3} + (x+h)^{1/3}x^{1/3} + x^{2/3}} = \frac{1}{3x^{2/3}}$$

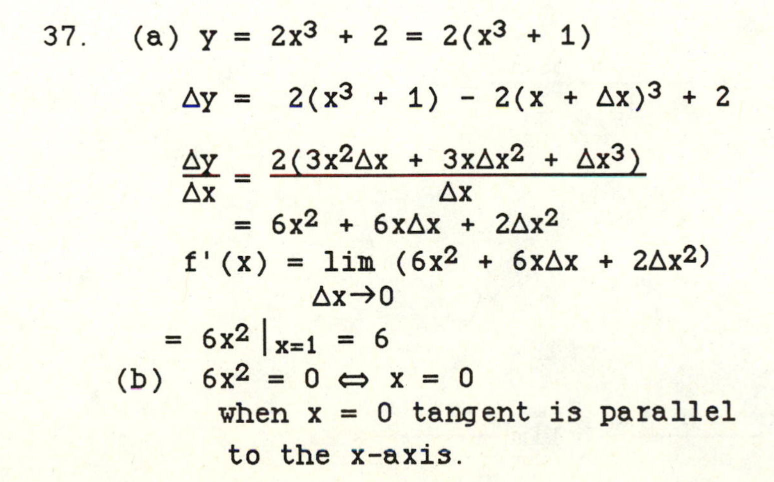

37. (a) $y = 2x^3 + 2 = 2(x^3 + 1)$

$$\Delta y = 2(x^3 + 1) - 2(x + \Delta x)^3 + 2$$

$$\frac{\Delta y}{\Delta x} = \frac{2(3x^2\Delta x + 3x\Delta x^2 + \Delta x^3)}{\Delta x}$$

$$= 6x^2 + 6x\Delta x + 2\Delta x^2$$

$$f'(x) = \lim_{\Delta x\to 0} (6x^2 + 6x\Delta x + 2\Delta x^2)$$

$$= 6x^2 \Big|_{x=1} = 6$$

(b) $6x^2 = 0 \Leftrightarrow x = 0$

when $x = 0$ tangent is parallel to the x-axis.

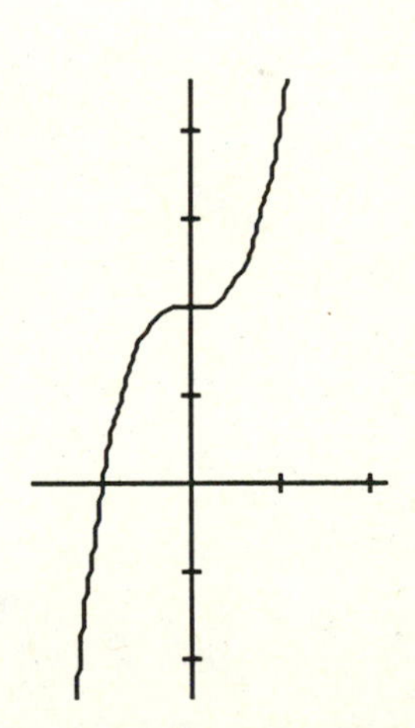

39. $y = 180x - 16x^2 = 4x(45 - 4x)$

$$f'(x_1) = \lim_{h\to 0} \frac{f(x_1+h) - f(x_1)}{h}$$
$$= \lim_{h\to 0} \frac{180(x_1 + h) - 16(x_1 + h)^2 - 180x_1 + 16x_1^2}{h}$$
$$= \frac{180h - 16(x_1^2 + 2x_1h + h^2) + 16x_1^2}{h}$$
$$= \frac{180h - 32x_1h - 16h^2}{h}$$
$$= 180 - 32x_1 - 16h$$

$$\left.\frac{dy}{dx}\right|_{x=x1} = 180 - 32x_1$$

$180 - 32x_1 = 0$ when $x_1 = \frac{45}{8}$

$$f(\tfrac{45}{8}) = 180\cdot\frac{45}{8} - 16(\tfrac{45}{8})^2 = \frac{8100}{8} - 16\cdot\frac{2045}{648} = \frac{2025}{4}.$$

Therefore, horizontal tangent occurs at $(\frac{45}{8}, \frac{2025}{4})$.

41. $s = 32t - 16t^2$; $s' = 32 - 32t = 0$ when $t = 1$. $s(1) = 16$

43. $$\lim_{x\to\infty} \frac{\sin x}{x} = 0$$

45. $$\lim_{x\to\infty} \frac{1 + \sin x}{x} = 0$$

47. $$\lim_{x\to 0} \frac{x}{\tan 3x} = \lim_{x\to 0} \frac{x\cos 3x}{\sin 3x} = \lim_{x\to 0} \frac{3x}{\sin 3x}\cdot\frac{\cos 3x}{3} = \frac{1}{3}$$

49. $$\lim_{x\to a} \frac{x^2 - a^2}{x - a} = \lim_{x\to a}(x + a) = 2a$$

51. $$\lim_{h\to 0} \frac{(x + h)^2 - x^2}{h} = \lim_{h\to 0} \frac{x^2 + 2xh + h^2 - x^2}{h} = \lim_{h\to 0} \frac{2xh + h^2}{h} = 2x$$

53. $$\lim_{\Delta x\to 0} \frac{\frac{1}{x + \Delta x} - \frac{1}{x}}{\Delta x} = \lim_{\Delta x\to 0} \frac{x - x - \Delta x}{(\Delta x)(x)(x + \Delta x)} = \lim_{\Delta x\to 0} \frac{-1}{x(x + \Delta x)} = -\frac{1}{x^2}$$

55. $\lim_{x\to 1}\frac{1-\sqrt{x}}{1-x}=\lim_{x\to 1}\frac{1-\sqrt{x}}{1-x}\cdot\frac{1+\sqrt{x}}{1+\sqrt{x}}=\lim_{x\to 0}\frac{1}{1+\sqrt{x}}=1$

57. $\lim_{x\to\infty}(1-x\cos x)$ does not exist.

59. $\lim_{x\to 0^+}\frac{|x|}{x}=1$

61. $\lim_{x\to 4^-}([x]-x)=3-4=-1$

63. $\lim_{x\to 3^+}\frac{[x]^2-9}{x^2-9}=0$

65. $\lim_{x\to 0}x[x]=0$

67. $f(x) = \frac{x-1}{2x^2-7x+5}$

(a) $\lim_{x\to\infty}\frac{x-1}{2x^2-7x+5} = 0$

(b) $\lim_{x\to 1}\frac{x-1}{(x-1)(2x-5)} = \lim_{x\to 1}\frac{1}{2x-5} = \frac{-1}{3}$

(c) $f(\frac{-1}{x}) = \frac{\frac{-1}{x}-1}{\frac{2}{x^2}+\frac{7}{x}+5} = \frac{-1-x}{\frac{2+7x+5x^2}{x}} = \frac{-x(x+1)}{5x^2+7x+2}$

$f(0) = \frac{-1}{5}$

$\frac{1}{f(x)} = \frac{2x^2-7x+5}{x-1} = 2x-5,\ x \neq 1, 5/2$

69. (a) $\lim_{n\to\infty}\sqrt{n^2+1}-h = \lim_{n\to\infty}\sqrt{n^2+1}-h\cdot\frac{\sqrt{n^2+1}+n}{\sqrt{n^2+1}+n}$

$= \lim_{n\to\infty}\frac{n^2+1-n^2}{\sqrt{n^2+1}+n} = \lim_{n\to\infty}\frac{1}{\sqrt{n^2+1}+n} = 0$

(b) $\lim_{n\to\infty}\sqrt{n^2+n}-n = \lim_{n\to\infty}\sqrt{n^2+n}-n\cdot\frac{\sqrt{n^2+n}+n}{\sqrt{n^2+n}+n}$

$= \lim_{n\to\infty}\frac{n^2+n-n^2}{\sqrt{n^2+n}+n} = \lim_{n\to\infty}\frac{1}{\sqrt{1+\frac{1}{n}}+1} = \frac{1}{2}$

71. We must find M such that $\left|\frac{t^2 + t}{t^2 - 1} - 1\right| = \left|\frac{t^2 + t - t^2 + 1}{t^2 - 1}\right| =$ $\left|\frac{t + 1}{t^2 - 1}\right| = \frac{1}{|t - 1|} < \varepsilon$ for all $t > M$. Let $t > 1$. Then $t - 1 > \frac{1}{\varepsilon}$ or $t > \frac{1}{\varepsilon} + 1$. Let $M = \frac{1}{\varepsilon} + 1$.

73. We are given that $F(t) \leq M$ and that $F(t) \to L$ as $t \to c$. Suppose that $L > M$ and take $\varepsilon = \frac{L - M}{2}$. There exists $\delta > 0$ for which $0 < |t - c| < \delta \Rightarrow |F(t) - L| < \frac{L - M}{2}$. But then we have $F(t) - L > L - \frac{L - M}{2} = \frac{L}{2} + \frac{M}{2} > \frac{M}{2} + \frac{M}{2} > M$, which is a contradiction. Therefore, $L \leq M$.

75. $f(x) = \frac{x^2 - 16}{|x - 4|}$

$$\lim_{x \to 4^+} \frac{x^2 - 16}{|x - 4|} = \lim_{x \to 4^+} \frac{x^2 - 16}{x - 4} = \lim_{x \to 4^+} (x + 4) = 8$$

$$\lim_{x \to 4^-} \frac{x^2 - 16}{4 - x} = \lim_{x \to 4^-} - \frac{(16 - x^2)}{4 - x} = \lim_{x \to 4^-} - (4 + x) = -8$$

f x) cannot be defined so as to make f continuous at since $\lim_{x \to 4} f(x)$ does not exist

77. $y = \frac{1}{[x]}$ is discontinuous when $0 \leq x < 1$, and at every integer.

79. The function $f(x) = |x|$ is continuous at $x = 0 \Rightarrow$ $\lim_{x \to 0} |x| = 0 \Rightarrow ||x| - 0|$ can be made to be as small as needed. In particular, $||x| - 0| = |x| = |x - 0| < \varepsilon$. Therefore, take $\delta \leq \varepsilon$.

81. Let f be a continuous function and suppose that $f(c) > 0$. Take $\varepsilon = \frac{f(c)}{2}$. Since the limit of $f(x) = f(c)$ as $x \to c$, there is $\delta > 0$ for which

$$|x - c| < \delta \Rightarrow |f(x) - f(c)| < \frac{f(c)}{2}.$$

But then we have $\frac{f(c)}{2} < f(x) < \frac{3f(c)}{2}$.

83. (a) $a^2 < b^2 \Rightarrow b^2 - a^2 > 0$. But

$b^2 - a^2 = (|b| + |a|)(|b| - |a|)$

and $|b| + |a|$ is positive.

Therefore $b^2 - a^2 > 0$

$\Leftrightarrow |b| - |a| > 0$ or $|b| > |a|$.

(b) $|a + b| < (|a| + |b|) \Leftrightarrow (a + b)^2 < (|a| + |b|)^2$, from part (a). But

$(|a| + |b|)^2 = |a|^2 + |b|^2 + 2|a||b|$ and

$(a + b)^2 = a^2 + b^2 + 2ab$.

Since $2|a||b| > 2ab$, the result follows.

(c) Since $a = b + (a - b)$,

$|a| = |b + (a - b)|$

$\leq |b| + |a - b|$, so

$|a| - |b| \leq |a - b|$.

This is true for all a,b so

$|b| - |a| \leq |b - a| = |a - b|$.

Together, these imply that

$||a| - |b|| \leq |a - b|$.

(d) From part (b), $|a_1 + a_2| \leq |a_1| + |a_2|$. Suppose that

$|a_1 + a_2 + \ldots. + a_k| \leq |a_1| + |a_2| + \ldots. + |a_k|$.

Then $|(a_1 + a_2 + \ldots. + a_k) + a_{k+1}| \leq$

$|(a_1 + a_2 + \ldots. + a_k)| + |a_{k+1}| \leq$

$|a_1| + |a_2| + \ldots. + |a_k| + |a_{k+1}|$.

85. Let $\Phi_i(x) = K_i(x - x_1)\ldots(x - x_{i-1})(x - x_{i+1})\ldots(x - x_n)$.

To find K, let $\Phi_i(x_i) = 1$. Then

$K = \dfrac{1}{(x_i - x_1)\ldots(x_i - x_{i-1})(x_i - x_{i+1})\ldots(x_i - x_n)}$. Thus,

$$\Phi_i(x) = \frac{(x - x_1)\ldots(x - x_{i-1})(x - x_{i+1})\ldots(x - x_n)}{(x_i - x_1)\ldots(x_i - x_{i-1})(x_i - x_{i+1})\ldots(x_i - x_n)}$$

CHAPTER 2

DERIVATIVES

2.1 POLYNOMIAL FUNCTIONS AND THEIR DERIVATIVES

1. $y = x \Rightarrow \frac{dy}{dx} = 1$ and $\frac{d^2y}{dx^2} = 0$

3. $y = x^2 \Rightarrow \frac{dy}{dx} = 2x$ and $\frac{d^2y}{dx^2} = 2$

5. $y = -x^2 + 3 \Rightarrow \frac{dy}{dx} = -2x$ and $\frac{d^2y}{dx^2} = -2$

7. $y = 2x + 1 \Rightarrow \frac{dy}{dx} = 2$ and $\frac{d^2y}{dx^2} = 0$

9. $y = \frac{x^3}{3} + \frac{x^2}{2} + x \Rightarrow \frac{dy}{dx} = x^2 + x + 1$ and $\frac{d^2y}{dx^2} = 2x + 1$

11. $s = 16t^2 + 3 \Rightarrow v = \frac{ds}{dt} = 32\,t$ and $a = \frac{d^2s}{dt^2} = 32$

13. $s = 16t^2 - 60t \Rightarrow v = \frac{ds}{dt} = 32\,t - 60$ and $a = \frac{d^2s}{dt^2} = 32$

15. $s = \frac{gt^2}{2} + v_0t + s_0$ $(g, v_0, s_0$ constants$) \Rightarrow v = \frac{ds}{dt} = gt + v_0$ and $a = \frac{d^2s}{dt^2} = g$

17. $y = 5x^3 - 3x^5 \Rightarrow \frac{dy}{dx} = 15x^2 - 15x^4 \Rightarrow \frac{d^2y}{dx^2} = 30x - 60x^3$

19. $y = \frac{x^4}{4} - \frac{x^3}{3} + \frac{x^2}{2} - x + 3 \Rightarrow \frac{dy}{dx} = x^3 - x^2 + x - 1$ and $\frac{d^2y}{dx^2} = 3\,x^2 - 2x + 1$

21. $12y = 6x^4 - 18x^2 - 12x \Rightarrow y = \frac{1}{2}x^4 - \frac{3}{2}x^2 - x$, so

$\frac{dy}{dx} = 2x^3 - 3x - 1$ and $\frac{d^2y}{dx^2} = 6x^2 - 3$

23. $y = x^2(x^3 - 1) = x^5 - x^2 \Rightarrow \frac{dy}{dx} = 5x^4 - 2x$ and $\frac{d^2y}{dx^2} = 20x^3 - 2$

25. $y = (3x - 1)(2x + 5) = 6x^2 + 13x - 5 \Rightarrow \frac{dy}{dx} = 12x + 13$ and $\frac{d^2y}{dx^2} = 12$

27. (c) If $y = x^2 + 5x$, then $y' = 2x + 5$ and $y'(3) = 11$

29. $y = x^3 - 3x + 1 \Rightarrow y' = 3x^2 - 3$. The slope of tangent is $y'(2) = 9$.
$m_\perp = -\frac{1}{9}$. At $(2, 3)$ the normal is $y - 3 = -\frac{1}{9}(x - 2)$ or $9y + x = 29$.

31. $y = x^3 + x \Rightarrow y' = 3x^2 + 1$. Since the slope is 4, we have

$y(1) = 2$ and $y(-1) = -2$. $\therefore$ Tangent line at $(1,2)$ with slope 4 is: $y - 2 = 4(x - 1)$ or $y = 4x - 2$

The tangent line at $(-1,-2)$ with slope 4 is: $y + 2 = 4(x + 1)$ or $y = 4x + 2$. Since $y = 3x^2 + 1$ is an upward opening parabola the smallest value will occur at the vertex $(0,1)$. Thus the smallest slope is 1 and this occurs when $x = 0$.

33. $y = x^3 \Rightarrow y' = 3x^2$. $y'(-2) = 12$. Tangent at $(-2,-8)$ is: $y + 8 = 12(x + 2)$ or $y = 12x + 16$. $x = 0 \Rightarrow y = 16$.

y-intercept is $(0,16)$. $y = 0 \Rightarrow x = -\frac{4}{3}$. x-intercept is $\left(-\frac{4}{3}, 0\right)$.

35. The curve $y = ax^2 + bx + c$ is tangent to $y = x$ at the origin $\Rightarrow (0,0)$ is a point on the curve. $\therefore 0 = 0a + 0b + c$ or $c = 0$. If $y = ax^2 + bx$, then $y' = 2ax + b$ measures the slope of the tangent to the curve at any point. In particular, $y = x$ has slope 1 at $x = 0$. $\therefore 1 = 2a(0) + b \Rightarrow b = 1$ and $y = ax^2 + 1$. Since the point $(1,2)$ is also given on the curve, $2 = (1)^2 a + 1$ or $a = 1$. $\therefore$ The equation is $y = x^2 + x$.

37. Mars: $s = 1.86t^2$ m
$v = 3.72t$ m/sec
$3.72t = 16.6$
$t = 4.46$ sec

Jupiter: $s = 11.44t^2$ m
$v = 22.88t$ m/sec
$22.88t = 16.6$
$t = .73$ sec

39. $s = 24t - 4.9t^2$ m $\Rightarrow v = 24 - 9.8t$ m/sec. The maximum height will occur when the rock ceases to rise, i.e. when the velocity is zero. $\therefore 24 - 9.8t = 0$ when $t = 2.45$ sec. $s(2.45) = 24(2.45) - 4.9(2.45)^2 = 29.39$ m.

41. $s = t^3 - 4t^2 - 3t \Rightarrow v = 3t^2 - 8t - 3.$

$3t^2 - 8t - 3 = 0 \Leftrightarrow (3t + 1)(t - 3) = 0 \Leftrightarrow t = -\frac{1}{3}$ or 3

$a = 6t - 8. \therefore\ a(3) = 10$ and $a(-\frac{1}{3}) = -10.$

43. $y = x^n \Rightarrow y' = nx^{n-1}$. Therefore $y'(x_1) = nx_1^{n-1}$ is the slope of the tangent at $P(x_1, y_1)$. Since this tangent contains the point $T(t, 0)$, its slope is also given by $m = \frac{y_1 - 0}{x_1 - t}$. Equating the two expressions for slope, we have

$$nx_1^{n-1} = \frac{y_1}{x_1 - t} \Leftrightarrow nx_1^{n-1} = \frac{x_1^n}{x_1 - t} \Leftrightarrow x_1 - t = \frac{x_1^n}{nx_1^{n-1}} \Leftrightarrow$$

$$x_1 - t = \frac{1}{n} x_1^{n-n+1} \Leftrightarrow t = x_1 - \frac{1}{n} x_1 = x_1(1 - \frac{1}{n}).$$

$\therefore$ The tangent to any other point $Q(x_2, y_2)$ has x-intercept $T(x_2(1-\frac{1}{n}), 0)$. To construct this tangent, connect these two points.

2.2 PRODUCTS, POWERS AND QUOTIENTS

1. $y = \frac{x^3}{3} - \frac{x^2}{2} + x - 1 \Rightarrow \frac{dy}{dx} = x^2 - x + 1$

3. $y = (x^2 + 1)^5$

$$\frac{dy}{dx} = 5(x^2 + 1)^4 \frac{d}{dx}(x^2 + 1) = 5(x^2 + 1)(2x) = 10x\ (x^2 + 1)^4$$

5. $y = (x + 1)^2 (x^2 + 1)^{-3}$

$$\frac{dy}{dx} = (x + 1)^2 \left[-3(x^2 + 1)^{-4} \frac{d}{dx}(x^2 + 1)\right] + (x^2 + 1)^{-3}[2(x + 1)]$$

$$= (x + 1)^2(-6x)(x^2 + 1)^{-4} + 2(x + 1)(x^2 + 1)^{-3}$$

$$= -2(x + 1)(x^2 + 1)^{-4}(2x^2 + 3x - 1)$$

7. $y = \dfrac{2x + 5}{3x - 2}$

$$\frac{dy}{dx} = \frac{(3x - 2)\dfrac{d}{dx}(2x + 5) - (2x + 5)\dfrac{d}{dx}(3x - 2)}{(3x - 2)^2}$$

$$= \frac{2(3x - 2) - 3(2x + 5)}{(3x - 2)^2}$$

$$= \frac{-19}{(3x - 2)^2}$$

9. $y = (1 - x)(1 + x^2)^{-1}$

$$\frac{dy}{dx} = (1 - x)\left[-(1 + x^2)^{-2}(2x)\right] + (1 + x^2)^{-1}(-1)$$

$$= (1 + x^2)^{-2}\left[-2x(1 - x) - (1 + x^2)\right]$$

$$= (1 + x^2)^{-2}(x^2 - 2x - 1)$$

11. $y = \dfrac{5}{(2x - 3)^4} = 5(2x - 3)^{-4} \Rightarrow \dfrac{dy}{dx} = -20(2x - 3)^{-5}(2) = \dfrac{-40}{(2x - 3)^5}$

13. $y = (5 - x)(4 - 2x) \Rightarrow \dfrac{dy}{dx} = -2(5 - x) - (4 - 2x) = 4x - 14$

15. $y = (2x - 1)^3(x + 7)^{-3}$

$$\frac{dy}{dx} = (2x - 1)^3\left[-3(x + 7)^{-4}\right] + (x + 7)^{-3}\left[3(2x - 1)^2(2)\right]$$

$$= (2x - 1)^2(x + 7)^{-4}\left[-3(2x - 1) + 6(x + 7)\right]$$

$$= 45(2x - 1)^2(x + 7)^{-4}$$

17. $y = (2x^3 - 3x^2 + 6x)^{-5} \Rightarrow \dfrac{dy}{dx} = -5(2x^3 - 3x^2 + 6x)^{-6}(6x^2 - 6x + 6)$

19. $y = \dfrac{(x - 1)^2}{x^2} = \dfrac{x^2 - 2x + 1}{x^2} = 1 - 2x^{-1} + x^{-2}$

$$\frac{dy}{dx} = 2x^{-2} - 2x^{-3}$$

21. $y = \dfrac{12}{x} - \dfrac{4}{x^3} + \dfrac{3}{x^4} = 12x^{-1} - 4x^{-3} + 3x^{-4}$

$$\frac{dy}{dx} = -12x^{-2} + 12x^{-4} - 12x^{-5}$$

23. $y = \dfrac{x^2 - 1}{x^2 + x - 2} = \dfrac{(x + 1)(x - 1)}{(x + 2)(x - 1)} = \dfrac{x + 1}{x + 2} \Rightarrow \dfrac{dy}{dx} = \dfrac{x + 2 - x - 1}{(x + 2)^2} = \dfrac{1}{(x + 2)^2}$

25. $s = \dfrac{t}{t^2 + 1} \Rightarrow \dfrac{ds}{dt} = \dfrac{t^2 + 1 - t(2t)}{(t^2 + 1)^2} = \dfrac{1 - t^2}{(t^2 + 1)^2}$

27. $s = (t^2 - t)^{-2} \Rightarrow \dfrac{ds}{dt} = -2(t^2 - t)^{-3}(2t - 1) = \dfrac{2 - 4t}{(t^2 - t)^3}$

29. $s = \dfrac{2t}{3t^2 + 1} \Rightarrow \dfrac{ds}{dt} = \dfrac{2(3t^2 + 1) - 2t(6t)}{(3t^2 + 1)^2} = \dfrac{2 - 6t^2}{(3t^2 + 1)^2}$

31. $s = (t^2 + 3t)^3 \Rightarrow \dfrac{ds}{dt} = 3(t^2 + 3t)^2(2t + 3)$

33. Each of the following is evaluated at $x = 0$:

(a) $\dfrac{d}{dx}(uv) = uv' + vu' = (5)(2)+(-1)(-3) = 13$

(b) $\dfrac{d}{dx}(\dfrac{u}{v}) = \dfrac{vu'-uv'}{v^2} = \dfrac{(-1)(-3)-(5)(2)}{(-1)^2} = -7$

(c) $\dfrac{d}{dx}(\dfrac{v}{u}) = \dfrac{uv'-vu'}{u^2} = \dfrac{(5)(2)-(-1)(-3)}{5^2} = \dfrac{7}{25}$

(d) $\dfrac{d}{dx}(7v-2u) = 7v'-2u' = 7(2) - 2(-3) = 20$

(e) $\dfrac{d}{dx}(u^3) = 3u^2u' = 3(5)^2(-3) = -225$

(f) $\dfrac{d}{dx}(5v^{-3}) = -3(5)(v^{-4})v' = -15(-1)^{-4}(2) = -30$

35. $y = x + \dfrac{1}{x} = x + x^{-1} \Rightarrow y' = 1 - x^{-2}$. When $x = 2$,

$y = 2 + \dfrac{1}{2} = \dfrac{5}{2}$ and $y' = 1 - \dfrac{1}{4} = \dfrac{3}{4}$.

$\therefore\ y - \dfrac{5}{2} = \dfrac{3}{4}(x - 2)$ or $4y - 3x = 4$.

37. $y = (3-2x)^{-1} \Rightarrow \dfrac{dy}{dx} = -(3-2x)^{-2}(-2) = 2(3-2x)^{-2}$

$\dfrac{d^2y}{dx^2} = -4(3-2x)^{-3}(-2) = 8(3-2x)^{-3}$

39. $y = x(x-1)(x+1)$

$\dfrac{dy}{dx} = x(x-1)\dfrac{d}{dx}(x+1) + x(x+1)\dfrac{d}{dx}(x-1) + (x+1)(x-1)\dfrac{d}{dx}(x)$

$= x(x-1) + x(x+1) + (x+1)(x-1) = 3x^2 - 1$

Alternatively, $y = x(x^2 - 1) = x^3 - x \Rightarrow \dfrac{dy}{dx} = 3x^2 - 1$

41. $y = (1 - x)(x + 1)(3 - x^2)$

$$\frac{dy}{dx} = (1 - x)(x + 1)\frac{d}{dx}(3 - x^2) + (1 - x)(3 - x^2)\frac{d}{dx}(x + 1) + (x + 1)(3 - x^2)\frac{d}{dx}(1 - x)$$

$$= (1 - x^2)(-2x) + (3 - 3x - x^2 + x^3) + (3 + 3x - x^2 - x^3)(-1) = 4x^3 - 8x$$

43. If $y = uv$ then $\frac{dy}{dt} = u\frac{dv}{dt} + v\frac{du}{dt} = u(.05v) + v(.04u) = .09uv = .09y$

45. $p(t) = \frac{a^2kt}{akt + 1} \Rightarrow \frac{dp}{dt} = \frac{(akt + 1)(a^2k) - (a^2kt)(ak)}{(akt + 1)^2}$

$= \frac{a^2k}{(akt + 1)^2}$. Since the numerator is constant and the denominator is increasing, $\frac{dp}{dt}$ is decreasing. ∴ The largest value will occur when $t = 0$.

∴ The maximum value is $\frac{dp}{dt} = a^2k$.

47. If $y = u_1u_2u_3...u_n$, then

$$\frac{dy}{dx} = u_2u_3...u_n\frac{d}{dx}(u_1) + ... + u_1u_2...u_{n-1}\frac{d}{dx}(u_n)$$

Let $u = u_1 = u_2 = ... = u_n$. Then $y = u^n$ and

$$\frac{dy}{dx} = uu..u\frac{du}{dx} + ...+ uu...u\frac{du}{dx}$$

$$= u^{n-1}\frac{du}{dx} + ... + u^{n-1}\frac{du}{dx} = nu^{n-1}\frac{du}{dx}.$$

2.3 IMPLICIT DIFFERENTIATION AND FRACTIONAL POWERS

1. $x^2 + y^2 = 1 \Rightarrow 2x + 2y\frac{dy}{dx} = 0 \Rightarrow 2y\frac{dy}{dx} = -2x \Rightarrow \frac{dy}{dx} = -\frac{x}{y}$

3. $x^2 - xy = 2 \Rightarrow 2x - x\frac{dy}{dx} - y = 0 \Rightarrow -x\frac{dy}{dx} = y - 2x \Rightarrow$

$$\frac{dy}{dx} = \frac{y - 2x}{x}$$

5. $y^2 = x^3 \Rightarrow 2y\frac{dy}{dx} = 3x^2 \Rightarrow \frac{dy}{dx} = \frac{3x^2}{2y}$

7. $x^{\frac{1}{2}} + y^{\frac{1}{2}} = 1 \Rightarrow \frac{1}{2}x^{-\frac{1}{2}} + \frac{1}{2}y^{-\frac{1}{2}}\frac{dy}{dx} = 0$

$y^{-\frac{1}{2}}\frac{dy}{dx} = -x^{-\frac{1}{2}} \Rightarrow \frac{dy}{dx} = -x^{-\frac{1}{2}}y^{\frac{1}{2}} = -\sqrt{\frac{y}{x}}$

9. $x^2 = \frac{x-y}{x+y} \Rightarrow x^3 + x^2y = x - y \Rightarrow 3x^2 + x^2\frac{dy}{dx} + 2xy = 1 - \frac{dy}{dx}$

$(x^2 + 1)\frac{dy}{dx} = 1 - 2xy - 3x^2 \Rightarrow \frac{dy}{dx} = \frac{1 - 2xy - 3x^2}{x^2 + 1}$

11. $y = x\sqrt{x^2 + 1} = x(x^2 + 1)^{\frac{1}{2}}$

$\frac{dy}{dx} = (x^2 + 1)^{\frac{1}{2}} + x\left[\frac{1}{2}(x^2 + 1)^{-\frac{1}{2}}(2x)\right] = \frac{\sqrt{x^2 + 1}}{1} + \frac{x^2}{\sqrt{x^2 + 1}} = \frac{2x^2 + 1}{\sqrt{x^2 + 1}}$

13. $2xy + y^2 = x + y \Rightarrow 2x\frac{dy}{dx} + 2y + 2y\frac{dy}{dx} = 1 + \frac{dy}{dx}$

$2x\frac{dy}{dx} + 2y\frac{dy}{dx} - \frac{dy}{dx} = 1 - 2y$

$\frac{dy}{dx} = \frac{1 - 2y}{2x + 2y - 1}$

15. $y^2 = \frac{x^2 - 1}{x^2 + 1} \Rightarrow 2y\frac{dy}{dx} = \frac{(x^2 + 1)(2x) - (x^2 - 1)(2x)}{(x^2 + 1)^2}$

$2y\frac{dy}{dx} = \frac{2x^3 + 2x - 2x^3 + 2x}{(x^2 + 1)^2} = \frac{4x}{(x^2 + 1)^2} \Rightarrow \frac{dy}{dx} = \frac{2x}{y(x^2 + 1)^2}$

17. $(3x + 7)^5 = 2y^3 \Rightarrow 5(3x + 7)^4(3) = 6y^2\frac{dy}{dx}$

$\frac{dy}{dx} = \frac{5(3x + 7)^4}{2y^2}$

19. $\frac{1}{x} + \frac{1}{y} = 1 \Rightarrow x^{-1} + y^{-1} = 1 \Rightarrow -x^{-2} - y^{-2}\frac{dy}{dx} = 0$

$\therefore \frac{dy}{dx} = -x^{-2}y^2 = -\left[\frac{y}{x}\right]^2$

21. $y^2 = x^2 - x \Rightarrow 2y\frac{dy}{dx} = 2x - 1 \Rightarrow \frac{dy}{dx} = \frac{2x - 1}{2y}$

23. $y = \dfrac{\sqrt[3]{x^2+3}}{x} = \dfrac{(x^2+3)^{\frac{1}{3}}}{x} \Rightarrow \dfrac{dy}{dx} = \dfrac{x\left[\frac{1}{3}(x^2+3)^{-\frac{2}{3}}(2x)\right] - (x^2+3)^{\frac{1}{3}}}{x^2}$

$= \dfrac{\frac{2}{3}x^2(x^2+3)^{-\frac{2}{3}} - (x^2+3)^{\frac{1}{3}}}{x^2} = \dfrac{2x^2 - 3(x^2+3)}{3x^2(x^2+3)^{\frac{2}{3}}} = \dfrac{-x^2-9}{3x^2(x^2+3)^{\frac{2}{3}}}$

25. $x^3 + y^3 = 18xy \Rightarrow 3x^2 + 3y^2\dfrac{dy}{dx} = 18\left(x\dfrac{dy}{dx} + y\right)$

$y^2\dfrac{dy}{dx} - 6x\dfrac{dy}{dx} = 6y - x^2 \Rightarrow \dfrac{dy}{dx} = \dfrac{6y - x^2}{y^2 - 6x}$

27. $y = (2x + 5)^{-\frac{1}{5}} \Rightarrow \dfrac{dy}{dx} = -\dfrac{2}{5}(2x + 5)^{-\frac{6}{5}}$

29. $y = \sqrt{1 - \sqrt{x}} = (1 - x^{\frac{1}{2}})^{\frac{1}{2}} \Rightarrow \dfrac{dy}{dx} = \dfrac{1}{2}(1 - x^{\frac{1}{2}})^{-\frac{1}{2}}\dfrac{d}{dx}(1 - x^{\frac{1}{2}})$

$= \dfrac{1}{2}(1 - x^{\frac{1}{2}})^{-\frac{1}{2}}(-\dfrac{1}{2}x^{-\frac{1}{2}}) = \dfrac{-1}{4\sqrt{x}\sqrt{1 - \sqrt{x}}} = \dfrac{-1}{4\sqrt{x - x\sqrt{x}}}$

31. $y = x^{\frac{1}{2}} \Rightarrow y' = \dfrac{1}{2}x^{-\frac{1}{2}}$; $y(4) = 2$ and $y'(4) = \dfrac{1}{4}$.

$\therefore$ Tangent line is: $y - 2 = \dfrac{1}{4}(x - 4)$ or $4y - x = 4$.

The x-intercept is $(-4,0)$ and y-intercept is $(0,1)$.

33. $b = a^{\frac{2}{3}} \Rightarrow \dfrac{db}{da} = \dfrac{2}{3}a^{-\frac{1}{3}}$. $\dfrac{db}{da}$ does not exist if $a = 0$.

35. $x^2 + y^2 = 1 \Rightarrow 2x + 2y\dfrac{dy}{dx} = 0 \Rightarrow \dfrac{dy}{dx} = -\dfrac{x}{y}$

$\dfrac{d^2y}{dx^2} = -\dfrac{y - x\frac{dy}{dx}}{y^2} = -\dfrac{y - x\left(-\frac{x}{y}\right)}{y^2} = -\dfrac{y^2 + x^2}{y^3} = -\dfrac{1}{y^3}$

37. $x^{\frac{2}{3}} + y^{\frac{2}{3}} = 1 \Rightarrow \dfrac{2}{3}x^{-\frac{1}{3}} + \dfrac{2}{3}y^{-\frac{1}{3}}\dfrac{dy}{dx} = 0 \Rightarrow \dfrac{dy}{dx} = -x^{-\frac{1}{3}}y^{\frac{1}{3}} = -\left(\dfrac{y}{x}\right)^{\frac{1}{3}}$

$\dfrac{d^2y}{dx^2} = -\left[x^{-\frac{1}{3}}\left(\dfrac{1}{3}y^{-\frac{2}{3}}\dfrac{dy}{dx}\right) + y^{\frac{1}{3}}\left(-\dfrac{1}{3}x^{-\frac{4}{3}}\right)\right] =$

$-\dfrac{1}{3}[(x^{-\frac{1}{3}}y^{-\frac{2}{3}})(-x^{-\frac{1}{3}}y^{\frac{1}{3}}) - y^{\frac{1}{3}}x^{-\frac{4}{3}}] = \dfrac{1}{3}x^{-\frac{4}{3}}y^{-\frac{1}{3}}$

39. $y^2 + 2y = 2x + 1 \Rightarrow 2y\frac{dy}{dx} + 2\frac{dy}{dx} = 2 \Rightarrow \frac{dy}{dx} = \frac{1}{y + 1}$

$$\frac{d^2y}{dx^2} = -(y + 1)^{-2}\frac{dy}{dx} = -\frac{1}{(y + 1)^2}\cdot\frac{1}{y + 1} = -\frac{1}{(y + 1)^3}$$

41. $y + 2\sqrt{y} = x \Rightarrow \frac{dy}{dx} + 2(\frac{1}{2}y^{-\frac{1}{2}})\frac{dy}{dx} = 1 \Rightarrow \frac{dy}{dx} + \frac{1}{\sqrt{y}}\frac{dy}{dx} = 1$

$$\frac{dy}{dx} = \frac{1}{1 + \frac{1}{\sqrt{y}}} = (1 + y^{-\frac{1}{2}})^{-1}$$

$$\frac{d^2y}{dx^2} = -(1 + y^{-1/2})^{-2}\left(-\frac{1}{2}y^{-3/2}\frac{dy}{dx}\right) = \frac{1}{2}y^{-3/2}(1 + y^{-1/2})^{-2}(1 + y^{-1/2})^{-1}$$

$$= \frac{1}{2}y^{-3/2}(1 + y^{-1/2})^{-3} = \frac{1}{2}[y^{1/2}(1 + y^{-1/2})]^{-3} = \frac{1}{2}(\sqrt{y} + 1)^{-3}$$

43. $x^2 + xy - y^2 = 1 \Rightarrow 2x + x\frac{dy}{dx} + y - 2y\frac{dy}{dx} = 0.$

At P(2,3): $4 + 2\frac{dy}{dx} + 3 - 6\frac{dy}{dx} = 0$ or $\frac{dy}{dx} = \frac{7}{4}$

Slope of normal is $m = -\frac{4}{7}$

(a) $y - 3 = \frac{7}{4}(x - 2) \Rightarrow 4y - 7x = -2$

(b) $y - 3 = -\frac{4}{7}(x - 2) \Rightarrow 7y + 4x = 29$

45. $x^2y^2 = 9 \Rightarrow 2x^2y\frac{dy}{dx} + 2xy^2 = 0.$ At P(−1, 3),

$$2(-1)^2(3)\frac{dy}{dx} + 2(-1)(3)^2 = 0 \Rightarrow \frac{dy}{dx} = 3$$

(a) Tangent: $y - 3 = 3(x + 1) \Rightarrow y - 3x = 6$

(b) Normal: $y - 3 = -\frac{1}{3}(x + 1) \Rightarrow 3y + x = 8.$

47. $y^2 - 2x - 4y - 1 = 0 \Rightarrow 2y\frac{dy}{dx} - 2 - 4\frac{dy}{dx} = 0.$ At P(−2,1)

$2\frac{dy}{dx} - 2 - 4\frac{dy}{dx} = 0 \Rightarrow \frac{dy}{dx} = -1$

(a) tangent: $y - 1 = -(x + 2) \Rightarrow y + x = -1$

(b) normal: $y - 1 = x + 2 \Rightarrow y - x = 3$

49. $xy + 2x - y = 0 \Rightarrow x\frac{dy}{dx} + y + 2 - \frac{dy}{dx} = 0 \Rightarrow \frac{dy}{dx} = \frac{y+2}{1-x}$. The line

$2x + y = 0$ has slope $m = -2$. To be parallel, the normal lines must also have slope $m = -2$. Since a normal is perpendicular to a tangent, the tangent line must have slope $m = \frac{1}{2}$. $\therefore$

$\frac{y+2}{1-x} = \frac{1}{2} \Leftrightarrow 2y + 4 = 1 - x \Leftrightarrow x = -3 - 2y$. Substituting,

$y(-3-2y) + 2(-3-2y) - y = 0 \Leftrightarrow y^2 + 4y + 3 = 0 \Rightarrow y = -3, -1$

$y = -3 \Rightarrow x = 3$ and $y + 3 = -2(x - 3) \Rightarrow y + 2x = 3$

$y = -1 \Rightarrow x = -1$ and $y + 1 = -2(x + 1) \Rightarrow y + 2x = -3$

51. $y = x^2+2x-3 \Rightarrow \frac{dy}{dx} = 2x + 2$. At $P(1,0)$, $\frac{dy}{dx} = 4$. $\therefore$ Normal has equation $y = -\frac{1}{4}(x - 1) = -\frac{1}{4}x + \frac{1}{4}$. To find point of intersection, equate $-\frac{1}{4}x + \frac{1}{4} = x^2 + 2x - 3 \Rightarrow$

$4x^2 + 9x - 13 = 0 \Rightarrow x = 1, -\frac{13}{4}$.

$y(-\frac{13}{4}) = -\frac{1}{4}(-\frac{13}{4}) + \frac{1}{4} = \frac{17}{16}$. $\therefore (-\frac{13}{4}, \frac{17}{16})$

53. $x^2 + xy + y^2 = 7$ crosses the x-axis when $y = 0$, i.e. when $x = \pm\sqrt{7}$

$2x + x\frac{dy}{dx} + y + 2y\frac{dy}{dx} = 0 \Rightarrow \frac{dy}{dx} = \frac{-2x - y}{x + 2y}$

At the points $(\pm\sqrt{7}, 0)$, $\frac{dy}{dx} = -2 \Rightarrow$ the tangents are parallel.

55. $s(t) = \sqrt{1 + 4t} \Rightarrow v(t) = \frac{1}{2}(1 + 4t)^{-\frac{1}{2}}(4) = \frac{2}{\sqrt{1 + 4t}}$

$v(6) = \frac{2}{\sqrt{25}} = \frac{2}{5}$ m/s

$a(t) = 2[-\frac{1}{2}(1 + 4t)^{-\frac{3}{2}}(4)] = \frac{-4}{(1 + 4t)^{\frac{3}{2}}}$

$a(6) = \frac{-4}{(25)^{\frac{3}{2}}} = -\frac{4}{125}$ m/s^2

57. $2x^2 + 3y^2 = 5 \Rightarrow 4x + 6y\frac{dy}{dx} = 0 \Rightarrow \frac{dy}{dx} = -\frac{2x}{3y}$.

$y^2 = x^3 \Rightarrow \frac{dy}{dx} = \frac{3x^2}{2y}$.

At $(1,1)$, $m_1 = -\frac{2}{3}$ and $m_2 = \frac{3}{2} \Rightarrow$ tangents are orthogonal

At $(1,-1)$, $m_1 = \frac{2}{3}$ and $m_2 = -\frac{3}{2} \Rightarrow$ tangents are orthogonal.

2.4 LINEAR APPROXIMATIONS

1. $f(x) = x^4$; $f'(x) = 4x^3$; $f'(1) = 4$ and $f(1) = 1$

   $L(x) = 1 + 4(x - 1) = 4x - 3$

   $L(1.01) = 4(1.01) - 3 = 1.04$

   By calculator, $f(1.01) = 1.04060$ to five decimal places.

3. $f(x) = x^{-1} \Rightarrow f'(x) = -x^{-2}$; $f'(2) = -\frac{1}{4}$ and $f(2) = \frac{1}{2}$

   $L(x) = \frac{1}{2} - \frac{1}{4}(x - 2) = -\frac{1}{4}x + 1$.

   $L(2.1) = -\frac{1}{4}(2.1) + 1 = 0.475$. By calculator, $f(2.1) = 0.47619$

5. $f(x) = x^3 - x \Rightarrow f'(x) = 3x^2 - 1$; $f(1) = 0$ and $f'(1) = 2$

   $L(x) = 0 + 2(x-1) = 2x - 2$. $L(1.1) = 2.2 - 2 = .2$

   By calculator, $f(1.1) = (1.1)^3 - 1.1 = 0.231$

7. $f(x) = x^3 - 2x + 3 \Rightarrow f'(x) = 3x^2 - 2$; $f(2) = 7$ and $f'(2) = 10$

   $L(x) = 7 + 10(x - 2) = 10x - 13$; $L(1.9) = 6$

   By calculator, $f(1.9) = 6.059$

9. $f(x) = \sqrt{x} \Rightarrow f'(x) = \frac{1}{2}x^{-\frac{1}{2}}$; $f(4) = 2$ and $f'(4) = \frac{1}{4}$.

   $L(x) = 2 + \frac{1}{4}(x - 4) = \frac{1}{4}x + 1$. $L(4.1) = 2.025$.

   By calculator, $f(4.1) = 2.02485$

11. $f(x) = \sqrt{x^2 + 9} \Rightarrow f'(x) = \frac{1}{2}(x^2 + 9)^{-\frac{1}{2}}(2x) = \frac{x}{\sqrt{x^2 + 9}}$

$L(x) = 5 - \frac{4}{5}(x + 4) = -\frac{4}{5}x + \frac{9}{5}$

$L(-4.2) = 5 - \frac{4}{5}(-4.2+4) = 5.16$

By calculator, $f(-4.2) = 5.16140$

13. $(1 + x)^2 \approx 1 + 2x$

15. $\frac{1}{(1 + x)^5} = (1 + x)^{-5} \approx 1 - 5x$

17. $\frac{2}{(1 - x)^4} = 2(1 - x)^{-4} \approx 2[1 - 4(-x)] = 2 + 8x$

19. $2\sqrt{1 + x} = 2(1 + x)^{\frac{1}{2}} \approx 2[1 + \frac{1}{2}x] = 2 + x$

21. $\frac{1}{1 + x} \approx 1 - x$

23. (a) $(1.0002)^{100} = (1 + .0002)^{100} \approx 1 + 100(.0002) = 1.02$

(b) $\sqrt[3]{1.009} = (1 + .009)^{\frac{1}{3}} \approx 1 + \frac{1}{3}(.009) = 1.003$

(c) $\frac{1}{0.999} = (.999)^{-1} = (1 - .001)^{-1} \approx 1 - (-.001) = 1.001$

25. $f(x) = x^2 + 2x$, $a = 0$, $\Delta x = 0.1$

$\Delta y = f(0.1) - f(0) = [(.1)^2 + 2(.1)] - 0 = .21$

$dy = f'(0)\Delta x = [2(0) + 2](.1) = .2$

$\text{Error} = \Delta y - dy = .21 - .2 = .01$

27. $f(x) = x^3 - x$, $a = 1$, $\Delta x = 0.1$

$\Delta y = f(1.1) - f(1) = [(1.1)^3 - 1] - 0 = .231$

$dy = f'(1)\Delta x = [3(1)^2 - 1](.1) = 0.2$

$\text{Error} = \Delta y - dy = 0.231 - 0.2 = 0.031$

29. $f(x) = x^{-1}$, $a = .5$, $\Delta x = .1$

$\Delta y = f(.6) - f(.5) = (.6)^{-1} - (.5)^{-1} = -.\overline{3}$

$dy = f'(.5)\Delta x = -(.5)^{-2}(.1) = -.4$

$\text{Error} = \Delta y - dy = -.\overline{3} - (-.4) = .0\overline{6}$

31. $V = \frac{4}{3}\pi r^3 \Rightarrow dV = 4\pi r^2 dr$

33. $V = x^3 \Rightarrow dV = 3x^2 dx$

35. $V = \pi r^2 h$ and h constant $\Rightarrow dV = 2\pi rhdr$

37. $V = \frac{1}{3}\pi r^2 h$ and h constant $\Rightarrow dV = \frac{2}{3}\pi rhdr$

39. $A = \pi r^2$, $dr = 0.02$, $r = 2$

(a) $dA = 2\pi rdr$. If $r = 2$ and $dr = 0.02$, $dA = 0.08\pi$

(b) % error $= \frac{dA}{A} \times 100 = \frac{.08\pi}{16\pi} \times 100 = 2\%$

41. If $x = 10$ cm, then $dx = (.01)(10) = .1$ cm.

$V = x^3 \Rightarrow dV = 3x^2 dx = 3(10)^2(.1) = 30\ cm^3$.

$\frac{dV}{V} = \frac{30}{1000} = .03$. $\therefore$ Error $= 3\%$

43. $A = x^2$. For A to be within 2% of true value, we need

$|\Delta A| \leq .02A = .02x^2$. $\Delta A \approx dA = (\frac{dA}{dx})dx = 2xdx$.

$\therefore\ |2xdx| \leq .02x^2 \Leftrightarrow |dx| \leq \frac{.02x^2}{2x} = .01x$.

The side x must be measured to within 1% of its true value.

45. $V = \frac{1}{3}\pi h^3$. We need $|\Delta V| \leq .01V$. Since $\Delta V \approx dV$ and $dV = \pi h^2 dh$,

$\pi h^2 |dh| \leq .01V \Rightarrow$

$$|dh| \leq \frac{.01(\frac{1}{3}\pi h^3)}{\pi h^2} = \frac{.01h}{3} = \frac{1}{3}\%\text{ of the true height.}$$

47. $r = \frac{D}{2} \Rightarrow V = \frac{1}{3}\pi r^3 = \frac{1}{6}\pi D^3$. $\Delta V \approx dv = \frac{1}{2}\pi D^2 dD$. To be within

3% of the Volume, we need $\frac{dV}{V} \leq .03$. $\Rightarrow$

$\frac{\frac{1}{2}\pi D^2 dD}{\frac{1}{6}\pi D^3} \leq .03 \Leftrightarrow \frac{dD}{D} \leq .01$. The diameter needs to be within 1%.

49. $W = kV = k\pi r^2 h = Kr^2$ where $K = k\pi h$ is constant.

We need $|\Delta W| \leq .001W$. $\Delta W \approx dW = 2Krdr$. $\therefore\ 2Kr|dr| \leq .001Kr^2 \Rightarrow$

$2|dr| \leq .001r \Rightarrow |dr| \leq .0005r$ or to within .05% of true value.

51. $y = x^3 + \Delta y = (x + \Delta x)^3 = x^3 + 3x^2\Delta x + 3x\Delta x^2 + \Delta x^3$

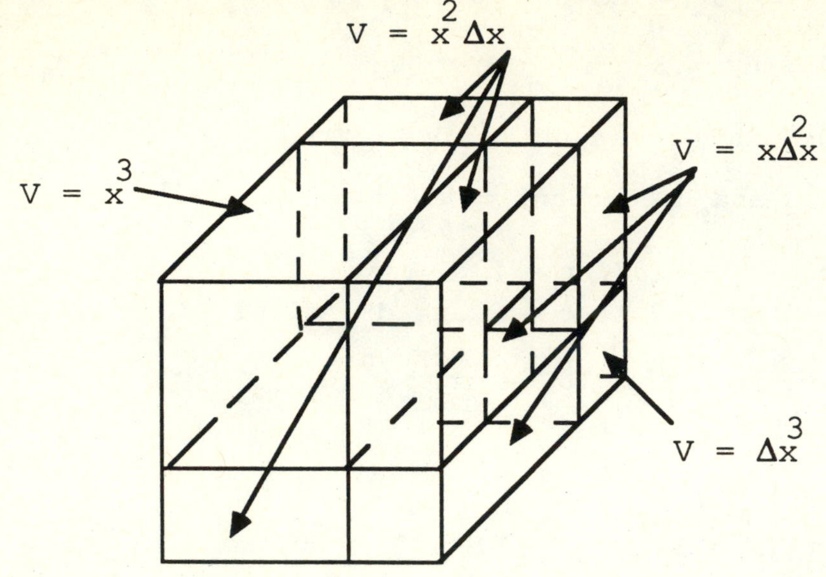

53. $f(x) = \dfrac{2}{1 - x} + \sqrt{1 + x} - 3.1$

(a) $f(0) = 2 + 1 - 3.1 = -.1$

$$f(\tfrac{1}{2}) = \frac{2}{1 - \frac{1}{2}} + \sqrt{1 + \frac{1}{2}} - 3.1 = 4 + \sqrt{\frac{3}{2}} - 3.1 > 0$$

$\therefore \exists\ x_1,\ 0 < x_1 < \frac{1}{2}$ for which $f(x_1) = 0$ (Intermediate Value Thm)

(b) Let $L_1(x) = 2 + 2x$ and $L_2(x) = 1 + \frac{1}{2}x$ be the linearizations of $f_1(x) = \dfrac{2}{1 - x}$ and $f_2(x) = \sqrt{1 + x}$ respectively.

Then $f(x) \approx L_1(x) + L_2(x) - 3.1 = 2 + 2x + 1 + \frac{1}{2}x - 3.1$

$= \frac{5}{2}x - .1 = 0 \Leftrightarrow x = .04$

(c) $f(0.04) = \dfrac{2}{1 - .04} + \sqrt{1 + .04} - 3.1 \approx 0.003137$

55. $M = \dfrac{M_o}{\sqrt{1 - (\frac{v}{c})^2}} = M_o\ [1 - (\frac{v}{c})^2)^{-\frac{1}{2}}$. The linearization of M is

$L(M) = M_o\ [1 + \frac{1}{2}(\frac{1}{c})^2 v^2] = M_o + \dfrac{M_o}{2c^2} v^2$. $\Delta M_o = M - M_o = \dfrac{M_o}{2c^2} v^2$.

$$\therefore \frac{\Delta M_o}{M_o} = \frac{\frac{M_o v^2}{2c^2}}{M_o} = \frac{1}{2c^2} v^2 \text{ will be equal to } 1\% = .01$$

$\Leftrightarrow v^2 = .02c^2 \Leftrightarrow v = c\sqrt{.02} = .14c.$ $\therefore$ an increase of velocity of 14% the speed of light is needed.

2.5 THE CHAIN RULE

1. $y = x^2 \Rightarrow \frac{dy}{dx} = 2x \qquad x = 2t - 5 \Rightarrow \frac{dx}{dt} = 2$
$\frac{dy}{dt} = \frac{dy}{dx} \bullet \frac{dx}{dt} = 2x(2) = 4x = 4(2t - 5) = 8t - 20$

3. $y = 8 - \frac{x}{3} \Rightarrow \frac{dy}{dx} = -\frac{1}{3}$ and $x = t^3 \Rightarrow \frac{dx}{dt} = 3t^2.$

$\frac{dy}{dt} = \frac{dy}{dx} \bullet \frac{dx}{dt} = -\frac{1}{3}(3t^2) = -t^2.$

5. $2x - 3y = 9 \Rightarrow 2 - 3\frac{dy}{dx} = 0 \Rightarrow \frac{dy}{dx} = \frac{2}{3}.$

$2x + \frac{t}{3} = 1 \Rightarrow 2\frac{dx}{dt} + \frac{1}{3} = 0 \Rightarrow \frac{dx}{dt} = -\frac{1}{6}.$

$\frac{dy}{dt} = \frac{dy}{dx} \bullet \frac{dx}{dt} = \frac{2}{3}(-\frac{1}{6}) = -\frac{1}{9}.$

7. $y = \sqrt{x + 2} \Rightarrow \frac{dy}{dx} = \frac{1}{2}(x + 2)^{-\frac{1}{2}}$
$x = \frac{2}{t} \Rightarrow \frac{dx}{dt} = -\frac{2}{t^2}$
$\frac{dy}{dt} = \frac{1}{2}(x + 2)^{-\frac{1}{2}}\left(\frac{-2}{t^2}\right) = -t^{-2}(2t^{-1} + 2)^{-\frac{1}{2}}$

9. $y = x^2 + 3x - 7 \Rightarrow \frac{dy}{dx} = 2x + 3. \quad x = 2t + 1 \Rightarrow \frac{dx}{dt} = 2$
$\frac{dy}{dt} = 2(2x + 3) = 2[2(2t + 1) + 3] = 8t + 10$

11. $z = w^2 - w^{-1} \Rightarrow \frac{dz}{dw} = 2w + w^{-2}. \quad w = 3x \Rightarrow \frac{dw}{dx} = 3.$

$\frac{dz}{dx} = \frac{dz}{dw} \bullet \frac{dw}{dx} = 3(2w + w^{-2}) = 3[6x + (3x)^{-2}] = 18x + \frac{1}{3x^2}$

13. $r = (s + 1)^{\frac{1}{2}} \Rightarrow \frac{dr}{ds} = \frac{1}{2}(s + 1)^{-\frac{1}{2}} \qquad s = 16t^2 - 20t \Rightarrow$

$\frac{ds}{dt} = 32t - 20. \quad \frac{dr}{dt} = \frac{dr}{ds} \cdot \frac{ds}{dt} = \frac{1}{2}(s + 1)^{-\frac{1}{2}}(32t - 20) =$

$(16t^2 - 20t + 1)^{-\frac{1}{2}}(16t - 10)$

15. $u = t + \frac{1}{t} \Rightarrow \frac{du}{dt} = 1 - t^{-2}$. $t = 1 - \frac{1}{v} \Rightarrow \frac{dt}{dv} = v^{-2}$

$\frac{du}{dv} = \frac{du}{dt} \bullet \frac{dt}{dv} = (1 - t^{-2})(v^{-2}) = [1 - (1 - \frac{1}{v})^{-2}]v^{-2} =$

$[1 - (\frac{v}{v - 1})^2][\frac{1}{v^2}] = \frac{(v - 1)^2 - v^2}{(v - 1)^2} \bullet \frac{1}{v^2} = \frac{1 - 2v}{v^2(v - 1)^2}$

17. $y = 3x^{\frac{2}{3}} \Rightarrow \frac{dy}{dx} = 2x^{-\frac{1}{3}}$. $x = 8t^3 \Rightarrow \frac{dx}{dt} = 24t^2$.

$\frac{dy}{dt} = (2x^{-\frac{1}{3}})(24t^2) = 48t^2 [8t^3]^{-\frac{1}{3}} = 24t$. By substitution,

$y = 3(8t^3)^{\frac{2}{3}} = 3(2t)^2 = 12t^2$ and $\frac{dy}{dt} = 24t$.

19. $y = \frac{1}{x^2 + 1} \Rightarrow \frac{dy}{dx} = -2x(x^2 + 1)^{-2}$. $x = \sqrt{4t - 1} = (4t - 1)^{\frac{1}{2}}$

$\Rightarrow \frac{dx}{dt} = \frac{1}{2}(4t - 1)^{-\frac{1}{2}}(4) = 2(4t - 1)^{-\frac{1}{2}}$.

$\frac{dy}{dt} = [-2x(x^2 + 1)^{-2}][2(4t - 1)^{-\frac{1}{2}}] =$

$= [-2\sqrt{4t - 1}(4t - 1 + 1)^{-2}][\frac{2}{\sqrt{4t - 1}}] =$

$= -4(\frac{1}{16t^2}) = -\frac{1}{4t^2}$. By substitution, $y = \frac{1}{(\sqrt{4t - 1})^2 + 1} =$

$(4t)^{-1}$ and $\frac{dy}{dt} = -(4t)^{-2}(4) = -\frac{1}{4t^2}$.

21. $g(x) = x^2 + 1 \Rightarrow g'(x) = 2x$. $x = f(t) = \sqrt{t + 1} \Rightarrow$

$f'(t) = \frac{1}{2}(t + 1)^{-\frac{1}{2}}$. $t = 0 \Rightarrow f(0) = 1$ and $f'(0) = \frac{1}{2}$.

$\frac{d}{dt}(g \circ f)\Big|_{x=0} = g'(f(0))f'(0) = g'(1)f'(0) = (2)(\frac{1}{2}) = 1$

23. $g(x) = \sqrt{1 + x^3} \Rightarrow g'(x) = \frac{1}{2}(1 + x^3)^{-\frac{1}{2}}(3x^2)$. $x = f(t) = t^{\frac{1}{3}} \Rightarrow$

$f'(t) = \frac{1}{3}t^{-\frac{2}{3}}$. $t = 1 \Rightarrow f(1) = 1$ and $f'(1) = \frac{1}{3}$.

$$\frac{d}{dt}(g\circ f)\Big|_{t=1} = g'(f(1))f'(1) = g'(1)f'(1) = \frac{3}{2\sqrt{2}} \cdot \frac{1}{3} = \frac{\sqrt{2}}{4}$$

25. $g(x) = \frac{2x}{x^2 + 1} \Rightarrow g'(x) = \frac{(x^2 + 1)(2) - 2x(2x)}{(x^2 + 1)^2} = \frac{2 - 2x^2}{(x^2 + 1)^2}$

$x = f(t) = 10t^2 + t + 1 \Rightarrow f'(t) = 20t + 1$.

$$\frac{d}{dt}(g\circ f)|_{t=0} = g'(f(0))f'(0) = g'(1)f'(0) = 0$$

27. $v = k\sqrt{s}$ m/sec $\Rightarrow a = \frac{dv}{dt} = \frac{1}{2}ks^{-\frac{1}{2}}\frac{ds}{dt} = \frac{1}{2}ks^{-\frac{1}{2}} \cdot ks^{\frac{1}{2}} = \frac{1}{2}k^2 = K$

29. $T = 2\pi\sqrt{\frac{L}{g}} \Rightarrow \frac{dT}{dL} = 2\pi\left[\frac{1}{2}\left(\frac{L}{g}\right)^{-\frac{1}{2}}\frac{1}{g}\right]$. $\frac{dT}{d\theta} = \frac{dT}{dL} \cdot \frac{dL}{d\theta} =$

$\pi\left(\frac{L}{g}\right)^{-\frac{1}{2}}\frac{1}{g}KL = \pi K\sqrt{\frac{L}{g}} = \frac{KT}{2}$

2.6 A BRIEF REVIEW OF TRIGONOMETRY

1.

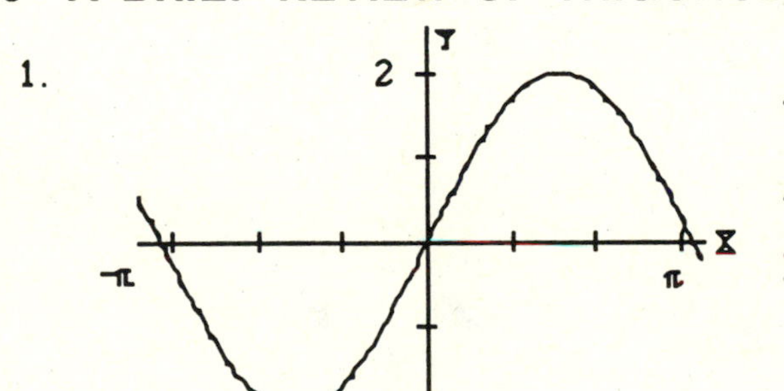

$y = 2\sin x$

Amplitude = 2

Period = 2π

3. $y = \sin(-x)$

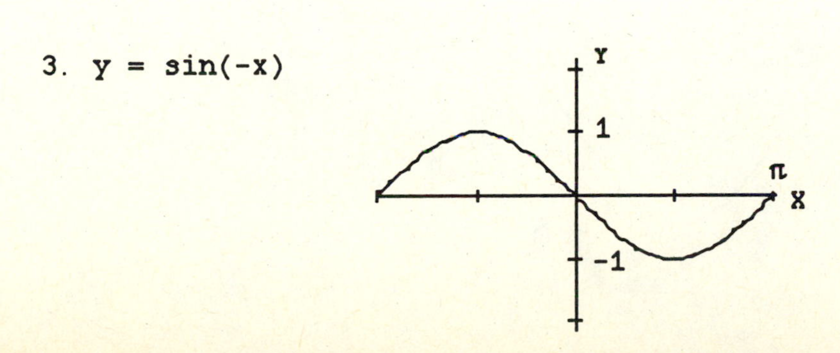

5. $y = 2\cos(3x)$

Amplitude = 2

Period = $\frac{2\pi}{3}$

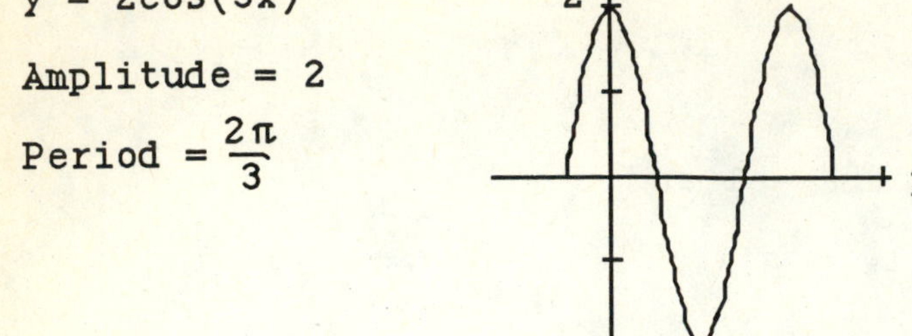

7. $y = \sin(x + \pi/2)$

$= -\cos(x)$

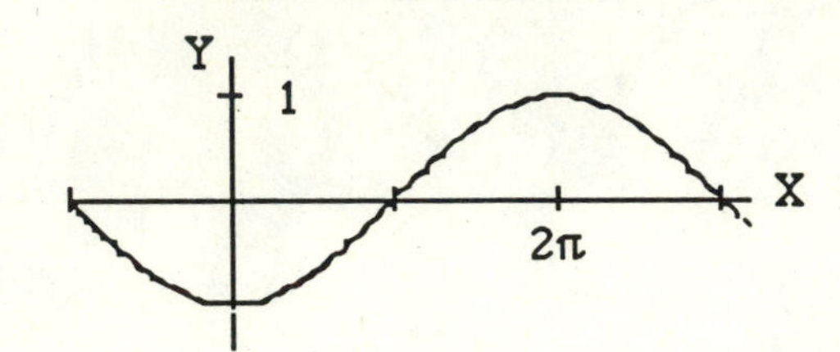

9. $y = |\cos x|$

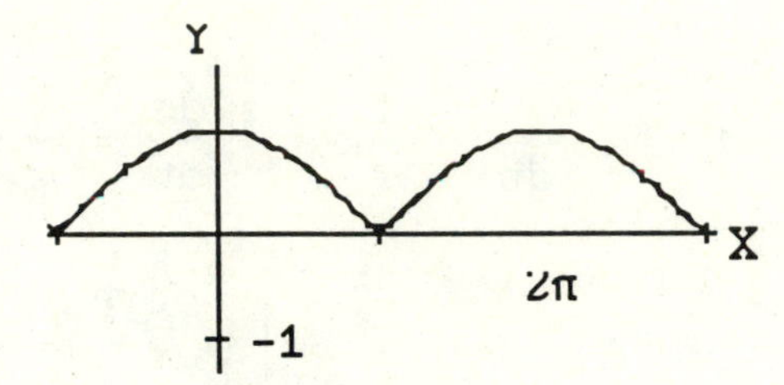

11. $y = .5(|\sin x| - \sin x)$

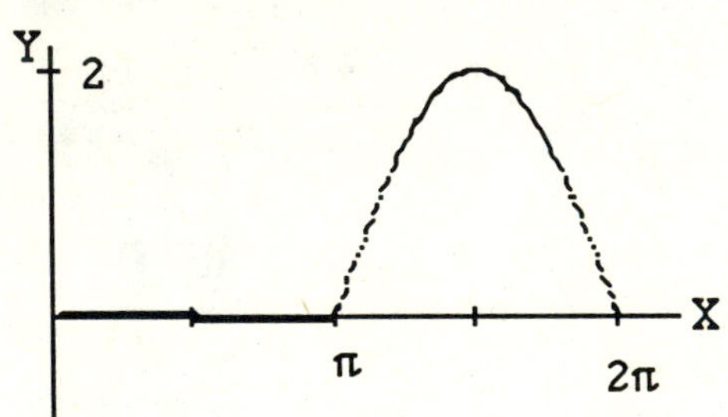

13. $y = \cos^2(x)$

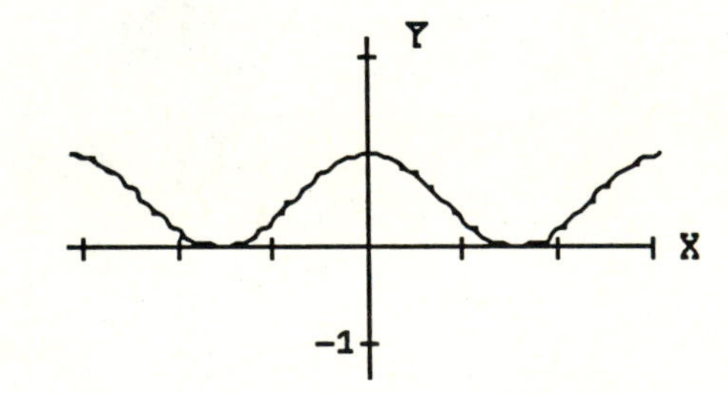

15. $y = \sin x - \cos x$

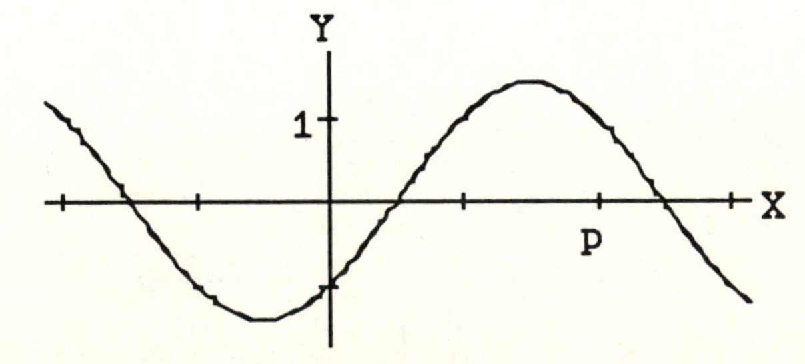

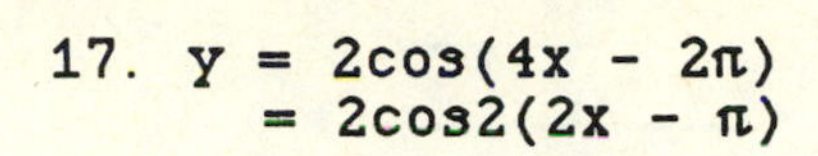

17. $y = 2\cos(4x - 2\pi)$
$= 2\cos 2(2x - \pi)$

Period = π
Amplitude = 2

Horizontal shift of π units to right is not visible because the period is the same.

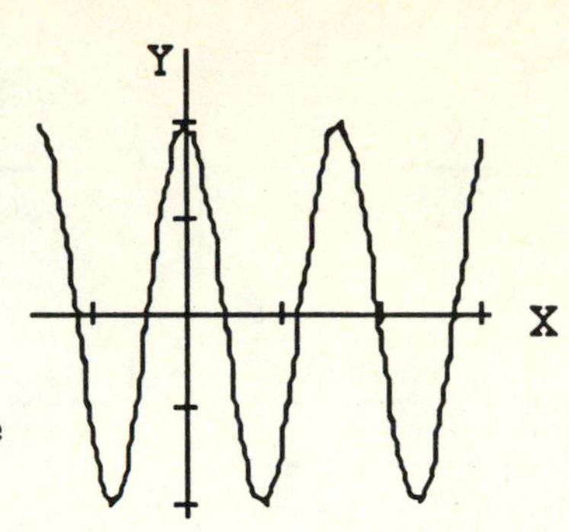

19. $y = \sec x$

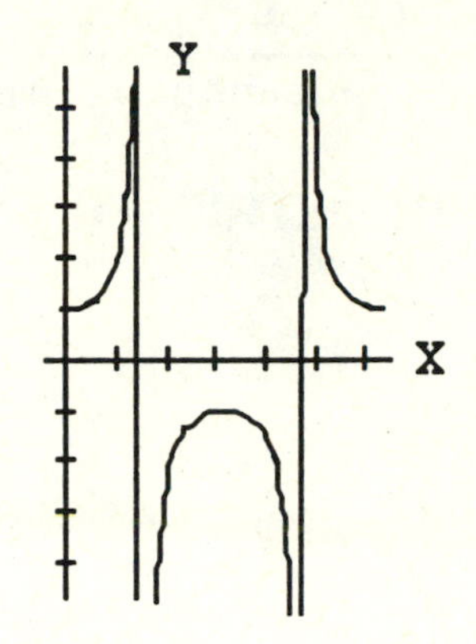

21. Let $f(x) = 37\left[\frac{2\pi}{365}(x - 101)\right] + 25.$

(a) Amplitude = [illegible] (b) Period = $2\pi \div \frac{2\pi}{365} = 365$

(c) Horizontal shift = 101 units to the right.

(d) Vertical shift = 25 units upward.

23. $$\lim_{h\to 0} \frac{(\sin h)(1 - \cos h)}{h^2} = \lim_{h\to 0} \frac{\sin h}{h} \bullet \frac{1 - \cos h}{h} = (1)(0) = 0$$

25. $$\lim_{x\to 0} \frac{1 - \cos x}{\sin x} = \lim_{x\to 0} \frac{x}{x} \bullet \frac{1 - \cos x}{\sin x} = \lim_{x\to 0} \frac{1 - \cos x}{x} \bullet \frac{x}{\sin x} = 0$$

27. In the formula (11d), let B = A. We then have

$\cos(A-A) = \cos A\cos A + \sin A\sin A$ or $\cos 2A + \sin 2A = \cos 0 = 1.$

29. $\sin(x - \frac{\pi}{2}) = (\sin x)(\cos\frac{\pi}{2}) - (\cos x)(\sin\frac{\pi}{2}) = (\sin x)(0) - (\cos x)(1) = -\cos x$

31. $\cos(x - \frac{\pi}{2}) = (\cos x)(\cos\frac{\pi}{2}) + (\sin x)(\sin\frac{\pi}{2}) = (\cos x)(0) + (\sin x)(1) = \sin x$

33. $\sin(\pi - x) = (\sin\pi)(\cos x) - (\cos\pi)(\sin x) = (0)(\cos x) - (-1)(\sin x) = \sin x$

35. $\tan(A+B) = \dfrac{\sin(A+B)}{\cos(A+B)} = \dfrac{\sin A\cos B + \cos A\sin B}{\cos A\cos B - \sin A\sin B} \bullet \dfrac{\frac{1}{\cos A\cos B}}{\frac{1}{\cos A\cos B}}$

$$\dfrac{\dfrac{\sin A\cos B}{\cos A\cos B} + \dfrac{\cos A\sin B}{\cos A\cos B}}{\dfrac{\cos A\cos B}{\cos A\cos B} - \dfrac{\sin A\sin B}{\cos A\cos B}} = \dfrac{\tan A + \tan B}{1 - \tan A\tan B}$$

$$\tan(A-B) = \tan(A+(-B)) = \dfrac{\tan A + \tan(-B)}{1 - \tan A\tan(-B)} = \dfrac{\tan A - \tan B}{1 + \tan A\tan B}$$

2.7 DERIVATIVES OF TRIGONOMETRIC FUNCTIONS

1. $y = \sin(x + 1) \Rightarrow y' = \cos(x + 1)$

3. $y = \sin\left(\frac{x}{2}\right) \Rightarrow y' = \frac{1}{2}\cos\left(\frac{x}{2}\right)$

5. $y = \cos 5x \Rightarrow y' = -5\sin 5x$

7. $y = \cos(-2x) = \cos(2x) \Rightarrow y' = -\sin(2x)(2) = -2\sin(2x)$

9. $y = \sin(3x + 4) \Rightarrow y' = 3\cos(3x + 4)$

11. $y = x\sin x \Rightarrow y' = x\cos x + \sin x$

13. $y = x\sin x + \cos x \Rightarrow y' = x\cos x + \sin x - \sin x = x\cos x$

15. $y = \dfrac{1}{\cos x} = \sec x \Rightarrow y' = \sec x\tan x$

17. $y = \sec(x - 1) \Rightarrow y' = \sec(x - 1)\tan(x - 1)$

19. $y = \sec(1 - x) \Rightarrow y' = -\sec(1 - x)\tan(1 - x)$

21. $y = \tan 2x \Rightarrow y' = 2\sec^2 2x$

23. $y = \sin^2 x \Rightarrow y' = 2\sin x\cos x = \sin 2x$

25. $y = \cos^2 5x \Rightarrow y' = -10\cos 5x\sin 5x$ $(=-5\sin 10x)$

27. $y = \tan(5x-1) \Rightarrow y' = 5\sec^2(5x-1)$

29. $y = 2\sin x\cos x = \sin 2x \Rightarrow y' = 2\cos 2x$

31. $y = \sqrt{2 + \cos 2x} \Rightarrow y' = \frac{1}{2}(2 + \cos 2x)^{-\frac{1}{2}}(-2\sin 2x) = \dfrac{-\sin 2x}{\sqrt{2 + \cos 2x}}$

33. $y = \cos\sqrt{x} \Rightarrow \frac{dy}{dx} = -\sin\sqrt{x}\left(\frac{1}{2}x^{-1/2}\right) = -\frac{\sin\sqrt{x}}{2\sqrt{x}}$

35. $y = \sqrt{\frac{1+\cos 2x}{2}} = \sqrt{\cos^2 x} = |\cos x| \Rightarrow \frac{dy}{dx} = -\sin x$ if $\cos x > 0$

and $\sin x$ if $\cos x < 0$.

37. $x = \tan y \Rightarrow 1 = \sec^2 y\frac{dy}{dx} \Rightarrow \frac{dy}{dx} = \frac{1}{\sec^2 y} = \cos^2 y$

39. $y^2 = \sin^4 2x + \cos^4 2x \Rightarrow$

$$2y\frac{dy}{dx} = 4(\sin^3 2x)(\cos 2x)(2) + 4(\cos^3 2x)(-\sin 2x)(2)$$

$$= 8\sin 2x\cos 2x(\sin^2 2x - \cos^2 2x) = 4\sin 4x(-\cos 4x) = -2\sin 8x$$

$$\frac{dy}{dx} = -\frac{\sin 8x}{y}$$

41. $x + \tan(xy) = 0 \Rightarrow 1 + \sec^2(xy)\left[y + x\frac{dy}{dx}\right] = 0$

$$1 + y\sec^2(xy) = -x\sec^2(xy)\frac{dy}{dx}$$

$$\frac{dy}{dx} = \frac{-(1 + y\sec^2(xy))}{x\sec^2(xy)}$$

or $\frac{dy}{dx} = -\frac{1}{x}[\cos^2(xy) + y]$

43. $x\sin 2y = y\cos 2x \Rightarrow 2x\cos 2y\frac{dy}{dx} + \sin 2y = -2y\sin 2x + \cos 2x\frac{dy}{dx}$

Evaluating at $(\frac{\pi}{4}, \frac{\pi}{2})$:

$$2(\frac{\pi}{4})(\cos\pi)\frac{dy}{dx} + \sin\pi = -2(\frac{\pi}{2})\sin\frac{\pi}{2} + (\cos\frac{\pi}{2})\frac{dy}{dx}$$

$$-\frac{\pi}{2}\frac{dy}{dx} = -\pi \Rightarrow \frac{dy}{dx} = 2$$

$y - \frac{\pi}{2} = 2(x - \frac{\pi}{4}) \Rightarrow y = 2x$ is the equation of the tangent.

45. $\lim_{x\to\frac{\pi}{4}} \frac{\sin x}{\cos x} = \lim_{x\to\frac{\pi}{4}} \tan x = 1$

47. $\lim_{x\to\pi} \sec(1+\cos x) = \sec(1+\cos\pi) = \sec(1-1) = \sec 0 = 1$

49. $\lim_{x\to 0} x\csc x = \lim_{x\to 0} \frac{x}{\sin x} = 1$

51. $\lim_{h\to 0} \frac{\cos(a+h) - \cos a}{h} = \frac{d}{dx}(\cos x)\Big|_{x=a} = -\sin a$

53. $y = \tan x$ and the linearization $L(x) = x$.

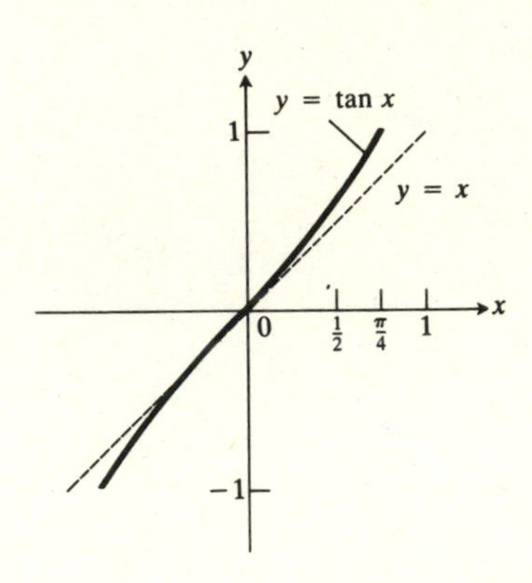

55. If $b = 1$, then the left and right hand limits of $f(x)$ at $x=0$ both exist and equal 1. Since $f(0)=1$, the function is continuous at $x = 0$.

57. $f(x) = \sqrt{1 + x} + \sin x \Rightarrow f'(x) = \frac{1}{2}(1 + x)^{-\frac{1}{2}} + \cos x.$

$f(x) \approx L(0) = f(0) + f'(0)x = 1 + \frac{3}{2}x.$

$\sqrt{1 + x} \approx L_1(0) = 1 + \frac{1}{2}x$ and $\sin x \approx L_2(0) = x$

Observe that $L(0) = L_1(0) + L_2(0)$.

2.8 PARAMETRIC FUNCTIONS

1. $x = \cos t$, $y = \sin t$, $0 \le t \le 2\pi$. The path is the unit circle, since $x^2 = \cos^2 t$, $y^2 = \sin^2 t \Rightarrow x^2 + y^2 = \cos^2 t + \sin^2 t = 1$. The particle begins at the point $P(\cos 0, \sin 0) = P(1,0)$ and travels in a counterclockwise direction to end at the point $Q(\cos 2\pi, \sin 2\pi) = Q(1,0) = P(1,0)$.

3. $x = \cos 2\pi t$, $y = \sin 2\pi t$, $0 \le t \le 1$, is the same path as that of problem 1 above.

5. $x = 3\cos t$, $y = 3\sin t$, $0 \le t \le 2\pi$, is a circle with center at the origin and radius 3, since $x^2 + y^2 = 9\cos^2 t + 9\sin^2 t = 9$. The particle starts and stops at the point $P(3,0)$ and travels in a counterclockwise direction.

7. $x = \cos^2 t$, $y = \sin^2 t$, $0 \le t \le \frac{\pi}{2}$. Since $x + y = 1$, the particle travels along a line, starting at $P(1,0)$ and stopping at $Q(0,1)$, in the direction from P to Q.

9. (a) clockwise: $x = 2\cos t$, $y = -2\sin t$, $0 \le t \le 2\pi$
   (b) counterclockwise: $x = 2\cos t$, $y = 2\sin t$, $0 \le t \le 2\pi$

11. $x = 2t - 5 \Rightarrow t = \dfrac{x+5}{2}$

$$y = 4t - 7 = 4\left(\frac{x+5}{2}\right) - 7$$

$$= 2x + 3$$

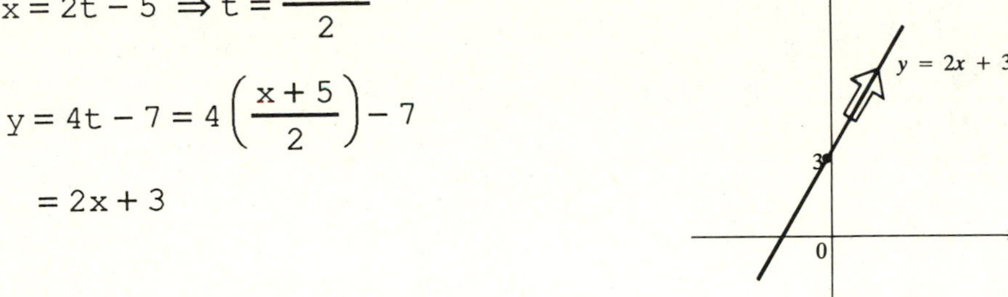

13. $x = 3t$, $y = 9t^2 \Rightarrow$

$y = (3t)^2 = x^2$

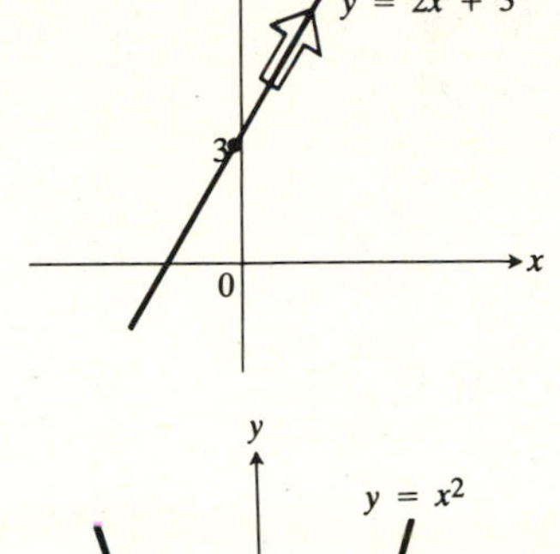

15. $x = t$, $y = \sqrt{1 - t^2} \Rightarrow$

$y = \sqrt{1 - x^2}$.

If $0 \le t \le 1$, then the curve goes from $(0,1)$ to $(1,0)$ only.

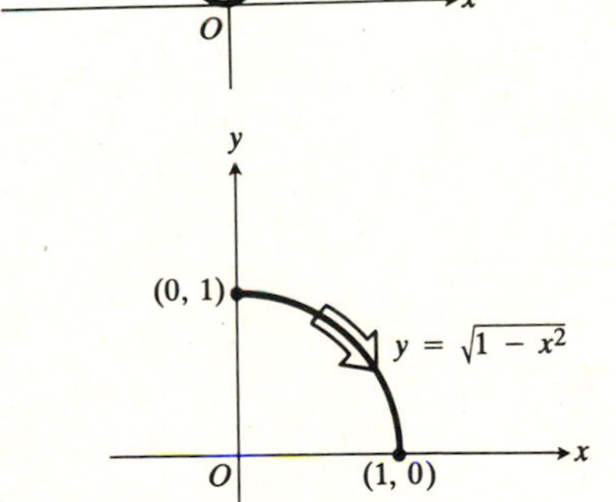

17. $x = t$, $y = \sqrt{t} \Rightarrow y = \sqrt{x}$

$(t \ge 0 \Rightarrow x \ge 0)$

19. $x = 3t$, $y = 2 - t$, $0 \le t \le 1$

$y = 2 - 2\left(\frac{x}{3}\right) = 2 - \frac{2}{3}x$

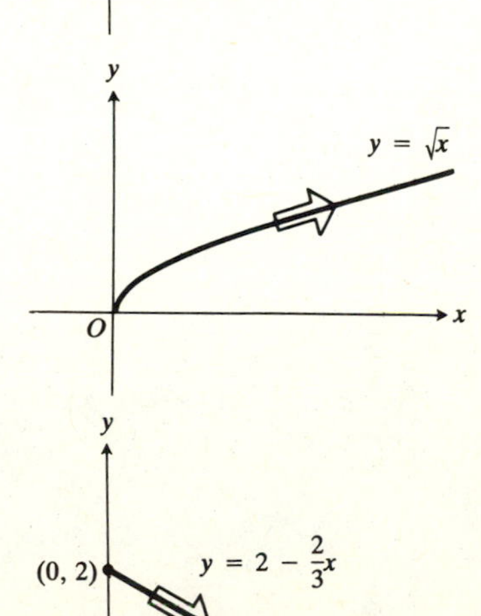

21. $x = 2t \Rightarrow \frac{dx}{dt} = 2 \qquad y = 1 + t \Rightarrow \frac{dy}{dt} = 1$

$\frac{dy}{dx} = \frac{\frac{dy}{dt}}{\frac{dx}{dt}} = \frac{1}{2}$. Eliminating t: $t = \frac{x}{2} \Rightarrow y = 1 + \frac{x}{2} \Rightarrow \frac{dy}{dx} = \frac{1}{2}$

23. $x = 5\cos t \Rightarrow \frac{dx}{dt} = -5\sin t \qquad y = 5\sin t \Rightarrow \frac{dy}{dt} = 5\cos t$

$\frac{dy}{dx} = \frac{\frac{dy}{dt}}{\frac{dx}{dt}} = \frac{5\cos t}{-5\sin t} = -\cot t$. **Elimintating t:**

$x^2 + y^2 = 25\cos^2 t + 25\sin^2 t = 25$

$\frac{dy}{dx} = -\frac{x}{y} = \frac{-5\cos t}{5\sin t} = -\cot$

25. $x = t^2 - \frac{\pi}{2}$. $y = \sin(t^2) = \sin(\frac{\pi}{2} + x) = \cos x$. $\frac{dy}{dx} = -\sin x$.
$\frac{dx}{dt} = 2t$ and $\frac{dy}{dt} = 2t\cos(t^2) \Rightarrow \frac{dy}{dx} = \frac{2t\cos(t^2)}{2t} = \cos(t^2) =$

$\cos(\frac{\pi}{2} + x) = -\sin x$.

27. $x = \cos t \Rightarrow \frac{dx}{dt} = -\sin t$. $y = 1 + \sin t \Rightarrow \frac{dy}{dt} = \cos t$.

$\frac{dy}{dx} = \frac{\frac{dy}{dt}}{\frac{dx}{dt}} = \frac{\cos t}{-\sin t} = -\cot t$. Eliminating t: $y - 1 = \sin t$

and $x = \cos t \Rightarrow (y-1)^2 + x^2 = \sin^2 t + \cos^2 t = 1$. $2(y-1)\frac{dy}{dx} + 2x = 0$

or $\frac{dy}{dx} = -\frac{x}{y-1} = -\cot t$.

29. (c)
$x = 4t - 5 \Rightarrow \frac{dx}{dt} = 4$. $\quad y = t^2 \Rightarrow \frac{dy}{dt} = 2t$.

$\frac{dy}{dx} = \frac{\frac{dy}{dt}}{\frac{dx}{dt}} = \frac{2t}{4}\Big|_{t=2} = 1.$

31. $x = t + \frac{1}{t} \Rightarrow \frac{dx}{dt} = 1 - \frac{1}{t^2}\Big|_{t=2} = \frac{3}{4}$.

$y = t - \frac{1}{t} \Rightarrow \frac{dy}{dt} = 1 + \frac{1}{t^2}\Big|_{t=2} = \frac{5}{4} \therefore \frac{dy}{dx} = \frac{\frac{5}{4}}{\frac{3}{4}} = \frac{5}{3}$.

$y(2) = \frac{3}{2}$ and $x(2) = \frac{5}{2}$. Tangent line: $y - \frac{3}{2} = \frac{5}{3}(x - \frac{5}{2})$

$\Rightarrow 6y - 9 = 10x - 25 \Rightarrow 3y - 5x + 8 = 0$.

33. $x = t\sqrt{2t + 5} \Rightarrow \frac{dx}{dt} = (2t + 5)^{\frac{1}{2}} + t\left[\frac{1}{2}(2t + 5)^{-\frac{1}{2}}(2)\right]$

$\frac{dx}{dt}\Big|_{t=2} = \sqrt{9} + \frac{2}{\sqrt{9}} = \frac{11}{3}$. $y = (4t)^{\frac{1}{3}} \Rightarrow \frac{dy}{dt} = \frac{1}{3}(4t)^{-\frac{2}{3}}(4)$.

$\frac{dy}{dt}\Big|_{t=2} = \frac{1}{3}$. $\frac{dy}{dx} = \frac{\frac{1}{3}}{\frac{11}{3}} = \frac{1}{11}$. $x(2) = 6$ and $y(2) = 2$.

$y - 2 = \frac{1}{11}(x - 6) \Rightarrow 11y - x = 16$.

35. $x = t^{-2} \Rightarrow \frac{dx}{dt} = -2t^{-3}\Big|_{t=2} = -\frac{1}{4}$. $y = \sqrt{t^2 + 12} \Rightarrow$

$\frac{dy}{dt} = \frac{1}{2}(t^2 + 12)^{-\frac{1}{2}}(2t)\Big|_{t=2} = \frac{1}{2}$. $x(2) = \frac{1}{4}$ and $y(2) = 4$.

$\frac{dy}{dx} = \frac{\frac{1}{2}}{\frac{-1}{4}} = -2$. Tangent line: $y - 4 = -2(x - \frac{1}{4}) \Rightarrow 2y + 4x = 9$.

37. $x = 80t \Rightarrow \frac{dx}{dt} = 80$. $y = 64t - 16t^2 \Rightarrow \frac{dy}{dt} = 64 - 32t$.

$\frac{dy}{dx} = \frac{64 - 32t}{80} = 0 \Leftrightarrow 64 - 32t = 0 \Rightarrow t = 2$.

39. $x^2 + y^2 = 25 \Rightarrow 2x\frac{dy}{dx} + 2y = 0$

$(3)(4) + (4)\frac{dy}{dx} = 0$ or $\frac{dy}{dx} = -3$

41. $x = 2t - 5 \Rightarrow \frac{dx}{dt} = 2.$ $y = 4t - 7 \Rightarrow \frac{dy}{dt} = 4.$ $\frac{dy}{dx} = \frac{4}{2} = 2.$

$$\frac{d^2y}{dx^2} = \frac{\frac{dy'}{dt}}{\frac{dx}{dt}} = \frac{\frac{d}{dt}\left(\frac{dy}{dx}\right)}{\frac{dx}{dt}} = \frac{0}{2} = 0.$$

43. $x = t \Rightarrow \frac{dx}{dt} = 1.$ $y = \sqrt{t} \Rightarrow \frac{dy}{dt} = \frac{1}{2\sqrt{t}}.$ $\frac{dy}{dx} = \frac{1}{2\sqrt{t}}.$

$$\frac{d^2y}{dx^2} = \frac{\frac{dy'}{dt}}{\frac{dx}{dt}} = \frac{\frac{d}{dt}\left(\frac{dy}{dx}\right)}{\frac{dx}{dt}} = \frac{-\frac{1}{4}t^{-3/2}}{1} = -\frac{1}{4}t^{-3/2}.$$

45. $x = t,\ y = t^{-1} \Rightarrow \frac{dx}{dt} = 1$ and $\frac{dy}{dt} = -t^{-2}$

$$\frac{dy}{dx} = \frac{\frac{dy}{dt}}{\frac{dx}{dt}} = \frac{-t^{-2}}{1}. \quad \frac{d^2y}{dx^2} = \frac{\frac{dy'}{dt}}{\frac{dx}{dt}} = \frac{2t^{-3}}{1}.$$

47. $x = \cos t,\ y = 1 + \sin t \Rightarrow \frac{dy}{dx} = -\cot$ (See Problem 27).

$$\frac{d^2y}{dx^2} = \frac{\frac{dy'}{dt}}{\frac{dx}{dt}} = \frac{\frac{d}{dt}(-\cot t)}{-\sin t} = \frac{\csc^2 t}{-\sin t} = -\frac{1}{\sin^3 t} \quad \left(= -\frac{1}{(1 - x^2)^{\frac{3}{2}}}\right)$$

49. $x = 80t,\ y = 64t - 16t^2 \Rightarrow \frac{dy}{dx} = \frac{64 - 32t}{80} = \frac{4}{5} - \frac{2}{5}t$ (See Problem 37).

$$\frac{d^2y}{dx^2} = \frac{\frac{d}{dt}\left(\frac{dy}{dx}\right)}{\frac{dx}{dt}} = \frac{\frac{-2}{5}}{80} = -\frac{1}{200}.$$

51. $$\frac{d^2y}{dx^2} = \frac{\frac{dy'}{dt}}{\frac{dx}{dt}} = \frac{3t^2 + 3}{\frac{dx}{dt}} = t^2 + 1 \Leftrightarrow \frac{dx}{dt} = \frac{3t^2 + 3}{t^2 + 1} = 3.$$

2.9 NEWTON'S METHOD

1. $f(x) = x^2 + x - 1.$ $f(0) = -1$ and $f(1) = 1.$ Take $x_1 = .5.$
$f'(x) = 2x + 1.$ $x_2 = .5 - \frac{f'(.5)}{f(.5)} = 0.625000.$ Similarly,
$x_3 = 0.618056$, $x_4 = 0.618034$. The root is 0.618.

3. $f(x) = x^4 + x - 3$. $f(1) = -1$ and $f(2) = 15$. The root is much closer to 1. Take $x_1 = 1$. $f'(x) = 4x^3 + 1$. $x_2 = 1 - \frac{f'(1)}{f(1)} =$ 1.20000, $x_3 = 1.165420$, $x_4 = 1.164037$, $x_5 = 1.164035$. The root is 1.164.

5. $f(x) = 2 - x^4$. $f(-2) = -14$ and $f(-1) = 1$. Take $x_1 = -1$. $f'(x) = -4x^3$. $x_2 = -1 - \frac{f(-1)}{f'(-1)} = -1.250000$. $x_3 = -1.193500$. $x_4 = -1.189230$. $x_5 = -1.189207$. The root is -1.189

7. If $f(x_o) = 0$ then $x_1 = x_o - \frac{f(x_o)}{f'(x_o)} = x_o - \frac{0}{f'(x_o)} = x_o$.
If $f(x_n) = 0$ then $x_{n+1} = x_n - \frac{f(x_n)}{f'(x_n)} = x_n$. The iterations are constant.

9. $f(x) = x^3 + 2x - 4$. $f(1) = -1$ and $f(2) = 8$. Take $x_1 = 1$. $f'(x) = 3x^2 + 2$. $x_2 = -1 - \frac{f(-1)}{f'(-1)} = 1.20000000$. $x_3 = 1.1797468$. $x_4 = 1.17950906$. $x_5 = 1.17950902$. The root is 1.179509

11. (i) $\Rightarrow$ (ii) A root of the equation would be such that

$x^3 - 3x - 1 = 0 \Rightarrow x^3 = 3x + 1$

(ii) $\Rightarrow$ (iii) If $x^3 = 3x + 1$ then $x^3 - 3x = 1$

(iii) $\Rightarrow$ (iv) If $x^3 - 3x = 1$ then $g'(x) = 0$ since

$$g(x) = \frac{1}{4}x^4 - \frac{3}{2}x^2 - x + 5 \Rightarrow g'(x) = x^3 - 3x - 1$$

(iv) $\Rightarrow$ (i) $g'(x) = f(x)$

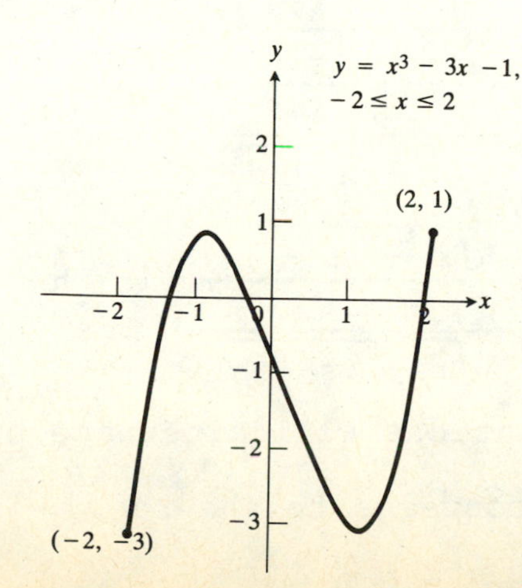

(c) 1.87939

(d) −0.34730
−1.53209

13. The point of intersection of $y = \cos x$ with $y = x$ is value of x for which $\cos x = x$ and hence a root of $\cos x - x = 0$. The two graphs cross between $x = 0$ and $x = \frac{\pi}{2}$. Take $x_1 = 1$.

$f(x) = \cos x - x$ and $f'(x) = -\sin x - 1$. $x_2 = 1 - \frac{\cos(1) - 1}{-\sin(1) - 1} =$ 0.7503639. $x_3 = 0.7391129$. $x_4 = 0.7390851$. $x_5 = 0.73908513$.

The root is 0.73909.

15. Divide $f(x) = x^4 - 2x^3 - x^2 - 2x + 2$ synthetically by successive integers until a sign change of the remainder indicates a root. The work is arranged in a table:

	1	-2	-1	-2	2
-2	1	-4	7	-16	34
-1	1	-3	2	-4	6
0	1	-2	-1	-2	2
1	1	-1	-2	-4	-2
2	1	0	-1	-4	-6
3	1	1	2	4	14

Between 0 and 1 and between 2 and 3

$f'(x) = 3x^3 - 6x^2 - 2x - 2$. Let $x_1 = .5$. $x_2 = 0.64062500$. $x_3 = 0.63017159$. $x_4 = 0.63011540$. $x_5 = 0.63011540$. The root between 0 and 1 is 0.630115. For the second root, let $x_1 = 2.5$. $x_2 = 2.5798611$. $x_3 = 2.5733190$. $x_4 = 2.5732719$. $x_5 = 2.5732719$. The root between 2 and 3 is 2.573272.

17.

$$f(x) = \sqrt{x - r} \text{ if } x \geq r \Rightarrow f'(x) = \frac{1}{2}(x - r)^{-\frac{1}{2}}$$

$$f(x) = \sqrt{r - x} \text{ if } x \leq r \Rightarrow f'(x) = \frac{1}{2}(r - x)^{-\frac{1}{2}}.$$

If $x_o = r + h$, then $x_1 = r + h - \dfrac{\sqrt{r + h - r}}{\dfrac{1}{2\sqrt{r + h - r}}} = r + h - 2h = r - h$

If $x_o = r - h$. then $x_1 = r - h - \dfrac{\sqrt{r - (r - h)}}{\dfrac{1}{2\sqrt{r - (r - h)}}} = r - h + 2h = r + h$

Thus, if $x_o \neq r$, successive iterations will alternate between $r + h$ and $r - h$ and will never converge to the root.

2.10 DERIVATIVE FORMULAS IN DIFFERENTIAL NOTATION

1. $y = x^3 - 3x \Rightarrow dy = (3x^2 - 3)\,dx$

3. $y = x\sqrt{1-x^2} \Rightarrow dy = \left[\sqrt{1-x^2} + \frac{1}{2}x(1-x^2)^{-1/2}(-2x)\right]dx = \frac{1-2x^2}{\sqrt{1-x^2}}dx$

5. $y + xy - x = 0 \Rightarrow dy + xdy + ydx - dx = 0$

$(1 + x)dy = (1 - y)dx$

$$dy = \frac{1 - y}{1 + x}\,dx$$

7. $y = \sin 5x \Rightarrow dy = 5\cos 5x\,dx$

9. $y = 4\tan\frac{x}{2} \Rightarrow dy = 4\sec^2\frac{x}{2}\cdot\frac{1}{2}\,dx = 2\sec^2\frac{x}{2}\,dx$

11. $y = 3\csc(1 - \frac{x}{3}) \Rightarrow dy = -3\csc(1 - \frac{x}{3})\cot(1 - \frac{x}{3})(-\frac{1}{3})dx$

$= \csc(1 - \frac{x}{3})\cot(1 - \frac{x}{3})\,dx$

13. $x = t + 1 \Rightarrow dx = dt. \quad y = t + \frac{t^2}{2} \Rightarrow dy = (1 + t)dt$

$$\frac{dy}{dx} = \frac{(1 + t)dt}{dt} = 1 + t$$

15. $x = \cos t \Rightarrow dx = -\sin t\,dt \quad y = 1 + \sin t \Rightarrow dy = \cos t\,dt$

$$\frac{dy}{dx} = \frac{\cos t\,dt}{-\sin t\,dt} = -\cot t$$

2.M MISCELLANEOUS PROBLEMS

1. $y = \frac{x}{\sqrt{x^2 - 4}} = x(x^2 - 4)^{-\frac{1}{2}} \Rightarrow$

$$\frac{dy}{dx} = (x^2 - 4)^{-\frac{1}{2}} + x\left[-\frac{1}{2}(x^2 - 4)^{-\frac{3}{2}}(2x)\right] = -4(x^2 - 4)^{-\frac{3}{2}}$$

3. $xy + y^2 = 1 \Rightarrow x\frac{dy}{dx} + y + 2y\frac{dy}{dx} = 0 \Rightarrow \frac{dy}{dx} = -\frac{y}{x + 2y}$.

5. $x^2y + xy^2 = 10 \Rightarrow x^2\frac{dy}{dx} + 2xy + 2xy\frac{dy}{dx} + y^2 = 0$

$$(x^2 + 2xy)\frac{dy}{dx} = -2xy - y^2 \Rightarrow \frac{dy}{dx} = -\frac{2xy + y^2}{2xy + x^2}$$

7. $y = \cos(1 - 2x) \Rightarrow \frac{dy}{dx} = (-\sin(1 - 2x))(-2) = 2\sin(1 - 2x)$

9. $y = \frac{x}{x+1} = x(x+1)^{-1} \Rightarrow \frac{dy}{dx} = (x+1)^{-1} - x(x+1)^{-2} = \frac{1}{(x+1)^2}$

11. $y = x^2\sqrt{x^2 - a^2} = x^2(x^2 - a^2)^{\frac{1}{2}}$

$\frac{dy}{dx} = 2x(x^2 - a^2)^{\frac{1}{2}} + x^2 \left[\frac{1}{2}(x^2 - a^2)^{-\frac{1}{2}}(2x)\right]$

$= 2x\sqrt{x^2 - a^2} + \frac{x^3}{\sqrt{x^2 - a^2}} = \frac{3x^3 - 2xa^2}{\sqrt{x^2 - a^2}}$

13. $y = \frac{x^2}{1 - x^2} \Rightarrow \frac{dy}{dx} = \frac{(1 - x^2)(2x) - x^2(-2x)}{(1 - x^2)^2} = \frac{2x}{(1 - x^2)^2}$

15. $y = \sec^2(5x) \Rightarrow \frac{dy}{dx} = 2(\sec 5x)(\sec 5x \tan 5x)(5) = 10\sec^2 5x \tan 5x$

17. $y = \frac{(2x^2 + 5x)^{\frac{3}{2}}}{3} \Rightarrow \frac{dy}{dx} = \frac{1}{3}(\frac{3}{2})(2x^2 + 5x)^{\frac{1}{2}}(4x + 5) = \frac{(4x + 5)\sqrt{2x^2 + 5x}}{2}$

19. $xy^2 + \sqrt{xy} = 2 \Rightarrow y^2 + 2xy\frac{dy}{dx} + x^{\frac{1}{2}}\left[\frac{1}{2}y^{-\frac{1}{2}}\frac{dy}{dx}\right] + \frac{1}{2}x^{-\frac{1}{2}}y^{\frac{1}{2}}$

$(2xy + \frac{\sqrt{x}}{2\sqrt{y}})\frac{dy}{dx} = -y^2 - \frac{\sqrt{y}}{2\sqrt{x}}$

$\frac{dy}{dx} = \frac{\dfrac{-2y^2\sqrt{x} - \sqrt{y}}{2\sqrt{x}}}{\dfrac{4xy\sqrt{y} + \sqrt{x}}{2\sqrt{y}}} = \frac{-2y^2\sqrt{xy}. - y}{4xy\sqrt{xy} + x}$

21. $x^{\frac{2}{3}} + y^{\frac{2}{3}} = a^{\frac{2}{3}} \Rightarrow \frac{2}{3}x^{-\frac{1}{3}} + \frac{2}{3}y^{-\frac{1}{3}}\frac{dy}{dx} = 0 \Rightarrow \frac{dy}{dx} = -\sqrt[3]{\frac{y}{x}}$

23. $xy = 1 \Rightarrow x\frac{dy}{dx} + y = 0 \Rightarrow \frac{dy}{dx} = -\frac{y}{x}$

25. $(x + 2y)^2 + 2xy^2 = 6 \Rightarrow 2(x + 2y)(1 + 2\frac{dy}{dx}) + 2y^2 + 4xy\frac{dy}{dx} = 0$

$$4(x + 2y)\frac{dy}{dx} + 4xy\frac{dy}{dx} = -2x - 4y - 2y^2$$

$$\frac{dy}{dx} = -\frac{x + 2y + y^2}{2x + 4y + 2xy}$$

27. $y^2 = \frac{x}{x + 1} \Rightarrow 2y\frac{dy}{dx} = \frac{x + 1 - x}{(x + 1)^2} \Rightarrow \frac{dy}{dx} = \frac{1}{2y(x + 1)^2}$

29. $xy + 2x + 3y = 1 \Rightarrow x\frac{dy}{dx} + y + 2 + 3\frac{dy}{dx} = 0$

$$(x + 3)\frac{dy}{dx} = -y - 2 \Rightarrow \frac{dy}{dx} = -\frac{y + 2}{x + 3}$$

31. $x^3 - xy + y^3 = 1 \Rightarrow 3x^2 - x\frac{dy}{dx} - y + 3y^2\frac{dy}{dx} = 0$

$$(3y^2 - x)\frac{dy}{dx} = y - 3x^2 \Rightarrow \frac{dy}{dx} = \frac{y - 3x^2}{3y^2 - x}$$

33. $y = \sqrt{\frac{1 + x}{1 - x}} \Rightarrow \frac{dy}{dx} = \frac{1}{2}\left(\frac{1 + x}{1 - x}\right)^{-\frac{1}{2}}\frac{d}{dx}\left(\frac{1 + x}{1 - x}\right)$

$$= \frac{1}{2}\sqrt{\frac{1 - x}{1 + x}} \cdot \frac{1 - x + 1 + x}{(1 - x)^2} = \sqrt{\frac{1 - x}{1 + x}} \cdot \frac{1}{(1 - x)^2}$$

$$= \frac{1}{(1 + x)^{\frac{1}{2}}(1 - x)^{\frac{3}{2}}}$$

35. $y = (x^3 + 1)^{\frac{1}{3}} \Rightarrow \frac{dy}{dx} = \frac{1}{3}(x^3 + 1)^{-\frac{2}{3}}(3x^2) = x^2(x^3 + 1)^{-\frac{2}{3}}$

37. $y = \cot 2x \Rightarrow \frac{dy}{dx} = -2\csc^2 2x$

39. $y = \frac{\sin x}{\cos^2 x} \Rightarrow \frac{dy}{dx} = \frac{\cos^2 x(\cos x) - \sin x(2\cos x)(-\sin x)}{\cos^4 x}$

$$= \frac{\cos^3 x + 2\sin^2 x\cos x}{\cos^4 x} = \frac{\cos^2 x + 2\sin^2 x}{\cos^3 x}$$

(Also: $\sec x + 2\tan^2 x\sec x$)

41. $y = x^2\cos 8x \Rightarrow \frac{dy}{dx} = -8x^2\sin 8x + 2x\cos 8x$

43. $y = \frac{\sin x}{1 + \cos x} \Rightarrow \frac{dy}{dx} = \frac{(1 + \cos x)(\cos x) - \sin x(-\sin x)}{(1 + \cos x)^2} =$

$\frac{\cos x + \cos^2 x + \sin^2 x}{(1 + \cos x)^2} = \frac{1 + \cos x}{(1 + \cos x)^2} = \frac{1}{1 + \cos x}$

45. $y = \csc x \Rightarrow \frac{dy}{dx} = -\csc x \cot x$

47. $y = \cos(\sin^2 x) \Rightarrow \frac{dy}{dx} = -\sin(\sin^2 x)\frac{d}{dx}(\sin^2 x)$

$= -\sin(\sin^2 x)(2\sin x\cos x) = -2\sin x\cos x\sin(\sin^2 x)$

49. $y = \sec^2 x \Rightarrow \frac{dy}{dx} = 2\sec x\sec x\tan x = 2\sec^2 x\tan x$

51. $y = \cos(\sin^2 3x) \Rightarrow \frac{dy}{dx} = -\sin(\sin^2 3x)(2\sin 3x)(\cos 3x)(3)$

$= -3\sin 6x\sin(\sin^2 3x)$ $\quad [2\sin 3x\cos 3x = \sin 6x]$

53. $y = \sqrt{2t + t^2} \Rightarrow \frac{dy}{dt} = \frac{1}{2}(2t + t^2)^{-\frac{1}{2}}(2 + 2t)$

$t = 2x + 3 \Rightarrow \frac{dt}{dx} = 2.$ $\quad \frac{dy}{dx} = \frac{dy}{dt} \bullet \frac{dt}{dx} = \frac{2(1 + t)}{\sqrt{2t + t^2}}$

55. $t = \frac{x}{1 + x^2}$ $\quad y = x^2 + t^2 \Rightarrow y = x^2 + \left(\frac{x}{1+ x^2}\right)^2$

$\frac{dy}{dx} = 2x + 2\left(\frac{x}{1 + x^2}\right)\left[\frac{1 + x^2 - x(2x)}{(1 + x^2)^2}\right] = 2x + \frac{2x(1 - x^2)}{(1 + x^2)^3}$

57. $x = t^2 + t \Rightarrow \frac{dx}{dt} = 2t + 1$ $\quad y = t^3 - 1 \Rightarrow \frac{dy}{dt} = 3t^2$

$\frac{dy}{dx} = \frac{\frac{dy}{dt}}{\frac{dx}{dt}} = \frac{3t^2}{2t + t}$

59. $y = \frac{x}{x^2 + 1} \Rightarrow \frac{dy}{dx} = \frac{x^2 + 1 - 2x^2}{(x^2 + 1)^2} = \frac{1 - x^2}{(x^2 + 1)^2}$

$\left.\frac{dy}{dx}\right|_{x=0} = 1.$ $\quad y(0) = 0.$ $\quad \therefore$ Tangent line is $y = x$.

61. $y = 2x^2 - 6x + 3 \Rightarrow y' = 4x - 6.$ $y(2) = -1,\ y'(2) = 2.$

Tangent line is: $y + 1 = 2(x - 2)$ or $y - 2x = -5$

63. $V = \pi[10 - \frac{x}{3}]x^2 = \pi[10x^2 - \frac{x^3}{3}] \Rightarrow \frac{dV}{dx} = \pi(20x - x^2)$

65. $s = at - 16t^2 \Rightarrow v = \frac{ds}{dt} = a - 32t.$ Maximum height will occur when $v = 0$, i.e. $a - 32t = 0$ or $t = \frac{a}{32}$. At this moment, $s=49$. That is, $a(\frac{a}{32}) - 16(\frac{a}{32})^2 = 49 \Rightarrow a = 56$ ft/sec.

67.
$$\lim_{\Delta x\to 0} \frac{[2 - 3(x + \Delta x)]^2 - [2 - 3x]^2}{\Delta x} =$$
$$\lim_{\Delta x\to 0} \frac{4 - 12(x + \Delta x) + 9(x + \Delta x)^2 - 4 + 12x - 9x^2}{\Delta x} =$$
$$\lim_{\Delta x\to 0} \frac{-12x - 12\Delta x + 9x^2 + 18x\Delta x + 9\Delta x^2 + 12x - 9x^2}{\Delta x} =$$
$$\lim_{\Delta x\to 0} \frac{-12\Delta x + 18x\Delta x + 9\Delta x^2}{\Delta x} = -12 + 18x.$$

This is the derivative of $f(x) = (2 - 3x)^2$

69. $x^2y + xy^2 = 6 \Rightarrow x^2\frac{dy}{dx} + 2xy + y^2 + 2xy\frac{dy}{dx} = 0.$ At $P(1,2)$,

$$(1)^2\frac{dy}{dx} + 2(1)(2) + (2)^2 + 2(1)(2)\frac{dy}{dx} = 0$$

$$5\frac{dy}{dx} + 8 = 0 \Rightarrow \frac{dy}{dx} = -\frac{8}{5}.$$

71. $y^3 + y = x \Rightarrow 3y^2\frac{dy}{dx} + \frac{dy}{dx} = 1.$ At $P(2,1)$, $3(1)^2\frac{dy}{dx} + \frac{dy}{dx} = 1$

$4\frac{dy}{dx} = 1 \Rightarrow \frac{dy}{dx} = \frac{1}{4}$. To find the $\frac{d^2y}{dx^2}$:

$(3y^2)\frac{d^2y}{dx^2} + \frac{dy}{dx} \bullet 6y\frac{dy}{dx} + \frac{d^2y}{dx^2} = 0.$ Evaluating at $P(2,1)$ and $\frac{dy}{dx} = \frac{1}{4}$:

$$3(1)^2\frac{d^2y}{dx^2} + 6(1)(\frac{1}{4})^2 + \frac{d^2y}{dx^2} = 0 \Rightarrow \frac{d^2y}{dx^2} = -\frac{3}{32}.$$

73. $x^2 = 4y \Rightarrow 2x = 4\dfrac{dy}{dx} \Rightarrow \dfrac{dy}{dx} = \dfrac{x}{2}$. At (2,1), $\dfrac{dy}{dx} = 1$, so the slope of the normal is -1. The normal is: $y - 1 = -(x - 2)$ or $y + x = 3$.

75. (a) $y = \sqrt{2x - 1} \Rightarrow \dfrac{dy}{dx} = \dfrac{1}{2}(2x - 1)^{-\frac{1}{2}}(2) = (2x - 1)^{-\frac{1}{2}}$

$$\frac{d^2y}{dx^2} = -\frac{1}{2}(2x - 1)^{-\frac{3}{2}}(2) = -(2x - 1)^{-\frac{3}{2}}$$

$$\frac{d^3y}{dx^3} = \frac{3}{2}(2x - 1)^{-\frac{5}{2}}(2) = 3(2x - 1)^{-\frac{5}{2}}$$

(b) $y = (3x + 2)^{-1} \Rightarrow \dfrac{dy}{dx} = -(3x + 2)^{-2}(3) = -3(3x + 2)^{-2}$

$$\frac{d^2y}{dx^2} = 6(3x + 2)^{-3}(3) = 18(3x + 2)^{-3}$$

$$\frac{d^3y}{dx^3} = -54(3x + 2)^{-4}(3) = -162(3x + 2)^{-4}$$

(c) $y = ax^3 + bx^2 + cx + d$ $\quad y' = 3ax^2 + 2bx + c$

$y'' = 6ax + 2b$ $\quad y''' = 6a$

77. $y = x^3 \Rightarrow y' = 3x^2$. $y'(a) = 3a^2$. The tangent line through (a, a^3) with slope $3a^2$ is: $y - a^3 = 3a^2(x - a)$ or $y = 3a^2x - 2a^3$. To find points of intersection with $y = x^3$, equate these:

$x^3 = 3a^2x - 2a^3$. Since we know $x = a$ is one root:

$$\begin{array}{r|rrrr} a & 1 & 0 & -3a^2 & 2a^3 \\ & & a & a^2 & -2a^3 \\ \hline & 1 & a & -2a^2 & \end{array} \qquad \begin{array}{r|rrr} a & 1 & a & -2a^2 \\ & & a & 2a^2 \\ \hline & 1 & 2a & \end{array} \qquad x = -2a$$

Observe that $y'(-2a) = 12a^2$ is 4 times the slope at $y'(a) = 3a^2$.

79. (a) The circle $(x-h)^2+(y-k)^2=a^2$ is tangent to $y=x^2+1$ at $(1,2)$ $\Rightarrow$ there is a mutual tangent line there. We find the equation of this tangent: $y'=2x$ and $y'(1)=2$ so $y=2x$. This tangent is $\perp$ the radius through (h,k) and $(1,2)$. Therefore

$$\frac{k-2}{h-1}=-\frac{1}{2} \Rightarrow h=5-2k. \ \therefore \text{ The locus is } \{(h,k) \mid h+2k=5\}$$

(b) $(x-h)^2+(y-k)^2=a^2 \Rightarrow (y-k)\frac{dy}{dx}=-x+h$ or $\frac{dy}{dx}=\frac{h-x}{y-k}$.

$$\frac{d^2y}{dx^2}=\left.\frac{(y-k)(-1)-(h-x)\frac{dy}{dx}}{(y-k)^2}\right]_{(x,y)=(1,2)}=\frac{-(2-k)-2(h-1)}{(2-k)^2}.$$

Since $\frac{d^2y}{dx^2}=2$ for the parabola, we have $\frac{-(2-k)-2(h-1)}{(2-k)^2}=2$ or $k-2h=2(2-k)^2$. Substituting $h=5-2k$ from part (a), we have $2k^2-13k+18=0$ or $k=\frac{9}{2}$ or 2. If $k=\frac{9}{2}$, $h=-4$

and the circle is $(x+4)^2+\left(y-\frac{9}{2}\right)^2=a^2$. Substituting the point

$(1,2)$ gives $a^2=\frac{125}{4}$, so $a=\frac{5\sqrt{5}}{2}$.

81. $x^2y+xy^2=6 \Rightarrow x^2\frac{dy}{dx}+2xy+y^2+2xy\frac{dy}{dx}=0$ or $\frac{dy}{dx}=-\frac{2xy+y^2}{x^2+2xy}$.

If $x=1$, then $y^2+y-6=0$ or $y=-3, 2$. At $(1,2)$, $\frac{dy}{dx}=-\frac{8}{5}$ and the tangent line is $y-2=-\frac{8}{5}(x-1)$ or $5y+8x=18$. At $(1,-3)$, $\frac{dy}{dx}=\frac{3}{5}$ and the tangent line is $y+3=\frac{3}{5}(x-1)$ or $5y-3x=-18$.

83. Let the coordinates of the upper right-hand corner of the gondola be the point $P(x,y)$. Observe that x is one-half the width of the gondola. Since the radius of the balloon is 15 ft. plus 8 ft. makes $y=-23$.

$x^2 + y^2 = 225 \Rightarrow 2x + 2y\frac{dy}{dx} = 0 \Rightarrow \frac{dy}{dx} = -\frac{x}{y}$. At (12,-9)

$\frac{dy}{dx} = -\frac{12}{-9} = \frac{4}{3}$. The slope of the tangent line through

(x,-23) and (12,-9) must have slope $\frac{4}{3}$. Therefore

$\frac{-23 - (-9)}{x - 12} = \frac{4}{3} \Rightarrow x = \frac{3}{2}$.

The width of the gondola is then 3 feet.

85. $y = 2x^2 - 3x + 5$. $\Delta y = y(3.1) - y(3) = 14.92 - 14 = .92$

$dy = f'(x)\Delta x = (4x - 3)\Delta x$. At $x = 3$, $dy = (9)(.1) = .9$

87. $\frac{d}{dt}(f(g(t)) = f'(g(t))g'(t)$. For $t = 1$,

$= f'(g(1))g'(1) = f'(3)g'(1) = (5)(6) = 30$

89. $x = y^2 + y \Rightarrow \frac{dx}{dy} = 2y + 1$. $u = (x^2 + x)^{\frac{3}{2}} \Rightarrow$

$\frac{du}{dx} = \frac{3}{2}(x^2 + x)^{\frac{1}{2}}(2x + 1)$. $\frac{dx}{dy} \cdot \frac{du}{dx} = \frac{du}{dy}$ and $\frac{dy}{du} = \frac{1}{\frac{du}{dy}} =$

$\frac{1}{\frac{3}{2}x^2 + x)^{\frac{1}{2}}(2x + 1)(2y + 1)}$.

91. $f'(x) = \sin(x^2)$ and $y = f\left(\frac{2x - 1}{x + 1}\right)$.

$\frac{dy}{dx} = f'\left(\frac{2x - 1}{x + 1}\right) \cdot \frac{d}{dx}\left(\frac{2x - 1}{x + 1}\right) = \sin\left(\frac{2x - 1}{x + 1}\right)^2 \cdot \frac{2(x + 1) -2x + 1}{(x + 1)^2} =$

$\frac{3}{(x + 1)^2} \sin\left(\frac{2x - 1}{x + 1}\right)^2$

93. $y = \sqrt{x^2 + 16} \Rightarrow \frac{dy}{dx} = \frac{1}{2}(x^2 + 16)^{-\frac{1}{2}}(2x) = \frac{x}{\sqrt{x^2 + 16}}$

$t = \frac{x}{x - 1} \Rightarrow \frac{dt}{dx} = \frac{x - 1 - x}{(x - 1)^2} = \frac{-1}{(x - 1)^2}$

$\frac{dy}{dt} = \frac{\frac{dy}{dx}}{\frac{dt}{dx}} = \frac{x}{\sqrt{x^2 + 16}} \cdot \frac{(x - 1)^2}{-1}$. At x=3, $\frac{dy}{dt} = \frac{3}{5}(-4) = -\frac{12}{5}$

95. Differentiate $\cos y = y \sin z$ with respect to z:

$$(-\sin y - \sin z)\frac{dy}{dz} = y\cos z$$

$$\frac{dy}{dz} = \frac{-y\cos z}{\sin y + \sin z}$$

Differentiate $z = x\sin y - y^2$ with respect to z:

$$1 = \sin y\frac{dx}{dz} + x\cos y\frac{dy}{dz} - 2y\frac{dy}{dz}$$

$$\sin y\frac{dx}{dz} = 1 - x\cos y\frac{dy}{dz} + 2y\frac{dy}{dz}$$

$$\sin y\frac{dx}{dz} = 1 - (x\cos y - 2y)\left(\frac{-y\cos z}{\sin y + \sin z}\right)$$

$$\frac{dx}{dz} = \frac{\sin y + \sin z + xy\cos y\cos z - 2y^2\cos z}{\sin y(\sin y + \sin z)}$$

97. $f(x) = (1 + \tan x)^{-1} \Rightarrow f'(x) = -(1 + \tan x)^{-2}(\sec^2 x)$.

$f(0) = 1$ and $f'(0) = -1$. $L(x) = f(0) + f'(0)x = 1 - x$

101. $x = \cos 3t \Rightarrow \frac{dx}{dt} = -3\sin 3t$. $y = \sin^2 3t \Rightarrow \frac{dy}{dt} = 6\sin 3t\cos 3t$

$$\frac{dy}{dx} = \frac{\frac{dy}{dt}}{\frac{dx}{dt}} = \frac{6\sin 3t\cos 3t}{-3\sin 3t} = -2\cos 3t.$$

$$\frac{d^2y}{dx^2} = \frac{\frac{d}{dt}\left(\frac{dt}{dx}\right)}{\frac{dx}{dt}} = \frac{6\sin 3t}{-3\sin 3t} = -2.$$

Also: $y = \sin^2 3t = 1 - \cos^2 3t = 1 - x^2 \Rightarrow \frac{dy}{dx} = -2x$ and $\frac{d^2y}{dx^2} = -2$

103. $x = t - t^2 \Rightarrow \frac{dx}{dt} = 1 - 2t \qquad y = t - t^3 \Rightarrow \frac{dy}{dt} = 1 - 3t^2$

$$\frac{dy}{dx} = \frac{\frac{dy}{dy}}{\frac{dx}{dt}} = \frac{1 - 3t^2}{1 - 2t}\Bigg|_{t=1} = 2.$$

$$\frac{d}{dt}\left(\frac{dy}{dx}\right) = \frac{(1 - 2t)^2(-6t) - (1 - 3t^2)(-2)}{(1 - 2t)^2} = \frac{6t^2 - 6t + 3}{(1 - 2t)^2}\Bigg|_{t=1} = 2$$

$$\frac{d^2y}{dx^2} = \frac{\frac{d}{dt}\left(\frac{dt}{dx}\right)}{\frac{dx}{dt}}\Bigg|_{t=1} = \frac{2}{-1} = -2$$

105. Let $y = u_1u_2\ldots u_n$. We want to prove that

$$\frac{dy}{dx} = \frac{du_1}{dx}u_2u_3\ldots u_n + u_1\frac{du_2}{dx}\ldots u_n + \ldots + u_1u_2\ldots u_{n-1}\frac{du_n}{dx}$$

The statement is true for $n = 1$, since

$$\frac{dy}{dx} = \frac{du_1}{dx} \text{ is true if } y = u_1$$

Assume, for $n = k$, that

$$\frac{dy}{dx} = \frac{du_1}{dx}u_2\ldots u_k + u_1\frac{du_2}{dx}u_3\ldots u_k + \ldots\ldots + u_1u_2\ldots u_{k-1}\frac{du_k}{dx}.$$

Then, for $n = k+1$, $y = u_1u_2\ldots u_ku_{k+1} = (u_1u_2\ldots u_k)u_{k+1}$ and

$$\frac{dy}{dx} = \frac{d}{dx}(u_1u_2\ldots u_k)u_{k+1} + u_1u_2\ldots u_k\frac{d}{dx}(u_{k+1})$$

107. (a) $\dfrac{d^2}{dx^2}(uv) = \dfrac{d}{dx}\left(\dfrac{d}{dx}(uv)\right) = \dfrac{d}{dx}\left[u\dfrac{dv}{dx} + v\dfrac{du}{dx}\right] =$

$$\frac{d}{dx}\left[u\frac{dv}{dx}\right] + \frac{d}{dx}\left[v\frac{du}{dx}\right] = u\frac{d^2v}{dx^2} + \frac{dv}{dx}\cdot\frac{du}{dx} + \frac{dv}{dx}\cdot\frac{du}{dx} + v\frac{d^2u}{dx^2} =$$

$$u\frac{d^2v}{dx^2} + 2\frac{dv}{dx}\cdot\frac{du}{dx} + v\frac{d^2u}{dx^2}$$

(b) $\dfrac{d^3}{dx^3}(uv) = \dfrac{d}{dx}\left(\dfrac{d^2}{dx^2}(uv)\right) = \dfrac{d}{dx}\left[u\dfrac{d^2v}{dx^2} + 2\dfrac{dv}{dx}\cdot\dfrac{du}{dx} + v\dfrac{d^2u}{dx^2}\right]$

$$= \frac{d}{dx}\left(u\frac{d^2v}{dx^2}\right) + 2\,\frac{d}{dx}\left(\frac{dv}{dx}\cdot\frac{du}{dx}\right) + \frac{d}{dx}\left(v\frac{d^2u}{dx^2}\right)$$

$$= \frac{du}{dx}\cdot\frac{d^2v}{dx^2} + u\frac{d^3v}{dx^3} + 2\left[\frac{dv}{dx}\cdot\frac{d^2u}{dx^2} + \frac{d^2v}{dx^2}\cdot\frac{du}{dx}\right] + \frac{dv}{dx}\cdot\frac{d^2u}{dx^2} + v\frac{d^3u}{dx^3}$$

$$= u\frac{d^3v}{dx^3} + 3\frac{du}{dx}\cdot\frac{d^2v}{dx^2} + 3\frac{d^2u}{dx^2}\cdot\frac{dv}{dx} + v\frac{d^3u}{dx^3}$$

Note: to begin to see the pattern, remember that

$(a + b)^3 = a^3 + 3a^2b + 3ab^2 + b^3$ which can be written as

$b^0a^3 + 3a^2b + 3ab^2 + b^3a^0$.

Define $\dfrac{d^0u}{dx^0} = u$ and $\dfrac{d^0v}{dx^0} = v$ and observe that the previous

could be obtained by regarding $\dfrac{d^3}{dx^3}(uv)$ as $\left(\dfrac{du}{dx} + \dfrac{dv}{dx}\right)^3$.

To prove the general formula, the following is also needed:

$${}_mC_k + {}_mC_{k+1} = {}_{m+1}C_{k+1} \quad \text{where} \quad {}_mC_k = \frac{m!}{k!\,(m-k)!}\,.$$

(c) We prove this by mathematical induction. The case for $n = 2$ is proved in part (a).

Assume the truth of the statement for $n = m$, that is:

$$\frac{d^m(uv)}{dx^m} = \frac{d^m u}{dx^m}v + m\frac{d^{m-1}u}{dx^{m-1}}\frac{dv}{dx} + \binom{m}{2}\frac{d^{m-2}u}{dx^{m-2}}\frac{dv^2}{dx^2} = \ldots + \binom{m}{m-1}\frac{du}{dx}\frac{d^{m-1}v}{dx^{m-1}} + u\frac{d^m v}{dx^m}$$

$$\frac{d^{m+1}(uv)}{dx^{m+1}} = \frac{d}{dx}\left(\frac{d^m(uv)}{dx^m}\right) = \left[v\frac{d^{m+1}u}{dx^{m+1}} + \frac{d^m u}{dx^m}\frac{dv}{dx}\right] + \left[m\frac{d^m u}{dx^m}\frac{dv}{dx} + m\frac{d^{m-1}u}{dx^{m-1}}\frac{d^2v}{dx^2}\right] + \ldots +$$

$$\left[\binom{m}{m-1}\frac{d^2u}{dx^2}\frac{d^{m-1}v}{dx^{m-1}} + \binom{m}{m-1}\frac{du}{dx}\frac{d^m u}{dx^m}v\right] + u\frac{d^{m+1}v}{dx^{m+1}}$$

$$= v\frac{d^{m+1}u}{dx^{m+1}} + \left[\frac{d^m u}{dx^m}\frac{dv}{dx} + m\frac{d^m u}{dx^m}\frac{dv}{dx}\right] + \left[m\frac{d^{m-1}u}{dx^{m-1}}\frac{d^2v}{dx^2} + \binom{m}{2}\frac{d^{m-1}u}{dx^{m-1}}\frac{d^2v}{dx^2}\right] + \ldots$$

$$+ u\frac{d^{m+1}v}{dx^{m+1}}$$

$$= v\frac{d^{m+1}u}{dx^{m+1}} + (m+1)\frac{d^m u}{dx^m}\frac{dv}{dx} + \binom{m+1}{2}\frac{d^{m-1}u}{dx^{m-1}}\frac{d^2v}{dx^2} + \ldots + u\frac{d^{m+1}v}{dx^{m+1}}$$

109. (a) To be continuous at $x = \pi$ requires that $y(\pi) = \lim_{x\to\pi} y(x)$.

From the right, $\lim_{x\to\pi^+}(mx + b) = m\pi + b = y(\pi)$. From the left, $\lim_{x\to\pi^-}\sin x = 0$. We require that $m\pi + b = 0$ or $m = -\frac{b}{\pi}$.

(b) $y' = \begin{cases} \cos x & \text{if } x < \pi \\ m & \text{if } x > \pi \end{cases}$. $y'(\pi) = m \Rightarrow \lim_{x\to\pi^-}\cos x = m$. Since $\cos\pi = -1$, we have $m = -1$. Then $b = \pi$.

111. (a) $f'(0) = \lim_{h\to 0}\frac{f(0+h) - f(0)}{h} = \lim_{h\to 0}\frac{h^2\sin\frac{1}{h}}{h} = \lim_{h\to 0} h\sin\frac{1}{h} = 0$ since $|\sin\frac{1}{h}| \le 1$.

(b) $f'(x) = x^2\cos\frac{1}{x}\left(-\frac{1}{x^2}\right) + 2x\sin\frac{1}{x} = 2x\sin\frac{1}{x} - \cos\frac{1}{x}$

(c) No, $\lim_{x\to 0}\cos\frac{1}{x}$ does not exist.

113. Define $g(x) = m(x - a) + c$, m and c constants.

$$e(x) = f(x) - g(x).$$

Further suppose that (1) $e(a) = 0$

$$(2) \lim_{x \to 0} \frac{e(x)}{x - a} = 0.$$

$e(a) = 0 \Rightarrow f(a) = g(a)$. But $g(a) = c$. $\therefore f(a) = c$.

$$0 = \lim_{x \to a} \frac{e(x)}{x - a} = \lim_{x \to a} \frac{f(x) - g(x)}{x - a} = \lim_{x \to a} \frac{f(x) - f(a) + f(a) - g(x)}{x - a}$$

$$= \lim_{x \to a} \frac{f(x) - f(a)}{x - a} + \lim_{x \to a} \frac{g(a) - g(x)}{x - a} \quad (\text{since } f(a) = g(a))$$

$$= f'(a) - g'(a) = f'(a) - m \Rightarrow f'(a) = m.$$

$\therefore g(x) = f(a) + f'(a)(x - a).$

CHAPTER 3

APPLICATIONS OF DERIVATIVES

3.1 CURVE SKETCHING WITH THE FIRST DERIVATIVE

1. $y = x^2 - x + 1$

   $y' = 2x - 1$

   $2x - 1 = 0 \Leftrightarrow x = \frac{1}{2}$

   f decreases on $(-\infty, \frac{1}{2})$

   f increases on $(\frac{1}{2}, \infty)$

   $f(\frac{1}{2}) = \frac{3}{4}$ is local minimum

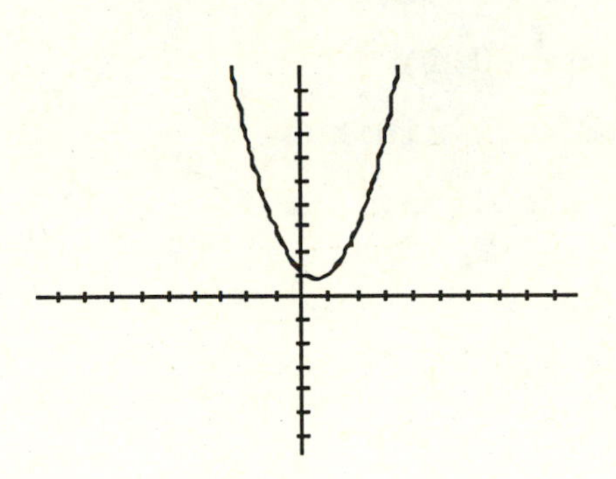

3. $y = \frac{x^3}{3} - \frac{x^2}{2} - 2x + \frac{1}{3}$

   $y' = x^2 - x - 2$

   $(x - 2)(x + 1) = 0 \Leftrightarrow x = 2$ or $x = -1$

   f increases on $(-\infty, -1) \cup (2, \infty)$

   f decreases on $(-1, 2)$

   $f(-1) = \frac{3}{2}$ is local maximum

   $f(2) = -3$ is local minimum

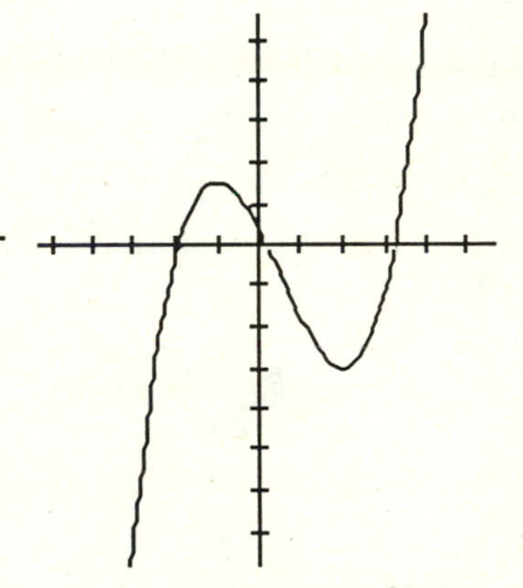

5. $y = x^3 - 27x + 36$

   $y' = 3x^2 - 27$

   $3x^2 - 27 = 0 \Leftrightarrow x \pm 3$

   f increases on $(-\infty, -3) \cup (3, \infty)$

   f decreases on $(-3, 3)$

   $f(-3) = 90$ is local maximum

   $f(3) = -18$ is local minimum

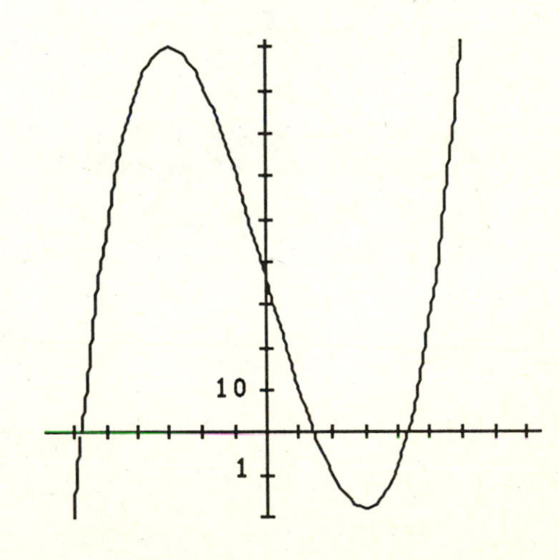

7. 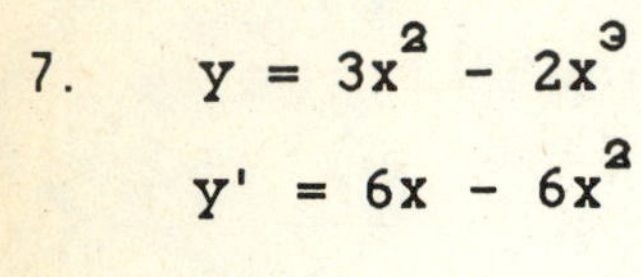

$y = 3x^2 - 2x^3$

$y' = 6x - 6x^2$

$6x(1 - x) = 0 \Leftrightarrow x = 0$ or $x = 1$

f decreases on $(-\infty, 0) \cup (1, \infty)$

f increases on $(0, 1)$

$f(0) = 0$ is local minimum

$f(1) = 1$ is local maximum

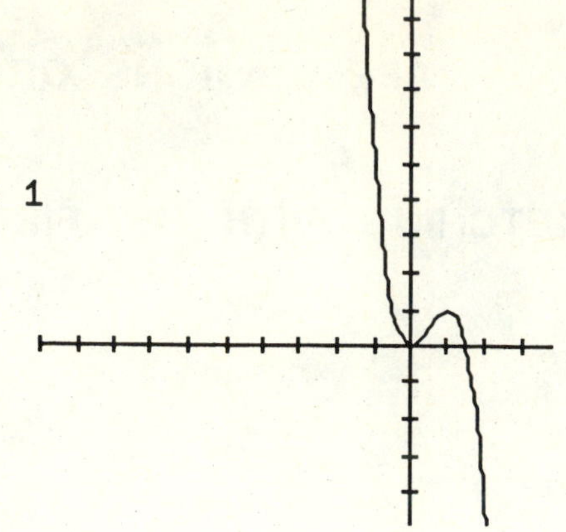

9. $y = x^4$

$y' = 4x^3$

$4x^3 = 0 \Leftrightarrow x = 0$

f decreases for $(-\infty, 0)$

f increases for $(0, \infty)$

$f(0) = 0$ is local minimum

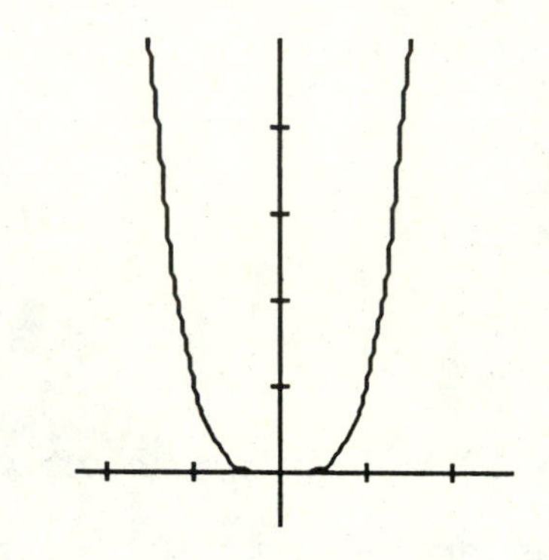

11. $y = \frac{1}{x^3}$

$y' = -3x^{-4} = -\frac{3}{x^4}$

$y' < 0$ for all $x \neq 0$.

f decreases for all $x \neq 0$
($x = 0$ is vertical asymptote)

There is no local maximum
or minimum value.

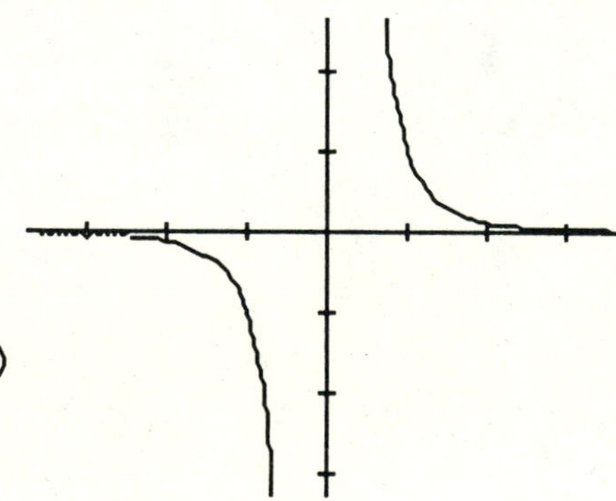

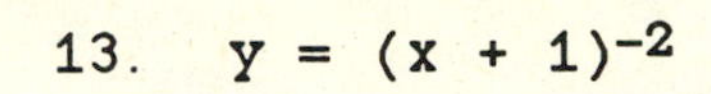

13. $y = (x + 1)^{-2}$

$$y' = -\frac{2}{(x + 1)^3}$$

Increasing on $(-\infty, -1)$

Decreasing on $(-1, \infty)$

No extreme value.

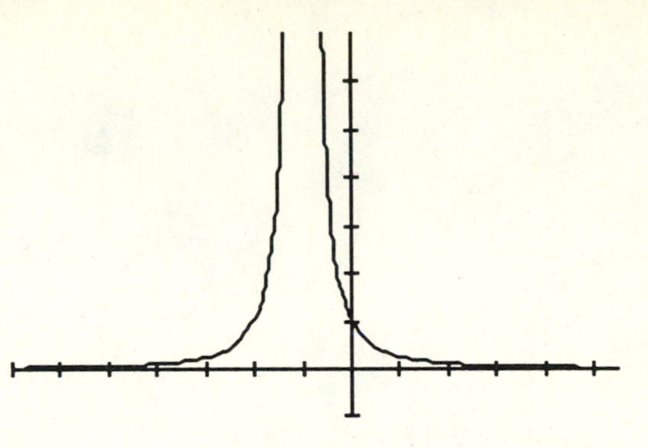

15. $y = \cos x, \ -\frac{3\pi}{2} \le x \le \frac{3\pi}{2}$

$y' = -\sin x$

$-\sin x = 0 \Leftrightarrow x = 0, -\pi, \pi$

f increases on $(-\pi, 0) \cup (\pi, \frac{3\pi}{2})$

f decreases on $(-\frac{3\pi}{2}, -\pi) \cup (0, \pi)$

$f(0)=1$ is local maximum; $f(-\pi) = f(\pi) = -1$ is local minimum

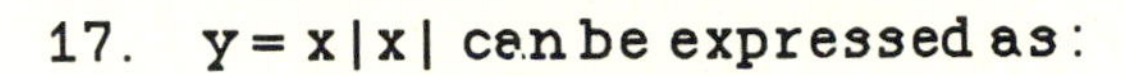

17. $y = x|x|$ can be expressed as:

$$y = \begin{cases} x^2 & \text{if } x \ge 0 \\ -x^2 & \text{if } x < 0 \end{cases}$$

$$y' = \begin{cases} 2x & \text{if } x > 0 \\ -2x & \text{if } x < 0 \end{cases}$$

The function is always increasing; there are no extrema

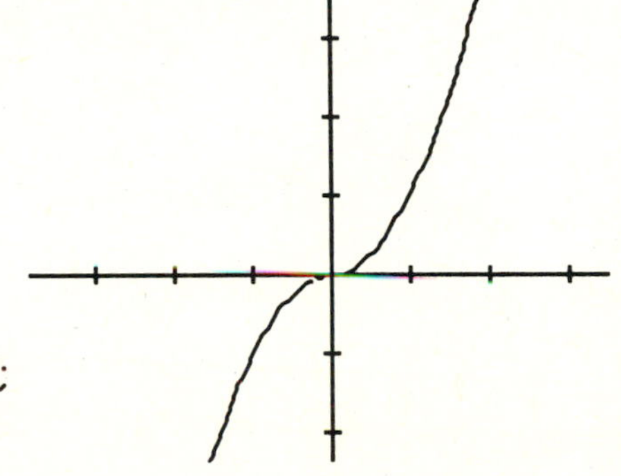

19. Since $f(x) = \cos x$ is an even function, $\cos|x| = \cos x$. Therefore, this problem is very similar to problem 15.

21. $y = \frac{x}{x+1} \Rightarrow y' = \frac{x + 1 - x}{(x+1)^2} = \frac{1}{(x+1)^2}$. Since $y' > 0$ for all x,

the functions increases on every interval in its domain.

23. There is no unique answer to this problem. Any function with properties like the one below is correct.

$$f(x) = \begin{cases} 1 - x & \text{if } x \le 0 \\ 1 - 2x & \text{if } x > 0 \end{cases} \quad f'(x) = \begin{cases} -1 & \text{if } x < 0 \\ -2 & \text{if } x > 0 \end{cases}$$

25. $y = x - 2\sin x$

$y' = 1 - 2\cos x$

$1 - 2\cos x = 0$

$\cos x = \frac{1}{2}$

$x = \frac{\pi}{3}$

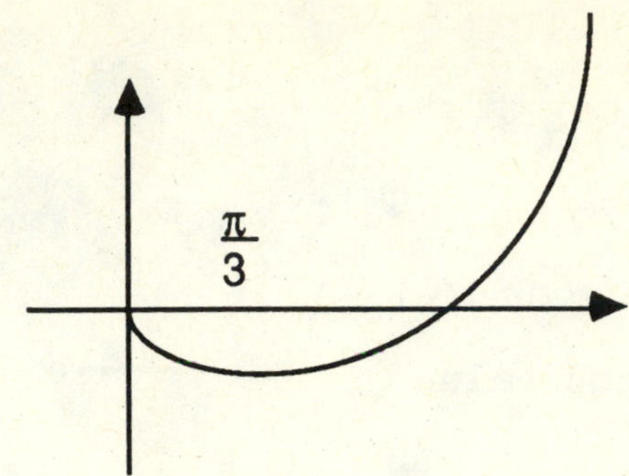

Use Newton's Method: $f(\pi) = \pi$ and $f(\frac{\pi}{2}) = \frac{\pi}{2} - 2 < 0 \Rightarrow$ there is a root r_1 such that $\frac{\pi}{2} < r_1 < \pi$. Take $x_0 = 1.5 \approx \frac{\pi}{2}$. Using the iterative formula

$$x_{n+1} = x_n - \frac{f(x_n)}{f'(x_n)},$$

we have $x_1 = 2.0765$, $x_2 = 1.9105$, $x_3 = 1.8956$, $x_4 = 1.8954$.

To two decimal places, $r_1 = 1.90$.

3.2 CONCAVITY AND POINTS OF INFLECTION

1. $y = x^2 - 4x + 3$

$y' = 2x - 4$ $\quad y'' = 2$

$y' = 0 \Leftrightarrow x = 2$

curve rises for $(2, \infty)$
curve falls for $(-\infty, 2)$
everywhere concave up

no inflection points
$f(2) = -1$ is minimum.

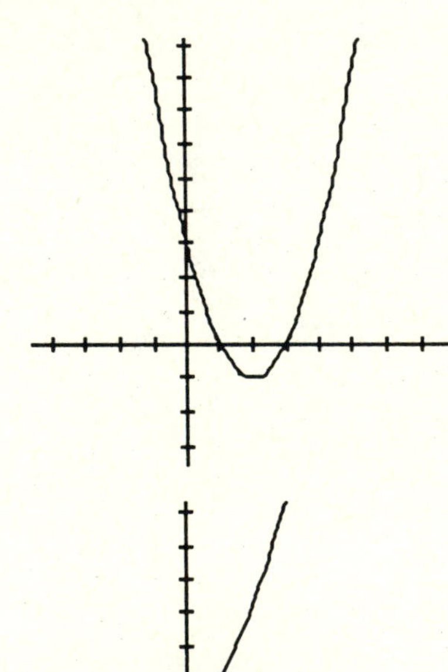

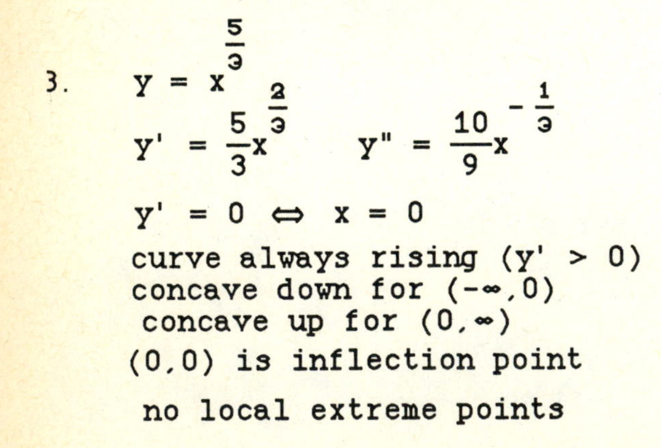

3. $y = x^{\frac{5}{3}}$

$y' = \frac{5}{3}x^{\frac{2}{3}}$ $\quad y'' = \frac{10}{9}x^{-\frac{1}{3}}$

$y' = 0 \Leftrightarrow x = 0$

curve always rising ($y' > 0$)
concave down for $(-\infty, 0)$
concave up for $(0, \infty)$
$(0,0)$ is inflection point

no local extreme points

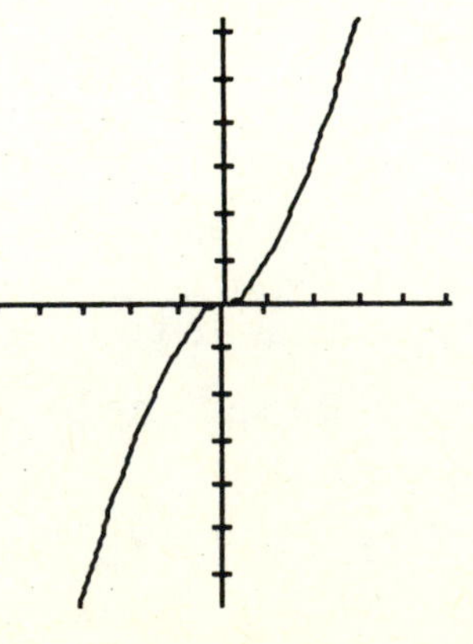

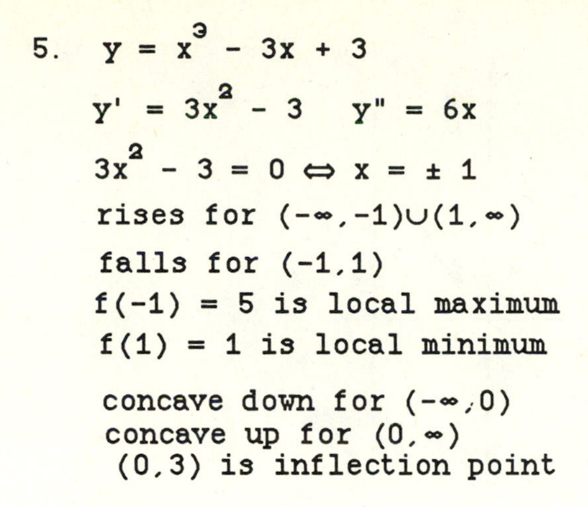

5. $y = x^3 - 3x + 3$

$y' = 3x^2 - 3 \quad y'' = 6x$

$3x^2 - 3 = 0 \Leftrightarrow x = \pm 1$

rises for $(-\infty,-1)\cup(1,\infty)$

falls for $(-1,1)$

$f(-1) = 5$ is local maximum

$f(1) = 1$ is local minimum

concave down for $(-\infty,0)$
concave up for $(0,\infty)$
$(0,3)$ is inflection point

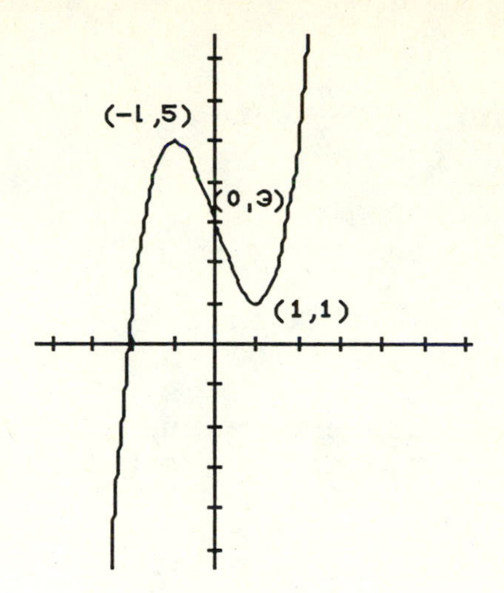

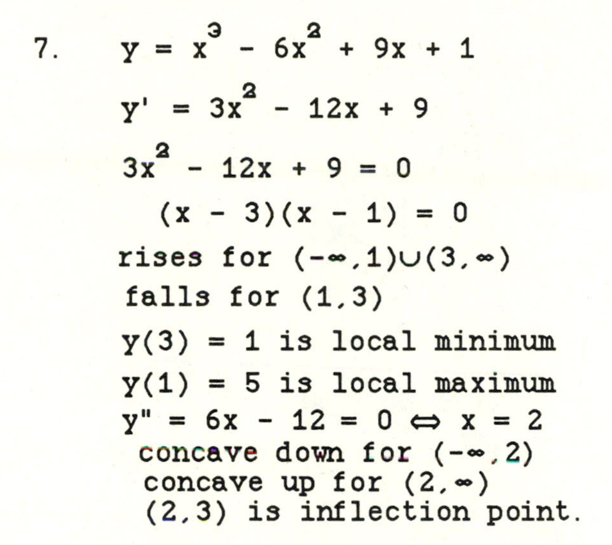

7. $y = x^3 - 6x^2 + 9x + 1$

$y' = 3x^2 - 12x + 9$

$3x^2 - 12x + 9 = 0$

$(x - 3)(x - 1) = 0$

rises for $(-\infty,1)\cup(3,\infty)$

falls for $(1,3)$

$y(3) = 1$ is local minimum

$y(1) = 5$ is local maximum

$y'' = 6x - 12 = 0 \Leftrightarrow x = 2$
concave down for $(-\infty,2)$
concave up for $(2,\infty)$
$(2,3)$ is inflection point.

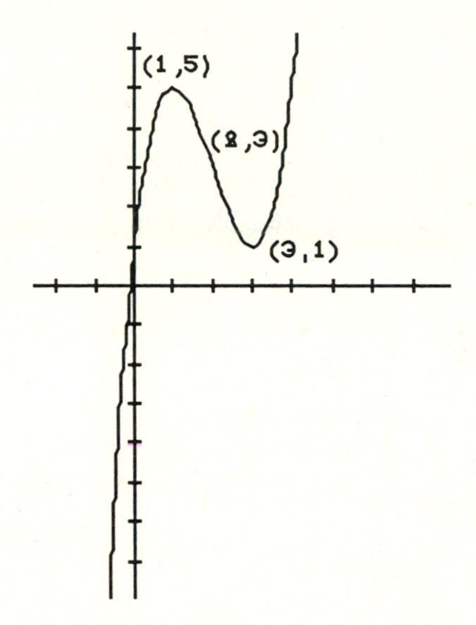

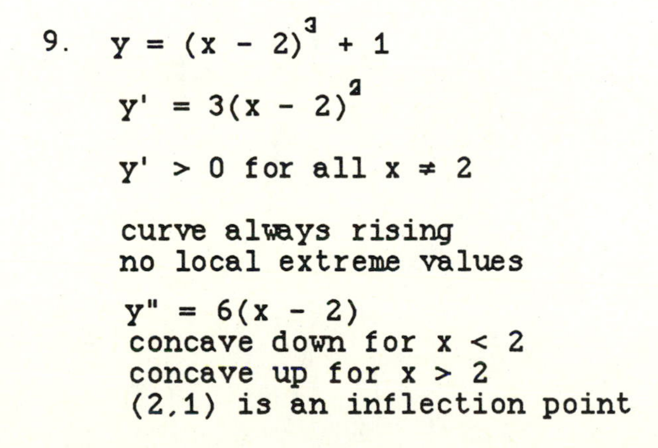

9. $y = (x - 2)^3 + 1$

$y' = 3(x - 2)^2$

$y' > 0$ for all $x \neq 2$

curve always rising
no local extreme values

$y'' = 6(x - 2)$
concave down for $x < 2$
concave up for $x > 2$
$(2,1)$ is an inflection point

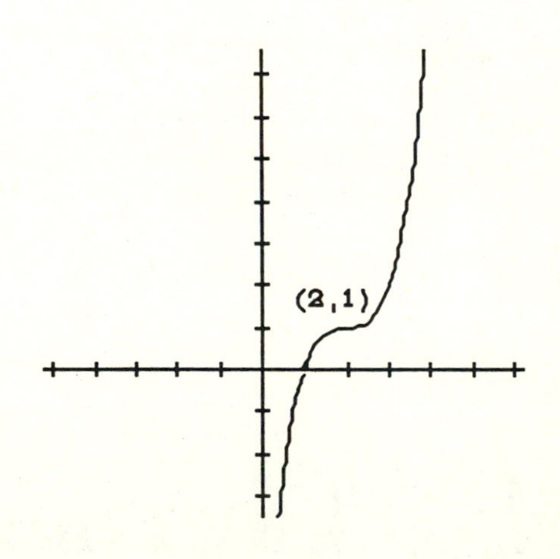

11. 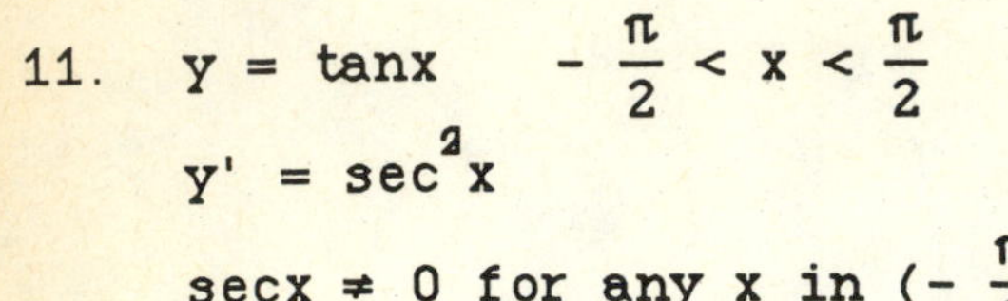
$y = \tan x \qquad -\frac{\pi}{2} < x < \frac{\pi}{2}$

$y' = \sec^2 x$

$\sec x \neq 0$ for any x in $(-\frac{\pi}{2}, \frac{\pi}{2})$

$y' > 0$ for all x in $(-\frac{\pi}{2}, \frac{\pi}{2})$
curve always rising

no local extreme values
$y'' = 2\sec x(\sec x \tan x) = 0$ if $x = 0$
concave down for $x < 0$
concave up for $x > 0$
$(0,0)$ inflection point

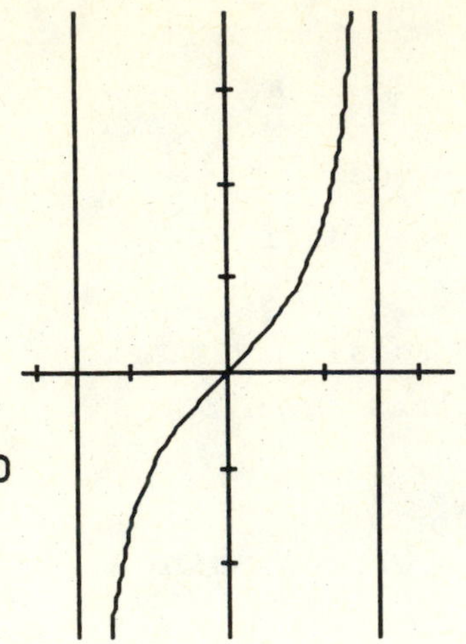

13. $y = -x^4 \Rightarrow y' = -4x^3$
The curve rises for
$x < 0$ and falls for $x > 0$.
$y(0) = 0$ is a maximum.
$y'' = -12x^2$. $y'' = 0$ if $x = 0$ but
$y'' < 0$ for all $x \neq 0$. The curve is
always concave down.

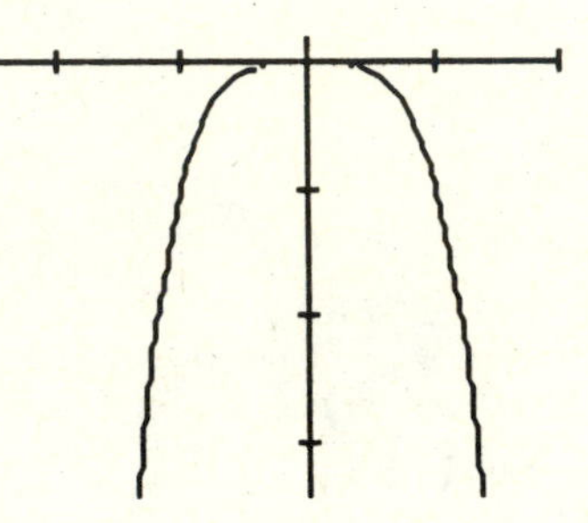

15. The answer is not unique. The curve $y = (x - 1)^2$ is an example. This is an upward opening parabola with vertex at $(1,0)$.

17. $y = 6 - 2x - x^2 \Rightarrow y' = -2(1 + x)$ and $y'' = -2$. There is a local maximum of $y = 7$ at $x = -1$. There are no inflection points. The curve is a downward opening parabola with vertex at the point $(-1,7)$.

19. $y = x(6 - 2x)^2$

$y' = (6 - 2x)^2 + 2x(6 - 2x)(-2)$

$= 12(3 - x)(1 - x)$

$y'' = 12[(3 - x)(-1) + (1 - x)(-1)]$

$= -12(4 - 2x)$

$y(1) = 16$ is local maximum

$y(3) = 0$ is local minimum.

$(2,8)$ is an inflection point.

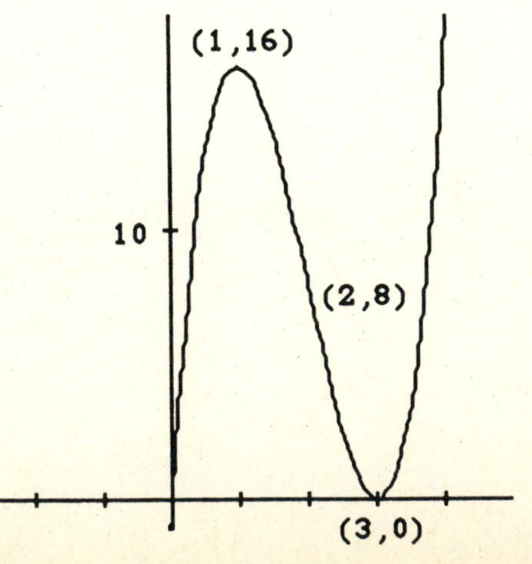

21. $y = 12 - 12x + x^3 \Rightarrow y' = -12 + 3x^2$ and $y'' = 6x$. $y' = 0$ if $x = \pm 2$. $y(-2) = 28$ is local maximum; $y(2) = 4$ is local minimum. The point (0,12) is an inflection point. The graph is similar to that of Problem 5 or 7.

23. $y = x^3 + 3x^2 + 3x + 2 \Rightarrow y' = 3x^2 + 6x + 3 = 3(x + 1)^2$. $y' > 0$ for all $x \neq -1$ so the graph is always rising. There is no maximum or minimum at $x = -1$ because the derivative does not change sign there. $y'' = 6(x + 1)$. The point (-1,1) is an inflection point. The graph is similar to Problem 9.

25. $y = x^3 - 6x^2 - 135x \Rightarrow y' = 3x^2 - 12x - 135 = 3(x - 9)(x + 5)$. $y(-5) = 400$ is local maximum and $y(9) = -972$ is local minimum. $y'' = 3(2x - 4)$. The point (2,-286) is an inflection point. The graph is similar in shape to the other cubics but greatly elongated.

27. $y = x^4 - 2x^2 \quad 2 \Rightarrow y' = 4x^3 - 4x = 4(x^2 - 1)$. The graph decreases on $(-\infty,-1)$, increases on $(-1,0)$, decreases on $(0,1)$ and increases on $(1,\infty)$. $y(-1) = y(1) = 1$ are local minimum values. $y(0) = 2$ is a local maximum.

$y'' = 4(3x^2 - 1) = 0$ when $x = \pm \frac{1}{\sqrt{3}}$

$y(\pm \frac{1}{\sqrt{3}}) = \frac{13}{9}$ so the points $(\pm\frac{1}{\sqrt{3}}, \frac{9}{13})$ are inflection points.

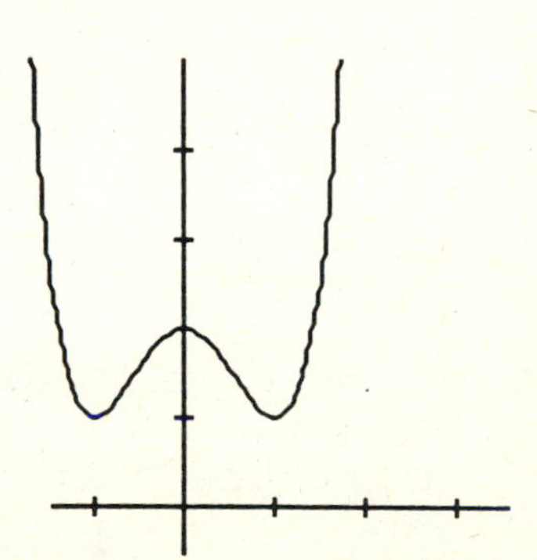

29. $y = 3x^4 - 4x^3 \Rightarrow y' = 12x^3 - 12x^2 = 12x^2(x - 1)$. $y' < 0$ for $x < 1$ and $y' > 0$ for $x > 1 \Rightarrow y(1) = -1$ is a local minimum. There is no local extreme value at $x = 0$ because y' does not change sign there. $y'' = 36x^2 - 24x = 12x(3x - 2)$.

$y'' > 0$ for $x < 0$ and $x > \frac{2}{3}$, but $y'' < 0$ for $0 < x < \frac{2}{3}$. $(0,0)$ and $(\frac{2}{3}, -\frac{16}{27})$ are inflection points.

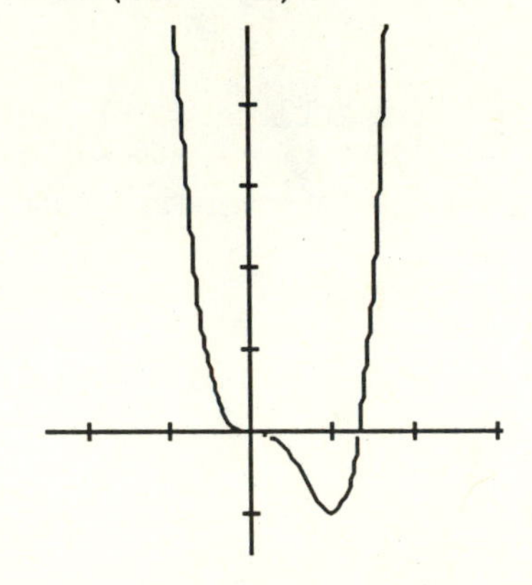

31. $y = \sin x + \cos x \Rightarrow y' = \cos x - \sin x$. $y' = 0$ if $\sin x = \cos x$. This occurs when $x = \frac{\pi}{4}, \frac{5\pi}{4}, -\frac{3\pi}{4}, -\frac{7\pi}{4}, \ldots$ The values $y(\frac{\pi}{4}) = y(-\frac{7\pi}{4}) = \ldots = \sqrt{2}$ are local maxima. The values

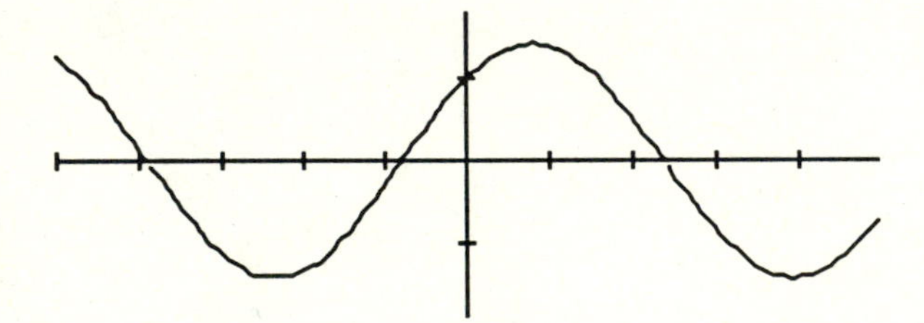

$y(\frac{5\pi}{4}) = y(-\frac{3\pi}{4}) = \ldots. = -\sqrt{2}$ are local minima

$y'' = -\sin x - \cos x = 0$ when $y = 0$. This occurs when $x = \frac{3\pi}{4}, -\frac{5\pi}{4}, \frac{7\pi}{4}, \ldots.$ Thus there are inflection points at each x-intercept.

33. (a) f' and f'' both negative state that the curve is both decreasing and concave down. Point T

(b) f' negative and f'' positive state that the curve is decreasing and concave up. Point P.

35.

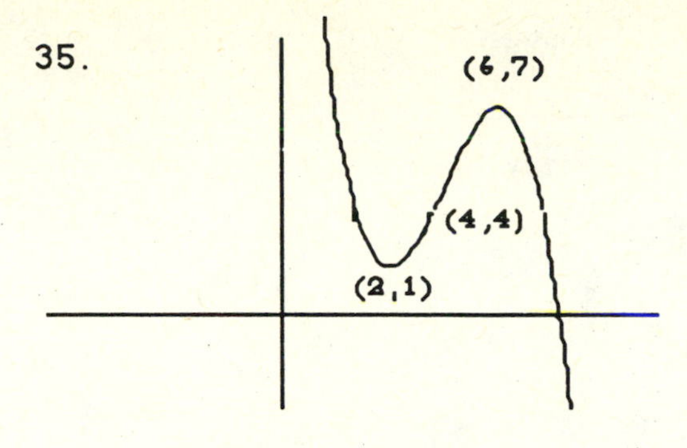

37.

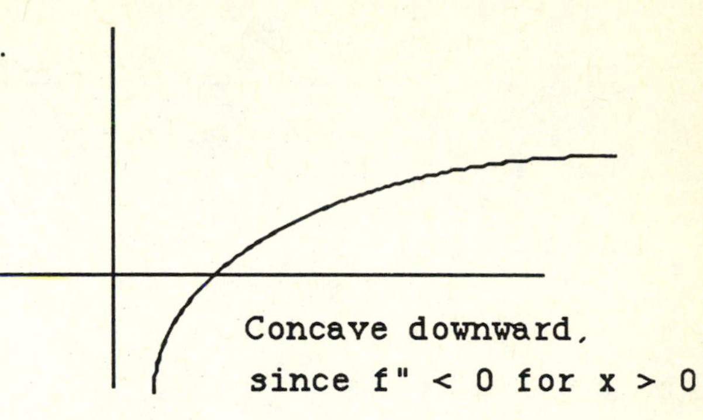

Concave downward, since $f'' < 0$ for $x > 0$

39. $y = 2x^3 + 2x^2 - 2x - 1$

$y' = 6x^2 + 4x - 2$
$= 2(3x - 1)(x + 1)$

$y'' = 12x + 4 = 4(3x + 1)$

$y(-1) = 1$ is local maximum

$y(1/3) = -37/27$ is local minimum.

$(-1/3, -5/27)$ is inflection point.

(a) three times

(b) once

(c) once

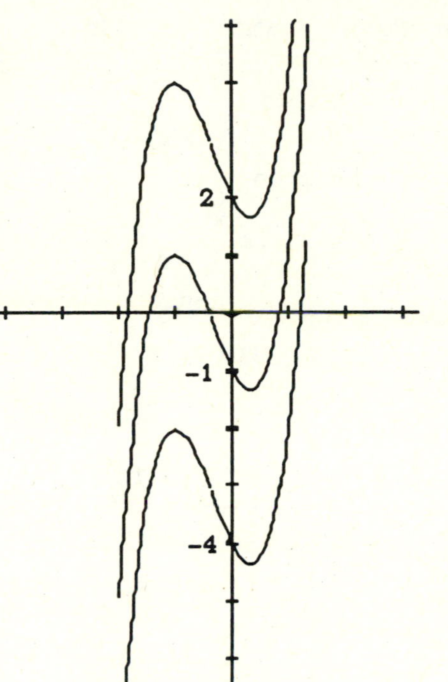

41. $y = x^4 + 8x^3 - 270x^2 \Rightarrow y' = 4x^3 + 24x^2 - 540x =$

$4x(x - 9)(x + 15)$. $y'' = 12x^2 + 48x - 540 = 12(x + 9)(x - 5)$.

x	y	curve
$x < -15$		falling, concave up
-15	$-37,125$	local minimum
$-15 < x < -9$		rising, convave up
-9	$-21,141$	inflection point
$-9 < x < 0$		rising, concave down,
0	0	local maximum
$0 < x < 5$		falling, concave down
5	$-5,125$	inflection point
$5 < x < 9$		falling, concave up
9	$-9,477$	local minimum
$9 < x$		rising, concave up

43. (a) At $x = 2$ because y' changes from positive to negative.
(b) At $x = 4$ because y' changes from negative to positive.

3.3 ASYMPTOTES AND SYMMETRY

1. Odd. 3. Odd 5. Neither 7. Even 9. Even

11. Odd. $f(-x) = \dfrac{-x}{(-x)^2 - 1} = \dfrac{-x}{x^2 - 1} = -\dfrac{x}{x^2 - 1} = -f(x)$

13. $y = \dfrac{1}{2x - 3}$

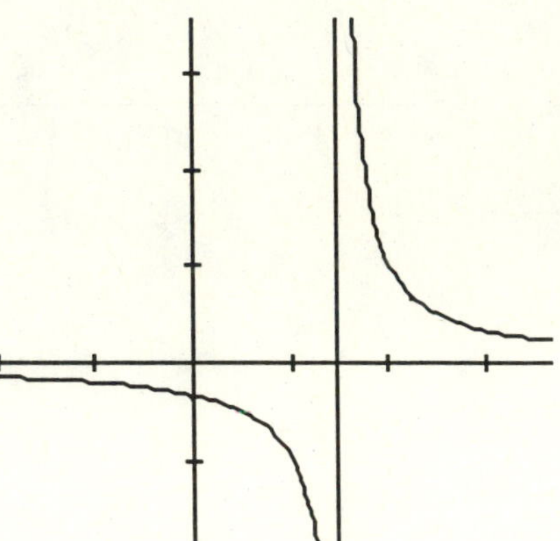

(a) No symmetry

(b) $x = 0 \Rightarrow y = -\dfrac{1}{3}$.

No x-intercept because y cannot be zero.

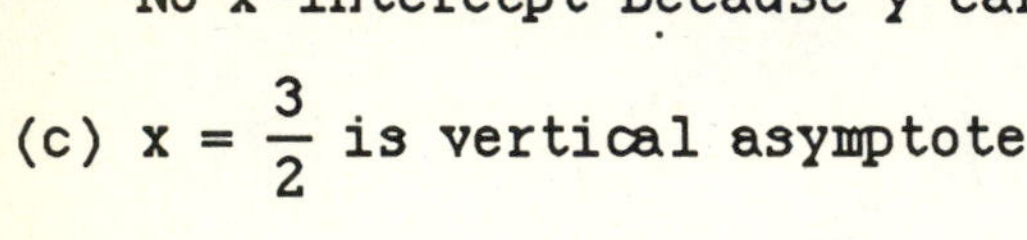

(c) $x = \dfrac{3}{2}$ is vertical asymptote

$y = 0$ is horizontal asymptote

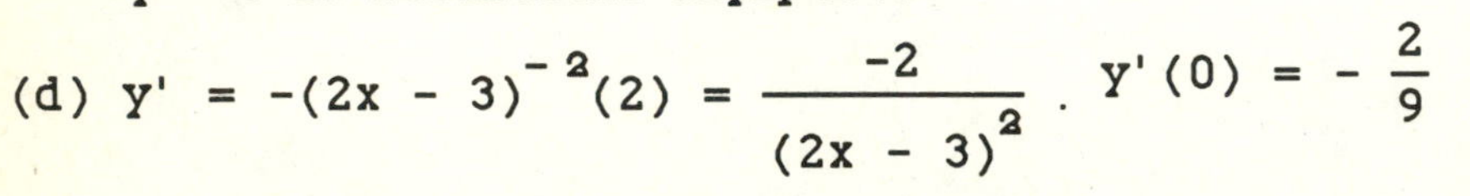

(d) $y' = -(2x - 3)^{-2}(2) = \dfrac{-2}{(2x - 3)^2}$. $y'(0) = -\dfrac{2}{9}$

(e) $y' < 0$ for all $x \neq \dfrac{3}{2} \Rightarrow$ graph always falling.

(f) $y'' = 2(2x - 3)^{-3} = \dfrac{4}{(2x - 3)^3}$. Concave down for x in $(-\infty, \frac{3}{2})$ and concave up for x in $(\frac{3}{2}, \infty)$.

15. $y = x - \dfrac{1}{x}$

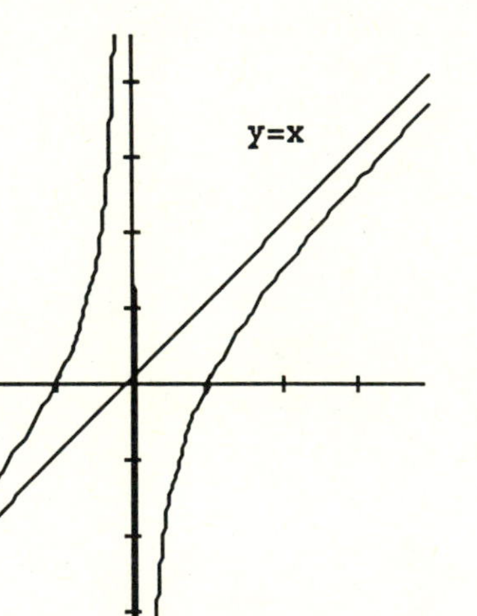

(a) odd. Symmetry to the origin

(b) No y-intercepts (x cannot be zero)
$y = 0 \Rightarrow x = \pm 1$

(c) $x = 0$: $x \to 0^+ \Rightarrow y \to -\infty$; $x \to 0^- \Rightarrow y \to +\infty$

$y = x$: $x \to \pm\infty \Rightarrow y \to x$

(d) $y' = 1 + \dfrac{1}{x^2}$. $y(\pm 1) = 2$

(e) $y' > 0$ for all $x \neq 0 \Rightarrow$

curve always rising

(f) $y'' = -\dfrac{2}{x^3}$. Concave up $(-\infty, 0)$ and down for $(0, \infty)$

(g) $y \approx x$ for large x and $y \approx -\dfrac{1}{x}$ for small x.

17. $y = \dfrac{x+3}{x+2} = 1 + \dfrac{1}{x+2}$

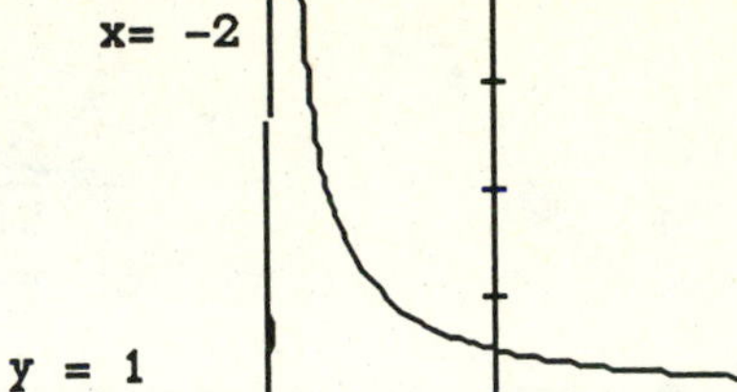

(a) No symmetry
(b) $x = 0 \Rightarrow y = \dfrac{3}{2}$;

$y = 0 \Rightarrow x = -3$

(c) Vertical: $x = -2$

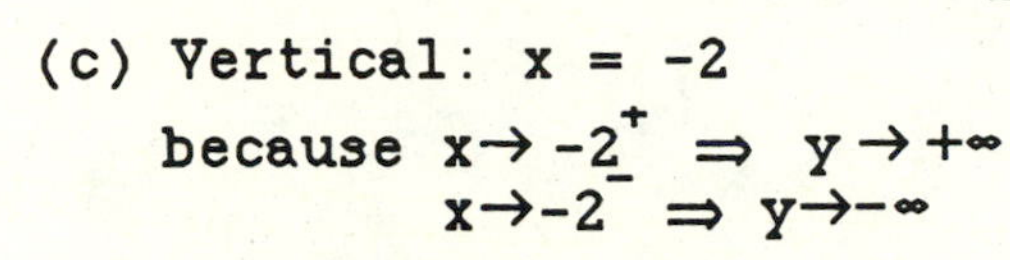

because $x \to -2^+ \Rightarrow y \to +\infty$
$x \to -2^- \Rightarrow y \to -\infty$

Horizontal: $y = 1$
because $x \to \pm\infty \Rightarrow y \Rightarrow 1$.

(d) $y' = -\dfrac{1}{(x+2)^2}$

$y'(0) = -\dfrac{1}{4}$ $y'(-3) = -1$

(e) $y' < 0$ for all $x \neq -2 \Rightarrow$ always falling

(f) $y''(x) = \dfrac{2}{(x+2)^3}$ concave up on $(-2, \infty)$
concave down on $(-\infty, -2)$

(g) $y \approx 1$ for large x; $y \approx \dfrac{1}{x+2}$ near $x = -2$.

19. $y = \dfrac{x+1}{x-1} = 1 + \dfrac{2}{x-1}$

(a) No symmetry
(b) $x = 0 \Rightarrow y = -1$ $y = 0 \Rightarrow x = -1$

(c) $x = 1$: $x \to 1^+ \Rightarrow y \to +\infty$

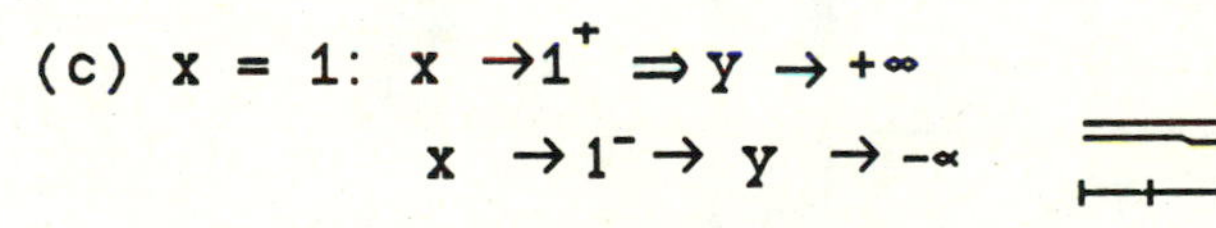

$x \to 1^- \to y \to -\infty$

$y = 1$: $x \to \infty \Rightarrow y \to 1$

y=1

x=1

(d) $y'(x) = -\dfrac{2}{(x-1)^2} < 0$ for all x.

$y'(0) = -2$; $y'(-1) = -\dfrac{1}{2}$

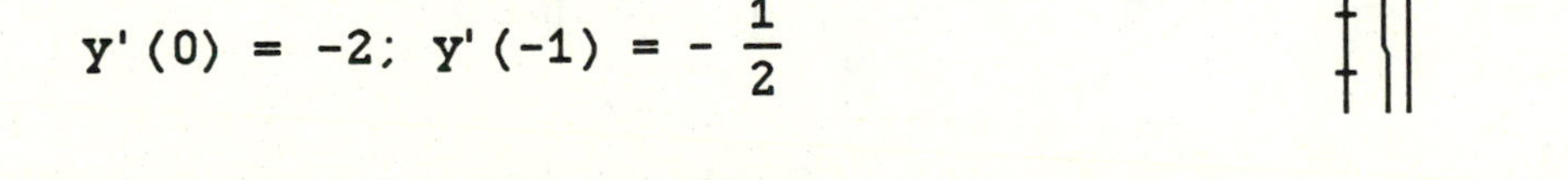

(e) graph is always falling.

(f) $y''(x) = \dfrac{4}{(x-1)^3}$ Convave up for $(1, \infty)$
Concave down for $(-\infty, 1)$

(g) $y \approx 1$ for large x; $y \approx \dfrac{2}{x-1}$ near $x = 1$.

21. $y = \dfrac{1}{x^2 + 1} = (x^2 + 1)^{-1}$

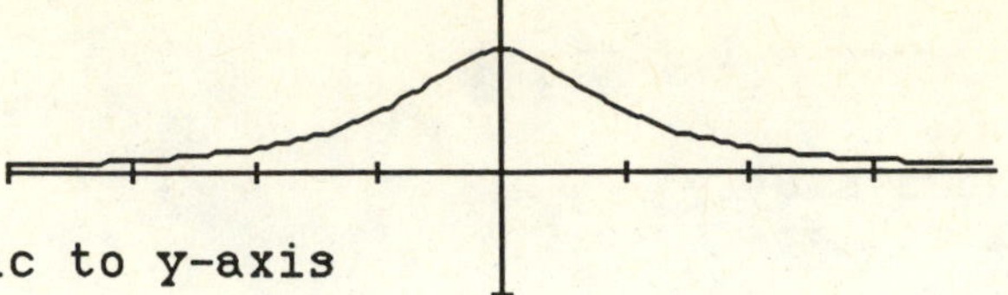

(a) even function - symmetric to y-axis

(b) $x = 0 \Rightarrow y = 1$. No x-intercept (y cannot be zero)

(c) No vertical asymptote. Horizontal: $y = 0$ since $x \to \pm\infty \Rightarrow y \to 0$

(d) $y'(x) + -(x^2 + 1)^{-2}(2x) = \dfrac{-2x}{(x^2 + 1)^2}$ $y'(0) = 0$

(e) $y' > 0$ if $x < 0$ rising
$y' < 0$ if $x > 0$ falling
$y(0) = 1$ is local maximum

(f) $y''(x) = (-2)(x^2 + 1)^{-2} + -2x[-2(x^2 + 1)^{-3}(2x)]$
$= \dfrac{2(3x^2 - 1)}{(x^2 + 1)^3}$. $y'' = 0 \Leftrightarrow x = \pm\dfrac{1}{\sqrt{3}}$.

concave up for $(-\infty, \frac{-1}{\sqrt{3}}) \cup (\frac{1}{\sqrt{3}}, \infty)$ concave down $(-\frac{1}{\sqrt{3}}, \frac{1}{\sqrt{3}})$

23. $y = \dfrac{x}{x^2 - 1}$

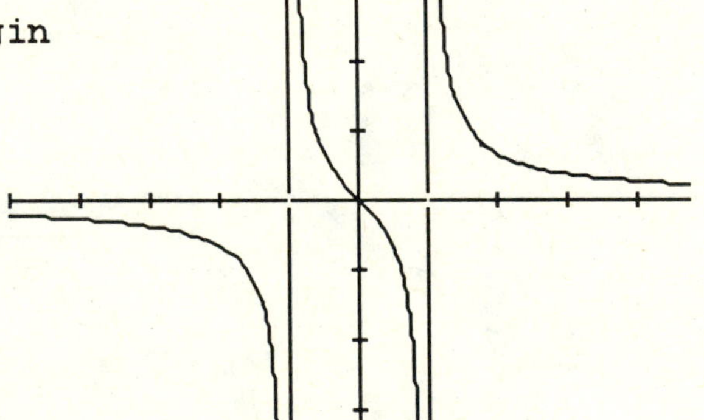

(a) Odd function
- symmetry to origin

(b) $x = 0 \Rightarrow y = 0$

(c) Vertical: $x = \pm 1$
$x \to 1^+$ or -1^-
$\Rightarrow y \to +\infty$
$x \to 1^-$ or -1^+
$\Rightarrow y \to -\infty$

Horizontal: $y = 0$
$x \to \pm\infty \Rightarrow y \to 0$

(d) $y' = \dfrac{x^2 - 1 - 2x^2}{(x^2 - 1)^2} = -\dfrac{1 + x^2}{(x^2 - 1)^2}$ $y'(0) = 1$

(e) $y' < 0$ for all $x \neq \pm 1 \Rightarrow$ always decreasing

(f) $y'' = -\left[\dfrac{(x^2 - 1)^2(2x) - (1 + x^2)[2(x^2 - 1)(2x)]}{(x^2 - 1)^4}\right]$

$= \dfrac{2x(x^2 + 3)}{(x + 1)^3(x - 1)^3}$ concave up on $(-1, 0) \cup (1, \infty)$
concave down on $(-\infty, -1) \cup (0, 1)$

(g) $y \approx \dfrac{1}{2(x - 1)}$ near $x = 1$; $y \approx \dfrac{1}{2(x + 1)}$ near $x = -1$

25. $y = \dfrac{x^2}{x^2 - 1} = 1 + \dfrac{1}{x^2 - 1}$

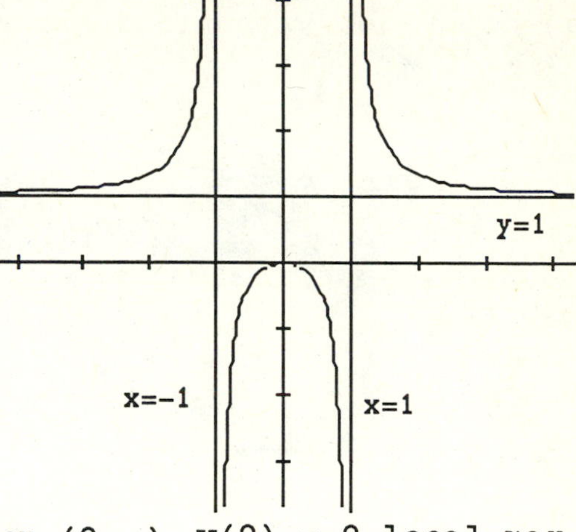

(a) Even - symmetric to y-axis

(b) $x = 0 \Rightarrow y = 0$

(c) Vertical: $x = \pm 1$

$x \to 1^+$ or $-1^- \Rightarrow y \to -\infty$

$x \to 1^-$ or $-1^+ \Rightarrow y \to +\infty$

Horizontal: $y = 1$

$x \to \pm\infty \Rightarrow y \to 1$

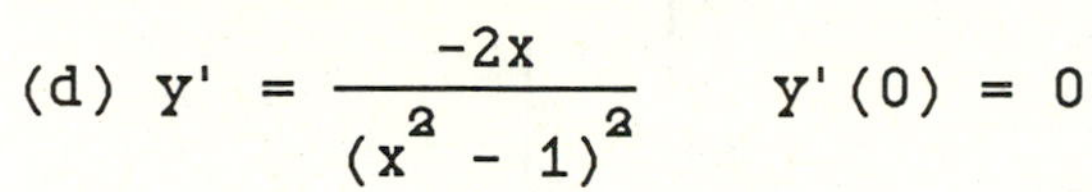

(d) $y' = \dfrac{-2x}{(x^2 - 1)^2}$ $\quad y'(0) = 0$

(e) rising on $(-\infty, 0)$ and falling on $(0, \infty)$ $y(0) = 0$ local max

(f) $y'' = \dfrac{(x^2 - 1)^2(-2) + 2x(2)(x^2 - 1)(2x)}{(x^2 - 1)^4} = \dfrac{2(1 + 3x^2)}{(x^2 - 1)^3}$

y'' does not exist at $x = \pm 1$, and changes sign there. concave down $(-1, 1)$ and concave up on $(-\infty, -1) \cup (1, \infty)$

(g) $y \approx 1$ for x large; $y \approx \dfrac{1}{2(x - 1)}$ near $x = 1$;

$y \approx \dfrac{1}{2(x + 1)}$ near $x = -1$.

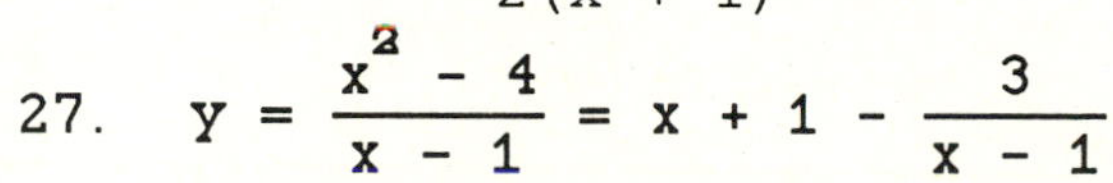

27. $y = \dfrac{x^2 - 4}{x - 1} = x + 1 - \dfrac{3}{x - 1}$

(a) No symmetry

(b) $x = 0 \Rightarrow y = 4$ $\quad y = 0 \Rightarrow x = \pm 2$

(c) Vertical: $x = 1$ $\quad x \to 1^+ \Rightarrow y \to +\infty$

$x \to 1^- \Rightarrow y \to -\infty$

Slant: $y \to x + 1$ as $x \to \infty$

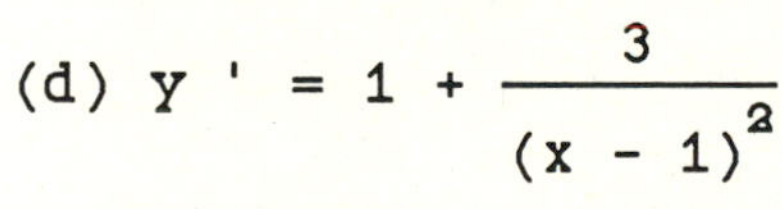

(d) $y' = 1 + \dfrac{3}{(x - 1)^2}$

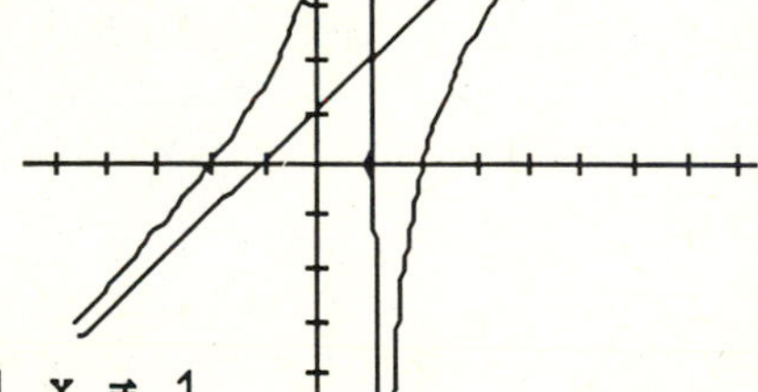

$y'(0) = 4$; $y'(2) = 4$; $y'(-2) = \dfrac{4}{3}$

(e) Always rising since $y' > 0$ for all $x \neq 1$.

(f) $y'' = \dfrac{-6}{(x - 1)^3}$ $\quad$ concave up $(-\infty, 1)$ and concave down $(1, \infty)$

(g) x large $\Rightarrow y \approx x + 1$.

$y \approx -\dfrac{3}{x - 1}$ near $x = 1$

29. $y = \dfrac{x^2 - 1}{2x + 4} = \dfrac{1}{2}x - 1 + \dfrac{3}{2x + 4}$

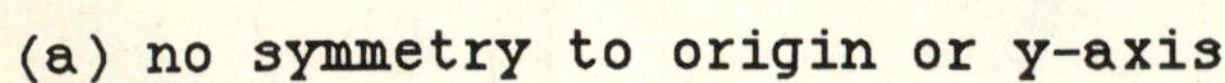

(a) no symmetry to origin or y-axis

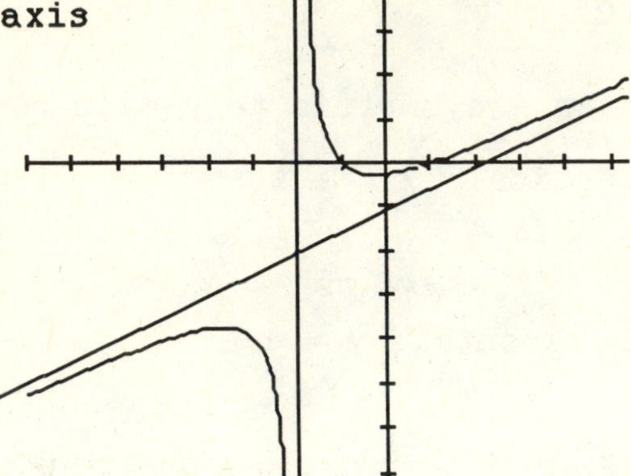

(b) $(0, -\frac{1}{4})$ $(\pm 1, 0)$

(c) vertical: $x = -2$

$x \to -2^+ \Rightarrow y \to \infty$
$x \to -2^- \Rightarrow y \to -\infty$

slant: $y = \frac{1}{2}x - 1$

(d) $y' = \dfrac{1}{2} - \dfrac{3}{2(x + 2)^2}$

$y'(1) = \frac{1}{3}$; $y'(-1) = -1$; $y'(0) = \frac{1}{8}$

(e) $y' = 0 \Leftrightarrow x = -2 \pm \sqrt{3}$. rising on $(-\infty, -2-\sqrt{3}) \cup (-2+\sqrt{3}\ \infty)$
falling on $(-2-\sqrt{3}, -2) \cup (-2, -2+\sqrt{3})$

(f) $y' = \dfrac{3}{(x + 2)^3}$ concave up $(-2, \infty)$ concave down $(-\infty, -2)$

(g) $y \approx \frac{1}{2}x - 1$ for x large; $y \approx \dfrac{3}{2x + 4}$ near $x = -2$

31. $y = \dfrac{2}{x} + 6x^2$.

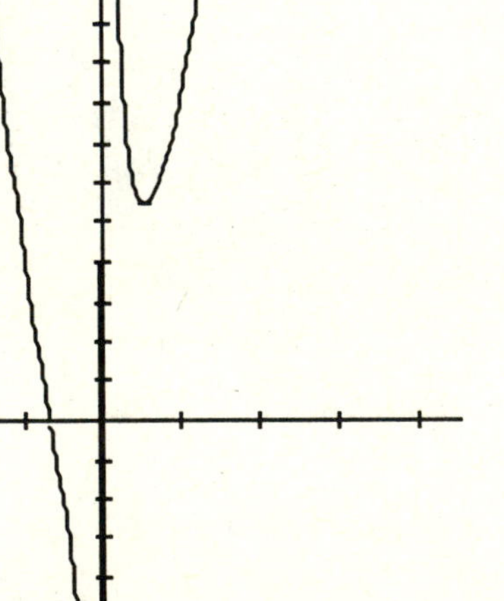

(a) No symmetry to origin or y-axis

(b) $(-\sqrt[3]{\frac{1}{3}}, 0)$

(c) Vertical: $x = 0$

$x \to 0^+ \Rightarrow y \to \infty$; $x \to 0^- \Rightarrow y \to -\infty$

$x \to \infty \Rightarrow y \to 6x^2$

(d) $y' = -\dfrac{2}{x^2} + 12x$.

$y'(-\sqrt[3]{\frac{1}{3}}) = -6\sqrt[3]{9}$.

(e) falling on $(-\infty, 0) \cup (0, \sqrt[3]{\frac{1}{6}})$

Rising on $(\sqrt[3]{\frac{1}{6}}, \infty)$

$y(\sqrt[3]{\frac{1}{6}}) = 3\sqrt[3]{6}$ is local minimum.

(f) Concave up on $(-\infty, -\sqrt[3]{\frac{1}{3}}) \cup (0,\infty)$; concave down on $(-\sqrt[3]{\frac{1}{3}}, 0)$

(g) $y \approx 6x^2$ for x large; $y \approx \frac{2}{x}$ near $x = 0$.

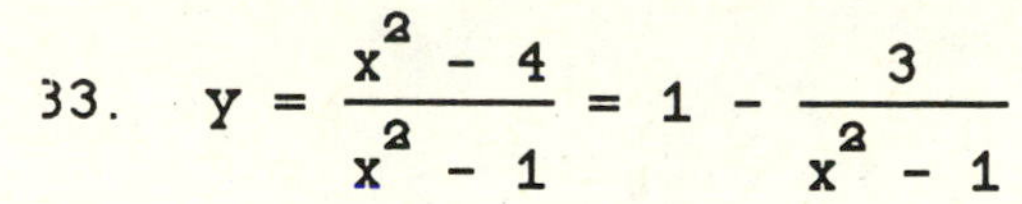

33. $y = \dfrac{x^2 - 4}{x^2 - 1} = 1 - \dfrac{3}{x^2 - 1}$

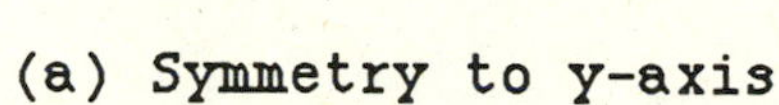

(a) Symmetry to y-axis

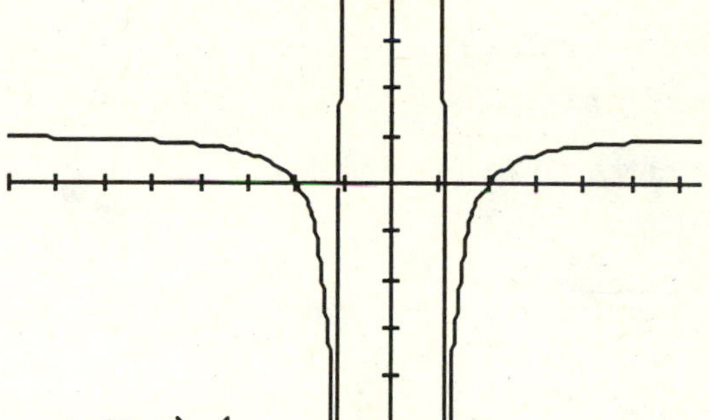

(b) (0,4) and (±2,0)

(c) Vertical: $x = \pm 1$

$x \to 1^+ \Rightarrow y \to -\infty$
$x \to 1^- \Rightarrow y \to \infty$

$x \to -1^+ \Rightarrow y \to \infty$
$x \to -1^- \Rightarrow y \to -\infty$

Horizontal: $y = 1$ $x \to \pm\infty \Rightarrow y \to 1$

(d) $y' = \dfrac{6x}{(x^2 - 1)^2}$ $y'(0) = 0$; $y'(2) = \frac{4}{3}$; $y'(-2) = -\frac{4}{3}$

(e) falling on $(-\infty, 0)$ rising on $(0,\infty)$ $y(0) = 4$ local max

(g) $y \approx 1$ for large x; $y \approx -\dfrac{3}{2(x - 1)}$ near $x = 1$;

$y \approx \dfrac{3}{2(x + 1)}$ near $x = -1$.

35. $y = \dfrac{x^2 + 1}{x^2 - 4x + 3} = 1 + \dfrac{4x - 2}{x^2 - 4x + 3}$

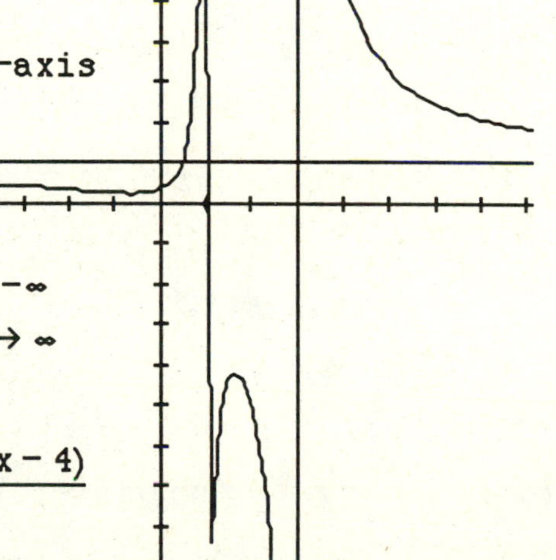

(a) No symmetry to origin or to y-axis

(b) $\left(0, \frac{1}{3}\right)$

(c) Horizontal: $x = 3$, $x = 1$

$x \to 3^+ \Rightarrow y \to \infty$; $x \to 3^- \Rightarrow y \to -\infty$

$x \to 1^+ \Rightarrow y \to -\infty$, $x \to 1^- \Rightarrow y \to \infty$

Vertical: $y = 1$

(d) $y' = \dfrac{(x^2 - 4x + 1)(4) - (4x - 2)(2x - 4)}{(x^2 - 4x + 3)^2}$

$= \dfrac{-4(x^2 - x - 1)}{(x^2 - 4x + 3)^2}$; $y'(0) = \dfrac{4}{9}$

(e) $y' = 0 \Leftrightarrow x = \dfrac{1 \pm \sqrt{5}}{2}$. Rising on $\left(\dfrac{1-\sqrt{5}}{2}, \dfrac{1+\sqrt{5}}{2}\right) \approx (-.6, 1.6)$;

Falling on $(-\infty, -.6) \cup (1.6, \infty)$. $y(1.6) = -4.24$ local maximum

$y(-.6) = .24$ local minimum

(f) $y'' = \dfrac{4(2x^3 - 3x^2 + 6x + 11)}{(x^2 - 4x + 3)^3}$. y'' changes sign at $x = 1$, $x = 3$

and $x = -1$

Concave down on $(-\infty, -1)$, concave up $(-1, 1)$, concave down on $(1, 3)$

and concave up on $(3, \infty)$

37. 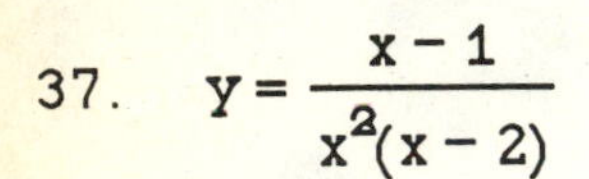
$y = \dfrac{x-1}{x^2(x-2)}$

(a) No symmetry to origin or y-axis

(b) $(1, 0)$

(c) Vertical: $x = 0$, $x = 2$

$x \to 2^+ \Rightarrow y \to \infty$

$x \to 2^- \Rightarrow y \to -\infty$

$x \to 0 \Rightarrow y \to \infty$

Horizontal: $y = 0$

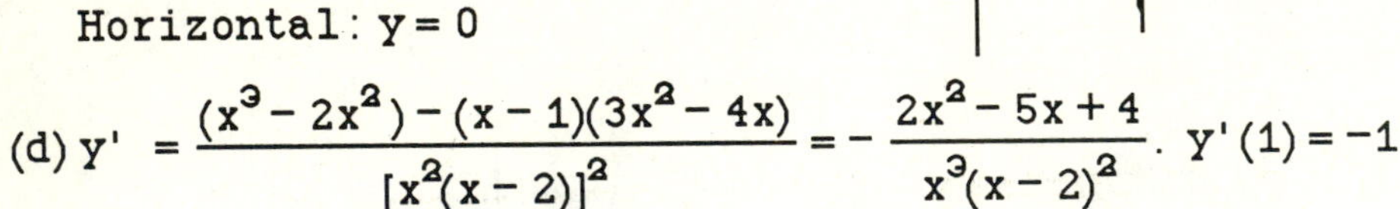
(d) $y' = \dfrac{(x^3 - 2x^2) - (x-1)(3x^2 - 4x)}{[x^2(x-2)]^2} = -\dfrac{2x^2 - 5x + 4}{x^3(x-2)^2}$. $y'(1) = -1$

(e) zeros of numerator are complex and do not affect signs.

Rising on $(-\infty, 0)$ and falling on $(0, 2) \cup (2, \infty)$

(f) $y'' = -\dfrac{x^3(x-2)^2 - (2x^2 - 5x + 4)[2x^3(x-2) + 3x^2(x-2)^2]}{[x^3(x-2)^2]^2}$

$= \dfrac{2(3x^3 - 12x^2 + 20x - 12)}{x^4(x-2)^3}$. The numerator has a zero

between $x = 1$ and $x = 2$ (specifically at $x = 1.22$).

The graph is concave up on $(-\infty, 0) \cup (0, 1.22)$,

concave down on $(1,22, 2)$ and concave up on $(2, \infty)$.

39. If $f(x) = 2 + \dfrac{\sin x}{x}$, $x > 0$, then $f'(x) = \dfrac{\cos x}{x} - \dfrac{\sin x}{x^2}$.

$\lim\limits_{x \to \infty} \dfrac{\cos x}{x} - \dfrac{\sin x}{x^2} = 0 =$ slope of line $y = 2$.

41. odd 43. even - is the reciprocal of an even function

45. odd 47. odd 49. even - it is squared

51. odd - the product of an odd function with an even one

53. even - the product of two odd functions (see 59 d)

55. neither

57. (a) even (b) even

59. Let u and v be odd functions, i.e. $u(-x) = -u(x)$ and $v(-x) = -v(x)$.

(a) $(u+v)(-x) = u(-x) + v(-x) = -u(x) - v(x) = -[u(x) + v(x)]$

$= -(u+v)(x)$. Therefore, $u + v$ is odd.

(b) $(u-v)(-x) = u(-x) - v(-x) = -u(x) + v(x) = -[u(x) - v(x)]$

$= -(u-v)(x)$. Therefore, $u - v$ is odd.

(c) $\left(\frac{u}{v}\right)(-x) = \frac{u(-x)}{v(-x)} = \frac{-u(x)}{-v(x)} = \left(\frac{u}{v}\right)(x)$. Therefore u/v is even.

(d) $(uv)(-x) = u(-x)v(-x) = (-u(x))(-v(x)) = u(x)v(x) = uv(x)$.

Therefore uv is even.

61. (a) It must be increasing because of the symmetry about the origin.
(b) It must be decreasing because of the symmetry about the y-axis.
(c) Concave down, as in part (a).
(d) Concave down, as in part (b).

3.4 MAXIMA AND MINIMA: THEORY

1. $y = x - x^2$ on $[0,1]$. $y' = 1 - 2x$. $y' = 0 \Leftrightarrow x = \frac{1}{2}$. $y' > 0$ if $x < \frac{1}{2}$ and $y' < 0$ if $x > \frac{1}{2} \Rightarrow y\left(\frac{1}{2}\right) = \frac{1}{4}$ is local maximum which absolute, since $y(0) = 0$ and $y(1) = 0$ are absolute minimimum values.

3. $y = x - x^3$ on $[0,1]$. $y' = 1 - 3x^2$. $y' = 0 \Leftrightarrow x = \pm\frac{1}{\sqrt{3}}$. The value $x = -\frac{1}{\sqrt{3}}$ is out of the domain. $y\left(\frac{1}{\sqrt{3}}\right) = \frac{2\sqrt{3}}{9}$ is local maximum which is absolute, since $y(0) = 0$ and $y(1) = 0$ are absolute minimum values.

5. $y = x^3 - 147x$ on $(-\infty, \infty)$. $y' = 3x^2 - 147$. $y' = 0 \Leftrightarrow x = \pm 7$.
$y' > 0$ if $x < -7$, $y' < 0$ if $-7 < x < 7$, and $y' > 0$ if $x > 7$.
Therefore, $y(-7) = 686$ is local maximum, and $y(7) = -686$
is local minimum. There are no absolute extrema.

7. $y = x^2 - 4x + 3$ on $(0, 3)$. $y' = 2x - 4$. $y' = 0 \Leftrightarrow x = 2$. $y'' = 2$
$\Rightarrow y(2) = -1$ is absolute minimum. No absolute maximum

values since the domain interval is open.

9. $y = x - x^2$ on $(0, 1)$. $y' = 1 - 2x$. $y' = 0 \Leftrightarrow x = \frac{1}{2}$. $y'' = -2 \Rightarrow$
$y\left(\frac{1}{2}\right) = \frac{1}{4}$ is a local maximum which is absolute. There are

no other extrema since the domain interval is open.

11. $y = 2x$ on $[0, 3]$. $y' = 2 \neq 0 \Rightarrow$ no local extrema. $y(0) = 0$ is
absolute minimun and $y(3) = 6$ is absolute maximum.

13. $y = x^2 + \frac{2}{x}$, $x > 0$. $y' = 2x - \frac{2}{x^2}$. $y' = 0 \Leftrightarrow 2x^3 - 2 = 0 \Leftrightarrow x = 1$.
$y'' = 2 + \frac{4}{x^3} > 0$ if $x = 1 \Rightarrow y(1) = 3$ is local minimum which is
absolute on $x > 0$. There are no other extreme values.

15. $y = x^3 + 3x^2 + 3x + 2$, $-\infty < x < \infty$. $y' = 3x^2 + 6x + 3$. $y' = 0 \Leftrightarrow$
$3(x + 1)^2 = 0 \Leftrightarrow x = -1$. There is no sign change in y'.
Therefore there are no extreme values.

17. $y = \sqrt{x} - x$, $x \geq 0$. $y' = \frac{1}{2\sqrt{x}} - 1$. $y' = 0 \Leftrightarrow 1 - 2\sqrt{x} = 0 \Leftrightarrow x = \frac{1}{4}$.

$y'' = -\frac{1}{4}x^{-\frac{3}{2}} < 0 \Rightarrow y\left(\frac{1}{4}\right) = \frac{1}{4}$ is local maximum which is

absolute. There is no local minimum.

19. $y = x^4 - 4x$ on $[0,2]$. $y' = 4x^3 - 4$. $y' = 0 \Leftrightarrow x = 1$.

$y'' = 12x^2$. $y''(1) > 0 \Rightarrow y(1) = -3$ is a relative minimum. $y(0) = 0$ and $y(2) = 8$. Therefore 8 is the absolute maximum, and -3 is an absolute minimum.

21. $y = \tan x$ on $\left[0, \frac{\pi}{2}\right]$. $y' = \sec^2 x > 0$. Therefore $y = \tan x$ is always rising; $y(0) = 0$ is absolute minimum, and there is no maximum.

23. $y = 2\sin x + \cos 2x$ on $\left[0, \frac{\pi}{2}\right]$. $y' = 2\cos x - 2\sin 2x$ $= 2\cos x - 4\sin x\cos x = 2\cos x(1 - 2\sin x)$. $y' = 0 \Leftrightarrow \cos x = 0$ or $\sin x = \frac{1}{2} \Leftrightarrow x = \frac{\pi}{2}$ or $\frac{\pi}{6}$. $y'' = -2\sin x - 4\cos 2x$. $y''\left(\frac{\pi}{6}\right) =$ $-2\left(\frac{1}{2}\right) - 4\left(\frac{1}{2}\right) < 0 \Rightarrow y\left(\frac{\pi}{6}\right) = 2\left(\frac{1}{2}\right) + \frac{1}{2} = \frac{3}{2} =$ local maximum. $y(0) = 1$ and $y\left(\frac{\pi}{2}\right) = 1$ are absolute minima, and the local maximum becomes absolute.

25. $y = x^4 - 8x^3 - 270x^2$ on $-\infty < x < \infty$. $y' = 4x^3 - 24x^2 - 540x$. $y' = 0 \Leftrightarrow$ $4x(x - 15)(x + 9) = 0 \Leftrightarrow x = -9, 0, 15$. $y(-9) = -9477$ is a local minimum and $y(15) = -37125$ is an absolute minimum. $y(0) = 0$ is a local maximum.

27. $y = (x - x^2)^{-1}$ on $(0,1)$. $y' = -\dfrac{1 - 2x}{(x - x^2)^2}$. $y' = 0 \Leftrightarrow x = \frac{1}{2}$.

$y' < 0$ if $x < \frac{1}{2}$ and $y' > 0$ if $x > \frac{1}{2} \Rightarrow y\left(\frac{1}{2}\right) = 4$ is a local minimum which is absolute on $(0,1)$. $y \to \infty$ if $x \to 0^-$ or $x \to 1^+ \Rightarrow$ no maximum values.

31. $y = \frac{x}{1+|x|}$ can be expressed as

$$y = \begin{cases} \frac{x}{1-x} & \text{for } x \le 0 \\ \frac{x}{1+x} & \text{for } x > 0 \end{cases} \quad \text{Then } y' = \begin{cases} \frac{1}{(1-x)^2} & \text{for } x \le 0 \\ \frac{1}{(1+x)^2} & \text{for } x > 0 \end{cases}$$

Note that each point for which y' does not exist is not in the domain of that piece. Therefore, $y' > 0$ always $\Rightarrow$ there are no extreme values.

33. $y = \sin|x|$, $-2\pi \le x \le 2\pi$.

$$y = \begin{cases} \sin x & \text{for } 0 \le x \le 2\pi \\ -\sin x & -2\pi \le x < 0 \end{cases} \quad y' = \begin{cases} \cos x & \text{for } 0 < x \le 2\pi \\ -\cos x & \text{for } -2\pi \le x < 0 \end{cases}$$

$x \to 0^+ \Rightarrow y' \to 1$ and $x \to 0^- \Rightarrow y' \to -1$. Therefore, $y'(0)$ does not exist and 0 is a critical point of y.

$\cos x = 0$ for $x = \frac{\pi}{2}$ and $\frac{3\pi}{2}$ in $(0, 2\pi]$ and $-\cos x = 0$ for

$x = -\frac{\pi}{2}$ and $-\frac{3\pi}{2}$ in $[-2\pi, 0)$. $y(0) = 0$ is local minimum.

$y\left(\pm\frac{\pi}{2}\right) = 1$ are absolute maximum values. $y\left(\pm\frac{3\pi}{2}\right) = -1$

are absolute minimum values.

35. $y = |x^2 - 1|$, $-1 \le x \le 2$.

$$y = \begin{cases} 1 - x^2 & \text{for } -1 \le x \le 1 \\ x^2 - 1 & \text{for } 1 < x \le 2 \end{cases} \quad y' = \begin{cases} -2x & \text{for } -1 < x < 1 \\ 2x & \text{for } 1 < x \le 2 \end{cases}$$

$y' = 0 \Leftrightarrow -2x = 0 \Leftrightarrow x = 0$. $y'' = -2 < 0 \Rightarrow y(0) = 1$ is local maximum.

y' does not exist if $x = 1$. $y(-1) = 0$, $y(1) = 0$ are absolute minimum values. $y(2) = 3$ is absolute maximum.

37. $(x-1)^2 \ge 0 \Leftrightarrow x^2 - 2x + 1 \ge 0 \Leftrightarrow x^2 + 1 \ge 2x$. If $x > 0$, then

$x + \frac{1}{x} \ge 2$

39. $y = \dfrac{x}{x^2+1}$. $y' = \dfrac{(x^2+1) - x(2x)}{(x^2+1)^2} = \dfrac{1-x^2}{(x^2+1)^2}$.

$y' = 0 \Leftrightarrow x = \pm 1$. $y' < 0$ if $x < -1$ or $x > 1$.

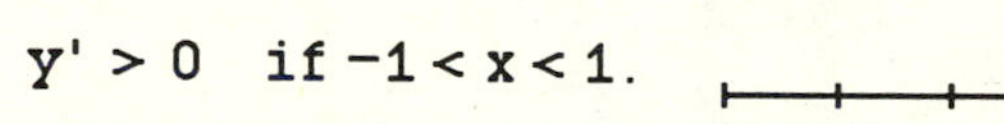

$y' > 0$ if $-1 < x < 1$.

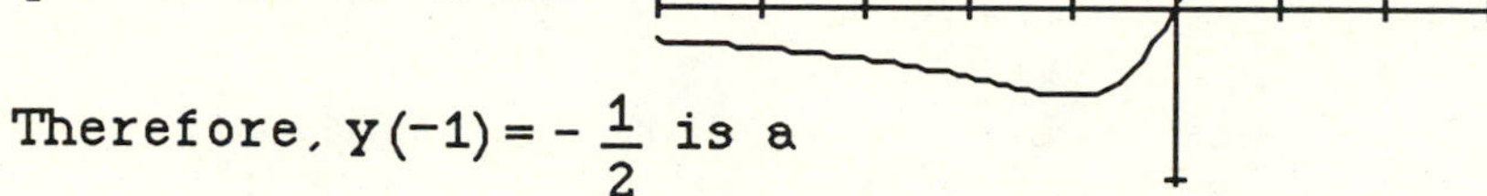

Therefore, $y(-1) = -\dfrac{1}{2}$ is a local minimum and $y(1) = \dfrac{1}{2}$ is a local maximum.

$y = 0$ is a horizontal asymptote. $y'' = \dfrac{2x(x^2-3)}{(x^2+1)^3}$.

$y'' > 0$ if $x > \sqrt{3}$, $y'' < 0$ if $0 < x < \sqrt{3}$, $y'' > 0$ if $-\sqrt{3} < x < 0$ and $y'' < 0$ if $x < -\sqrt{3}$. Therefore inflection points are $\left(-\sqrt{3}, -\dfrac{\sqrt{3}}{4}\right)$, $(0, 0)$ and $\left(\sqrt{3}, \dfrac{\sqrt{3}}{4}\right)$

41. (b), f' must change sign at $x = c$, since it exists. The function $f(x) = x^2$ serves as a counterexample for (a) and (c).

43. $y = 4\sin^2 x - 3\cos^2 x$. $y' = 8\sin x\cos x + 6\cos x\sin x =$ $14\sin x\cos x$. $y' = 0 \Leftrightarrow \sin x = 0$ or $\cos x = 0 \Leftrightarrow$ $x = 0, \pm\dfrac{\pi}{2}, \pm\pi, \pm\dfrac{3\pi}{2}, \ldots$. $y(0) = y(\pm\pi) = \ldots = -3$.

$y\left(\pm\dfrac{\pi}{2}\right) = y\left(\pm\dfrac{3\pi}{2}\right) = \ldots = 4$. Maximum height is 4 units.

3.5 MAXIMA AND MINIMA: PROBLEMS

1. (a) Let x and $20 - x$ represent the numbers. Then the expression to be maximized is

$$f(x) = x^2 + (20 - x)^2, \quad 0 \le x \le 20.$$
$$f'(x) = 4x - 40 = 0 \text{ if } x = 10.$$
$$f''(x) = 4 \qquad f(10) = 200 \text{ is a minimum.}$$

Therefore, the maximum must occur at the endpoints, or when the numbers are 0 and 20, with maximum of 400.

(b) $f(x) = x^3(20 - x)^2 = x^3(400 - 40x + x^2) = 400x^3 - 40x^4 + x^5$.

$f'(x) = 1200x^2 - 160x^3 + 5x^4$. $f'(x) = 0 \Leftrightarrow 5x^2(x - 20)(x - 12) = 0$

$\Leftrightarrow x = 12$ or 20. $f''(x) = 2400x - 480x^2 + 20x^3$. $f''(20) = 207920$

$f''(12) = -5760$. Therefore $x = 12$ gives maximum. Dimensions are 12 by 8.

(c) $g(x) = 20 - x + \sqrt{x}$. $g'(x) = -1 + \dfrac{1}{2\sqrt{x}}$. $g'(x) = 0 \Leftrightarrow x = \dfrac{1}{4}$.

$g''(x) = -\dfrac{1}{4}x^{-\frac{3}{2}}$. $g''\left(\dfrac{1}{4}\right) < 0 \Rightarrow g\left(\dfrac{1}{4}\right)$ is a maximum.

3. Let l = length and w = width. Then $lw = 16 \Rightarrow l = 16w^{-1}$.
$p = 2l + 2w = 2(16w^{-1}) + 2w$. $p' = -32w^{-2} + 2$. $p' = 0 \Leftrightarrow$
$w = 4$. $p'' = 64w^{-3}$. $p''(4) > 0 \Rightarrow p(4) = 16$ is a minimum.

5. $d = \sqrt{(1 - \cos t)^2 + (\sqrt{3} - \sin t)^2}$. It is sufficient to minimize the expression under the radical.
$D = 1 - 2\cos t + \cos^2 t + 3 - 2\sqrt{3}\sin t + \sin^2 t = 5 - 2\cos t - 2\sqrt{3}\sin t$.
$D' = 2\sin t - 2\sqrt{3}\cos t$. $D' = 0 \Leftrightarrow 2\sin t = 2\sqrt{3}\cos t \Leftrightarrow \tan t = \sqrt{3}$.
$\Leftrightarrow t = \dfrac{\pi}{3}$. $D'' = 2\cos t + 2\sqrt{3}\sin t$. $D''\left(\dfrac{\pi}{3}\right) > 0 \Rightarrow \left(\cos\dfrac{\pi}{3}, \sin\dfrac{\pi}{3}\right) =$
$\left(\dfrac{1}{2}, \dfrac{\sqrt{3}}{2}\right)$ is closest point.

7. Let (x, y) be a point on the parabola. Then $A = 2xy =$
$2x(12 - x^2)$. $A' = 24 - 6x^2$. $A' = 0 \Leftrightarrow x = 2$. $A'' = -12 < 0$
so $A(2) = 32$ is largest area.

9. Let y = dimension parallel to river and x the other two sides.
Then $2x + y = 800 \Rightarrow y = 800 - 2x$. $A = xy = x(800 - 2x) =$
$800x - 2x^2$. $A' = 800 - 4x$. $A' = 0 \Leftrightarrow x = 200$ m. $A'' = -4 < 0 \Rightarrow$
$A(200) = 200(400) = 80{,}000$ m^2 is maximum area.

11. Let x be the base dimension and y the side. Then

$x^2 + 4xy = 108 \Rightarrow y = \dfrac{108 - x^2}{4x} = 27x^{-1} - \dfrac{1}{4}x$. $V = x^2y =$

$x^2\left(27x^{-1} - \dfrac{1}{4}x\right) = 27x - \dfrac{1}{4}x^3$. $V' = 27 - \dfrac{3}{4}x^2$. $V' = 0 \Leftrightarrow$

$x = 6$. $V'' = -\dfrac{3}{2}x$. $V''(6) = -9 < 0 \Rightarrow V(6)$ is maximum. The

dimensions are 6 by 6 by 3.

13. $A = \dfrac{1}{2}ab$, where $a^2 + b^2 = 20^2$. $A = \dfrac{1}{2}a(20^2 - a^2)^{\frac{1}{2}}$.

$$A' = \frac{1}{2}\left[(20^2 - a^2)^{\frac{1}{2}} + \frac{1}{2}a(20^2 - a^2)^{-\frac{1}{2}}(-2a)\right] = \frac{20^2 - 2a^2}{2\sqrt{20^2 - a^2}}.$$

$A' = 0 \Leftrightarrow a = 10\sqrt{2}$. $A' > 0$ if $a < 10\sqrt{2}$ and $A' < 0$ if $a > 10\sqrt{2}$

$\Rightarrow A(10\sqrt{2})$ is a maximum. Then $b = 20^2 - 200 = 10\sqrt{2} = a$.

15. $x = (t-1)(t-4)^4$. $x'(t) = (t-1)^4 + 4(t-1)(t-4)^3$

$= (t-4)^3(t - 4 + 4t - 4) = (t-4)^3(5t-8)$.

(a) The particle is at rest when $t = 4$ or $t = \dfrac{8}{5}$.

(b) The particle moves to the left for t in $\left(\dfrac{8}{5}, 4\right)$.

(c) $a(t) = x''(t) = 5(t-4)^3 + 3(5t-8)(t-4)^2 = 4(t-4)^2(5t-11)$.

$a < 0$ for $\dfrac{8}{5} < t < \dfrac{11}{5}$. Since v and a have the same sign,

the velocity is increasing. $a > 0$ for $\dfrac{11}{5} < t < 4$. Since

v and a have opposite signs, the velocity is decreasing.

Therefore, the maximum velocity occurs when $t = \dfrac{11}{5}$,

and is $v\left(\dfrac{11}{5}\right) = -\dfrac{2187}{125}$.

17. Let x and y be the dimensions of the written material. Then

$xy = 50 \Rightarrow y = 50x^{-1}$. The total area is $A = (8 + x)(4 + y)$

$= 32 + 4x + 400x^{-1} + 50$. $A' = 4 - 400x^{-2}$. $A' = 0 \Leftrightarrow x = 10$.

$A'' = 800x^{-3}$. $A''(10) > 0 \Rightarrow$ maximum area. Overall dimensions

are 18 in. by 9 in.

19. Let x and y be the dimensions. Then $xy = 216 \Rightarrow y = 216x^{-1}$.

$p = 3x + 2y = 3x + 432x^{-1}$. $p' = 3 - 432x^{-2}$. $p' = 0 \Leftrightarrow x = 12$.

$p'' = 864x^{-3}$. $p''(12) > 0 \Rightarrow$ minimum. The dimensions are 12 by 18.

21. $V = \pi r^2 h$. $\pi r^2 h = 1000 \Rightarrow h = \frac{1000}{\pi} r^{-2}$. $S = \pi r^2 + 2\pi rh =$

$\pi r^2 + 2\pi r\left(\frac{1000}{\pi r^2}\right) = \pi r^2 + 2000r^{-1}$. $S' = 2\pi r - 2000r^{-2}$.

$S' = 0 \Leftrightarrow 2\pi r = 2000r^{-2} \Leftrightarrow r^3 = \frac{2000}{2\pi} \Leftrightarrow r = \frac{10}{\sqrt[3]{\pi}}$.

$S'' = 2\pi + 6000r^{-3} > 0$ for $r > 0 \Rightarrow$ minimum. $h = \frac{10}{\sqrt[3]{\pi}}$.

23. Let x and y be the other dimensions. Then $xy = 2 \Rightarrow y = 2x^{-1}$.

The cost is $C = 10x + 5(2y) + 5(2xy) = 10x + 20x^{-1} + 20$.

$C' = 10 - 20x^{-2}$. $C' = 0 \Leftrightarrow x = \sqrt{2}$. $C'' = 40x^{-3}$. $C''(\sqrt{2}) > 0$

$\Rightarrow$ cost is minimum. Dimensions are $\sqrt{2}$ by $\sqrt{2}$ by 1.

25. Let $D = (x - 2)^2 + \left(x + \frac{1}{2}\right)^2$. $D' = 2(x - 2) + 2\left(x + \frac{1}{2}\right)(2x)$.

$= 4(x^3 + x - 1)$. $D' = 0 \Leftrightarrow x^3 + x - 1 = 0 \Leftrightarrow x(x^2 + 1) = 1 \Leftrightarrow$

$\frac{1}{x^2 + 1} = x$. $D'' = 4(3x^2 + 1) > 0 \Rightarrow$ minimum value.

27. Let x and y be the legs of the right triangle, and H be the fixed hypotenuse. Then $x^2 + y^2 = H^2$. $V = \frac{1}{3}\pi x^2 y =$

$\frac{\pi}{3} y(H^2 - y^2) = \frac{\pi H^2}{3} y - \frac{\pi}{3} y^3$. $V' = \frac{\pi H^2}{3} - \pi y^2$. $V' = 0 \Leftrightarrow$

$\pi y^2 = \frac{\pi H^2}{3} \Leftrightarrow y = \sqrt{\frac{H^2}{3}} = \frac{H}{\sqrt{3}}$. $V'' = -2\pi y < 0 \Rightarrow$ maximum.

$x^2 = H^2 - \frac{H^2}{3} = \frac{2H^2}{3} \Rightarrow x = \sqrt{\frac{2}{3}}\, H$.

29. $F(x) = x^2 + \frac{a}{x} \Rightarrow f'(x)$

(a) For a local minimum at $x = 2$, we need $f'(2) = 0$ or

$4 - \frac{a}{4} = 0$ or $a = 16$.

(b) For a local minimum at $x = -3$, we need $f'(-3) = 0$ or

$-6 - \frac{a}{9} = 0$ or $a = -54$

(c) $f''(x) = 2 + \frac{2a}{x^3}$. For an inflection point at $x = 1$ we need $f''(1) = 0$,

or $2 + 2a = 0$ or $a = -1$.

(d) $f'(x) = 0$ if $2x^3 - a = 0$ or if $x = \left(\frac{a}{2}\right)^{\frac{1}{3}}$. For this value of x,

$f'' = 2 + \frac{2a}{\frac{a}{2}} = 6 > 0$ for all values of a. Therefore, there

are no local maximum values.

31. Let x be a side of the square and y be a side of the triangle.

(a) Then $4x + 3y = L \Rightarrow y = \frac{L - 4x}{3}$. $A = x^2 + \frac{y^2\sqrt{3}}{4} = x^2 + \frac{\sqrt{3}}{4}\left(\frac{L - 4x}{3}\right)^2$.

$A' = 2x + 2\left(\frac{L - 4x}{3}\right)\left(-\frac{4}{3}\right)\left(\frac{\sqrt{3}}{4}\right)$. $A' = 0 \Leftrightarrow 2x - \frac{2\sqrt{3}}{3}\left(\frac{L - 4x}{3}\right) = 0$

$\Leftrightarrow 2x + \frac{8\sqrt{3}}{9}x = \frac{2L\sqrt{3}}{9} \Leftrightarrow x = \frac{L}{3\sqrt{3} + 4}$. $A''(x) = 2 + \frac{8\sqrt{3}}{9} > 0 \Rightarrow$

local minimum value. Cut piece $\frac{4L}{3\sqrt{3} + 4}$ for square.

(b) If $y = 0$, $x = \frac{L}{4}$ and $a = \frac{L^2}{16}$. If $x = 0$, $y = \frac{L}{3}$ and $A = \frac{L^2}{12\sqrt{3}}$.

The maximum occurs if all of the area is in the square.

33. $V = \frac{1}{3}\pi x^2 (y + r) = \frac{1}{3}\pi (r^2 - y^2)(r + y) = \frac{\pi}{3}(r^3 - ry^2 + r^2y - y^3)$

$V' = \frac{\pi}{3}(-2ry + r^2 - 3y^2)$. $V' = 0 \Leftrightarrow r^2 - 2ry - 3y^2 = 0 \Leftrightarrow$

$(r - 3y)(r + y) = 0 \Leftrightarrow y = \frac{r}{3}$. $V'' = \frac{\pi}{3}(-2r - 6y)$. $V''\left(\frac{r}{3}\right) < 0$

$\Rightarrow V$ is a maximum. $V = \frac{32\pi r^3}{81}$.

35. Let x = radius and y = height of inscribed cylinder.

Then $\frac{h}{r} = \frac{y}{r-x} \Rightarrow y = \frac{hr-hx}{r}$. $V_{cyl} = \pi x^2 y = \pi x^2 \left(\frac{hr-hx}{r}\right) =$

$\frac{\pi h}{r}(rx^2 - x^3)$. $V' = \frac{\pi h}{r}(2rx - 3x^2)$. $V' = 0 \Leftrightarrow 2rx - 3x^2 = 0$

$\Leftrightarrow x = 0$ or $x = \frac{2r}{3}$. $V'' = 2r - 6x$. $V''\left(\frac{2r}{3}\right) = 2r - 4r < 0 \Rightarrow$

maximum value. $y = \dfrac{hr - h\left(\frac{2r}{3}\right)}{r} = \frac{h}{3}$. $V_{cyl} = \pi\left(\frac{2r}{3}\right)^2\left(\frac{h}{3}\right) =$

$\frac{4\pi r^2 h}{27}$. $V_{cone} = \frac{\pi}{3} r^2 h$. $\dfrac{V_{cyl}}{V_{cone}} = \dfrac{\frac{4\pi r^2 h}{27}}{\frac{\pi r^2 h}{3}} = \frac{4}{9}$.

37. Let x and y be the dimensions of the beam. Then $S = Kxy^3$.

$x^2 + y^2 = D^2 \Rightarrow S = Ky^3(D^2 - y^2)^{\frac{1}{2}}$.

$S' = \frac{1}{2}Ky^3(D^2 - y^2)^{-\frac{1}{2}}(-2y) + 3Ky^2(D^2 - y^2)^{\frac{1}{2}} = \dfrac{3D^2Ky^2 - 4Ky^4}{\sqrt{D^2 - y^2}}$.

$S' = 0 \Leftrightarrow Ky^2(3D^2 - 4y^2) = 0 \Leftrightarrow y = 0$ or $y = \frac{\sqrt{3}}{2}D$. $S' < 0$ if

$y > \frac{\sqrt{3}}{2}D$ and $S' > 0$ if $y < \frac{\sqrt{3}}{2}D \Rightarrow$ stiffness is maximum if

$y = \frac{\sqrt{3}}{2}D$ and $x = D^2 - y^2 = \frac{1}{2}D$.

39. Let 2x and y be the dimensions of the rectangle. Then

$p = 2y + 2x + \pi x$. $2\frac{dy}{dx} + 2 + \pi = 0 \Rightarrow \frac{dy}{dx} = -\frac{2+\pi}{2}$.

$L = \frac{1}{2}\left(\frac{\pi x^2}{2}\right) + 2xy$. $L' = \frac{\pi x}{2} + 2x\frac{dy}{dx} + 2y = \frac{\pi x}{2} + 2x\left(-\frac{2+\pi}{2}\right) + 2y$.

$L' = 0 \Leftrightarrow \frac{\pi x}{2} - 2x - \pi x + 2y = 0 \Leftrightarrow x\left(\frac{\pi}{2} + 2\right) = 2y \Leftrightarrow \frac{x}{y} = \frac{4}{4+\pi}$.

$L'' = \frac{\pi}{2} + 2x\frac{d^2y}{dx^2} + 2\frac{dy}{dx} + 2\frac{dy}{dx} = \frac{\pi}{2} + 0 + 4\left(-\frac{2+\pi}{2}\right) = -4 - \frac{3\pi}{2} < 0$

$\Rightarrow$ Light is maximum when $\frac{2x}{y} = \frac{8}{4+\pi}$.

41. $S = \pi r^2 + 2\pi rh + 2\pi r^2 = 3\pi r^2 + 2\pi rh$. $h = \frac{S - 3\pi r^2}{2\pi r}$.

$V = \pi r^2 h + \frac{2}{3}\pi r^3 = \pi r^2\left(\frac{S - 3\pi r^2}{2\pi r}\right) + \frac{2\pi r^3}{3} = \frac{rS}{2} - \frac{5\pi r^3}{6}$.

$V' = \frac{S}{2} - \frac{5\pi r^2}{2} = 0 \Leftrightarrow r = \sqrt{\frac{S}{5\pi}}$. $V'' = -5\pi r < 0$ for $r > 0 \Rightarrow$ maximum value.

$h = \frac{S - 3\pi\left(\frac{S}{5\pi}\right)}{2\pi\sqrt{\frac{S}{5\pi}}} = \frac{\frac{S}{5}}{\pi\sqrt{\frac{S}{5\pi}}} = \sqrt{\frac{S}{5\pi}}$. Maximum occurs when $r = h$

43. (a) $f(x) = x^2 - x + 1$ is never negative, because $f'(x) = 2x - 1$ $= 0$ if $x = \frac{1}{2}$. $f''(x) = 2 \Rightarrow f\left(\frac{1}{2}\right) = \frac{3}{4}$ is a local minimum which is absolute since there are no other critical points.

(b) $f(x) = 3 + 4\cos x + \cos 2x$ is never negative because $f'(x) = -4\sin x - 2\sin 2x = -4\sin x - 4\sin x\cos x = 0 \Leftrightarrow$ $\sin x(1 + \cos x) = 0 \Leftrightarrow \sin x = 0$ or $\cos x = -1 \Leftrightarrow x = 0$ or π (over one period). $f(0) = 8$ and $f(\pi) = 0$.

45. If $y = \sin x + \cos x$, then $y' = \cos x - \sin x$ and $y' = 0 \Leftrightarrow \cos x = \sin x$ or

$x = \frac{\pi}{4}$ or $\frac{5\pi}{4}$. $y\left(\frac{\pi}{4}\right) = \frac{\sqrt{2}}{2} + \frac{\sqrt{2}}{2} = \sqrt{2}$ is a maximum and

$y\left(\frac{5\pi}{4}\right) = -\frac{\sqrt{2}}{2} - \frac{\sqrt{2}}{2} = -\sqrt{2}$ is a minimum value.

47. Let x = the radius of the hemisphere, and y = height of the cylinder. Then $V = \pi x^2 y + \frac{2}{3}\pi x^3 \Rightarrow y = \frac{V}{\pi x^2} - \frac{2x}{3}$. $C = 2\pi xy + 2(2\pi x^2)$

$$= 2\pi x\left(\frac{V}{\pi x^2} - \frac{2x}{3}\right) + 4\pi x^2 = \frac{2V}{x} + \frac{8\pi}{3}x^2.$$

$C' = -2Vx^{-2} + \frac{16\pi}{3}x = 0 \Leftrightarrow x^3 = \frac{3V}{8\pi}$ or $x = \left(\frac{3V}{8\pi}\right)^{\frac{1}{3}}$.

$C'' = 4Vx^{-3} + \frac{16\pi}{3} > 0$ for $x > 0 \Rightarrow$ this gives a minimum value.

If $x = \left(\frac{3V}{8\pi}\right)^{\frac{1}{3}}$ then $y = \left(\frac{3V}{\pi}\right)^{\frac{1}{3}}$.

49. $(x+y)^2 + (b-a)^2 = c^2 \Rightarrow 2(x+y)\left(1 + \frac{dy}{dx}\right) = 0 \Rightarrow \frac{dy}{dx} = -1$

$d_1 = (x^2 + b^2)^{\frac{1}{2}}$ and $d_2 = (y^2 + a^2)^{\frac{1}{2}}$. Then $t = d_1 + d_2$ is the distance to be minimized.

$$\frac{dt}{dx} = \frac{1}{2}(x^2 + b^2)^{-\frac{1}{2}}(2x) + \frac{1}{2}(y^2 + a^2)^{-\frac{1}{2}}(2y)\left(\frac{dy}{dx}\right)$$

$$= x(x^2 + b^2)^{-\frac{1}{2}} - y(y^2 + a^2)^{-\frac{1}{2}}$$

$$\frac{dt}{dx} = 0 \Leftrightarrow x(y^2 + a^2)^{-\frac{1}{2}} = y(x^2 + b^2)^{-\frac{1}{2}} \Leftrightarrow x^2(y^2 + a^2) = y^2(x^2 + b^2)$$

$$\Leftrightarrow x^2a^2 = y^2b^2 \Leftrightarrow y = \frac{a}{b}x.$$

$$\frac{d^2t}{dx^2} = (x^2 + b^2)^{-\frac{1}{2}} - \frac{x}{2}(x^2 + b^2)^{-\frac{3}{2}}(2x) - (y^2 + b^2)^{-\frac{1}{2}}\frac{dy}{dx} - + \frac{y}{2}(y^2 + a^2)^{-\frac{3}{2}}(2y)\frac{dy}{dx}$$

$$= \frac{x^2 + b^2 - x^2}{(x^2 + b^2)^{\frac{3}{2}}} + \frac{y^2 + b^2 - y^2}{(y^2 + b^2)^{\frac{3}{2}}} > 0 \Rightarrow \text{minimum distance.}$$

$$t = \sqrt{x^2 + b^2} + \sqrt{\frac{a^2x^2}{b^2} + a^2} = \left(1 + \frac{a}{b}\right)\sqrt{x^2 + b^2}.$$

To get x in terms of c we use: $c^2 = (b - a)^2 + \left(x + \frac{ax}{b}\right)^2 =$

$$(b - a)^2 + x^2\left(1 + \frac{a}{b}\right)^2 \Rightarrow x^2 = \frac{c^2 - (b - a)^2}{\left(1 + \frac{a}{b}\right)^2}.$$

$$t = \left(1 + \frac{a}{b}\right)\sqrt{\frac{c^2 - (b - a)^2 + b^2\frac{(b + a)^2}{b^2}}{\left(1 + \frac{a}{b}\right)^2}} = \sqrt{c^2 + 4ab}$$

51. The distance between the graphs is $D(x) = f(x) - g(x)$.
$D'(x) = f'(x) - g'(x) = 0 \Leftrightarrow f'(x) = g'(x)$.
$D''(x) = f''(x) - g''(x) < 0$ since, from the graph, $f''(x) < 0$ and $g''(x) > 0$ for $a < x < b$. Therefore, the critical point occurs where $f'(c) = g'(c)$, so that the tangent lines are parallel.

53. Marginal cost $= \frac{dy}{dx}$. Let the revenue be $R = xP$, so that

$$\frac{dR}{dx} = x\frac{dP}{dx} + P.$$

(a) Profit $T = xP - y = R - y$ would be maximized at a point where $\frac{dT}{dx} = \frac{dR}{dx} - \frac{dy}{dx} = 0$ or where $\frac{dR}{dx} = \frac{dy}{dx}$.

(b) For T to be maximal, $\frac{d^2T}{dx^2} < 0$, or $\frac{d^2R}{dx^2} - \frac{d^2y}{dx^2} < 0$ or $\frac{d^2y}{dx^2} > \frac{d^2R}{dx^2}$.

55. $2c + 2x + y = 1 \Rightarrow y = 1 - 2x - 2c$. Also $\sin\alpha = \frac{c}{x}$.

$z = y + \sqrt{x^2 - c^2} = 1 - 2x - 2x + \sqrt{x^2 - c^2}$.

$\frac{dz}{dx} = -2 + \frac{1}{2}(x^2 - c^2)^{-\frac{1}{2}}(2x) = 0 \Leftrightarrow \frac{x}{\sqrt{x^2 - c^2}} = 2 \Leftrightarrow x^2 = 4(x^2 - c^2)$

$\Leftrightarrow x = \frac{2}{\sqrt{3}}c$ or $\frac{c}{x} = \frac{\sqrt{3}}{2} = \sin\alpha \Rightarrow \alpha = \frac{\pi}{3}$ or $2\alpha = \frac{2\pi}{3}$.

$\frac{d^2z}{dx^2} = \frac{-c^2}{(x^2 - c^2)^{\frac{3}{2}}} < 0 \Rightarrow$ this angle gives maximal distance.

3.6 RELATED RATES OF CHANGE

1. $A = \pi r^2 \Rightarrow \frac{dA}{dt} = 2\pi r\frac{dr}{dt}$

3. $V = s^3 \Rightarrow \frac{dV}{dt} = 3s^2\frac{ds}{dt}$

5. $V = IR \Rightarrow$ (a) $\frac{dV}{dt} = 1$ (b) $\frac{dI}{dt} = -\frac{1}{3}$

   (c) $R = VI^{-1} \Rightarrow \frac{dR}{dt} = I^{-1}\frac{dV}{dt} + V(-I^{-2}\frac{dI}{dt})$

   (d) $\frac{dR}{dt} = \frac{1}{2}(1) - (\frac{12}{4})(-\frac{1}{3}) = \frac{3}{2}$

7. Let x = distance from 2nd base

   y = distance from 3rd base.

   Then $y = \sqrt{x^2 + 90^2} = \sqrt{60^2 + 90^2} = 30\sqrt{13}$

   $y^2 = x^2 + 90^2 \Rightarrow 2y\frac{dy}{dt} = 2x\frac{dx}{dt}$. Now $\frac{dx}{dt} = -16$ ft/s so

   $30\sqrt{13}\frac{dy}{dt} = (60)(-16) \Rightarrow \frac{dy}{dt} = -\frac{32}{\sqrt{13}}$ ft/s

9. $A = \frac{1}{2}xy \Rightarrow \frac{dA}{dt} = \frac{1}{2}\left(x\frac{dy}{dt} + y\frac{dx}{dt}\right)$

From Ex. 3, $x = 10 \Rightarrow y = 24$ and $\frac{dy}{dt} = -\frac{5}{3}$

$\frac{dA}{dt} = \frac{1}{2}\left[(10)\left(-\frac{5}{3}\right) + (24)(4)\right] = \frac{119}{3}$

$A = \frac{1}{2}x\sqrt{26^2 - x^2} \Rightarrow \frac{da}{dx} = \frac{1}{2}(26^2 - x^2)^{\frac{1}{2}} + \frac{1}{2}x\left(\frac{1}{2}\right)(26^2 - x^2)^{-\frac{1}{2}}(-2x)$

$= \frac{\sqrt{26^2 - x^2}}{2} - \frac{x^2}{\sqrt{26^2 - x^2}} = \frac{26^2 - 2x^2}{2\sqrt{26^2 - x^2}}$

$\frac{dA}{dx} = 0 \Leftrightarrow t = \frac{x}{4} = \frac{13\sqrt{2}}{4}$ sec. Since $\frac{dA}{dx} > 0$ for $0 < x < 13\sqrt{2}$ and $\frac{dA}{dx} < 0$ for $13\sqrt{3} < x < 26$, this give a maximum value.

11. An accumulation of moisture is a change in volume. $\therefore \frac{dV}{dt} = kS$, where k is the constant of proportionality. $S = \frac{4}{3}\pi r^3 \Rightarrow \frac{dV}{dt} = 4\pi r^2\frac{dr}{dt}$. Equating, $4\pi r^2\frac{dr}{dt} = kS = k(4\pi r^2) \Rightarrow \frac{dr}{dt} = k$.

13. $V = \frac{4}{3}\pi r^3 \Rightarrow \frac{dV}{dt} = 4\pi r^2\frac{dr}{dt}$. $\therefore 100 = 4\pi(3)^2\frac{dr}{dt} \Rightarrow \frac{dr}{dt} = \frac{25}{9\pi}$ ft/min.

Since $S = 4\pi r^2$, $\frac{dS}{dt} = 8\pi r\frac{dr}{dt} = 8\pi(3)\frac{25}{9\pi} = \frac{200}{30}$ ft^2/min.

15. $s^2 = x^2 + (200 + y)^2 \Rightarrow 2s\frac{ds}{dt} = 2x\frac{dx}{dt} + 2(200 + y)\frac{dy}{dt} \Rightarrow$

$\frac{ds}{dt} = \frac{(66)(66) + (215)(15)}{\sqrt{215^2 + 66^2}} \approx 33.7$ ft/sec

17. (a) $3x^2 - y^2 = 12 \Rightarrow 6x\frac{dx}{dt} - 2y\frac{dy}{dt} = 0 \Rightarrow \frac{dx}{dt} = \frac{y}{3x}\frac{dy}{dt}$

$\therefore \frac{dx}{dt}\Big|_{x=4} = \left(\frac{\pm 6}{12}\right)(6) = \pm 3$

(b) $\frac{dy}{dx} = \frac{\frac{dy}{dt}}{\frac{dx}{dt}} = \frac{6}{\pm 3} = \pm 2$

19. $\frac{dx}{dt} = 10$ m/sec. $\tan\theta = \frac{y}{x} = \frac{x^2}{x} = x$. $\sec^2\theta \frac{d\theta}{dt} = \frac{dx}{dt}$.

$x = 3 \Rightarrow \sec^2\theta = 1 + \tan^2\theta = 10$ and $\frac{d\theta}{dt} = 1$.

$x = 103 \Rightarrow \sec^2\theta = 1 + 103^2 = 10610$ and $\frac{d\theta}{dt} = \frac{10}{10610} \approx 0.00094$

21. (a) Length of shadow: $\frac{6}{y} = \frac{16}{y+z} \Rightarrow y = \frac{3}{5}z$

$\frac{dy}{dt} = \frac{3}{5}\frac{dz}{dt} = \frac{3}{5}(5) = 3$ ft/sec.

(b) Tip of shadow: $\frac{6}{y} = \frac{16}{x} \Rightarrow x = \frac{8}{3}y \Rightarrow \frac{dx}{dt} = \frac{8}{3}\frac{dy}{dt} = \frac{8}{3}(3) = 8$ ft/sec.

25. $s^2 = 1 + (x-z)^2 \Rightarrow 2s\frac{ds}{dt} = 2(x-z)\left(\frac{dx}{dt} - \frac{dz}{dt}\right)$

$(1.5)(-136) = \frac{\sqrt{5}}{2}\left(\frac{dx}{dt} - 120\right) \Rightarrow \frac{dx}{dt} = 62.4$ miles per hour.

27. (a) At noon, ship A is due north of ship B $\Rightarrow x = s \Rightarrow \frac{ds}{dt} = 12$.

(b) At 1 p.m. ship B is due east of ship B $\Rightarrow y = s \Rightarrow \frac{ds}{dt} = 8$.

(c) $s^2 = x^2 + y^2$, $x = 12 - 12t$ and $y = 8t$.

$\therefore s^2 = (12-12t)^2 + (8t)^2 \Rightarrow 2s\frac{ds}{dt} = 2(12-12t)(-12) + 128t$

$= 0 \Leftrightarrow t = \frac{9}{13}$. Then $x = \frac{48}{13}$, $y = \frac{72}{13}$ and $s \approx 6.5$.

The minimum distance is greater than 5, and the ships do not see each other.

3.7 THE MEAN VALUE THEOREM

1. $f(x) = x^4 + 3x + 1$, $[-2,-1]$. $f(-2) = 11$ and $f(-1) = -1 \Rightarrow$ by the Intermediate Value Theorem that f has at least one root between -2 and -1. $f'(x) = 4x^3 + 3 = 0 \Leftrightarrow x = -\left(\frac{3}{4}\right)^{\frac{1}{3}} > -1$.

$\therefore$ f' has no zero between -2 and -1, so f has exactly one root.

3. $f(x) = 2x^3 - 3x^2 - 12x - 6$, $[-1, 0]$. $f(-1) = 1$, $f(0) = -6 \Rightarrow$ by The Intermediate Value Theorem that f has at least one root between -1 and 0. $f'(x) = 6x^2 - 6x - 12 = 6(x-2)(x+1)$. $f'(x) = 0$ if $x = 2$ or -1. $\therefore$ f has exactly one root on $[-1, 0]$.

5.(a) (i) $y = x^2 - 4 = 0 \Leftrightarrow x = \pm 2$. $y' = 2x = 0 \Leftrightarrow x = 0$. Note $-2 < 0 < 2$.

(ii) $y = x^2 + 8x + 15 = (x+3)(x+5) = 0 \Leftrightarrow x = -3$ or -5.

$y' = 2x + 4 = 0 \Leftrightarrow x = -4$. Note $-5 < -4 < -3$.

(iii) $y = (x+1)(x-2)^2 = 0 \Leftrightarrow x = -1$ or $x = 2$. $y' = 3x(x-2) = 0$ $\Leftrightarrow x = 0$ or $x = 2$. Note that $x = 2$ is a double root of y and that y' has a zero at $x = 2$ also.

(iv) $y = x(x-9)(x-24) = 0 \Leftrightarrow x = 0$, 9 or 24. $y' = 3(x-4)(x-18)$ $= 0 \Leftrightarrow x = 4$ or $x = 18$. Note that $0 < 4 < 9$ and $9 < 18 < 24$.

(b) Let r_1 and r_2 be zeros of the polynomial $p(x) = x^n + a_{n-1}x^{n-1} + \ldots + a_1x + a_0$. Then $p(r_1) = p(r_2) = 0$. Since polynomials are everywhere continuous and differentiable, by Rolle's Theorem, $p'(r) = 0$ for some $r_1 < r < r_2$. But $p'(x) = nx^{n-1} + (n-1)a_{n-1}x^{n-2} + \ldots + a_1$.

7. $f(x) = x^2 + 2x - 1$, $a = 0$, $b = 1$. $f'(x) = 2x + 2$

$$2c + 2 = \frac{f(1) - f(0)}{1 - 0} \Rightarrow 2c + 2 = 3 \Leftrightarrow c = \frac{1}{2}$$

9. $f(x) = x + \frac{1}{x}$, $a = \frac{1}{2}$, $b = 2$. $f'(x) = 1 - x^{-2}$.

$$1 - \frac{1}{c^2} = \frac{f(2) - f\left(\frac{1}{2}\right)}{2 - \frac{1}{2}} \Rightarrow 1 - \frac{1}{c^2} = 0 \Leftrightarrow c = \pm 1. \ \therefore c = 1$$

11. By Mean Value Theorem, $\exists$ $a < c < x$ for which $f'(c) = \frac{f(x) - f(a)}{x - a}$

Since $|f'(c)| \le 1$, $\left|\frac{f(x) - f(a)}{x - a}\right| = \frac{|f(x) - f(a)|}{|x - a|} \le 1 \Rightarrow |f(x) - f(a)| \le |x - a|$.

13. Let $f(x) = \sin x$. Then f is continuous on [a,b] and differentiable on (a,b) $\Rightarrow \exists$ c for which $f'(c) = \cos c = \dfrac{\sin b - \sin a}{b - a}$.

Since $|\cos c| \le 1$, $\dfrac{|\sin b - \sin a|}{|b - a|} \le 1 \Rightarrow |\sin b - \sin a| \le |b - a|$.

15. There is no contradiction. The converse of the Mean Value Theorem is not true. The theorem states sufficient conditions for there to be a horizontal tangent line, but not necessary conditions. That is, there can be a horizontal tangent line on an interval without differentiability on that interval.

17. The function $f(x) = [x]$ is continuous on [0,1) but not continuous on [0,1] as the theorem requires.

19. This is precisely what the Mean Value Theorem asserts. The expression on the right side is the average value of a function on an interval (in this case from 0 miles to 30 miles in 1 hour). The expression on the left is the instantaneous rate of change at some moment. The Mean Value Theorem states that these must be equal at least once.

21. Consider the given function on [−3,0] and [0,3].

On [−3,0], we have $\dfrac{f(0) - f(-3)}{0 - (-3)} \le 1 \Rightarrow f(0) + 3 \le 3 \Rightarrow f(0) \le 0$

On [0,3], we have $\dfrac{f(3) - f(0)}{3 - 0} \le 1 \Rightarrow 3 - f(0) \le 3 \Rightarrow f(0) \ge 0$.

Since f is differentiable for all x, f must be continuous for all x, so $f(0) = 0$.

23. No. Corollary 3 states that functions with the same derivatives differ from each other by at most a constant. Therefore, the family of functions $f(x) = 3x + b$ describes all functions for which $f'(x) = 3$.

25. Let x_1 and x_2 be any two values such that $a \leq x_1 < x_2 \leq b$.

By Mean Value Theorem, $\exists$ c for which $x_1 < c < x_2$ and

$\dfrac{f(x_2) - f(x_1)}{x_2 - x_1} = f'(c)$. Since $f'(x) = 0$ for all x in (a,b),

$f(x_2) - f(x_1) = 0 \Rightarrow f(x_2) = f(x_1)$. $\therefore$ f(x) is constant on (a,b).

3.8 INDETERMINATE FORMS AND L'HOPITAL'S RULE

1. $$\lim_{x\to 2} \frac{x-2}{x^2-4} = \lim_{x\to 2} \frac{1}{2x} = \frac{1}{4}$$

3. $$\lim_{x\to\infty} \frac{5x^2-3x}{7x^2+1} = \lim_{x\to\infty} \frac{10x-3}{14x} = \lim_{x\to\infty} \frac{10}{14} = \frac{5}{7}$$

5. $$\lim_{t\to 0} \frac{\sin t^2}{t} = \lim_{t\to 0} \frac{2\sin t\cos t}{1} = 0$$

7. $$\lim_{x\to 0} \frac{\sin 5x}{x} = \lim_{x\to 0} \frac{5\cos 5x}{1} = 5$$

9. $$\lim_{\theta\to\pi} \frac{\sin\theta}{\pi-\theta} = \lim_{\theta\to\pi} \frac{\cos\theta}{-1} = \frac{-1}{-1} = 1$$

11. $$\lim_{x\to\frac{\pi}{4}} \frac{\sin x - \cos x}{x - \frac{\pi}{4}} = \lim_{x\to\frac{\pi}{4}} \frac{\cos x + \sin x}{1} = \sqrt{2}$$

13. $$\lim_{x\to\frac{\pi}{2}} \left[-\left(x-\frac{\pi}{2}\right)\tan x\right] = \lim_{x\to\frac{\pi}{2}} \left[\frac{-\left(x-\frac{\pi}{2}\right)}{\cot x}\right] = \lim_{x\to\frac{\pi}{2}} \left(\frac{-1}{-\csc^2 x}\right) = 1$$

15. $$\lim_{x\to 1} \frac{2x^2-(3x+1)\sqrt{x}+2}{x-1} = \lim_{x\to 1} \frac{4x - 3\sqrt{x} - \frac{3x+1}{2\sqrt{x}}}{1} = -1$$

17. $$\lim_{x\to 0} \frac{\sqrt{a(a+x)}-a}{x} = \lim_{x\to 0} \frac{\frac{1}{2}[a(a+x)]^{-\frac{1}{2}}(a)}{1} = \frac{1}{2}$$

19. $$\lim_{x\to 0} \frac{x(\cos x - 1)}{\sin x - x} = \lim_{x\to 0} \frac{-x\sin x + \cos x - 1}{\cos x - 1} =$$

$$\lim_{x\to 0} \frac{-x\cos x - \sin x - \sin x}{-\sin x} = \lim_{x\to 0} \frac{x\sin x - \cos x - 2\cos x}{-\cos x} = 3$$

21. $$\lim_{r\to 1} \frac{a(r^n-1)}{r-1} \text{ (n a positive integer)} = \lim_{r\to 1} \frac{nar^{n-1}}{1} = na$$

23. $\lim_{x\to\infty}\left(x-\sqrt{x^2+x}\right)=\lim_{x\to\infty}\frac{x-\sqrt{x^2+x}}{1}\cdot\frac{x+\sqrt{x^2+x}}{x+\sqrt{x^2+x}}$

$$\lim_{x\to\infty}\frac{-x}{x+\sqrt{x^2+x}}=\lim_{x\to\infty}\frac{-\frac{x}{x}}{\frac{x}{x}+\sqrt{\frac{x^2}{x^2}+\frac{x}{x^2}}}=\lim_{x\to\infty}\frac{-1}{1+\sqrt{1+\frac{1}{x}}}=-\frac{1}{2}$$

25. $\lim_{x\to\infty}\frac{\sqrt{10x+1}}{\sqrt{x+1}}=\lim_{x\to\infty}\frac{\sqrt{10+\frac{1}{x}}}{\sqrt{1+\frac{1}{x}}}=\sqrt{10}$

27. (a) $y=\sec x+\tan x,\ -\frac{\pi}{2}<x<\frac{\pi}{2}.\ y=\frac{1+\sin x}{\cos x}\geq 0$

since $|\sin x|\leq 1\Rightarrow 1+\sin x>0$, and $\cos x>0$ for $-\frac{\pi}{2}<x<\frac{\pi}{2}$.

$\therefore\ y'=\sec x\tan x+\sec^2 x=\frac{\sin x+1}{\cos^2 x}\geq 0$ also.

$y''=\sec x(\sec^2 x)+\tan x(\sec x\tan x)+2\sec x(\sec x\tan x)$

$$=\frac{1}{\cos^3 x}+\frac{\sin^2 x}{\cos^3 x}+\frac{2\sin x}{\cos^3 x}=\frac{(1+\sin x)^2}{\cos^3 x}\geq 0.$$

(b) $\lim_{x\to\left(-\frac{\pi}{2}\right)^-}(\sec x+\tan x)=\lim_{x\to\left(-\frac{\pi}{2}\right)^-}\frac{1+\sin x}{\cos x}=\lim_{x\to\left(-\frac{\pi}{2}\right)^-}\frac{\cos x}{-\sin x}=0$

(c)

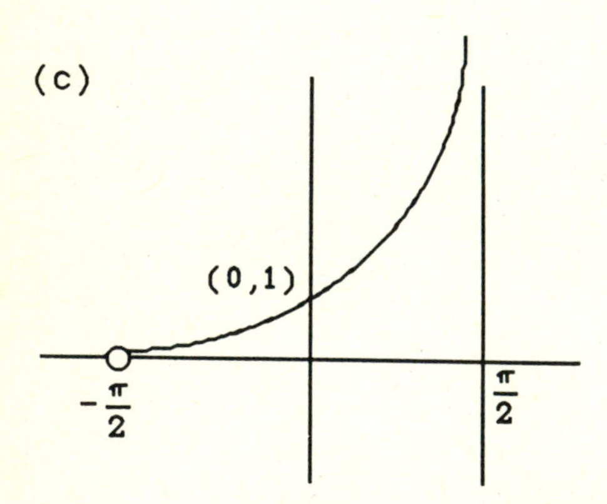

3.9 QUADRATIC APPROXIMATIONS AND APPROXIMATION ERRORS

1. $\sin x\approx x \quad\Rightarrow\quad x\sin x\approx x^2$

3. $\cos x\approx 1-\frac{x^2}{2} \quad\Rightarrow\quad \cos\sqrt{1+x}\approx 1-\frac{1+x}{2}=\frac{1}{2}-\frac{1}{2}x$ since $\sqrt{1+x}\approx 0$ if $x\approx -1$.

5. $\sqrt{x}=\sqrt{1+(x-1)}\approx 1+\frac{x-1}{2}-\frac{(x-1)^2}{8}=\frac{3}{8}+\frac{3}{4}x-\frac{1}{8}x^2.$

(b) $|f''(x)| \le 1$ so take $M = 1$. $|e_1(x)| < \frac{1}{2}(.1)^2 < 0.005$

9. $y = \tan x$, $y' = \sec^2 x$ and $y'' = 2\sec^2 x \tan x$. $y(0) = 0$, $y'(0) = 1$ and $y''(0) = 0$. Therefore, $\tan x \approx x$. $f'''(x) = 4\sec^2 x \tan^2 x + 2\sec^4 x$. If one takes $x = \frac{\pi}{6}$, then $|e_2(x)| \le \frac{1}{6}\left(\frac{48}{9}\right)(0.1)^3 \approx 0.00089$.

If one uses a calculator and $x = 0.1$, $|e_2(x)| \le 0.0003$

11. $f(x) = \sin x$ and $f\left(\frac{\pi}{2}\right) = 1$; $f'(x) = \cos x$ and $f'\left(\frac{\pi}{2}\right) = 0$.

$f''(x) = -\sin x$ and $f''\left(\frac{\pi}{2}\right) = -1$. Therefore, $\sin x \approx 1 - \frac{1}{2}\left(x - \frac{\pi}{2}\right)^2$.

$|e_2(x)| \le \frac{1}{6}(0.1)^3 \approx 0.000167$

13. $f(x) = \cos x$ and $f\left(\frac{\pi}{2}\right) = 0$; $f'(x) = -\sin x$ and $f'\left(\frac{\pi}{2}\right) = -1$.

$f''(x) = -\cos x$ and $f''\left(\frac{\pi}{2}\right) = 0$. Therefore, $\cos x \approx -\left(x - \frac{\pi}{2}\right)$

$|e_2(x)| \le \frac{1}{6}(0.1)^3 \approx 0.000167$

15. $|e_2(x)| \le \max\left|\frac{f'''(x)}{6}\right| x^2 = \frac{1}{6}x^2$. Therefore,

(a) $\frac{1}{6}x^2 \le 0.01 \Leftrightarrow |x| \le \frac{\sqrt{6}}{10} \approx 0.245$

(b) $\frac{1}{6}x^2 \le 0.01\,|x| \Leftrightarrow |x| \le 0.06$

17. If $y = (1+x)^k$, then $y' = k(1+x)^{k-1}$ and $y'' = k(k-1)(1+x)^{k-2}$. Therefore, $y(0) = 1$, $y'(0) = k$ and $y''(0) = k(k-1)$, and

$$y \approx 1 + kx + \frac{k(k-1)}{2}x^2.$$

19. $f(x) = x^3 + 5x - 7$, $f'(x) = 3x^2 + 5$, $f''(x) = 6x$, and $f'''(x) = 6$. $f(1) = -1$, $f'(1) = 8$, $f''(1) = 6$ and $f'''(1) = 6$. Therefore

$$f(x) = -1 + 8(x-1) + \frac{6}{2}(x-1)^2 + \frac{6}{6}(x-1)^3 = x^3 + 5x - 7 = f(x).$$

3.M MISCELLANEOUS

1. $y = 9x - x^2$; $y' = 9 - 2x$; $y'' = -2$

   (a) y increases on $\left(-\infty, \frac{9}{2}\right)$

   (b) y decreases on $\left(\frac{9}{2}, \infty\right)$

   (c) never concave up

   (d) always concave down

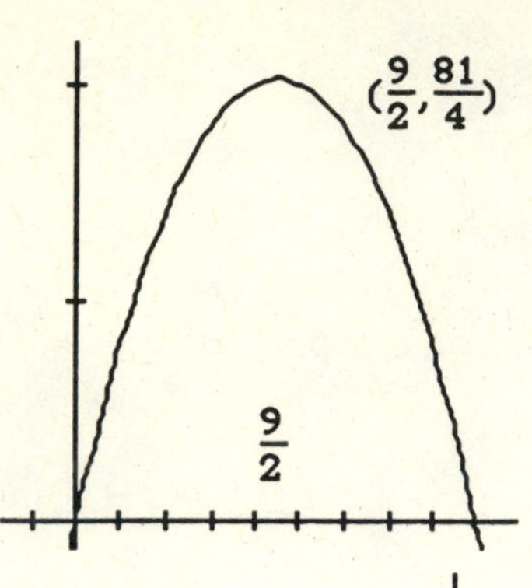

3. $y = 4x^3 - x^4$; $y' = 12x^2 - 4x^3$; $y'' = 24x - 12x^2$

   (a) y increases on $(-\infty, 3)$

   (b) y decreases on $(3, \infty)$

   (c) concave up on $(0, 2)$

   (d) concave down on $(-\infty, 0) \cup (2, \infty)$

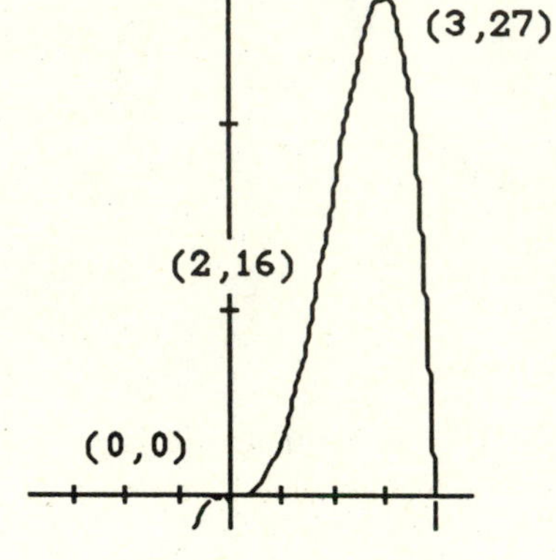

5. $y = x^2 + 4x^{-1}$; $y; = 2x - 4x^{-2}$; $y'' = 2 + 8x^{-3}$

   (a) y increases on $(\sqrt[3]{2}, \infty)$

   (b) y decreases on $(-\infty, \sqrt[3]{2})$

   (c) concave up on $(-\infty, -\sqrt[3]{4}) \cup (0, \infty)$

   (d) concave down on $(-\sqrt[3]{4}, 0)$

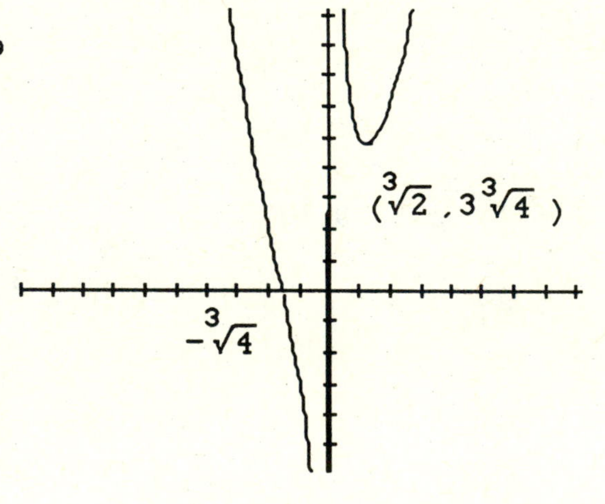

7. $y = 5 - x^{2/3}$; $y' = -\frac{2}{3}x^{-1/3}$; $y'' = \frac{1}{9}x^{-4/3}$

   (a) y increases on $(-\infty, 0)$

   (b) y decreases on $(0, \infty)$

   (c) concave up on for all $x \neq 0$

   (d) for no x

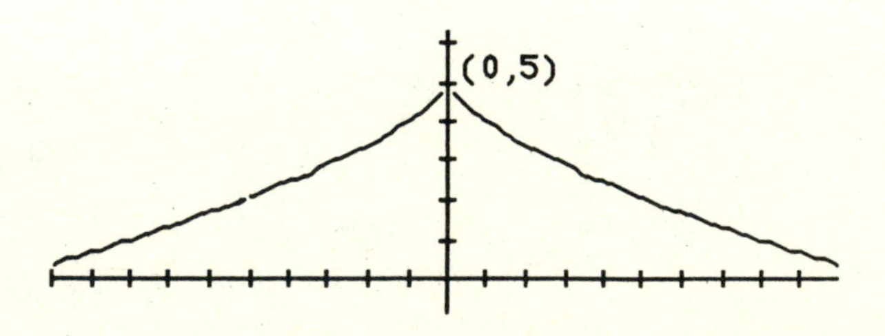

11. $y = \dfrac{x^2}{ax+b}$; $y' = \dfrac{ax^2+2bx}{(ax+b)^2}$

$$y'' = \frac{2(2a^2x^2+4axb+b^2)}{(ax+b)^3}$$

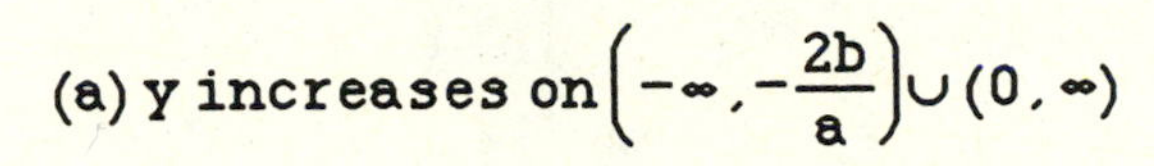

(a) y increases on $\left(-\infty, -\dfrac{2b}{a}\right) \cup (0, \infty)$

(b) y decreases on $\left(-\dfrac{2b}{a}, 0\right)$

(c) concave up on $\left(-\dfrac{b}{a}, \infty\right)$

(d) concave down on $\left(-\infty, -\dfrac{b}{a}\right)$

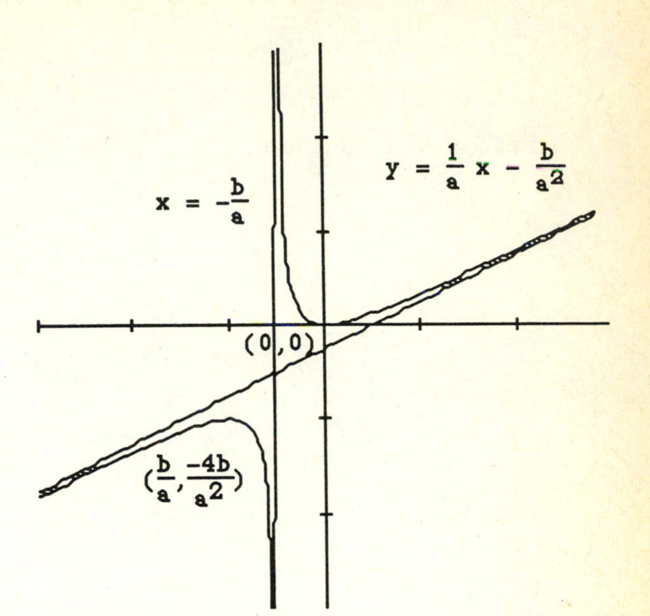

13. $y = (x-1)(x+1)^2$; $y' = (x+1)(3x-1)$; $y'' = 6x+2$

(a) y increases on $(-\infty, -1) \cup \left(\dfrac{1}{3}, \infty\right)$

(b) y decreases on $\left(-1, \dfrac{1}{3}\right)$

(c) concave up on $\left(-\dfrac{1}{3}, \infty\right)$

(d) concave down on $\left(-\infty, -\dfrac{1}{3}\right)$

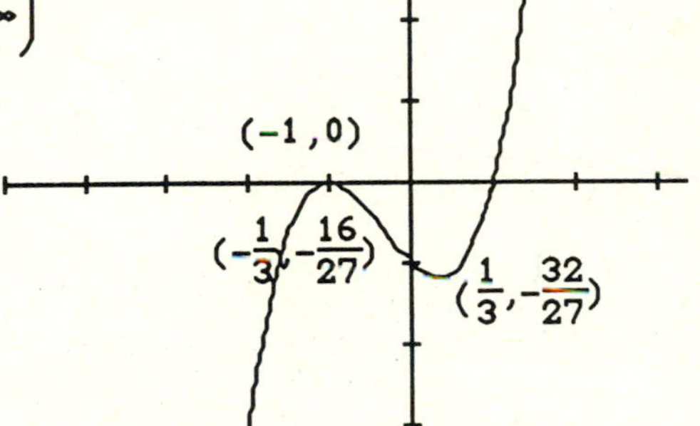

15. $y = 2x^3 - 9x^2 + 12x - 4$; $y' = 6x^2 - 18x + 12$; $y'' = 12x - 18$

(a) y increases on $(-\infty, 1) \cup (2, \infty)$

(b) y decreases on $(1, 2)$

(c) concave up on $\left(\dfrac{3}{2}, \infty\right)$

(d) concave down on $\left(-\infty, \dfrac{3}{2}\right)$

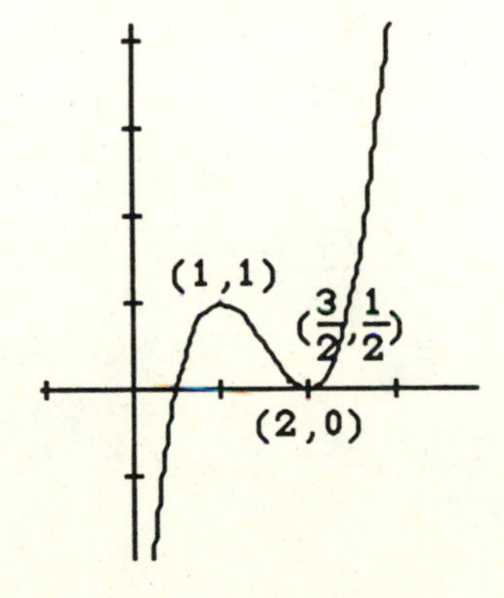

17. The relation $xy^2 = 3(1 - x)$ is symmetric to the x-axis, and not a function. We will solve it for y, graph the top half, and reflect this about the y-axis for the bottom half.

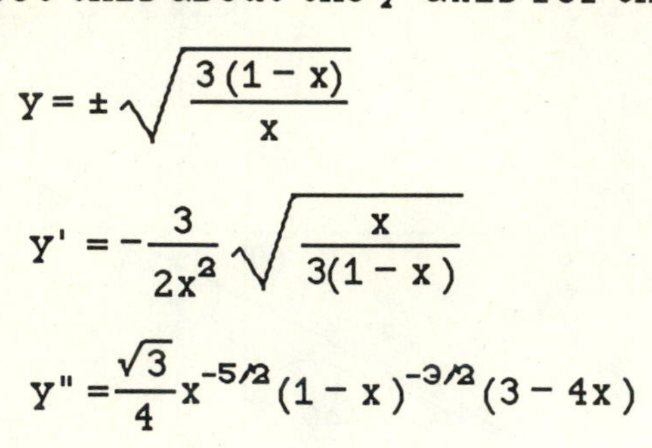

$$y = \pm\sqrt{\frac{3(1-x)}{x}}$$

$$y' = -\frac{3}{2x^2}\sqrt{\frac{x}{3(1-x)}}$$

$$y'' = \frac{\sqrt{3}}{4}x^{-5/2}(1-x)^{-3/2}(3-4x)$$

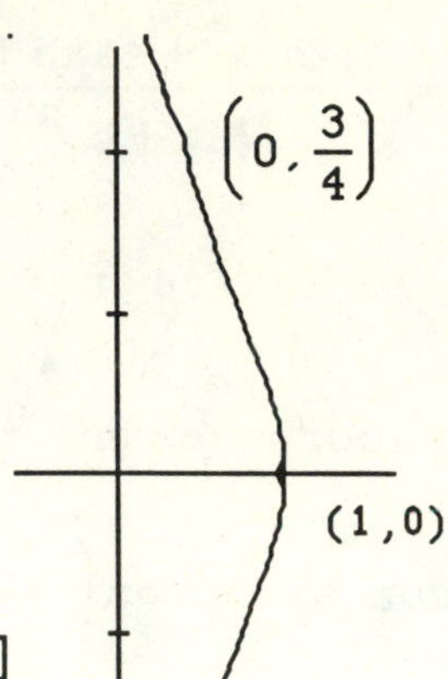

(a) y does not increase on its domain which is $(0,1]$

(b) y decreases on $(0,1]$

(c) concave up on $\left(0, \frac{3}{4}\right)$

(d) concave down on $\left(\frac{3}{4}, 1\right]$

19 (a). $y = \dfrac{x+1}{x^2+1}$; Domain is all reals

(a) No symmetry to origin or y-axis

(b) $(0,1)$ and $(-1,0)$

(c) No vertical asymptote

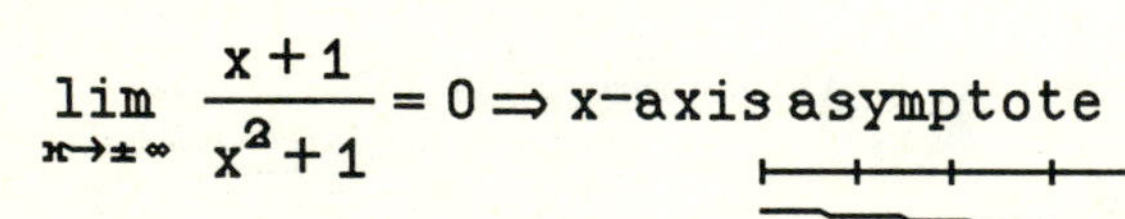

$$\lim_{x\to\pm\infty} \frac{x+1}{x^2+1} = 0 \Rightarrow \text{x-axis asymptote}$$

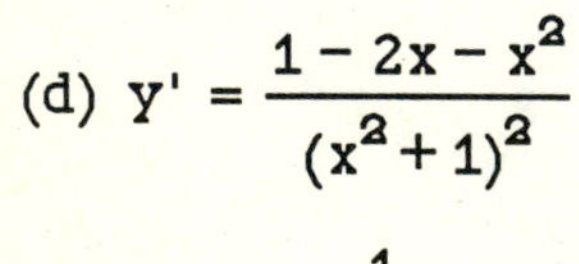

(d) $y' = \dfrac{1-2x-x^2}{(x^2+1)^2}$

$y'(-1) = \frac{1}{2}$; $y'(0) = 1$

(e) rising for x in $(-1-\sqrt{2}, -1+\sqrt{2})$

falling for x in $(-\infty, -1-\sqrt{2}) \cup (-1+\sqrt{2}, \infty)$

(f) $y'' = \dfrac{2(x^3+3x^2-3x-1)}{(x^2+1)^3}$; $x^3+3x^2-3x-1 = (x-1)x^2+4x+1)$

$y'' = 0$ if $x = 1, -2 \pm \sqrt{3}$.

Concave down if x in $(-\infty, -2-\sqrt{3}) \cup (-2+\sqrt{3}, 1)$

Concave up if x in $(-2-\sqrt{3}, (-2+\sqrt{3}) \cup (1, \infty)$

19(b). $y = \dfrac{x^2+1}{x+1} = x - 1 + \dfrac{2}{x+1}$; Domain is all $x \neq -1$

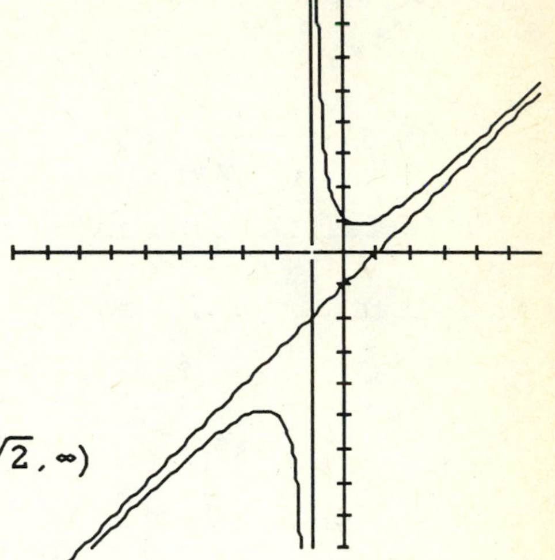

(a) No symmetry to origin or y-axis

(b) (0, 1)

(c) Vertical asymptote: $x = -1$

slant asymptote: $y = x - 1$

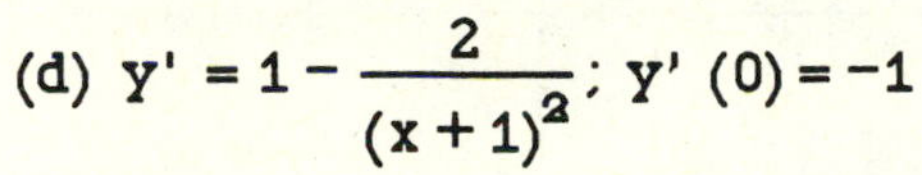

(d) $y' = 1 - \dfrac{2}{(x+1)^2}$; $y'(0) = -1$

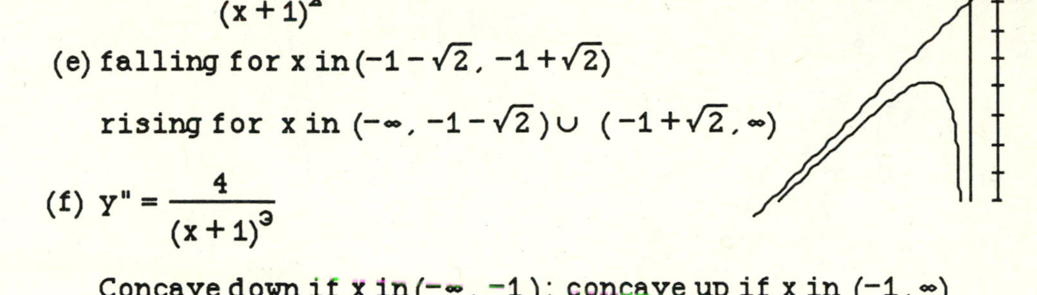

(e) falling for x in $(-1 - \sqrt{2}, -1 + \sqrt{2})$

rising for x in $(-\infty, -1 - \sqrt{2}) \cup (-1 + \sqrt{2}, \infty)$

(f) $y'' = \dfrac{4}{(x+1)^3}$

Concave down if x in $(-\infty, -1)$; concave up if x in $(-1, \infty)$

21. $y = x(x+1)(x-2)$; Domain is all x

(a) No symmetry to origin or y-axis

(b) (0, 0), (−1, 0) and (2, 0)

(c) no asymptotes

(d) $y' = 3x^2 - 2x - 2$

$y'(0) = -2$; $y'(-1) = 3$' $y'(2) = 6$

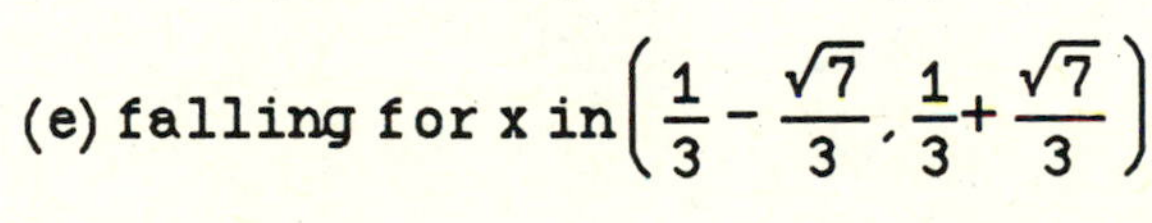

(e) falling for x in $\left(\dfrac{1}{3} - \dfrac{\sqrt{7}}{3}, \dfrac{1}{3} + \dfrac{\sqrt{7}}{3}\right)$

rising for x in $\left(-\infty, \dfrac{1}{3} - \dfrac{\sqrt{7}}{3}\right) \cup \left(\dfrac{1}{3} + \dfrac{\sqrt{7}}{3}, \infty\right)$

(f) $y'' = 6x - 2$

Concave down if x in $\left(-\infty, \dfrac{1}{3}\right)$; concave up if x in $\left(\dfrac{1}{3}, \infty\right)$

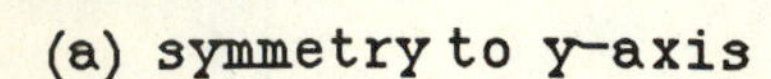

23. $y = \dfrac{8}{4 - x^2}$; Domain is all $x \neq \pm 2$

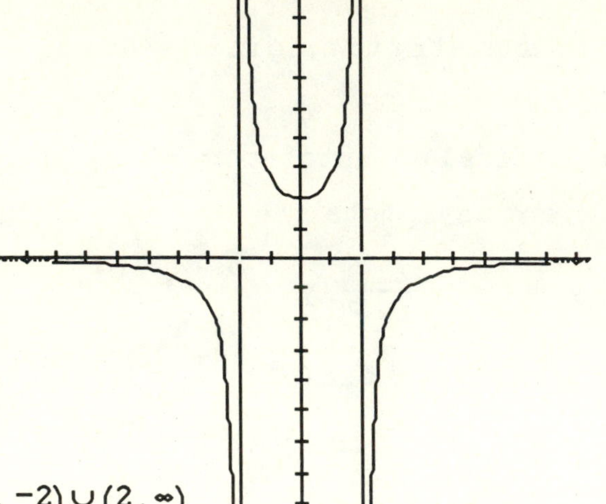

(a) symmetry to y-axis

(b) (0, 2) (c) $x = \pm 2$, $y = 0$

(d) $y' = \dfrac{16x}{(4 - x^2)^2}$; $y'(0) = 0$

(e) falling for x in $(-\infty, 0)$

rising for x in $(0, \infty)$

(f) $y'' = \dfrac{16(3x^2 + 4)}{(4 - x^2)^3}$

Concave down if x in $(-\infty, -2) \cup (2, \infty)$

Concave up if x in $(-2, 2)$

25. $x^2y - y = 4(x - 2) \Rightarrow y = \dfrac{4(x - 2)}{x^2 - 1}$; Domain is all $x \neq \pm 1$

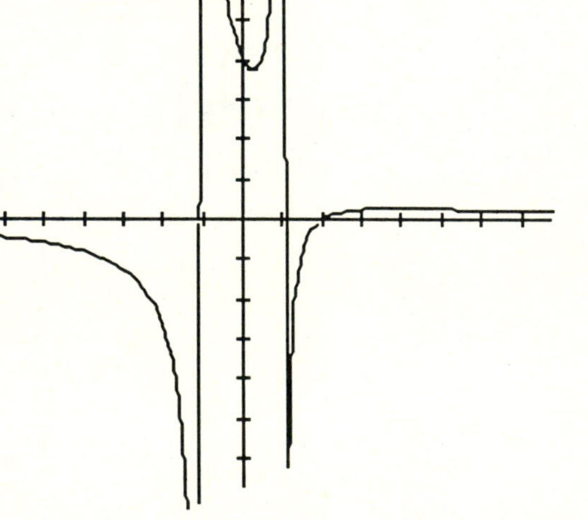

(a) No symmetry

(b) (2, 0), (0, 8) (c) $x = \pm 1$, $y = 0$

(d) $y' = \dfrac{-4(x^2 - 4x + 1)}{(x^2 - 1)^2}$; $y'(2) = \dfrac{4}{3}$; $y'(0) = -4$

(e) falling for x in $(-\infty, 2 - \sqrt{3}) \cup (2 + \sqrt{3}, \infty)$

rising for x in $(2 - \sqrt{3}, 2 + \sqrt{3})$

(f) $y'' = \dfrac{8(x^3 - 6x^2 + 3x - 2)}{(x^2 - 1)^3}$

Concave down if x in $(-\infty, -1) \cup (1, 5.5)$

Concave up if x in $(-1, 1) \cup (5.5, \infty)$

27. $y = \dfrac{ax+b}{cx^2+dx+e}$, a,b,c,d, and e are either 0 or 1.

(1) If there is no y − intercept, then $x \neq 0 \Rightarrow e = 0$

(2) x–axis an asymptote $\Rightarrow \lim\limits_{x\to\infty} \dfrac{ax+b}{cx^2+dx+e} = 0 \Rightarrow c \neq 0$ so $c = 1$

(3) x–intercept is $(-1,0) \Rightarrow 0 = \dfrac{-a+b}{c-d} \Rightarrow a = b$. If $a = b = 0$, the

funtion is $y = 0$. Therefore $a = b = 1$. The value of d must be

0, because otherwise the function is $y = \dfrac{1}{x}$.

29. If $\dfrac{dy}{dx} = 6(x-1)(x-2)^2(x-3)^3(x-4)^4$, then $\dfrac{dy}{dx}$ can only change

signs at x = 1 and x = 4. Moreover, there is a local maximum at

x = 1 because the change is from positive to negative, and a

local minimum at x = 3 because the change is from negative to positive.

31. $f(x) = 4x^3 - 8x^2 + 5x \Rightarrow f'(x) = 12x^2 - 16x + 5 = (2x-1)(6x-5)$

$f'(x) = 0 \Leftrightarrow x = \frac{1}{2}$ or $x = \frac{5}{6}$. We check $\{0, \frac{1}{2}, \frac{5}{6}, 2\}$.

$f(0) = 0 \quad f(\frac{1}{2}) = 1 \quad f(\frac{5}{6}) = \frac{25}{27} \quad f(2) = 10$. The largest value

of f on [0,2] is 10.

33. $y = ax^3 + bx^2 + cx + d \Rightarrow f'(x) = 3ax^2 + 2bx + c$ and

$f''(x) = 6ax + 2b$.

Inflection pt. at $(1,-6) \Rightarrow f''(1) = 0 \Rightarrow$ $6a + 2b = 0$
Local maximum at $(-1,10) \Rightarrow f'(-1) = 10 \Rightarrow$ $3a - 2b + c = 10$
$(1,-6)$ on curve $\Rightarrow f(1) = -6 \Rightarrow$ $a + b + c + d = -6$

$(-1,10)$ on curve $\Rightarrow f(-1) = 10 \Rightarrow$ $-a + b - c + d = 10$

$6a + 2b = 0 \Rightarrow b = -3a$. Substituting into second equation:

$3a - 2(-3a) + c = 0 \Rightarrow c = -9a$. Substituting for b and c:

$-a - 3a + 9a + d = 10 \Rightarrow 5a + d = 10 \Rightarrow d = 10 - 5a$

$a + (-3a) + (-9a) + (10 - 5a) = -6 \Rightarrow a = 1$.

Then $b = -3$, $c = -9$, $d = 5$ and $y = x^3 - 3x^2 - 9x + 5$

35. $p = 2r + s \Rightarrow 2r + s = 100 \Rightarrow s = 100 - 2r.$

$A = \frac{1}{2}rs = \frac{1}{2}r(100 - 2r) = 50r - r^2.$ $A' = 50 - 2r$

$\Rightarrow r = 25$ ft. $A'' = -2 \Rightarrow$ maximum area.

37. $V = \frac{1}{3}\pi r^2 h.$

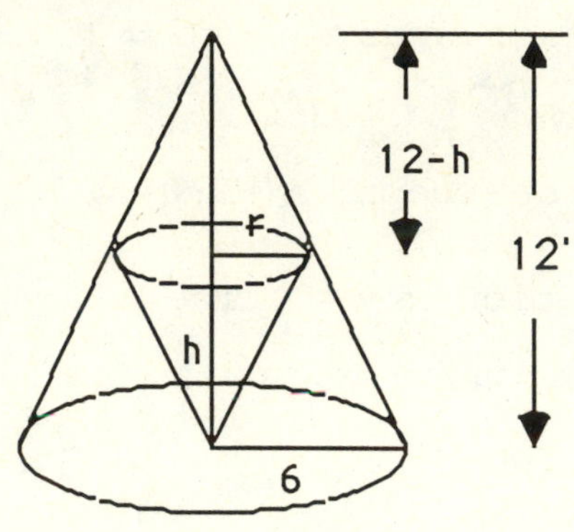

$\frac{r}{12 - h} = \frac{6}{12} \Rightarrow h = 2(6 - r)$

$V = \frac{2}{3}\pi r^2(6 - r)$

$= 4\pi r^2 - \frac{2}{3}\pi r^3, \quad 0 \le r \le 6$

$V' = 8\pi r - 2\pi r^2 = 2\pi r(4 - r) = 0 \Leftrightarrow r = 0$ or $r = 4$

$V(0) = V(6) = 0.$ $\therefore$ $r = 4$ and $h = 4$ give maximum volume.

39. $12y = 36 - x^2 \Rightarrow y = \frac{36 - x^2}{12}.$

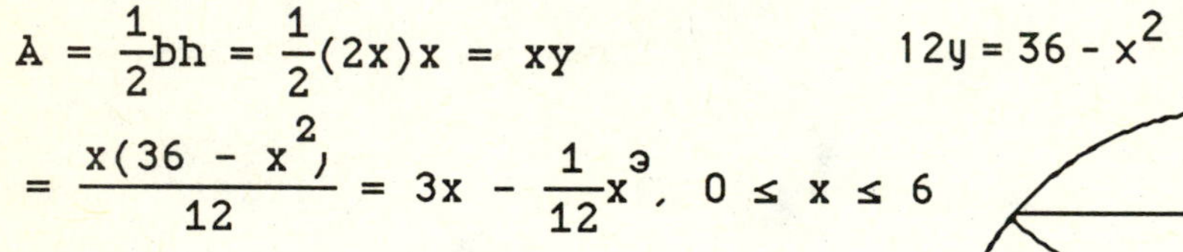

$A = \frac{1}{2}bh = \frac{1}{2}(2x)x = xy$

$= \frac{x(36 - x^2)}{12} = 3x - \frac{1}{12}x^3, \quad 0 \le x \le 6$

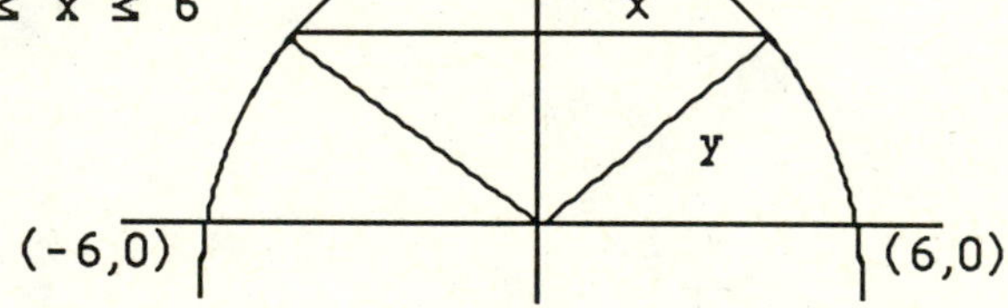

$A' = 3 - \frac{1}{4}x^2 \Rightarrow x = \pm 2\sqrt{3}.$

$A(0) = A(6) = 0.$

$\therefore A(2\sqrt{3}) = 4\sqrt{3}$ is largest triangle.

41. $x^2 - y^2 = 1$ is a hyperbola with vertices $(\pm 1, 0)$. A restraint on x is that $|x| \ge 1$. Let s be the distance from the point (x,y) to the point $P(a,0)$. Then we have

$s = \sqrt{(x - a)^2 + y^2} = \sqrt{(x - a)^2 + (x^2 - 1)}.$

$$\frac{ds}{dx} = \frac{1}{2}\left[(x-a)^2 + (x^2-1)\right]^{-\frac{1}{2}}(2(x-a)+2x) =$$

$$\frac{4x-2a}{2\sqrt{(x-a)^2+(x^2-1)}}. \quad \frac{ds}{dx} = 0 \Leftrightarrow x = \frac{a}{2}.$$

(a) If $a = 4$ then $x = 2$ and $y = \pm\sqrt{3}$. $\frac{ds}{dx} > 0$ if $x > 2$

and $\frac{ds}{dx} < 0$ if $x < 2 \Rightarrow$ distance is a minimum.

(b) If $a = 2$ then $x = 1$ and $y = 0$, which is the vertex and hence the closest.

(c) If $a = \sqrt{2}$ then $x = \frac{\sqrt{2}}{2} < 1$ (not on hyperbola) so again the vertex $(1,0)$ is closest

45. Let k and p be the fixed base and perimeter of the triangle. Let x and y denote the other two sides. Then $p = x + y + k$ and $s = \frac{p}{2}$. $A = \sqrt{s(s-x)(s-y)(s-k)} = \sqrt{\frac{p}{2}(\frac{p}{2}-x)(\frac{p}{2}-y)(\frac{p}{2}-k)} =$

$$\frac{1}{4}\sqrt{p(p-2x)(p-2y)(p-2k)} = \frac{\sqrt{p(p-2k)}}{4}\sqrt{(p-2x)(p-2(p-x-k))} =$$

$$K\sqrt{(p-2x)(2x+2k-p)}.$$

$$A' = \frac{K}{2} \bullet \frac{2(p-2x)-2(2x+2k-p)}{\sqrt{(p-2x)(2x+2k-p)}} = K\frac{2p-4x-2k}{\sqrt{(p-2x)(2x+2k-p)}}.$$

$$A' = 0 \Leftrightarrow 2p - 4x - 2k = 0 \Leftrightarrow x = \frac{p-k}{2} \Rightarrow y = \frac{p-k}{2}.$$

$0 < x < \frac{p-k}{2} \Rightarrow A' > 0$ and $\frac{p-k}{2} < x \Rightarrow$ maximum area

47. $|PQ| = \sqrt{|OP|^2 + |OQ|^2} = \sqrt{x^2 + y^2} = \sqrt{x^2 + \left(\frac{bx}{x - a}\right)^2}$

$$d'(x) = \frac{1}{2}\left(x^2 + \frac{b^2x^2}{(x - a)^2}\right)^{-\frac{1}{2}}\left[2x + 2\left(\frac{bx}{x - a}\right)\left(\frac{b(x - a) - bx}{(x - a)^2}\right)\right]$$

$$= \frac{1}{2\sqrt{\frac{x^2(x - a)^2 + b^2x^2}{(x - a)^2}}} \bullet \frac{2x(x - a)^3 - 2ab^2x}{(x - a)^3}$$

$$= \frac{x[(x - a)^3 - ab^2]}{\sqrt{x^2(x - a)^2 + b^2x^2}} = 0 \Leftrightarrow x = 0 \text{ or } (x - a)^3 - ab^2 = 0$$

$$(x - a)^3 = ab^2$$

$$x = a + (ab^2)^{\frac{1}{3}}.$$

$x = 0$ is out of the domain. $d' > 0$ for values to the right and negative for values to the left of this root, giving a minimum value. Then

$$y = b + \frac{ab}{x - a} = b + \frac{ab}{a + (ab^2)^{\frac{1}{3}} - a} = b + (a^2b)^{\frac{1}{3}}.$$

$$x^2 = a^2 + 2a(ab^2)^{\frac{1}{3}} + (ab^2)^{\frac{2}{3}} \text{ and } y^2 = b^2 + 2b(a^2b)^{\frac{1}{3}} + (a^2b)^{\frac{2}{3}}$$

$$a^2 + 2a^{\frac{4}{3}}b^{\frac{2}{3}} + a^{\frac{2}{3}}b^{\frac{4}{3}} \qquad b^2 + 2b^{\frac{4}{3}}a^{\frac{2}{3}} + a^{\frac{4}{3}}b^{\frac{2}{3}}$$

$$x^2 + y^2 = a^2 + 3a^{\frac{4}{3}}b^{\frac{2}{3}} + 3a^{\frac{2}{3}}b^{\frac{4}{3}} + b^2 = (a^{\frac{2}{3}} + b^{\frac{2}{3}})^{\frac{3}{2}}.$$

(b) $|OP| + |OQ| = x + y = x + \frac{bx}{x - a} = s(x)$.

$$s'(x) = 1 + \frac{b(x - a) - bx}{(x - a)^2} = \frac{(x - a)^2 - ab}{(x - a)^2} = 0 \text{ if } x = a + \sqrt{ab}.$$

$y = b + \frac{ab}{a + \sqrt{ab} - a} = b + \sqrt{ab}$. y' changes from negative to positive $\Rightarrow$ this is a minimum.

Then $x + y = a + 2\sqrt{ab} + b = (\sqrt{a} + \sqrt{b})^2$

(c) $|OP||OQ| = xy = x(\frac{bx}{x - a}) = \frac{bx^2}{x - a} = p(x)$.

$$p'(x) = \frac{(x - a)(2bx) - bx^2}{(x - a)^2} = \frac{2bx^2 - 2abx - bx^2}{(x - a)^2} = \frac{bx(x - 2a)}{(x - a)^2} = 0$$

if $x = 2a$. y' changes from negative to positive $\Rightarrow$ minimum.

$y = b + \frac{ab}{2a - a} = 2b$. $xy = 4ab$.

49. Let $S = (x_1 - x)^2 + (y_1 - y)^2$ be the square of the distance between $P_1(x_1, y_1)$ and any point $Q(x, y)$ on line L. Then

$\frac{dS}{dx} = -2\ (x_1 - x) - 2(y_1 - y)\frac{dy}{dx}$. Since P is on the line

$y = -\frac{a}{b}x - \frac{c}{b}$, $\frac{dy}{dx} = -\frac{a}{b}$ so that $\frac{dS}{dx} = -2\ (x_1 - x) + 2(y_1 - y)\frac{a}{b}$. Now,

let $P(x_0, y_0)$ be the point on L closest to P_1, so that $\frac{dS}{dx} = 0$.

Then $2(y_1 - y_0)\frac{a}{b} = 2(x_1 - x_0)$ or $\frac{y_1 - y_0}{x_1 - x_0} = \frac{b}{a}$. Thus, the slope of

PP_1 is the negative reciprocal of the slope of L, and $PP_1 \perp L$.

(b) $S = (x_1 - x_0)^2 + (y_1 - y_0)^2 = (x_1 - x_0)^2 + \frac{b^2}{a^2}(x_1 - x_0)^2$

$= (x_1 - x_0)^2 \left(\frac{a^2 + b^2}{a^2}\right)$. Now $P(x_0, y_0)$ satisfies both

the equation of L: $ax_0 + by_0 + c = 0$ and PP_1: $y_1 - y_0 = \frac{b}{a}(x_1 - x_0)$.

Substituting, $ax_0 + b\left[y_1 - \frac{b}{a}(x_1 - x_0)\right] + c = 0 \Rightarrow$

$a^2x_0 + [a^2x_1 - a^2x_1] + aby_1 - b^2(x_1 - x_0) + ac = 0 \Rightarrow$

$-a^2(x_1 - x_0) + a(ax_1 + by_1 + c) - b^2(x_1 - x_0) = 0 \Rightarrow$

$(a^2 + b^2)(x_1 - x_0) = a(ax_1 + by_1 + c)$. Hence

$S = \frac{a^2 + b^2}{a^2}\left[\frac{a(ax_1 + by_1 + c)}{a^2 + b^2}\right]^2 = \frac{ax_1 + by_1 + c}{a^2 + b^2}$. Since $S = d^2$,

$$d = \frac{|ax_1 + by_1 + c|}{\sqrt{a^2 + b^2}}.$$

51. Given that $ax + \frac{b}{x} \geq c$ where a,b, and c are positive constants. Let $f(x) = ax + \frac{b}{x} \Rightarrow f'(x) = a - \frac{b}{x^2}$ and $f''(x) = \frac{2b}{x^3}$. $f'(x) = 0 \Leftrightarrow x = \sqrt{\frac{b}{a}}$. $f'' > 0$ for $x > 0$ so this value gives a minimum, that is

$$f\left(\sqrt{\frac{b}{a}}\right) > ax + \frac{b}{x} \geq c \Rightarrow a\sqrt{\frac{b}{a}} + b\sqrt{\frac{a}{b}} \geq c \Rightarrow 2\sqrt{ab} \geq c \Rightarrow$$

$$4ab \geq c^2 \Rightarrow ab \geq \frac{c^2}{4}.$$

53. If $f(x) = ax^2 + 2bx + c$, then $f'(x) = 2ax + 2b = 0 \Leftrightarrow x = -\frac{b}{a}$. If $a > 0$, then $f'(x) = 2a > 0$ and there is a minimum value at $x = -\frac{b}{a}$.

Therefore, for any x, $f(x) \geq f\left(-\frac{b}{a}\right) = a\left(-\frac{b}{a}\right)^2 + 2b\left(-\frac{b}{a}\right) + c =$ $-b^2 + ac \geq 0 \Rightarrow b^2 - ac \leq 0$.

55. $B^2 = AC$ only if $F(x_0) = 0$ for some x_0, where $F(x)$ is as in Problem 54. Then $f(x_0) = (a_1x_0 + b_1)^2 + \ldots + (a_nx_0 + b_n)^2 = 0$, so that $ax_i + b_i = 0$ for each i.

57. By similar triangles, $\frac{12 - y}{12} = \frac{x}{36}$, where y is the height of the rectangle and x is the length of theside which lies on the base of the triangle. Then $A = xy = x\left(12 - \frac{1}{3}x\right)$, and $A' = 12 - \frac{2}{3}x$. $A' = 0$ if $x = 18$. $A'' = -\frac{2}{3} < 0 \Rightarrow A(18)$ is maximum.

The dimensions and 18 by 6.

59. If $d = |\sin t - \sin\left(t + \frac{\pi}{3}\right)|$, then $d = 0$ when $t = \frac{\pi}{3}, \frac{4\pi}{3}, \ldots$

Otherwise, $d' = \cos t - \cos\left(t + \frac{\pi}{3}\right) = 0$ when $t = -\frac{\pi}{6}, \frac{5\pi}{6}, \ldots$ and for these values, $d = 1$.

61. $A = \frac{1}{2}r^2\theta$ and $p = r\theta + 2r = \frac{2A}{r} + 2r$. $p' = -\frac{2A}{r^2} + 2 = 0 \Leftrightarrow$ $r = \sqrt{A}$. $p'' = \frac{4A}{r^3} > 0 \Rightarrow$ minimum value. If $= \sqrt{A}$, $\theta = 2$ radians.

63. (a) $A = \pi(r_2^2 - r_1^2) \Rightarrow \frac{dA}{dt} = 2\pi\left(r_2 \frac{dr_2}{dt} - r_1 \frac{dr_1}{dt}\right) = 2\pi[6(.01) - 4(.02)] = -0.04$

(b) $r_1 = 3 + at$, $r_2 = 5 + bt$ and we want t for which $\frac{dA}{dt} = 0$.

$2\pi[(5 + bt)(b) - (3 + at)(a)] = 0 \Leftrightarrow (b^2 - a^2)t = 3a - 5b$ or

$t = \frac{3a - 5b}{b^2 - a^2}$. $\frac{d^2A}{dt^2} = b^2 - a^2 < 0$ since $\frac{3}{5}a < b < a \Rightarrow$ a maximum value.

65. $s(t) = at - (1 + a^4)t^2 \Rightarrow \frac{ds}{dt} = a - 2t(1 + a^4) = 0 \Leftrightarrow t = \frac{a}{2(1 + a^4)}$.

For this t, $s = a\left(\frac{a}{2(1 + a^4)}\right) - (1 + t^4)\left(\frac{a}{2(1 + a^4)}\right)^2 = \frac{a^2}{4(1 + a^4)} > 0$.

For $t > \frac{a}{2(1 + a^4)}$, $\frac{ds}{dt} < 0$ and the particle moves back. If you

consider $s(a) = \frac{a^2}{4(1 + a^4)}$, then $\frac{ds}{da} = \frac{2a(1 + a^4) - 4a^5}{4(1 + a^4)^2} = 0 \Leftrightarrow$

$2a - 2a^5 = 0 \Leftrightarrow a(1 - a^4) = 0 \Leftrightarrow a = \pm 1, 0$ and $s(1) = \frac{1}{8}$.

67. If $f(x) = (c_1 - x)^2 + (c_2 - x)^2 + \ldots + (c_n - x)^2$, then

$f'(x) = -2(c_1 - x) - 2(c_2 - x) - \ldots - 2(c_n - x) = 0 \Leftrightarrow$

$nx = c_1 + c_2 + \ldots + c_n$ or $x = \frac{1}{n}(c_1 + c_2 + \ldots + c_n)$. $f''(x) = 2n$

$\Rightarrow$ minimum value. Notice that this is just the arithmetic mean.

69. Let $M = a_1 a_2 \ldots a_{n-1}$ and $N = a_1 + a_2 + \ldots + a_{n-1}$.

Define $f(x) = \frac{\frac{N + x}{n}}{Mx^{\frac{1}{n}}} = \frac{N + x}{Mn} x^{-\frac{1}{n}} = \frac{N}{Mn} x^{-\frac{1}{n}} + \frac{1}{Mn} x^{1 - \frac{1}{n}}$.

$f'(x) = \frac{1}{Mn}\left[-\frac{N}{n} x^{-\frac{1}{n} - 1} + \left(1 - \frac{1}{n}\right) x^{-\frac{1}{n}}\right] = \frac{1}{Mn^2} x^{-\frac{1}{n} - 1}\left[-N + (n-1)x\right]$

$f'(x) = 0$ if $-N + (n-1)x = 0 \Leftrightarrow x = \frac{N}{n-1}$.

That is, x is the arithmetic mean.

71. $V = \frac{dx}{dt} = f(x) \Rightarrow \frac{dv}{dx} = f'(x)$.

$a = \frac{d^2x}{dt^2} = \frac{dv}{dt}$. Since $\frac{dv}{dt} = \frac{dv}{dx} \bullet \frac{dx}{dt}$, we have $a = f'(x)f(x)$

73. $V = x^3 \Rightarrow \frac{dV}{dx} = 3x^2 \frac{dx}{dt}$. $300 = 3(20)^2 \frac{dx}{dt} \Rightarrow \frac{dx}{dt} = \frac{1}{4}$ in/min.

75. (a) $V = \frac{4}{3}\pi r^3 \Rightarrow \frac{dV}{dt} = 4\pi r^2 \frac{dr}{dt}$. $\frac{dr}{dt} = \frac{-12\pi}{4\pi(20)^2} = -\frac{3}{400}$ ft/min

$S = 4\pi r^2 \Rightarrow \frac{dS}{dt} = 8\pi r \frac{dr}{dt}$. $\frac{dS}{dt} = 8\pi(20)(-\frac{3}{400}) = -\frac{6\pi}{5}\frac{ft^2}{min}$

If $f(x) \geq 0$ for all x then $f(-\frac{b}{a}) \geq 0$. Conversely, suppose $b^2 - c \leq 0 \Rightarrow -b^2 + c \geq 0$. Then $f(x) \geq f(-\frac{b}{a}) \geq 0$.

(b) $\Delta r \approx dr = \frac{dr}{dt}\Delta r = -\frac{3}{400}\frac{ft}{min} \bullet \frac{6\ sec}{1} \bullet \frac{1\ min}{60\ sec} = -\frac{3}{4000}$ ft

$\Delta S \approx \frac{dS}{dt}\Delta t = -\frac{6\pi}{5}\frac{ft^2}{min} \bullet \frac{6\ sec}{1} \bullet \frac{1\ min}{60\ sec} = -\frac{3\pi}{25}$ ft^2

77. Let $s = \sqrt{x^2 + y^2} = \sqrt{x^2 + x^3}$. $\frac{ds}{dt} = \frac{1}{2}(x^2 + x^3)^{-\frac{1}{2}}(2x + 3x^2)\frac{dx}{dt}$

$\frac{dx}{dt}(2^2 + 3(2)^2)(\frac{1}{2})(4 + 8)^{-\frac{1}{2}} = 2$ $\frac{8}{\sqrt{12}}\frac{dx}{dt} = 2$ $\frac{dx}{dt} = \frac{\sqrt{3}}{2}$.

79. By similar triangles, $\frac{x}{y} = \frac{5}{10}$ or $y = \frac{1}{2}x$. The volume is then $V = \frac{1}{3}\pi x^2 y = \frac{1}{12}\pi y^3$ and $\frac{dV}{dt} = \frac{\pi}{4}y^2\frac{dy}{dt}$. The net change in volume is $\frac{dV}{dt} = c - 0.08\sqrt{y}$ ft^3/min. At the moment when $y = \frac{25}{4}$ and $\frac{dy}{dt} = .02$ ft/min,

$\frac{\pi}{4}\left(\frac{25}{4}\right)^2(.02) = c - .08\sqrt{\frac{25}{4}}$ or $c = \frac{25\pi}{128} + \frac{1}{5} \approx 0.8$. For $0 \leq y \leq 10$, $\frac{dV}{dt} > 0$, so the tank will fill.

81. Let the altitude fromA to BC = h, and |BC| = a. Let K = area of BCED. Then $K = \frac{1}{2}x(y - a)$. By similar triangles,

$\frac{y}{h - x} = \frac{a}{h}$, so $y = \frac{a}{h}(h - x)$ and $K = \frac{1}{2}x\left(\frac{a}{h}(h - x) - a\right) = \frac{a}{2h}(2hx - x^2)$.

Then $\frac{dK}{dx} = \frac{a}{2h}(2h - 2x) = y$.

83. Let x = distance ship A traveled, y = distance ship B traveled and s = distance between them. Then $s^2 = x^2 + y^2 - 2xy\cos 60^\circ$ or $s = \sqrt{x^2 + y^2 - xy}$. Then $\frac{ds}{dt} = \frac{2x\frac{dx}{dt} + 2y\frac{dy}{dt} - x\frac{dy}{dt} - y\frac{dx}{dt}}{2\sqrt{x^2 + y^2 - xy}}$. Evaluating, afer 2 hours, $\frac{ds}{dt} = \frac{65}{\sqrt{13}} = 5\sqrt{13}$ mph.

85. If $(y+1)^3 = x^2$, then $3(y+1)^2\frac{dy}{dx} = 2x$ or $\frac{dy}{dx} = \frac{2x}{3(y+1)^2}$.

When $x = 0$, $y = -1$ and $\frac{dy}{dx}$ does not exist. Therefore, Rolle's Theorem does not apply since $\frac{dy}{dx}$ does not exist for all x in $(-1, 1)$.

87. If $y = \sin x \sin(x+2) - \sin^2(x+1)$, $-\pi \le x \le \pi$, then

$$\frac{dy}{dx} = [\sin x\cos(x+2) + \sin(x+2)\cos x] - 2\sin(x+1)\cos(x+1)$$

$$= \sin(x + x + 2) - \sin 2(x+1) = 0.$$ Therefore y is a constant function and the graph is a horizontal line. If $x = \pi$, $y \approx 0.7$.

89. By the Mean Value Theorem, $\frac{f(6) - f(0)}{6} = f'(c)$ for some c in $(0, 6)$ or $f(6) - f(0) = 6f'(c) \le 6(2) = 12$.

91. (a) $(x+1)^2 \ge 0 \Rightarrow x^2 + 1 \ge -2x \Rightarrow 1 \ge -2\left(\frac{x}{x^2+1}\right) \Rightarrow -\frac{1}{2} \le \frac{x}{x^2+1}$.

$(x-1)^2 \ge 0 \Rightarrow x^2 + 1 \ge 2x \Rightarrow 1 \ge 2\left(\frac{x}{x^2+1}\right) \Rightarrow \frac{x}{x^2+1} \le \frac{1}{2}$

(b) By the Mean Value Theorem, $|f(b) - f(a)| \le |f'(c)|\,|(b-a)|$ for some c in $[a,b]$. By part (a), $f'(c) = \frac{c}{c^2+1} \le \frac{1}{2}$.

Therefore, $|f(b) - f(a)| \le \frac{1}{2}|b - a|$

93. (a) $\lim_{x\to\infty} \frac{\sqrt{x+5}}{\sqrt{x}+5} = \lim_{x\to\infty} \frac{\sqrt{1 + \frac{5}{x}}}{1 + \frac{5}{\sqrt{x}}} = 1$

(b) $\lim_{x\to\infty} \frac{2x}{x + 7\sqrt{x}} = \lim_{x\to\infty} \frac{2}{1 + \frac{7}{\sqrt{x}}} = 2$

95. (a) $F(b) = 0$ by definition of k. Direct substitution shows that $F(a) = f(a) - f(a) - (a-1)f'(a) - \ldots - k(a-a)^{n+1} = 0$

(b) $F^{(k)}(a) = f^{(k)}(a) - f^{(k)}(a) - f^{(k+1)}(a-a) - \ldots - k(n+1)(n)(n-1)\ldots(a-a)^{n+1-k}$, $= 0$, $k = 1, 2, \ldots, n$.

(c) Since $F(a) = F(b) = 0$, by Rolle's Theorem there exists a c_1 such that $a < c_1 < b$ and $F'(c_1) = 0$. Since $F'(a) = F'(c_1) = 0$, there exists a c_2 such that $a < c_2 < c_1$ and $F''(c_2) = 0$. In a there exists $c_3, c_4, \ldots, c_n$, $a < c_{n+1} < c_n < c_{n-1} < \ldots < c_1 < b$ such that $0 = F''(c_1) = F''(c_2) = \ldots = F^{(n+1)}(c_{n+1})$.

(d) $F^{(n+1)}(x) = f^{(n+1)}(x) - k(n+1)!$. Since $0 = F^{(n+1)}(c_{n+1})$,

$f^{(n+1)}(c_{n+1}) - k(n+1)!$, and $k = \dfrac{f^{(n+1)}(c_{n+1})}{(n+1)!}$.

97. (a) $$x_{n+1} = x_n - \frac{f(x_n)}{f'(x_n)} = x_n - \frac{\frac{1}{x_n} - a}{-\left(\frac{1}{x_n}\right)^2} = x_n + x_n - ax_n^2 = x_n(2 - ax_n)$$

(b) $x_1 = x_0(2 - ax_0) = ax_1\left(\frac{2}{a} - x_0\right)$. Therefore, $x_0 > \frac{2}{a}$ would make $x_1 < 0$ and $0 < x_0 < \frac{2}{a} \Rightarrow x_1 > 0$. Equality would make all of all of the iterations zero. The appropriate interval is $0 < x_0 < \frac{2}{a}$.

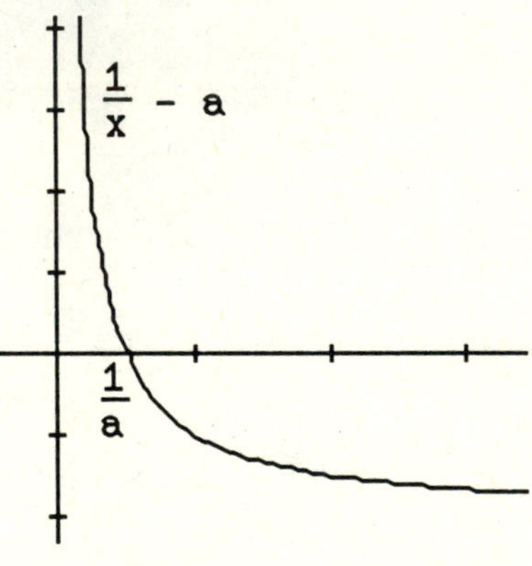

CHAPTER 4

INTEGRATION

4.1 INDEFINITE INTEGRALS

1. (a) $\int 2x\,dx = x^2 + C$ (b) $\int 3\,dx = 3x + C$ (c) $\int (2x+3)\,dx = x^2 + 3x + C$

3. (a) $\int 3x^2\,dx = x^3 + C$ (b) $\int x^2\,dx = \frac{1}{3}x^3 + C$ (c) $\int (x^2 + 2x)\,dx = \frac{1}{3}x^3 + x^2 + C$

5. (a) $\int -3x^{-4}\,dx = x^{-3} + C$ (b) $\int x^{-4}\,dx = -\frac{1}{3}x^{-3} + C$

   (c) $\int (x^{-4} + 2x + 3)\,dx = -\frac{1}{3}x^{-3} + x^2 + 3x + C$

7. (a) $\int \frac{3}{2}\sqrt{x}\,dx = x^{\frac{3}{2}} + C$ (b) $\int 4\sqrt{x}\,dx = \frac{8}{3}x^{\frac{3}{2}} + C$

   (c) $\int (x^2 - 4\sqrt{x})\,dx = \frac{1}{3}x^3 - \frac{8}{3}x^{\frac{3}{2}} + C$

9. (a) $\int (2x-1)\,dx = x^2 - x + C$ (b) $\int (2x-1)^2\,dx = \frac{1}{6}(2x-1)^3 + C$

   (c) $\int (2x-1)^3\,dx = \frac{1}{8}(2x-1)^4 + C$

11. (a) $\int (x^2-3)\,2x\,dx = (x^2-3)^2 + C$ (b) $\int (x^2-3)\,x\,dx = \frac{1}{4}(x^2-3) + C$

13. $\frac{dy}{dx} = 2x - 7 \Rightarrow y = \int (2x-7)\,dx = x^2 - 7x + C$

15. $\frac{dy}{dx} = x^2 + 1 \Rightarrow y = \int (x^2+1)\,dx = \frac{1}{3}x^3 + x + C$

17. $\frac{dy}{dx} = \frac{-5}{x^2}$, $x > 0$, $\Rightarrow y = \int \frac{-5}{x^2}\,dx = 5x^{-1} + C$

19. $\frac{dy}{dx} = (x-2)^4 \Rightarrow y = \int (x-2)^4\,dx = \frac{1}{5}(x-2)^5 + C$

21. $\frac{dy}{dx} = \frac{1}{x^2} + x$, $x > 0 \Rightarrow y = \int (x^{-2} + x)\,dx = -x^{-1} + \frac{1}{2}x^2 + C$

23. $\frac{dy}{dx} = \frac{x}{y}$, $y > 0 \Rightarrow y\,dy = x\,dx \Rightarrow \int y\,dy = \int x\,dx \Rightarrow \frac{1}{2}y^2 = \frac{1}{2}x^2 + C_1$

   $y^2 = x^2 + C$ or $y = \sqrt{x^2 + C}$, $y > 0$

25. $\frac{dy}{dx} = \frac{x+1}{y-1}$, $y > 1 \Rightarrow \int (y-1)\,dy = \int (x+1)\,dx \Rightarrow$

   $\frac{1}{2}(y-1)^2 = \frac{1}{2}(x+1)^2 + C_1 \Rightarrow (y-1)^2 = (x+1)^2 + C$

27. $\frac{dy}{dx} = \sqrt{xy}, x > 0, y > 0 \Rightarrow \int y^{-\frac{1}{2}}\,dy = \int x^{\frac{1}{2}}\,dx \Rightarrow$

$2y^{\frac{1}{2}} = \frac{2}{3}x^{\frac{3}{2}} + C_1 \Rightarrow y^{\frac{1}{2}} = \frac{1}{3}x^{\frac{3}{2}} + C$

29. $\frac{dy}{dx} = 2xy^2, y > 0 \Rightarrow y^{-2}\,dy = 2x\,dx \Rightarrow -y^{-1} = x^2 + C$ or $y = -\frac{1}{x^2 + C}$

31. $x^3 \frac{dy}{dx} = -2, x > 0 \Rightarrow dy = -2x^{-3}\,dx \Rightarrow y = x^{-2} + C$

33. $\frac{ds}{dt} = 3t^2 + 4t - 6 \Rightarrow \int ds = \int (3t^2 + 4t - 6)dt \Rightarrow s = t^3 + 2t^2 - 6t + C$

35. $\frac{dy}{dt} = (2t + t^{-1})^2, t > 0 \Rightarrow \int dy = \int (4t^2 + 4 + t^{-2})\,dt \Rightarrow$

$y = \frac{4}{3}t^3 + 4t - t^{-1} + C$

37. $a = \frac{dv}{dt} = 1.6 \text{ m/sec}^2 \Rightarrow v = \int 1.6dt = 1.6t + C$. $v_o = 0 \Rightarrow C = 0$.

$v(30) = 1.6(30) = 48$ m/sec.

4.2 SELECTING A VALUE FOR THE CONSTANT OF INTEGRATION

1. $v = 3t^2, s_o = 4 \Rightarrow s = t^3 + C$. $4 = 0 + C \Rightarrow C = 4$. $\therefore s = t^3 + 4$

3. $v = (t+1)^2, s_o = 0 \Rightarrow s = \frac{1}{3}(t+1)^3 + C$. $0 = \frac{1}{3}(1)^3 + C \Rightarrow C = -\frac{1}{3}$.

$\therefore s = \frac{1}{3}(t+1)^3 - \frac{1}{3}$

5. $v = (t+1)^{-2}, s_o = -5 \Rightarrow s = -(t+1)^{-1} + C$. $-5 = -(-1)^{-1} + C \Rightarrow C = -4$

$\therefore s = -(t+1)^{-1} - 4$

7. $a = 9.8, v_o = 20, s_o = 0$. $a = \frac{dv}{dt} \Rightarrow v = 9.8t + C_1$. $20 = 0 + C_1 \Rightarrow$

$C_1 = 20$ and $v = \frac{ds}{dt} = 9.8t + 20$. $s = 4.9t^2 + 20t + C_2$.

$0 = 0 + C_2 \Rightarrow C_2 = 0$. $\therefore s = 4.9t^2 + 20t$.

9. $a = 2t, v_o = 1, s_o = 1$. $a = \frac{dv}{dt} \Rightarrow v = t^2 + C_1$. $1 = 0 + C_1 \Rightarrow$

$C_1 = 1$ and $v = \frac{ds}{dt} = t^2 + 1$. $s = \frac{1}{3}t^3 + t + C_2$.

$1 = 0 + C_2 \Rightarrow C_2 = 1$. $\therefore s = \frac{1}{3}t^3 + t + 1$.

11. $a = 2t + 2$, $v_0 = 1$, $s_0 = 0$. $v = t^2 + 2t + C_1$. $1 = 0 + C_1 \Rightarrow C_1 = 1$.

$v = t^2 + 2t + 1 \Rightarrow s = \frac{1}{3}t^3 + t^2 + t + C_2$. $0 = 0 + C_2 \Rightarrow C_2 = 0$.

$\therefore s = \frac{1}{3}t^3 + t^2 + t$.

13. $\frac{dy}{dx} = 3x^2 + 2x + 1$, $y = 0$ when $x = 1$. $y = x^3 + x^2 + x\ C$.

$0 = 1 + 1 + 1 + C \Rightarrow C = -3$. $\therefore y = x^3 + x^2 + x - 3$

15. $\frac{dy}{dx} = 4(x - 7)^3$, $y = 10$ when $x = 8$. $y = (x - 7)^4 + C$

$10 = 1 + C \Rightarrow C = 9$. $\therefore y = (x - 7)^4 + 9$

17. $\frac{dy}{dx} = x\sqrt{y}$, $y = 1$ when $x = 0$. $y^{-\frac{1}{2}}\,dy = x\,dx \Rightarrow 2y^{\frac{1}{2}} = \frac{1}{2}x^2 + C$

$2 = 0 + C \Rightarrow C = 2$. $\therefore 2y^{\frac{1}{2}} = \frac{1}{2}x^2 + 2$ or $y = \left(\frac{x^2}{4} + 1\right)^2$

19. $\frac{dy}{dx} = 2xy^2$, $y = 1$ when $x = 1 \Rightarrow y^{-2}\,dy = 2x\,dx \Rightarrow -y^{-1} = x^2 + C$

$-1 = 1 + C \Rightarrow C = -2$ $\therefore$ $\frac{1}{y} = 2 - x^2$ or $y = \frac{1}{2 - x^2}$

21. $\frac{d^2y}{dx^2} = 2 - 6x$, $y = 1$, $\frac{dy}{dx} = 4$ when $x = 0$.

$\frac{dy}{dx} = 2x - 3x^2 + C$. $4 = 0 - 0 + C \Rightarrow C = 4$. $\frac{dy}{dx} = 2x - 3x^2 + 4$.

$y = x^2 - x^3 + 4x + C$. $1 = 0 + C \Rightarrow C = 1$. $y = x^2 - x^3 + 4x + 1$.

23. $\frac{d^2y}{dx^2} = \frac{3x}{8}$, through $(4, 4)$ with slope 3.

$\frac{dy}{dx} = \frac{3}{16}x^2 + C$. $3 = \frac{3}{16}(4)^2 + C \Rightarrow C = 0$. $\frac{dy}{dx} = \frac{3}{16}x^2$.

$y = \frac{1}{16}x^3 + C$. $4 = \frac{4^3}{16} + C \Rightarrow C = 0$. $y = \frac{1}{16}x^3$.

25. $\frac{dy}{dx} = 3\sqrt{x} \Rightarrow y = 2x^{\frac{3}{2}} + C$. $(9, 4)$ belongs to graph so

$4 = 2(9)^{\frac{3}{2}} + C \Rightarrow C = -50$ and $f(x) = 2x^{\frac{3}{2}} - 50$.

27. $a = \frac{dv}{dt} = 9.8 \Rightarrow v = 9.8t + C$. $v_0 = 0 \Rightarrow C = 0$.
$v = \frac{ds}{dt} = 9.8t \Rightarrow s = 4.9t^2 + C$. $s_0 = 0 \Rightarrow C = 0$. He will enter the water when $4.9t^2 = 30$, or when $t = 2.47$ sec. His velocity at that time is $v(2.47) = 24.25$ m/s.

29. It cannot be choice (a) because $y(1) \neq 0$. The slope of tangent through $(1, 1)$ and $\left(\frac{1}{2}, 0\right)$ is 2. In (b) and (c), $y'(1) \neq 2$, so it cannot be either of these. It is (d), because $y = x^2 + C$ and $1 = 1 + C \Rightarrow C = 0$.

31. $a = v\frac{dv}{ds}$ and $F = ma \Rightarrow -\frac{mgR^2}{s^2} = mv\frac{dv}{ds}$. Then
$v\,dv = -\frac{gR^2}{s^2}\,ds$ or $\frac{1}{2}v^2 = \frac{gR^2}{s} + C$. At the surface of the earth,
$s = R$ and $v_0 = \sqrt{2gR}$. So $\frac{2gR}{2} = \frac{gR^2}{R} + C \Rightarrow C = 0$. Then
$v = \sqrt{\frac{2gR^2}{s}} = \sqrt{2gR}\sqrt{\frac{R}{s}} = v_0\sqrt{\frac{R}{s}}$. Since $\frac{ds}{dt} = v_0\sqrt{\frac{R}{s}}$,
$\sqrt{s}\,ds = v_0\sqrt{R}\,dt$ or $\frac{2}{3}s^{\frac{3}{2}} = v_0\sqrt{R}\,t + C$. $s(0) = R \Rightarrow C = \frac{2}{3}R^{\frac{3}{2}}$.

Then $s^{\frac{3}{2}} = \frac{3}{2}v_0\sqrt{R}\,t + R^{\frac{3}{2}} = R^{\frac{3}{2}}\left(1 + \frac{3v_0 t}{2R}\right)$.

4.3 SUBSTITUTION METHOD OF INTEGRATION

1. $\int (x-1)^{243}\,dx = \frac{1}{244}(x-1)^{244} + C$

3. $\int \frac{1}{\sqrt{1-x}}\,dx = \int -u^{-\frac{1}{2}}\,du = -2u^{\frac{1}{2}} + C = -2(1-x)^{\frac{1}{2}} + C$

   Let $u = 1 - x \Rightarrow du = -dx$

5. $\int x\sqrt{2x^2 - 1}\,dx = \int u^{\frac{1}{2}}\left(\frac{1}{4}\,du\right) = \frac{1}{4}\left(\frac{2}{3}u^{\frac{3}{2}}\right) + C = \frac{1}{6}(2x^2 - 1)^{\frac{3}{2}} + C$

   Let $u = 2x^2 - 1 \Rightarrow du = 4x\,dx \Rightarrow x\,dx = \frac{1}{4}\,du$

7. $\int (2-t)^{\frac{2}{3}}\,dt = \int -u^{\frac{2}{3}}\,du = -\frac{3}{5}u^{\frac{5}{3}} + C = -\frac{3}{5}(2-t)^{\frac{5}{3}} + C$

9. $\int (1+x^3)^2\,dx = \int (1+2x^3+x^6)\,dx = x + \frac{1}{2}x^4 + \frac{1}{7}x^7 + C$

11. $\int x(x^2+1)^{10}\,dx = \int u^{10}\left(\frac{1}{2}du\right) = \frac{1}{2}\left(\frac{1}{11}u^{11}\right) + C = \frac{1}{22}(x^2+1)^{11} + C$

Let $u = x^2+1 \Rightarrow du = 2x\,dx \Rightarrow x\,dx = \frac{1}{2}du$

13. $\int \frac{x^2}{\sqrt{1+x^3}}\,dx = \int u^{-1/2}\left(\frac{1}{3}du\right) = \frac{1}{3}(2u^{1/2}) + C = \frac{2}{3}(1+x^3)^{1/2} + C$

Let $u = 1+x^3 \Rightarrow du = 3x^2\,dx \Rightarrow x^2\,dx = \frac{1}{3}du$

15. $\int \frac{dx}{(3x+2)^2} = \int u^{-2}\left(\frac{1}{3}du\right) = -\frac{1}{3}u^{-1} + C = -\frac{1}{3}(3x+2)^{-1} + C$

Let $u = 3x+2 \Rightarrow du = 3\,dx \Rightarrow dx = \frac{1}{3}\,du$

17. $\int \frac{3r\,dr}{\sqrt{1-r^2}} = 3\int u^{-\frac{1}{2}}\left(-\frac{1}{2}du\right) = -\frac{3}{2}(2u^{\frac{1}{2}}) + C = -3\sqrt{1-r^2} + C$

Let $u = 1-r^2 \Rightarrow du = -2r\,dr \Rightarrow r\,dr = -\frac{1}{2}\,du$

19. $\int x^4(7-x^5)^3\,dx = -\frac{1}{5}\int u^3\,du = -\frac{1}{5}\left(\frac{1}{4}u^4\right) + C = -\frac{1}{20}(7-x^5)^4 + C$

Let $u = 7-x^5 \Rightarrow du = -5x^4\,dx \Rightarrow x^4\,dx = -\frac{1}{5}\,du$

21. $\int \frac{ds}{(s+1)^3} = \int (s+1)^{-3}\,ds = -\frac{1}{2}(s+1)^{-2} + C$

23. $\int \frac{1}{x^2+4x+4}\,dx = \int (x+2)^{-2}\,dx = -(x+2)^{-1} + C$

25. $\int \frac{x+1}{2\sqrt{x+1}}\,dx = \frac{1}{2}\int \frac{u}{\sqrt{u}}\,du = \frac{1}{2}\int u^{\frac{1}{2}}\,du = \frac{1}{3}u^{\frac{3}{2}} + C = \frac{1}{3}(x+1)^{\frac{3}{2}} + C$

Let $u = x+1 \Rightarrow du = dx$

27. $\int (y^3+6y^2+12y+8)(y^2+4y+4)\,dy = \int (y+2)^3(y+2)^2\,dy$

$= \int (y+2)^5\,dy = \frac{1}{6}(y+2)^6 + C$

29. $\int \frac{1}{\sqrt{x}(1+\sqrt{x})^2}\,dx = \int \frac{1}{u^2}\cdot 2du = \int 2u^{-2}\,du = -2u^{-1} + C = -2(1+\sqrt{x})^{-1} + C$

Let $u = 1+\sqrt{x} \Rightarrow du = \frac{1}{2\sqrt{x}}\,dx \Rightarrow \frac{dx}{\sqrt{x}} = 2du$

31. $\frac{dy}{dx} = x\sqrt{1+x^2}$, $y = 0$ when $x = 0$. $y = \int x\sqrt{1+x^2}\,dx$

$= \frac{1}{3}(1+x^2)^{3/2} + C$. $0 = \frac{1}{3}(1)^{3/2} + C \Rightarrow C = -\frac{1}{3}$. $\therefore y = \frac{1}{3}(1+x^2)^{3/2} - \frac{1}{3}$.

33. $\frac{dr}{dz} = 24z(3z^2 - 1)^3$, $r = -3$ when $z = 0$. $r = \int 24z(3z^2 - 1)^3\,dz$

$= (3z^2 - 1)^4 + C$. $-3 = (-1)^4 + C \Rightarrow C = -4$. $\therefore r = (3z^2 - 1)^4 - 4$.

35. $2y\frac{dy}{dx} = 3x\sqrt{x^2+1}\sqrt{y^2+1}$, $y = 0$ when $x = 0$.

$[(y^2+1)^{-\frac{1}{2}} y]\,dy = \frac{3}{2}x\sqrt{x^2+1}\,dx \Rightarrow (y^2+1)^{\frac{1}{2}} = \frac{1}{2}(x^2+1)^{\frac{3}{2}} + C.$

$(1)^{\frac{1}{2}} = \frac{1}{2}(1)^{\frac{3}{2}} + C \Rightarrow C = \frac{1}{2}$. $\therefore\ 2(y^2+1)^{\frac{1}{2}} = (x^2+1)^{\frac{3}{2}} + 1$

37. Only (a) and (c). You are never allowed to factor a variable term out of the integal sign.

4.4 INTEGRALS OF TRIGONOMETRIC FUNCTIONS

1. $\int \sin 3x\,dx = -\frac{1}{3}\cos 3x + C$

3. $\int \sec^2(x+2)\,dx = \tan(x+2) + C$

5. $\int \csc\left(x+\frac{\pi}{2}\right)\cot\left(x+\frac{\pi}{2}\right)dx = -\csc\left(x+\frac{\pi}{2}\right) + C$

7. $\int x\sin(2x^2)\,dx = \frac{1}{4}\int \sin u\,du = -\frac{1}{4}\cos u + C = -\frac{1}{4}\cos(2x^2) + C$

Let $u = 2x^2\,dx \Rightarrow du = 4x\,dx \Rightarrow x\,dx = \frac{1}{4}\,du$

9. $\int \sin 2t\,dt = -\frac{1}{2}\cos 2t + C$

11. $\int 4\cos 3y\,dy = \frac{4}{3}\sin 3y + C$

13. $\int \sin^2 x\cos x\,dx = \int u^2\,du = \frac{1}{3}u^3 + C = \frac{1}{3}\sin^3 x + C$

Let $u = \sin x \Rightarrow du = \cos x\,dx$

15. $\int \sec^2 2\theta\,d\theta = \frac{1}{2}\tan 2\theta + C$

17. $\int \sec\frac{x}{2}\tan\frac{x}{2}\,dx = 2\sec\frac{x}{2} + C$

19. $\int \frac{d\theta}{\sin^2\theta} = \int \csc^2\theta\,d\theta = -\cot\theta + C$

21. $\int \cos^2 y\, dy = \int \left(\frac{1}{2} + \frac{1}{2}\cos 2y\right) dy = \frac{1}{2}y + \frac{1}{4}\sin 2y + C$

23. $\int (1 - \sin^2 3t)\cos 3t\, dt = \int \cos 3t\, dt - \int \sin^2 3t \cos 3t\, dt$

$= \frac{1}{3}\sin 3t - \frac{1}{9}\sin^3 3t + C$

25. $\int \frac{\cos x\, dx}{\sin^2 x} = \int \cot x \csc x\, dx = -\csc x + C$

27. $\int \frac{\sin 2t\, dt}{\sqrt{2 - \cos 2t}} = \int (2 - \cos 2t)^{-\frac{1}{2}} \sin 2t\, dt = (2 - \cos 2t)^{\frac{1}{2}} + C$

29. $\int \cos^2 \frac{2x}{3} \sin \frac{2x}{3}\, dx = -\frac{1}{2}\cos^3 \frac{2x}{3} + C$

31. $\int \sec\theta (\sec\theta + \tan\theta)\, d\theta = \int (\sec^2\theta + \sec\theta\tan\theta)\, d\theta =$

$\tan\theta + \sec\theta + C$

33. $\int (\sec^2 y + \csc^2 y)\, dy = \tan y - \cot y + C$

35. $\int (3\sin 2x + 4\cos 3x)dx = -\frac{3}{2}\cos 2x + \frac{4}{3}\sin 3x + C$

37. $\int \tan^2 x \sec^2 x\, dx = \frac{1}{3}\tan^3 x + C$

39. $\int \cot^3 x \csc^2 x\, dx = -\frac{1}{4}\cot^4 x + C$

41. $\int \sqrt{\tan x}\, \sec^2 x\, dx = \frac{2}{3}\tan^{\frac{3}{2}} x + C$

43. $\int (\sin x)^{\frac{3}{2}} \cos x\, dx = \frac{2}{5}\sin^{\frac{5}{2}} x + C$

45. $\int \cos^3 x\, dx = \int (\cos^2 x)\cos x\, dx = \int (1 - \sin^2 x)\cos x\, dx =$

$\int \cos x\, dx - \int \sin^2 x \cos x\, dx = \sin x - \frac{1}{3}\sin^3 x + C.$

47. $\int \cos^5 x\, dx = \int (\cos^2 x)^2 \cos x\, dx = \int (1 - \sin^2 x)^2 \cos x\, dx =$

$\int (1 - 2\sin^2 x + \sin^4 x)\cos x\, dx = \sin x - \frac{2}{3}\sin^3 x + \frac{1}{5}\sin^5 x + C.$

49. $\int \cos^{-4} 2x \sin 2x\, dx = \frac{1}{6} \cos^{-3} 2x + C$

51. $\int \frac{\tan^2 \sqrt{x}}{\sqrt{x}}\, dx = 2\int \tan^2 u\, du = 2\int (\sec^2 u - 1)\, du = 2 \tan u - 2u + C$

$= 2 \tan \sqrt{x} - 2\sqrt{x} + C$

Let $u = \sqrt{x} \Rightarrow du = \frac{1}{2\sqrt{x}} dx \Rightarrow \frac{1}{\sqrt{x}} dx = 2\, du$

53. $\int \frac{x \operatorname{cox} \sqrt{3x^2 - 6}}{\sqrt{3x^2 - 6}}\, dx = \frac{1}{3}\int \cos u\, du = \frac{1}{3} \sin u\, du + C = \frac{1}{3} \sin \sqrt{3x^2 - 6} + C$

Let $u = \sqrt{3x^2 - 6} \Rightarrow du = \frac{3x\, dx}{\sqrt{3x^2 - 6}} dx$

55. (d) because $\frac{d}{dx}\left(\sec x - \frac{\pi}{4}\right) = \sec x \tan x$

57. $2y \frac{dy}{dx} = 5x - 3 \sin x$, $y = 0$ when $x = 0$.

$2y\, dy = (5x - 3 \sin x)\, dx \Rightarrow y^2 = \frac{5}{2}x^2 + 3 \cos x + C$. $0 = 0 + 3 + C \Rightarrow C = -3$.

$\therefore\ y^2 = \frac{5}{2} x^2 + 3 \cos x - 3$.

59. $\frac{dy}{dx} = \frac{\pi \cos \pi x}{\sqrt{y}}$, $y = 1$ when $x = \frac{1}{2}$

$\sqrt{y}\, dy = \pi \cos \pi x\, dx \Rightarrow \frac{2}{3}y^{\frac{3}{2}} = \sin \pi x + C$. $\frac{2}{3} - \sin\frac{\pi}{2} + C \Rightarrow C = -\frac{1}{3}$

$2y^{\frac{3}{2}} = 3 \sin \pi x - 1$ or $y = \left(\frac{3\sin \pi x - 1}{2}\right)^{\frac{2}{3}}$

61. $v = \frac{ds}{dt} = 6 \sin 2t$, $s = 0$ when $t = 0$. $s = \int 6 \sin 2t\, dt =$

$-3 \cos 2t + C$. $0 = -3 \cos 0 + C \Rightarrow C = 3 \Rightarrow s = -3 \cos 2t + 3$.

$s\left(\frac{\pi}{2}\right) = -3 \cos \pi + 3 = 6$ m.

63. All three are correct. The double-angle cosine formulas

$$\frac{1}{2} - \frac{1}{2} \cos 2x = \sin^2 x \text{ and } \frac{1}{2} + \frac{1}{2} \cos 2x = \cos^2 x$$

state that $\cos 2x$, $\sin^2 x$ and $\cos^2 x$ differ from each other by a constant.

65. (a) Let $u = x - 1$ and $du = dx$. Then

$$\int \sqrt{1 + \sin^2(x-1)}\sin(x-1)\cos(x-1)\,dx = \int \sqrt{1+\sin^2 u}\,\sin u \cos u\,du.$$

Let $v = \sin u$ and $dv = \cos u\,du$. Then

$$\int \sqrt{1+\sin^2 u}\,\sin u\cos u\,du = \int \sqrt{1+v^2}\;v\,dv.$$

Let $w = 1 + v^2$ and $dw = 2v\,dv$ or $v\,dv = \frac{1}{2}dw$. Then

$$\int \sqrt{1+v^2}\;v\,dv = \int \frac{1}{2}\sqrt{w}\,dw = \frac{1}{3}w^{\frac{3}{2}} + C = \frac{1}{3}(1+v^2)^{\frac{3}{2}} + C$$

$$= \frac{1}{3}(1+\sin^2 u)^{\frac{3}{2}} + C = \frac{1}{3}[1+\sin^2(x-1)]^{\frac{3}{2}} + C.$$

(b) Let $v = \sin(x-1)$ and $dv = \cos(x-1)\,dx$. Then

$$\int \sqrt{1+\sin^2(x-1)}\sin(x-1)\cos(x-1)\,dx = \int \sqrt{1+v^2}\;v\,dv.$$

Let $w = 1 + v^2$ and $dw = 2v\,dv$ or $v\,dv = \frac{1}{2}dw$. Then

$$\int \sqrt{1+v^2}\;v\,dv = \int \frac{1}{2}\sqrt{w}\,dw = \frac{1}{3}w^{\frac{3}{2}} + C = \frac{1}{3}(1+v^2)^{\frac{3}{2}} + C$$

$$= \frac{1}{3}[1+\sin^2(x-1)]^{\frac{3}{2}} + C.$$

4.5 DEFINITE INTEGRALS. THE AREA UNDER A CURVE

1. $$L = \left[f(0) + f\left(\frac{1}{4}\right) + f\left(\frac{1}{2}\right) + f\left(\frac{3}{4}\right)\right]\cdot\frac{1}{4}$$
$$= \left\{1 + \frac{3}{2} + 2 + \frac{5}{2}\right]\cdot\frac{1}{4}$$
$$= \frac{7}{4}$$

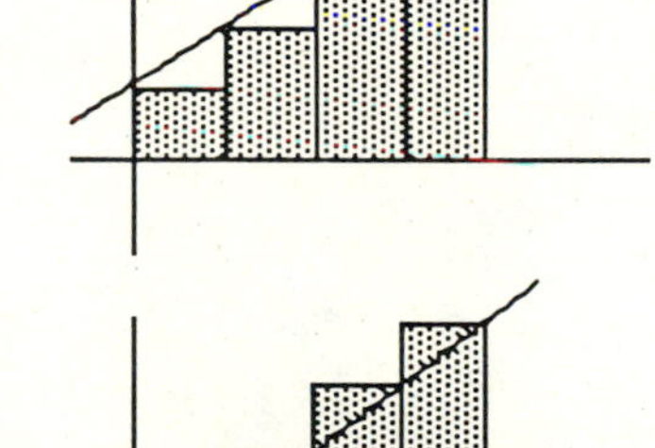

$$U = \left[\frac{3}{2} + 2 + \frac{5}{2} + 3\right]\cdot\frac{1}{4}$$
$$= \frac{9}{4}$$

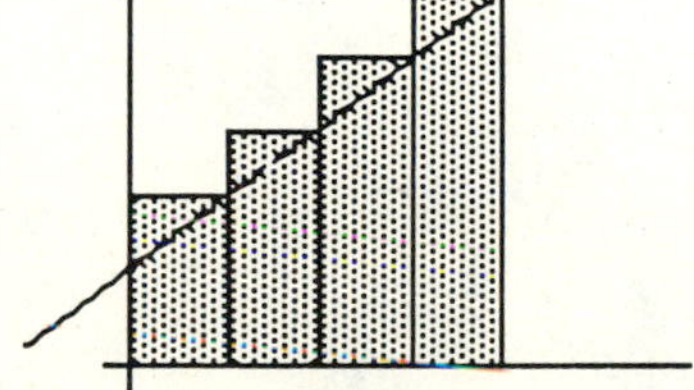

3. $L = \left(\sin 0 + \sin\frac{\pi}{4} + \sin\frac{3\pi}{4} + \sin\pi \right) \cdot \frac{\pi}{4}$

$= \left(\frac{\sqrt{2}}{2} + \frac{\sqrt{2}}{2} \right) \cdot \frac{\pi}{4} = \frac{\pi\sqrt{2}}{4}$

$U = \left(\sin\frac{\pi}{4} + \sin\frac{\pi}{2} + \sin\frac{\pi}{2} + \sin\frac{3\pi}{4} \right) \cdot \frac{\pi}{4}$

$= \frac{(2+\sqrt{2})\pi}{4}$

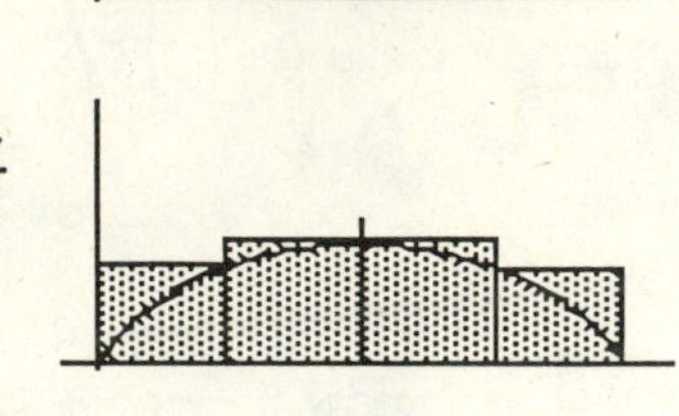

5. $L = (\sqrt{0} + \sqrt{1} + \sqrt{2} + \sqrt{3})(1)$

≈ 4.15

$U = (\sqrt{1} + \sqrt{2} + \sqrt{3} + \sqrt{4})(1)$

≈ 6.15

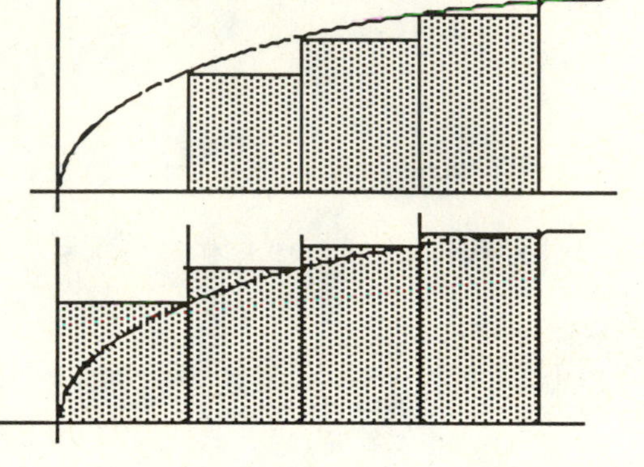

7. $\sum_{i=-1}^{3} 2^{i} = 2^{-1} + 2^{0} + 2^{1} + 2^{2} + 2^{3} = \frac{1}{2} + 1 + 2 + 4 + 8$

9. $\sum_{n=0}^{4} \frac{n}{4} = \frac{0}{4} + \frac{1}{4} + \frac{2}{4} + \frac{3}{4} + \frac{4}{4} = \frac{5}{2}$

11. $\sum_{m=0}^{5} \sin\frac{m\pi}{2} = \sin 0 + \sin\frac{\pi}{2} + \sin\pi + \sin\frac{3\pi}{2} + \sin 2\pi + \sin\frac{5\pi}{2} = 1$

13. $\sum_{i=0}^{3} (i^{2} + 5) = 5 + 6 + 9 + 14 = 34$

15. (a) $\sum_{j=2}^{7} 2^{j-2} = 2^{0} + 2^{1} + 2^{2} + 2^{3} + 2^{4} + 2^{5}$ (c) $\sum_{j=0}^{5} 2^{j} = 2^{0} + 2^{1} + 2^{2} + 2^{3} + 2^{4} + 2^{5}$

(b) $\sum_{k=0}^{5} 2^{k} = 2^{0} + 2^{1} + 2^{2} + 2^{3} + 2^{4} + 2^{5}$ (d) $\sum_{j=1}^{6} 2^{j-1} = 2^{0} + 2^{1} + 2^{2} + 2^{3} + 2^{4} + 2^{5}$

Therefore, all express the same sum.

17. (a) $\int_{1}^{9} -2f(x)\,dx = -2\int_{1}^{9} f(x)\,dx = -2\left(\int_{1}^{7} f(x)\,dx + \int_{7}^{9} f(x)\,dx \right) = -2\,(-1+5) = -8$

(b) $\int_{7}^{9} [\,2f(x) - h\,(x)\,]\,dx = 2\int_{7}^{9} f(x)\,dx - \int_{7}^{9} h\,(x)\,dx = 2\,(5) - 4 = 6$

(c) $\int_9^7 f(x)\,dx = -\int_7^9 f(x)\,dx = -5$ (d) $\int_7^7 [f(x) + h(x)]\,dx = 0$

19. On each subinterval $[x_{i-1}, x_i]$, let $L_i = f(x_{i-1})\Delta x$ and $U_i = f(x_i)\Delta x$.

Then $U_i - L_i = [f(x_i) - f(x_{i-1})]\Delta x$ and $U - L = \sum_{i=0}^{n} [f(x_i) - f(x_{i-1})]\Delta x$

$= [f(x_n) - f(x_0)]\Delta x = [f(b) - f(a)]\Delta x$

21. Use the notation of problem 19, and assume that f is increasing.

Then $U_i - L_i = [f(x_i) - f(x_{i-1})]\Delta x_i$ and $U - L = \sum_{i=0}^{n} [f(x_i) - f(x_{i-1})]\Delta x_i$

$\le \sum_{i=0}^{n} [f(x_i) - f(x_{i-1})]\Delta x_{max} = [f(x_n) - f(x_0)]\Delta x_{max} = [f(b) - f(a)]\Delta x_{max}$.

If f is decreasing, we have $L - U \le [f(b) - f(a)]\Delta x_{max}$. Therefore,

$|U - L| \le [f(b) - f(a)]\Delta x_{max}$.

4.6 CALCULATING DEFINITE INTEGRALS BY SUMMATION

1. Verify $\sum_{k=1}^{n} k^3 = \left(\frac{n(n+1)}{2}\right)^2$. For $n = 1$, $1^3 = \left(\frac{1\cdot 2}{2}\right)^2 = 1$.

For $n = 2$, $9 = 1^3 + 2^3 = \left(\frac{2\cdot 3}{2}\right)^2 = 9$. For $n = 3$, $36 = 1^3 + 2^3 + 3^3 = \left(\frac{3\cdot 4}{2}\right)^2 = 36$.

In general, $1^3 + 2^3 + .. + (n+1)^3 = \left(\frac{n(n+1)}{2}\right)^2 + (n+1)^3 =$

$\left(\frac{n(n+1)}{2}\right)^2 + \frac{4(n+1)^3}{4} = \frac{(n+1)^2}{4}[n^2 + 4(n+1)] = \left(\frac{(n+1)(n+2)}{2}\right)^2$.

3. $U = [m(a + \Delta x) + m(a + 2\Delta x) + ... + m(a + n\Delta x)$

$= m[na + (1 + 2 + ... + n)]\Delta x = m\left[na + \frac{n(n-1)}{2}\right]\Delta x$

$= m\left[a + \frac{n-1}{2}\right]n\Delta x = m\left[a + \frac{n-1}{2}\cdot\frac{b-a}{n}\right](b-a)$.

$\int_a^b mx\,dx = \lim_{n\to\infty} m\left[a + \frac{n-1}{n}\cdot\frac{b-a}{2}\right](b-a) = m\left(\frac{b^2 - a^2}{2}\right)$.

5. Let $\Delta x = \frac{b}{n}$, $x_i = \frac{ib}{n}$, and $f(x_i) = \frac{i^3b^3}{n^3}$. Then $\sum_{i=1}^{n} \frac{i^3b^3}{n^3}\cdot\frac{ib}{n}$

$$= \frac{b^4}{n^4}\sum_{i=1}^{n} i^3 = \frac{b^4}{n^4}\left(\frac{n(n+1)}{2}\right)^2 . \lim_{n\to\infty} \frac{b^4}{n^4}\left(\frac{n(n+1)}{2}\right)^2$$

$$= \frac{b^4}{4} \lim_{n\to\infty} \frac{n^4 + 2n^3 + n^2}{n^4} = \frac{b^4}{4}.$$

7. Prove that $\sum_{k=1}^{n} \frac{1}{k(k+1)} = \frac{n}{n+1}$. If $n = 1$, then $\frac{1}{1\cdot 2} = \frac{1}{1+1}$.

$$\sum_{k=1}^{n+1} \frac{1}{k(k+1)} = \sum_{k=1}^{n} \frac{1}{k(k+1)} + \frac{1}{(n+1)(n+2)} = \frac{n}{n+1} + \frac{1}{(n+1)(n+2)}$$

$$= \frac{n(n+2)+1}{(n+1)(n+2)} = \frac{(n+1)^2}{(n+1)(n+2)} = \frac{n+1}{n+2}.$$

9. $\int_0^1 (1 + x^2)\,dx = \int_0^1 1\,dx + \int_0^1 x^2\,dx = 1 + \frac{1}{3} = \frac{4}{3}$

11. Let $f(x) = x^2$ on $[0, 1]$. Let $\Delta x = \frac{1-0}{n} = \frac{1}{n}$. Then

$S_n = \frac{1}{n}\left[\left(\frac{1}{n}\right)^2 + \left(\frac{2}{n}\right)^2 + \ldots + \left(\frac{n-1}{n}\right)^2\right]$ is a lower sum and hence

$$\lim_{n\to\infty} S_n = \int_0^1 x^2\,dx = \frac{1}{3}.$$

4.7 THE FUNDAMENTAL THEOREM OF INTEGRAL CALCULUS

1. $\int_0^3 y\,dx = \int_0^3 (x^2 + 1)\,dx = \frac{1}{3}x^3 + x\Big]_0^3 = 9 + 3 = 12$

3. $\int_0^4 \sqrt{2x+1}\,dx = \frac{1}{3}(2x+1)^{3/2}\Big]_0^4 = \frac{26}{3}$

5. $\int_1^2 (2x+1)^{-2}\,dx = -\frac{1}{2}(2x+1)^{-1}\Big]_1^2 = \frac{1}{15}$

7. $\int_0^2 (x^3 + 2x + 1)\,dx = \frac{1}{4}x^4 + x^2 + x\Big]_0^2 = 10$

9. $\int_0^2 \frac{x\,dx}{\sqrt{2x^2+1}} = \frac{1}{2}\sqrt{2x^2+1}\Big]_0^2 = 1$

11. $\int_0^1 (1-x)\,dx = x - \frac{1}{2}x^2\Big]_0^1 = \frac{1}{2}$

13. $\int_0^1 \sqrt{1-x}\,dx = -\frac{2}{3}(1-x)^{\frac{3}{2}}\Big]_0^1 = \frac{2}{3}$

15. $\int_{-1}^1 (1-y^2)\,dy = 2\left(y - \frac{1}{3}y^3\right)\Big]_0^1 = \frac{4}{3}$

17. $\sin^2 3x = 0 \Leftrightarrow 3x = 0 \text{ or } \pi \Rightarrow x = 0 \text{ or } \frac{\pi}{3}$.

$$A = \int_0^{\frac{\pi}{3}} \sin^2 3x\,dx = \int_0^{\frac{\pi}{3}}\left(\frac{1}{2} - \frac{1}{2}\cos 6x\right)dx = \frac{1}{2}x - \frac{1}{12}\sin 6x\Big]_0^{\frac{\pi}{3}} = \frac{\pi}{6}$$

19. $2 - x = x^2 \Leftrightarrow (x+2)(x-1) = 0 \Leftrightarrow x = 1 \text{ or } -2$.

The point of intersection for the 'vertex' of the region is (1,1). Therefore,

$$A = \int_0^1 [(2-y) - \sqrt{y}]\,dy = 2y - \frac{1}{2}y^2 - \frac{2}{3}y^{\frac{3}{2}}\Big]_0^1 = \frac{5}{6}$$

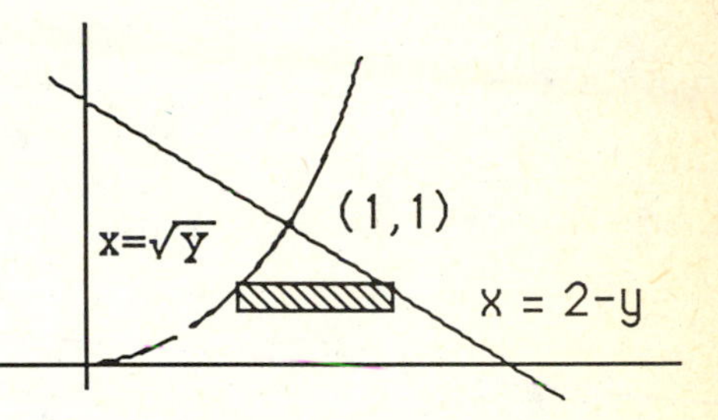

21. $\int_1^2 (2x+5)\,dx = x^2 + 5x\Big]_1^2 = 8$

23. $\int_{-1}^1 (x+1)^2\,dx = \frac{1}{3}(x+1)^3\Big]_{-1}^1 = \frac{8}{3}$

25. $\int_0^{\pi} \sin x\,dx = -\cos x\Big]_0^{\pi} = 2$

27. $\int_{\frac{\pi}{4}}^{\frac{\pi}{2}} \frac{\cos x\,dx}{\sin^2 x} = -(\sin x)^{-1}\Big]_{\frac{\pi}{4}}^{\frac{\pi}{2}} = \sqrt{2} - 1$

29. $\int_0^{\pi} \sin^2 x\,dx = \int_0^{\pi}\left(\frac{1}{2} - \frac{1}{2}\cos 2x\right)dx = \frac{1}{2}x - \frac{1}{4}\sin 2x\Big]_0^{\pi} = \frac{\pi}{2}$

31. $\int_0^1 \frac{dx}{(2x+1)^3} = -\frac{1}{4}(2x+1)^{-2}\Big]_0^1 = \frac{2}{9}$

33. $\int_0^1 \sqrt{5x+4}\,dx = \frac{2}{15}(5x+4)^{\frac{3}{2}}\Big]_0^1 = \frac{38}{15}$

35. $\int_{-1}^{0}\left(\frac{x^7}{2}-x^{15}\right)dx=\frac{1}{16}x^8-\frac{1}{16}x^{16}\Big]_{-1}^{0}=0$

37. $\int_{1}^{2}\frac{x^2+1}{x^2}\,dx=\int_{1}^{2}\left(1+\frac{1}{x^2}\right)dx=x-\frac{1}{x}\Big]_{1}^{2}=\frac{3}{2}$

39. $\int_{\frac{\pi}{6}}^{\frac{\pi}{2}}\cos^2\theta\sin\theta\,d\theta=-\frac{1}{3}\cos^3\theta\Big]_{\frac{\pi}{6}}^{\frac{\pi}{2}}=\frac{\sqrt{3}}{8}$

41. $\int_{-4}^{4}|x|\,dx=\int_{-4}^{0}(-x)\,dx+\int_{0}^{4}x\,dx=-\frac{1}{2}x^2\Big]_{-4}^{0}+\frac{1}{2}x^2\Big]_{0}^{4}=16$

43. $\int_{0}^{b}\left(-\frac{h}{b}x+h\right)dx=-\frac{h}{b}\left(\frac{x^2}{2}\right)+hx\Big]_{0}^{b}=\frac{1}{2}bh$

45. $\frac{d}{dx}\int_{0}^{x}\sqrt{1+t^2}\,dt=\sqrt{1+x^2}$

47. $\frac{d}{dx}\int_{x}^{1}\sqrt{1-t^2}\,dt=-\frac{d}{dx}\int_{1}^{x}\sqrt{1-t^2}\,dt=-\sqrt{1-x^2}$

49. $\frac{d}{dx}\int_{1}^{2x}\cos(t^2)\,dt=\left(\frac{d}{du}\int_{1}^{u}\cos(t^2)\,dt\right)\left(\frac{du}{dx}\right)$, where $u=2x$,

$=\cos(u^2)(2)=2\cos(2x)^2=2\cos 4x^2$

51. $\frac{d}{dx}\int_{\sin x}^{0}\frac{dt}{2+t}=-\frac{d}{dx}\int_{0}^{\sin x}\frac{dt}{2+t}=-\frac{1}{2+\sin x}\cdot\frac{d}{dx}(\sin x)=\frac{-\cos x}{2+\sin x}$

53. $\frac{d}{dx}\int_{\cos x}^{0}\frac{1}{1-t^2}\,dt=-\frac{d}{dx}\int_{0}^{\cos x}\frac{1}{1-t^2}\,dt=\frac{-1}{1-\cos^2 x}\cdot\frac{d}{dx}(\cos x)$

$=\frac{\sin x}{1-\cos^2 x}=\frac{1}{\sin x}=\csc x$

55. If $F(x)=\int_{0}^{x}f(t)\,dt=\sin x$, then $F'(x)=f(x)=\cos x$.

57. (a) If $f(x)=2+\int_{0}^{x}\frac{10}{1+t}dt$, then $f(0)=2$ and $f'(x)=\frac{10}{1+x}$

so $f'(0)=10$ and $L(x)=2+10x$.

(b) $f''(x)=-\frac{10}{(1+x)^2}$ and $f'(0)=-10$. Therefore $Q(x)=2+10x-5x^2$.

59. (a) Let $F(x)=\int_{0}^{x}f(t)\,dt=x\cos\pi x$. Then

$F'(x)=f(x)=\cos\pi x-\pi x\sin\pi x$ so $f(4)=\cos 4\pi-4\pi\sin 4\pi=1$.

(b) Let $F(x) = \int_0^{x^2} f(t)\,dt = x\cos\pi x$. Then

$F'(x) = 2x\,f(x^2) = \cos\pi x - \pi x\sin\pi x$. When $x = 2$,

$4f(4) = = \cos 2\pi - 2\pi\sin 2\pi = 1$ so $f(4) = \frac{1}{4}$.

(c) If $\int_0^{f(x)} t^2\,dt = x\cos\pi x$, then $\left.\frac{1}{3}t^3\right]_0^{f(x)} = x\cos\pi x$. Thus,

$\frac{1}{3}f^3(x) = x\cos\pi x$, $f(x) = (3x\,\text{cox}\,\pi x)^{\frac{1}{3}}$, and $f(4) = (12)^{\frac{1}{3}}$.

4.8 SUBSTITUTION IN DEFINITE INTEGRALS

1. (a) $\int_0^3 \sqrt{y+1}\,dy = \int_1^4 \sqrt{u}\,du = \left.\frac{2}{3}u^{\frac{3}{2}}\right]_1^4 = \frac{14}{3}$

(b) $\int_{-1}^0 \sqrt{y+1}\,dy = \int_0^1 \sqrt{u}\,du = \left.\frac{2}{3}u^{\frac{3}{2}}\right]_0^1 = \frac{2}{3}$

Let $u = y+1 \Rightarrow du = dy$. In (a) $y = 0 \Rightarrow u = 1$, $y = 3 \Rightarrow u = 4$.

In (b) $y = -1 \Rightarrow u = 0$, $y = 0 \Rightarrow u = 1$

3. (a) $\int_0^{\frac{\pi}{4}} \tan x\sec^2 x\,dx = \int_0^1 u\,du = \left.\frac{1}{2}u^2\right]_0^1 = \frac{1}{2}$

(b) $\int_{-\frac{\pi}{4}}^0 \tan x\sec^2 x\,dx = \int_{-1}^0 u\,du = \left.\frac{1}{2}u^2\right]_{-1}^0 = -\frac{1}{2}$

Let $u = \tan x \Rightarrow du = \sec^2 x\,dx \Rightarrow$. Then in (a), $x = 0 \Rightarrow u = 0$,

$x = \frac{\pi}{4} \Rightarrow u = 1$, and in (b) $x = -\frac{\pi}{4} \Rightarrow u = -1$, $x = 0 \Rightarrow u = 0$.

5. (a) $\int_0^1 \frac{x^3}{\sqrt{x^4+9}}\,dx = \int_9^{10} \frac{1}{4}u^{-\frac{1}{2}}\,du = \left.\frac{1}{2}u^{\frac{1}{2}}\right]_9^{10} = \frac{\sqrt{10}-3}{2}$

(b) $\int_{-1}^0 \frac{x^3}{\sqrt{x^4+9}}\,dx = \int_{10}^9 \frac{1}{4}u^{-\frac{1}{2}}\,du = \left.\frac{1}{2}u^{\frac{1}{2}}\right]_{10}^9 = \frac{3-\sqrt{10}}{2}$

Let $u = x^4 + 9 \Rightarrow du = 4x^3\,dx \Rightarrow x^3\,dx = \frac{1}{4}\,du$. Then in (a), $x = 0 \Rightarrow u = 9$,

$x = 1 \Rightarrow u = 10$, , and in (b) $x = -1 \Rightarrow u = 10$, $x = 0 \Rightarrow u = 9$.

7. (a) $\int_0^{\sqrt{7}} x(x^2+1)^{\frac{1}{3}}\,dx = \int_1^8 \frac{1}{2}u^{\frac{1}{3}}\,du = \frac{3}{8}u^{\frac{4}{3}}\Big]_1^8 = \frac{45}{8}$

(b) $\int_{-\sqrt{7}}^{0} x(x^2+1)^{\frac{1}{3}}\,dx = \int_8^1 \frac{1}{2}u^{\frac{1}{3}}\,du = \frac{3}{8}u^{\frac{4}{3}}\Big]_8^1 = -\frac{45}{8}$

Let $u = x^2+1 \Rightarrow du = 2x\,dx \Rightarrow x\,dx = \frac{1}{2}\,du$. Then in (a), $x = 0 \Rightarrow u = 1$, $x = \sqrt{7} \Rightarrow u = 8$, and in (b) $x = -\sqrt{7} \Rightarrow u = 8$, $x = 0 \Rightarrow u = 1$.

9. (a) $\int_0^{\frac{\pi}{6}} (1-\cos 3x)\sin 3x\,dx = \int_0^1 \frac{1}{3}u\,du = \frac{1}{6}u^2\Big]_0^1 = \frac{1}{6}$

(b) $\int_{\frac{\pi}{6}}^{\frac{\pi}{3}} (1-\cos 3x)\sin 3x\,dx = \int_1^2 \frac{1}{3}u\,du = \frac{1}{6}u^2\Big]_1^2 = \frac{1}{2}$

Let $u = 1-\cos 3x \Rightarrow du = 3\sin 3x\,dx \Rightarrow \sin 3x\,dx = \frac{1}{3}u\,du$

Then in (a), $x = 0 \Rightarrow u = 0$, $x = \frac{\pi}{6} \Rightarrow u = 1$, and in (b) $x = \frac{\pi}{3} \Rightarrow u = 2$.

11. (a) $\int_0^{2\pi} \frac{\cos x\,dx}{\sqrt{2+\sin x}} = 2\sqrt{2+\sin x}\Big]_0^{2\pi} = 0$

(b) $\int_{-\pi}^{\pi} \frac{\cos x\,dx}{\sqrt{2+\sin x}} = 2\sqrt{2+\sin x}\Big]_{-\pi}^{\pi} = 0$

13. (a) $\int_{-\pi}^{\pi} x\cos(2x^2)\,dx = \frac{1}{4}\sin(2x^2)\Big]_{-\pi}^{\pi} = 0$

(b) $\int_{-\pi}^{0} x\cos(2x^2)\,dx = \frac{1}{4}\sin(2x^2)\Big]_{-\pi}^{0} = -\frac{1}{4}\sin(2\pi^2)$

15. $\int_0^1 \sqrt{t^5+2t}\,(5t^4+2)\,dt = \frac{2}{3}(t^5+2t)^{\frac{3}{2}}\Big]_0^1 = 2\sqrt{3}$

17. $\int_0^{\frac{\pi}{2}} \cos^3 2x\sin 2x\,dx = \int_1^{-1} -\frac{1}{2}u^3\,du = -\frac{1}{8}u^4\Big]_1^{-1} = 0$

Let $u = \cos 2x \Rightarrow du = -2\sin 2x\,dx \Rightarrow \sin 2x\,dx = -\frac{1}{2}du$.

Then $x = 0 \Rightarrow u = 1$, $x = \frac{\pi}{2} \Rightarrow u = -1$.

19. (a)

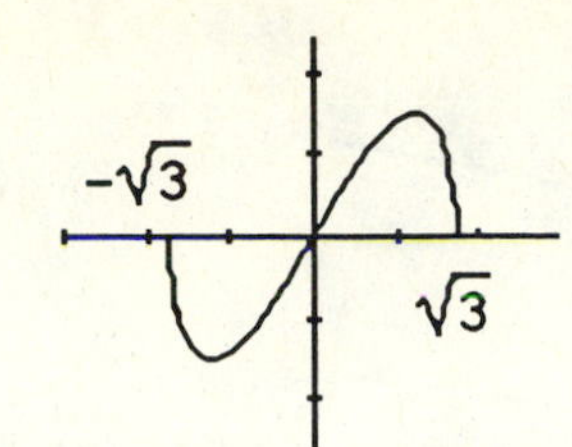

(b) $2\int_0^{\sqrt{3}} x\sqrt{3-x^2}\,dx = -\frac{2}{3}(3-x^2)^{3/2}\Big]_0^{\sqrt{3}} = 2\sqrt{3}$

21. $\int_1^3 \frac{\sin 2x}{x}\,dx = 2\int_1^3 \frac{\sin 2x}{2x}\,dx = \int_2^6 \frac{\sin u}{u}\,du$, where we let

$u = 2x \Rightarrow \frac{1}{2}du = dx$ and $x = 1 \Rightarrow u = 2$, $x = 3 \Rightarrow u = 6$.

Therefore, $\int_1^3 \frac{\sin 2x}{x}\,dx = F(6) - F(2)$.

23. (a) Let $F(x) = h(x)\sin x$. Then $F(-x) = h(-x)\sin(-x) = h(x)(-\sin x) = -h(x)\sin x = -F(x)$. Therefore, the fun[illegible]on i

(b) Let $u = -x$ so that $dx = -du$, and $x = 0 \Rightarrow u = 0$, $x = -a \Rightarrow u = a$.

Then $\int_{-a}^{0} h(x)\sin x\,dx = -\int_0^a h(x)\sin x\,dx = \int_a^0 h(-u)\sin(-u)(-du)$

$= \int_a^0 h(u)\sin(u)\,du = -\int_0^a h(x)\sin x\,dx$.

(c) $\int_{-a}^{a} h(x)\sin x\,dx = \int_{-a}^{0} h(x)\sin x\,dx + \int_0^a h(x)\sin x\,dx$

$= -\int_0^a h(x)\sin x\,dx + \int_0^a h(x)\sin x\,dx = 0$.

(d) $h(x) = \sec x$ is even, $a = \frac{\pi}{4}$, and $-a = -\frac{\pi}{4}$. By part (c)

$\int_{-\frac{\pi}{4}}^{\frac{\pi}{4}} \sec x \sin x\,dx = 0$.

25. $\int_0^1 x^2\,dx = \frac{1}{3}x^3\Big]_0^1 = \frac{1}{3}$

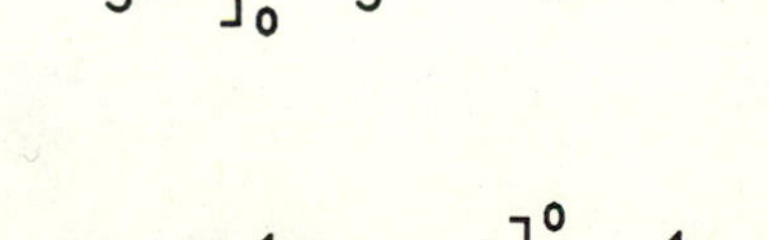

$\int_{-1}^{0} (x+1)^2\,dx = \frac{1}{3}(x+1)^3\Big]_{-1}^{0} = \frac{1}{3}$

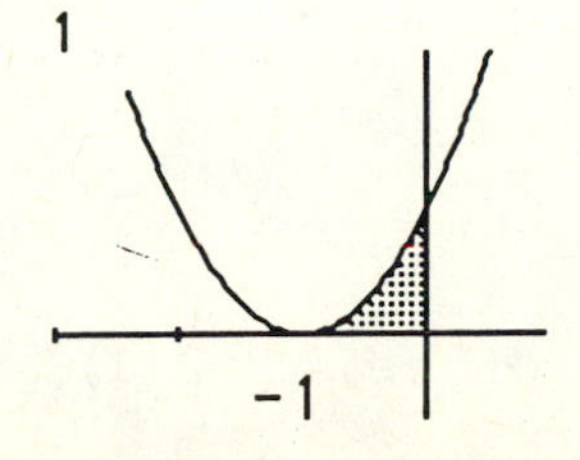

27. $\int_4^8 \sqrt{x-4}\,dx = \frac{2}{3}(x-4)^{\frac{3}{2}}\Big]_0^8 = \frac{16}{3}$

$\int_{-1}^3 \sqrt{x+1}\,dx = \frac{2}{3}(x+1)^{\frac{3}{2}}\Big]_{-1}^3 = \frac{16}{3}$

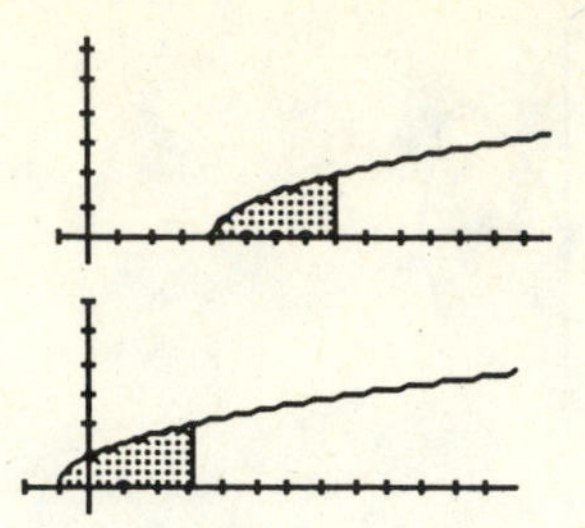

4.9 RULES FOR APPROXIMATING DEFINITE INTEGRALS

1.

i	0	1	2	3	4
x_i	0	0.5	1.0	1.5	2.0
y_i	0	0.5	1.0	1.5	2.0

(a) Trapezoidal Rule: $\frac{2-0}{2\cdot 4}(0 + 2(.5) + 2(1) + 2(1.5) + 2) = 2$

(b) Simpson's Rule: $\frac{2-0}{3\cdot 4}(0 + 4(.5) + 2(1) + 4(1.5) + 2) = 2$

(c) $\int_0^2 x\,dx = \frac{1}{2}x^2\Big]_0^2 = 2$

3.

i	0	1	2	3	4
x_i	0	0.5	1.0	1.5	2.0
y_i	0	0.125	1.0	3.375	8.0

(a) Trapezoidal Rule: $\frac{2-0}{2\cdot 4}(0 + 2(.125) + 2(1) + 2(3.375) + 8) = 4.25$

(b) Simpson's Rule: $\frac{2-0}{3\cdot 4}(0 + 4(.125) + 2(1) + 4(3.375) + 8) = 4$

(c) $\int_0^2 x^3\,dx = \frac{1}{4}x^4\Big]_0^2 = 4$

5. (a) $\int_0^4 \sqrt{x}\,dx \approx \frac{1}{2}(0 + 2\sqrt{1} + 2\sqrt{2} + 2\sqrt{3} + \sqrt{4}) \approx 5.146$

(b) $\int_0^4 \sqrt{x}\,dx \approx \frac{1}{3}(0 + 4\sqrt{1} + 2\sqrt{2} + 4\sqrt{3} + \sqrt{4}) \approx 5.252$

(c) $\int_0^4 \sqrt{x}\,dx = \frac{2}{3}x^{\frac{3}{2}}\Big]_1^4 = \frac{16}{3}$

7. (a) If $f(x) = \frac{1}{x}$, then $f''(x) = 2x^{-3}$ which is a decreasing function on $[1,2]$. Therefore, $M = 2(1)^{-3} = 2$.

$E_T \le \frac{b-a}{12}h^2 M = \frac{1}{12}\left(\frac{1}{10}\right)^2(2) = \frac{1}{600} \approx 0.0017$

(b) $f^{(4)}(x) = 24x^{-5}$ takes its maximum M at $x = 1$.

$$E_S \le \frac{b-a}{180}h^4 M = \left(\frac{1}{180}\right)\left(\frac{1}{10^4}\right)(24) \approx 1.33 \times 10^{-5}.$$

(c) $E_T \le \left(\frac{1}{12}\right)\left(\frac{1}{4}\right)^2 (2) = \frac{1}{96} \approx 0.0104$

$E_S \le \;= \left(\frac{1}{180}\right)\left(\frac{1}{4}\right)^4 (24) = \frac{1}{1920} \approx 0.0052$

9. $f''(x) = 0$, so (a) the Trapezoidal Rule is exact for any n, and (b) Simpson's Rule is exact for any even n.

11. (a) $\frac{2}{12}\left(\frac{2}{n}\right)^2 (12) < 10^{-4} \Leftrightarrow \frac{8}{n^2} < 10^{-4} \Leftrightarrow n^2 > 8 \times 10^4$

$\Leftrightarrow n > 2\sqrt{2} \times 10^2 \approx 282.3$. Therefore, $n \ge 283$.

(b) Simpson's Rule is exact for any even n.

13. (a) $f''(x) = \frac{-1}{4x\sqrt{x}}$ is an increasing function, and $|f''(1)| = \frac{1}{4}$ is maximum.

$E_T \le \frac{3}{12}\left(\frac{3}{n}\right)^2\left(\frac{1}{4}\right) < 10^{-4} \Leftrightarrow \frac{9}{16n^2} < 10^{-4} \Leftrightarrow n > \sqrt{\frac{9}{16} \times 10^4} = 75..$

Hence, any $n \ge 75$.

(b) $|f^{(4)}(x)| = |-\frac{15}{16}x^{-\frac{7}{2}}| \le \frac{15}{16}$ on $[1, 4]$.

$E_S \le \frac{3}{180}\left(\frac{3}{n}\right)^4\left(\frac{15}{16}\right) < 10^{-4} \Leftrightarrow \; < \; n^4 \ge \frac{3645}{2880} \times 10^4 \approx 10.6.$

Hence, any $n \ge 12$ (since n must be even).

15. $E_S \le \frac{1}{180}\left(\frac{1}{n^4}\right)(3) \le 10^{-5} \Rightarrow 60\,n^4 \ge 10^5$ or $n > 6$.

17. $S \approx \frac{1}{3}(1.5 + 4(1.6) + 2(1.8) + 4(1.9) + 2(2.0) + 4(2.1) + 2.1) \approx 11.2\ \text{ft}^2$.

$V = \frac{5000\ \text{lb}}{\frac{42\ \text{lb}}{\text{ft}^3}} = 119.05\ \text{ft}^3$. $l = \frac{V}{S} = \frac{119.05}{10.6} \approx 10.6$ ft.

19. (a) $E_S \le \frac{\pi}{180}\left(\frac{\pi}{4}\right)^4 \cdot 1 \approx 0.00664$

(b) $\frac{\pi}{12}\left(\frac{2}{\pi} + \frac{8\sqrt{2}}{\pi} + 2 + \frac{8\sqrt{2}}{\pi} + \frac{2}{\pi}\right) \approx 2.74$

(c) $\frac{.00664}{2.74} \cdot 100 = 0.24\ \%$

4.M MISCELLANEOUS

1. $\frac{dy}{dx} = xy^2 \Rightarrow y^{-2}\,dy = xdx \Rightarrow -y^{-1} = \frac{1}{2}x^2 + C$ or $y = \frac{-2}{C + x^2}$.

3. $\frac{dy}{dx} = \frac{x^2 - 1}{y^2 + 1} \Rightarrow (y^2 + 1)\,dy = (x^2 - 1)\,dx \Rightarrow \frac{1}{3}y^3 + y = \frac{1}{3}x^3 - x + C$

5. $\frac{dr}{ds} = \left(\frac{2 + r}{3 - s}\right)^2 \Rightarrow (2 + r)^{-2}\,dr = (3 - s)^{-2}\,ds \Rightarrow \frac{-1}{2 + r} = \frac{1}{3 - s} + C$

7. (a) $\frac{dy}{dx} = x\sqrt{x^2 - 4}$, $y = 3$ when $x = 2$. $y = \frac{1}{3}(x^2 - 4)^{\frac{3}{2}} + C$.

   $3 = \frac{1}{3}(0) + C \Rightarrow C = 3$. Therefore $y = \frac{1}{3}(x^2 - 4)^{\frac{3}{2}} + 3$.

   (b) $\frac{dy}{dx} = xy^3$, $y = 1$ when $x = 0$. $y^{-3}\,dy = x\,dx \Rightarrow -\frac{1}{2}y^{-2} = \frac{1}{2}x^2 + C$.

   $-\frac{1}{2}(1) = 0 + C \Rightarrow C = \frac{1}{2}$. Therefore $-\frac{1}{2y^2} = \frac{1}{2}x^2 - \frac{1}{2}$ or $y = \frac{1}{\sqrt{1 - x^2}}$.

   (c) $\frac{dy}{dx} = x^3y^2$, $y = 4$ when $x = 1$. $y^{-2}\,dy = x^3\,dx \Rightarrow -y^{-1} = \frac{1}{4}x^4 + C$.

   $-\frac{1}{4} = \frac{1}{4} + C \Rightarrow C = -\frac{1}{2}$. $\therefore -\frac{1}{y} = \frac{x^4}{4} - \frac{1}{2}$ or $y = \frac{4}{2 - x^4}$.

   (d) $\sqrt{y + 1}\frac{dy}{dx} = \frac{1}{x^2}$, $y = 3$ when $x = -3$. $\sqrt{y + 1}\,dy = x^{-2}\,dx \Rightarrow$

   $\frac{2}{3}(y + 1)^{\frac{3}{2}} = -x^{-1} + C$. $\frac{2}{3}(4)^{\frac{3}{2}} = \frac{1}{3} + C \Rightarrow C + 5$.

   $\frac{2}{3}(y + 1)^{\frac{3}{2}} = -\frac{1}{x} + 5$ or $y = \left(\frac{15}{2} - \frac{3}{2x}\right)^{\frac{2}{3}} - 1$.

9. $\frac{dy}{dx} = 3x^2 + 2 \Rightarrow y = x^3 + 2x + C$. The point $(1, 1)$ belongs to the curve, so $1 = 1 + 2 + C \Rightarrow C = -4$. The equation is $y = 3x^2 + 2x - 4$.

11. (a) $a = \frac{dv}{dt} = \sqrt{t} - \frac{1}{\sqrt{t}} \Rightarrow v = \frac{2}{3}t^{\frac{3}{2}} - 2t^{\frac{1}{2}} + C$. $v = \frac{4}{3}$ when $t = 0$,

    so $C = \frac{4}{3}$. $\therefore v(t) = \sqrt{t} - \frac{1}{\sqrt{t}} + \frac{4}{3}$.

(b) $v = \frac{ds}{dt} = \sqrt{t} - \frac{1}{\sqrt{t}} + \frac{4}{3} \Rightarrow s = \frac{4}{15}t^{\frac{5}{2}} - \frac{4}{3}t^{\frac{3}{2}} + \frac{4}{3}t + C.$

$S = -\frac{4}{15}$ when $t = 0$, so $C = -\frac{4}{15}$ and $s(t) = \frac{4}{15}t^{\frac{5}{2}} - \frac{4}{3}t^{\frac{3}{2}} + \frac{4}{3}t - \frac{4}{15}.$

13. We are given $\frac{d^2x}{dt^2} = 4x$. We must express the derivative as a function of x.

Since $a = \frac{dv}{dt} = \frac{dv}{dx} \cdot \frac{dx}{dt} = v \cdot \frac{dv}{dx}$, we may write $v \cdot \frac{dv}{dx} = 4x$ or $v\,dv = 4x\,dx$.

$\therefore \frac{1}{2}v^2 = -2x^2 + C$. $v = 0$ when $x = 5$, so $C = 50$. $\therefore \frac{1}{2}v^2 = -2x^2 + 50$.

$\frac{1}{2}v^2 = -2(3)^2 + 50 \Rightarrow v^2 = 64$ or $v = \pm 8$. Since it is moving left, $v = -8$.

15. $\frac{dy}{dx} = x\sqrt{1 + x^2}$, $y = -2$ when $x = 0$. $y = \frac{1}{3}(1 + x^2)^{\frac{3}{2}} + C.$

$-2 = \frac{1}{3} + C \Rightarrow C = -\frac{7}{3}$. $\therefore\ y = \frac{1}{3}(1 + x^2)^{\frac{3}{2}} - \frac{7}{3}.$

17. $\sqrt{x}\, y \frac{dy}{dx} = x + 1$, $y = 2$ when $x = 1$. $y\,dy = (x^{\frac{1}{2}} + x^{-\frac{1}{2}})\,dx \Rightarrow$

$\frac{1}{2}y^2 = \frac{2}{3}x^{\frac{3}{2}} + 2x^{\frac{1}{2}} + C$. $2 = \frac{2}{3} + 2 + C \Rightarrow C = -\frac{2}{3}$. $\frac{1}{2}y^2 = \frac{2}{3}x^{\frac{3}{2}} + 2x^{\frac{1}{2}} - \frac{2}{3}$ or

$y = 2\left(\frac{1}{3}x^{\frac{3}{2}} + x^{\frac{1}{2}} - \frac{1}{3}\right)^{\frac{1}{2}}.$

19. $\frac{dy}{dx} = x\sqrt{9y + x^2y} \Rightarrow y^{-\frac{1}{2}}\,dy = x\sqrt{9 + x^2}\,dx$ so $2y^{\frac{1}{2}} = \frac{1}{3}(9 + x^2)^{\frac{3}{2}} + C.$

$y = 36$ when $x = 0$ so $12 = \frac{1}{3}(9)^{\frac{3}{2}} + C \Rightarrow C = 3$. $2y^{\frac{1}{2}} = \frac{1}{3}(9 + x^2)^{\frac{3}{2}} + 3$ or

$y = \left(\frac{1}{6}(9 + x^2)^{\frac{3}{2}} + \frac{3}{2}\right)^2.$

21. $a = -32\ \text{ft/s}^2$ so $v = -32t + C$. $v_0 = 96$ when $t = 0$ so $C = 96$ and $v = -32t + 96$. It has reached its maximum height when $v = 0$ or $-32t + 96 = 0$ or when $t = 3$ seconds. $s = -16t^2 + 96t + C$. The initial height is 0 because it is thrown from the ground, so $C = 0$. Then $s(3) = -16(3)^2 + 96(3) = 144$ ft.

23. $a = \frac{dv}{dt} = -K\sqrt{v} \Rightarrow \frac{1}{\sqrt{v}}\,dv = -K\,dt \Rightarrow 2\sqrt{v} = -Kt + C$. $v = 16$ when $t = 0$

so $2\sqrt{16} = C$ or $C = 8$ and $2\sqrt{v} = -Kt + 8$. $V = 0$ when $t = 4 \Rightarrow 0 = -4K + 8$ or

$K = -2$. $\therefore 2\sqrt{v} = -2t + 8$, or $v = (4 - t)^2$.

(a) $v(2) = 4$ ft/s.

(b) $s = \int_0^4 |4 - t|^2\,dt = -\frac{1}{3}(4 - 7)^3\Big]_0^4 = \frac{64}{3}$ ft.

25. $\lim_{n\to\infty} \frac{1^5 + 2^5 + 3^5 + \ldots + n^5}{n^6} = \lim_{n\to\infty}\left[\left(\frac{1}{n}\right)^5 + \left(\frac{2}{n}\right)^5 + \ldots + \left(\frac{n}{n}\right)^5\right]\cdot\frac{1}{n}$

Take $f(x) = x^5$. Identify $\frac{1}{n}$ with the needed partition of $[0, 1]$

since $\Delta x = \frac{1}{n} = \frac{1 - 0}{n}$. Since f is increasing on $[0, 1]$, the set

$\left\{\frac{1}{n}, \frac{2}{n}, \ldots, \frac{n}{n}\right\}$ is the set of the right endpoints of the partition.

Then $\sum_{j=1}^{n}\left(\frac{1}{j}\right)^5 \cdot \frac{1}{n}$ is an upper Riemann Sum for $f(x) = x^5$ on $[0, 1]$

and $\lim_{n\to\infty} \sum_{j=1}^{n}\left(\frac{1}{j}\right)^5 \cdot \frac{1}{n} = \int_0^1 x^5\,dx = \frac{1}{6}x^6\Big]_0^1 = \frac{1}{6}$.

27. $\lim_{n\to\infty} \frac{1}{n}\left[f\left(\frac{1}{n}\right) + f\left(\frac{2}{n}\right) + \ldots + f\left(\frac{n}{n}\right)\right] = \int_0^1 f(x)\,dx$, where we have

identified $\Delta x = \frac{1}{n}$ as the partition of $[0, 1]$ into n equal parts.

29. (a) The polygon can be divided into n isosceles triangles by joining the vertices to the center of the circle. The vertex angle of each triangle is $\theta_n = \frac{2\pi}{n}$, and the area of the triangle is $A = \frac{1}{2}r^2 \sin\theta_n$.

Then the area of the polygon $A_n = nA = \frac{nr^2}{2}\sin\theta_n$.

(b) Let $x = \frac{\pi}{n}$, so that $x \to 0$ as $n \to \infty$. Then

$$\lim_{n\to\infty} \frac{nr^2}{2}\sin\frac{2\pi}{n} = \lim_{x\to 0} \frac{r^2}{2}\left[\frac{\pi}{x}\sin 2x\right] = \pi r^2 \lim_{x\to 0}\frac{\sin 2x}{2x} = \pi r^2.$$

31. $\int_0^\pi \cos 4x \sin 4x\,dx = -\frac{1}{8}\cos^2 4x\Big]_0^\pi = 0$

33. $\int_{-1}^0 \frac{12\,dx}{(2 - 3x)^2} = 4(2 - 3x)^{-1}\Big]_{-1}^0 = \frac{6}{5}$

35. $\int_0^{\frac{\pi}{2}} \frac{\sin x \cos x \, dx}{\sqrt{1+3\sin^2 x}} = \frac{1}{3}(1+3\sin^2 x)^{\frac{1}{2}}\Big]_0^{\frac{\pi}{2}} = \frac{1}{3}$

37. $\int_{-\frac{\pi}{2}}^{\frac{\pi}{2}} 15 \sin^4 3x \cos 3x \, dx = \sin^5 3x \Big]_{-\frac{\pi}{2}}^{\frac{\pi}{2}} = -2$

39. $\int_0^1 \frac{dr}{\sqrt[3]{(7-5r)^2}} = -\frac{3}{5}(7-5r)^{1/3}\Big]_0^1 = \frac{3}{5}(7^{1/3} - 2^{1/3})$

41. $\int_0^1 \pi x^2 \sec^2\left(\frac{\pi x^3}{3}\right) dx = \tan \frac{\pi x^3}{3}\Big]_0^1 = \sqrt{3}$

43. $\int_{-1}^3 f(x)\, dx = \int_{-1}^2 (3-x)\, dx + \int_2^3 \frac{x}{2}\, dx = 3x - \frac{1}{2}x^2\Big]_{-1}^2 + \frac{x^2}{4}\Big]_2^3 = \frac{35}{4}$

45. (a) Let $F(x) = \int_0^x \frac{du}{u+\sqrt{u^2+1}}$. Then $F'(x) = \lim_{h\to 0} \frac{F(x+h)-F(x)}{h} =$

$$\lim_{h\to 0} \frac{\int_0^{x+h} \frac{du}{u+\sqrt{u^2+1}} - \int_0^x \frac{du}{u+\sqrt{u^2+1}}}{h} = \lim_{h\to 0} \frac{1}{h}\int_x^{x+h} \frac{du}{u+\sqrt{u^2+1}}$$

$= \frac{1}{1+\sqrt{x^2+1}}$ by the Fundamental Theorem for Calculus.

(b) If $F(x) = \int_a^x f(t)\, dt$, then $F'(2) = \lim_{x\to 2} \frac{f(x)-f(2)}{x-2} =$

$$\lim_{x\to 2} \frac{\int_a^x f(t)\, dt - \int_a^2 f(t)\, dt}{x-2} = \lim_{x\to 2} \frac{1}{x-2}\int_2^x f(t)\, dt.$$

$\therefore \lim_{x\to 2} \frac{x}{x-2}\int_2^x f(t)\, dt = \lim_{x\to 2} x \cdot \lim_{x\to 2} \frac{x}{x-2}\int_2^x f(t)\, dt = 2F'(2) = 2f(2).$

(In this problem, we must assume that f is continuous)

47. The derivative of the left side of the equation is:

$$\frac{d}{dx}\int_0^x \left[\int_0^u f(t)\,dt\right] = \int_0^x f(t)\,dt.$$

The derivative of the right side of the equation is:

$$\frac{d}{dx}\left[\int_0^x f(u)\,(x-u)\,du\right] = \frac{d}{dx}\int_0^x x\,f(u)\,du - \frac{d}{dx}\int_0^x u\,f(u)\,du$$

$$= x\,\frac{d}{dx}\int_0^x f(u)\,du + \int_0^x f(u)\,du\,\frac{d}{dx}(x) - x\,f(x)$$

$$= x\,f(x) + \int_0^x f(u)\,du - x\,f(x) = \int_0^x f(u)\,du.$$

Since each side has the same derivative, they differ from each other by at most a constant. Since both sides equal zero when $x = 0$, the constant must be zero.

49. $S \approx \frac{15}{3}(0 + 4(36) + 2(54) + 4(51) + 2(49.5) + 4(54)\ 2(64.4) + 4(67.5) + 42)$

$= 6{,}059\ \text{ft}^2$. The cost is then $(2.10)(6{,}059) = \$12{,}723.90$. No, the job cannot be done for $11,000.

CHAPTER 5

APPLICATIONS OF DEFINITE INTEGRALS

5.1 DISTANCE TRAVELED BY A MOVING BODY

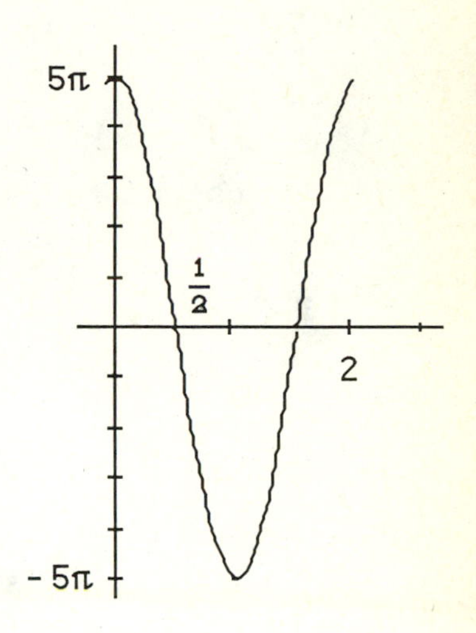

1. (b) Total distance $= 5\pi\int_0^{\frac{1}{2}} \cos \pi t\, dt - 5\pi\int_{\frac{1}{2}}^{\frac{3}{2}} \cos \pi t$

$$+ 5\pi\int_{\frac{3}{2}}^{2} \cos \pi t\, dt$$

$$= 4\cdot 5\pi\int_0^{\frac{1}{2}} \cos \pi t\, dt = 20 \sin \pi t\Big]_0^{\frac{1}{2}} = 20$$

(c) Net change $= 5\pi\int_0^2 \cos \pi t\, dt = 5 \sin \pi t\Big]_0^2 = 0$

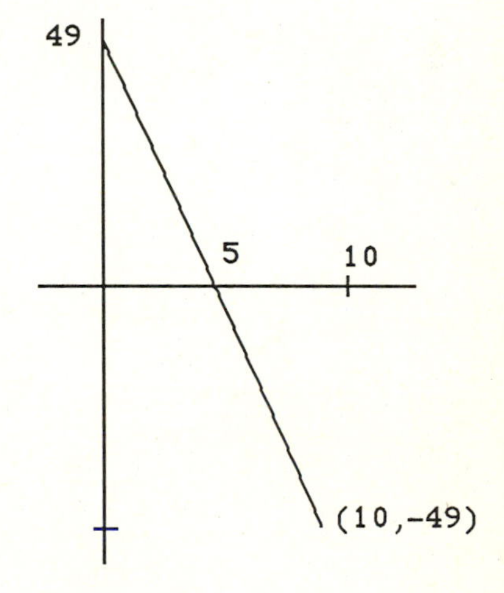

3. (b) Total distance $= \int_0^{10} |\,49 - 9.8t\,|\, dt$

$$= \int_0^5 (49 - 9.8\,t)\, dt - \int_5^{10} (49 - 9.8\,t)\, dt$$

$$= 245$$

(c) Net change $= \int_0^{10} (49 - 9.8\,t)\, dt = 0$

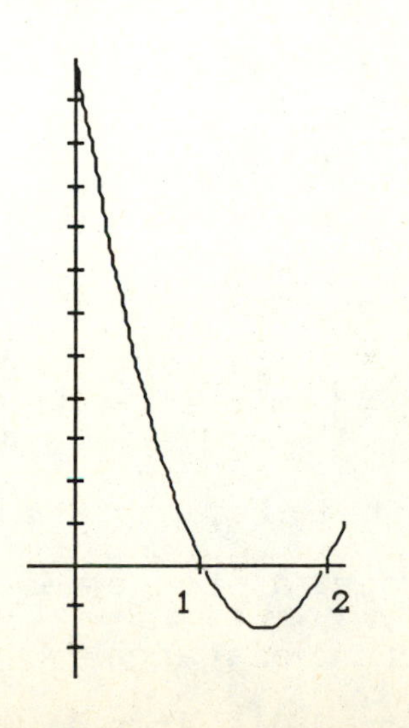

5. (b) Total distance $= \int_0^2 |\,6(t-1)(t-2)\,|\, dt$

$$= \int_0^1 6(t-1)(t-2)\, dt - \int_1^2 6(t-1)(t-2)\, dt$$

$$= 2t^3 - 9t^2 + 12\,t\Big]_0^1 - (2t^3 - 9t^2 + 12\,t)\Big]_1^2$$

$= 6$ m.

(c) Net change $= \int_0^2 6(t-1)(t-2)\, dt$

$$= 2t^3 - 9t^2 + 12\,t\Big]_0^2$$

$= 4$ m.

7. (b) Total distance = $6\int_0^{\frac{\pi}{2}} |\sin 3t|\, dt$

$= 6\int_0^{\frac{\pi}{3}} \sin 3t\, dt - 6\int_{\frac{\pi}{3}}^{\frac{\pi}{2}} \sin 3t\, dt$

$= 6$

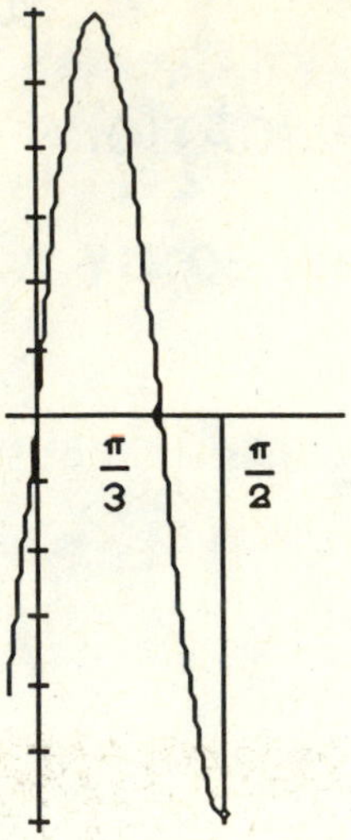

(c) Net change = $6\int_0^{\frac{\pi}{2}} \sin 3t\, dt$

$= 2$

9. (a) $s(t) = \int \cos t\, dt = \sin t + C.$ $s(0) = 0 \Rightarrow C = 0.$ $\therefore s(t) = \sin t$

(b) T.D. $= \int_0^{2\pi} |\cos t|\, dt = 4\int_0^{\frac{\pi}{2}} \cos t\, dt = 4 \sin t\Big]_0^{\frac{\pi}{2}} = 4$

(c) N.C. $= \int_0^{2\pi} \cos t\, dt = \sin t\Big]_0^{2\pi} = 0$ or N.C. $= s(2\pi) - s(0) = 0$

11. (a) $s(t) = \int 5\pi \cos \pi t\, dt = 5 \sin \pi t + C.$ $s(0) = 5 \Rightarrow C = 5.$ $\therefore s(t) = 5 \sin \pi t + 5$

(b) T.D. $= \int_0^{\frac{3}{2}} |5\pi \cos \pi t|\, dt = 5\pi\int_0^{\frac{1}{2}} \cos \pi t\, dt - 5\pi\int_{\frac{1}{2}}^{\frac{3}{2}} \cos \pi t\, dt = 15$

(c) N.C. $= \int_0^{\frac{3}{2}} 5\pi \cos \pi t\, dt = -5$ or N.C. $= s\left(\frac{3}{2}\right) - s(0) = -5$

13. $a = -4\pi^2 \cos 2\pi t,\ v(0) = 2$

$v = \int -4\pi^2 \cos 2\pi t\, dt = -2\pi \sin 2\pi t + C.$ $v(0) = 2 \Rightarrow C = 2.$

$v(t) = -2\pi \sin 2\pi t + 2$

N.C. $= \int_0^2 (-2\pi \sin 2\pi t + 2)\, dt = 4$

15. $a = g,\ v(0) = 0 \Rightarrow v = gt + C.$ $v(0) = 0 \Rightarrow C = 0$ and $v(t) = gt$

N.C. $= \int_0^2 gt\, dt = \frac{1}{2} gt^2\Big]_0^2 = 2g$

17. (a) Total distance = 2; net change = 2

(b) Total distance = 4; net change = 0

(c) Total distance = 4; net change = 4

(d) Total distance = 2; net change = 2

5.2 AREAS BETWEEN CURVES

1. The region is bounded by $y = x^2 - 2$ and $y = 2$. The limits are where $x^2 - 2 = 2 \Leftrightarrow x = \pm 2$. Using symmetry, we have that

$$A = 2\int_0^2 [2 - (x^2 - 2)]\,dx = 2\int_0^2 (-x^2 + 4)dx = -\frac{1}{3}x^3 + 4x\Big] = \frac{32}{3}$$

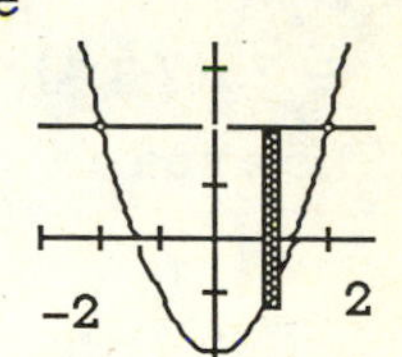

3. The region is bounded by $x = y^2 - y^3$ and the y-axis. The limits are the points where the graph crosses the y-axis: $y^2(1 - y) = 0 \Leftrightarrow y = 0,\ 1$. Then

$$A = \int_0^1 (y^2 - y^3)\,dy = \frac{1}{3}y^3 - \frac{1}{4}y^4\Big] = \frac{1}{12}$$

5. The region is bounded by $y = 2x - x^2$ and $y = -3$. They intersect where $2x - x^2 = -3$ or $x = 3, -1$. The area is:

$$A = \int_{-1}^3 [(2x - x^2) - (-3)]\,dx = x^2 - \frac{1}{3}x^3 + 3x\Big]_{-1}^3 = \frac{32}{3}$$

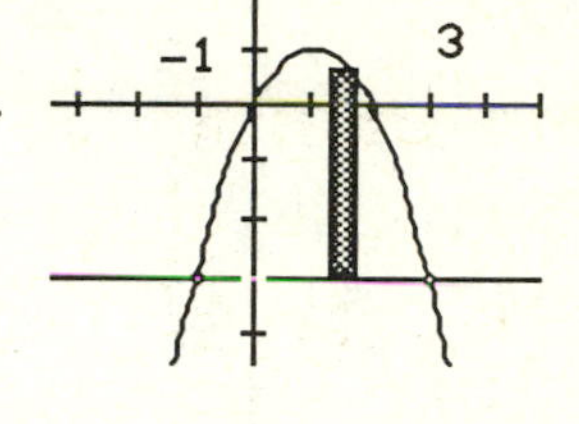

7. The region is bounded by $x = 3y - y^2$ and $x + y = 3$. They intersect where $3y - y^2 = 3 - y$ or $y = 3, 1$. The area is:

$$A = \int_1^2 [(3y - y^2) - (3 - y)]\,dy = 2y^2 - \frac{1}{3}y^3 - 3y\Big]_1^2 = \frac{4}{3}$$

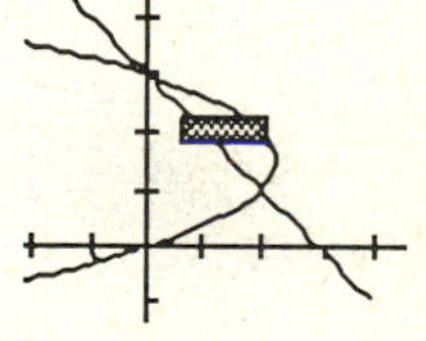

9. The region is bounded by $x = y^2$ and $x = y + 2$. They intersect where $y^2 - y - 2 = 0$ or $y = 2, -1$. The area is:

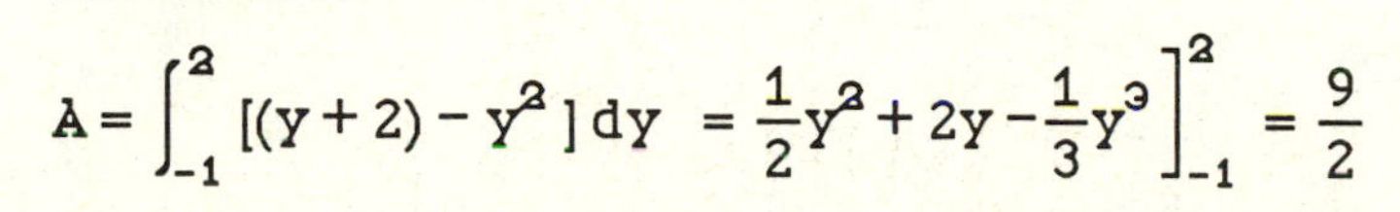

$$A = \int_{-1}^2 [(y + 2) - y^2]\,dy = \frac{1}{2}y^2 + 2y - \frac{1}{3}y^3\Big]_{-1}^2 = \frac{9}{2}$$

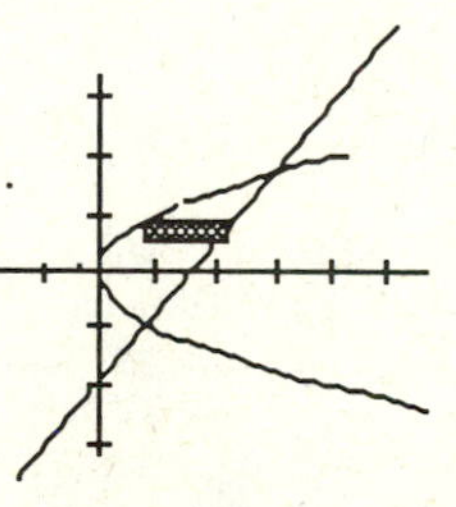

11. The region is bounded by $x = y^3$ and $x = y^2$. They intersect

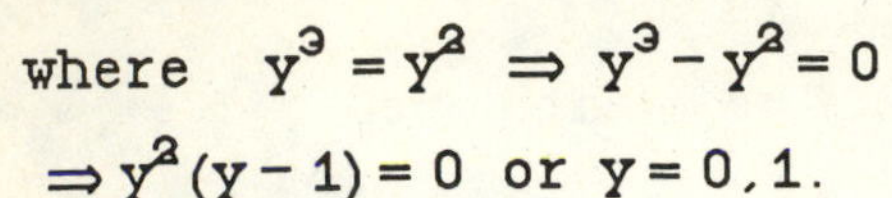

where $y^3 = y^2 \Rightarrow y^3 - y^2 = 0$

$\Rightarrow y^2(y-1) = 0$ or $y = 0, 1$.

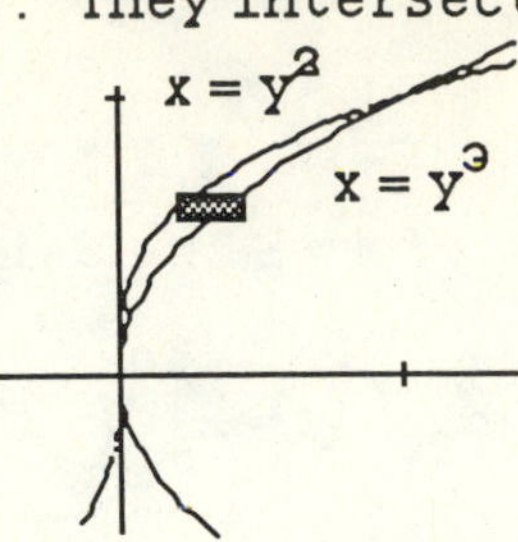

The area is:

$$A = \int_0^1 (y^3 - y^2)\,dy = \frac{1}{4}y^4 - \frac{1}{3}y^3\Big]_0^1 = \frac{1}{12}$$

13. The region is bounded by $y = x^2 - 2x$ and $y = x$.

They intersect where $x^2 - 2x = x$ or $x = 0, 3$.

The area is:

$$A = \int_0^3 [x - (x^2 - 2x)]\,dx = \frac{3}{2}x^2 - \frac{1}{3}x^3\Big]_0^3 = \frac{9}{2}$$

15. The region is bounded by $x = -2y^2 + 3$ and $x = y^2$. They intersect where

$-2y^2 + 3 = y^2$ or $y = \pm 1$. Using symmetry,

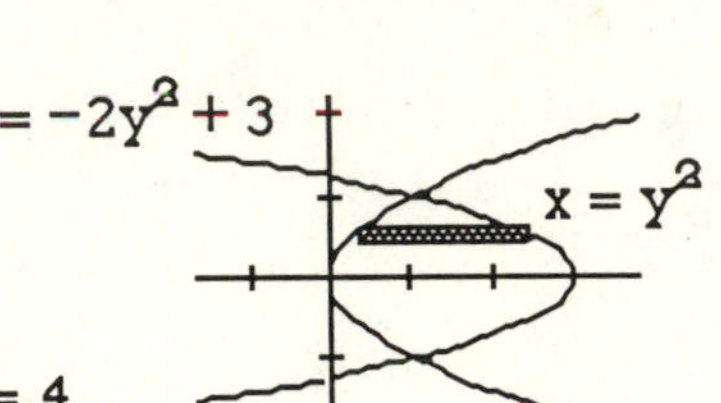

$$A = 2\int_0^1 [(-2y^2 + 3) - y^2]\,dy = 2\left[-\frac{2}{3}y^3 + 3y - \frac{1}{2}y^2\right]_0^1 = 4.$$

17. The region is bounded by $y = x$ and $y = 2 - (x-2)^2$.

They intersect where $2 - (x-2)^2 = x$ or $x = 1, 2$.

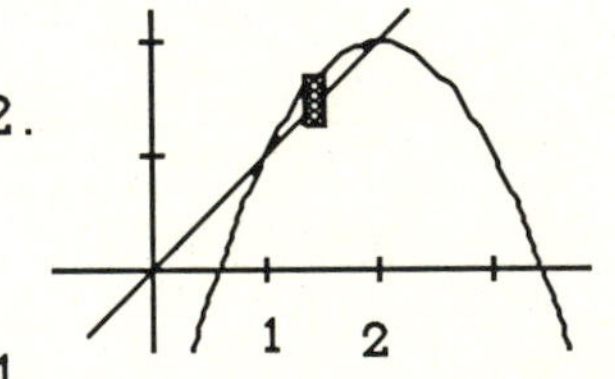

The area is:

$$A = \int_1^2 ([2 - (x-2)^2] - x)\,dx = -\frac{1}{3}x^3 + \frac{3}{2}x^2 - 2x\Big]_1^2 = \frac{1}{6}$$

19. The region is bounded by $y = 4 - 4x^2$ and $y = x^4 - 1$.

They intersect where $4 - 4x^2 = x^4 - 1$

or $x^4 + 4x^2 - 5 = 0 \Leftrightarrow x = \pm 1$. Using symmetry,

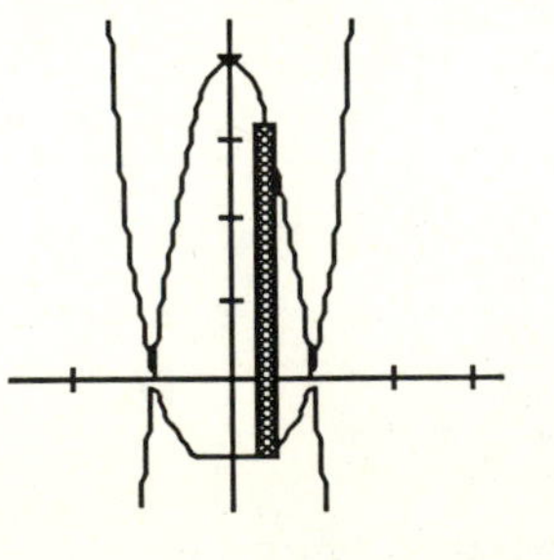

$$A = 2\int_0^1 [(4 - 4x^2) - (x^4 - 1)]\,dx = 2\left[5x - \frac{4}{3}x^3 - \frac{1}{5}x^5\right]_0^1 = \frac{104}{15}$$

21. Using symmetry, $A = 2\int_0^{\pi} (\cos x - (-1))\,dx = 2[\sin x + x]_0^{\pi} = 2\pi$

23. The region is bounded by $y = \sin\left(\frac{\pi x}{2}\right)$ and $y = x$. They intersect where $\sin\left(\frac{\pi x}{2}\right)$ or $x = -1, 0,$ or 1. Using symmetry,

$$A = 2\int_0^1 \left(\sin\left(\frac{\pi x}{2}\right) - x\right)dx = -\frac{4}{\pi}\cos\frac{\pi x}{2} - x^2\Big]_0^1 = \frac{4}{\pi} - 1.$$

25. The graphs intersect when $\sin x = \cos x$ or $x = \frac{\pi}{4}$.

$$A = \int_0^{\frac{\pi}{4}} (\cos x - \sin x)\,dx = \sin x + \cos x\Big]_0^{\frac{\pi}{4}} = \sqrt{2} - 1$$

27. The region is composed of two distinct parts, the left half bounded by $y = x^2$, $x = 1$ and x– axis with area given by $A = \int_0^1 x^2\,dx$, and the right piece bounded by $y = 2 - x$, x–axis and $x = 1$ with area given by $A = \int_1^2 (2 - x)\,dx$. The total area is:

$$A = \int_0^1 x^2\,dx + \int_1^2 (2 - x)\,dx = \frac{1}{3}x^3\Big]_0^1 + 2x - \frac{1}{2}x^2\Big]_1^2 = \frac{5}{6}$$

29. $y = 3 - x^2$ and $y = -1 \Rightarrow 3 - x^2 = -1 \Leftrightarrow x = \pm 2$

(a) $A = \int_{-2}^2 [3 - x^2 - (-1)]\,dx = 2\int_0^2 (4 - x^2)dx = 4x - \frac{1}{3}x^3\Big]_0^2 = \frac{32}{3}$

(b) $A = 3\int_{-1}^3 \sqrt{3 - y}\,dy = -\frac{4}{3}(3 - y)^{3/2}\Big]_{-1}^3 = \frac{32}{3}$

31. $A = \int_{\frac{\pi}{4}}^{\frac{3\pi}{4}} (\sin x - \cos x)dx = -\cos x - \sin x\Big]_{\frac{\pi}{4}}^{\frac{3\pi}{4}} = 2\sqrt{2}$

33. Area of the parabolic region is:

$$2\int_0^a (a^2 - x^2)\,dx = 2\,a^2 x - \frac{1}{3}x^3\Big]_0^a = \frac{4}{3}a^3$$

Area of the triangle is: $\frac{1}{2}[2a \cdot a^2] = a^3$.

The $\lim\limits_{a\to 0} \dfrac{a^3}{\frac{4a^3}{3}} = \dfrac{3}{4}$.

5.3 CALCULATING VOLUMES BY SLICING. VOLUMES OF REVOLUTION

1. $V=\pi\int_0^2 y^2\,dx = \pi\int_0^2 (2-x)^2\,dx=\pi\int_0^2 (4-4x+x^2)\,dx$

$=\pi\left[4x-2x^2+\frac{1}{3}x^3\right]_0^2=\frac{8\pi}{3}$

3. $x-x^2=0 \Leftrightarrow x=0$ or 1. Therefore, $V=\pi\int_0^1 y^2\,dx$

$=\pi\int_0^1 (x-x^2)^2\,dx=\pi\int_0^1 (x^2-2x^3+x^4)\,dx$

$=\pi\left[\frac{1}{3}x^3-\frac{1}{2}x^4+\frac{1}{5}x^5\right]_0^1=\frac{\pi}{30}$

5. $x^2-2x=0 \Leftrightarrow x=0$ or 2. Therefore, $V=\pi\int_0^2 y^2\,dx$

$=\pi\int_0^2 (x^2-2x)^2\,dx=\pi\int_0^2 (x^4-4x^3+4x^2)\,dx$

$=\pi\left[\frac{1}{5}x^5-x^4+\frac{4}{3}x^3\right]_0^2=\frac{16\pi}{15}$

7. $V=\pi\int_0^1 y^2\,dx=\pi\int_0^1 (x^4)^2\,dx=\pi\int_0^1 x^8\,dx=\pi\frac{1}{9}x^9\Big]_0^1=\frac{\pi}{9}$

9. $V=\pi\int_{-\frac{\pi}{4}}^{\frac{\pi}{4}} \sec^2 x\,dx=\pi\tan x\Big]_{-\frac{\pi}{4}}^{\frac{\pi}{4}}=2\pi$

11. $V=\pi\int_0^2 x^2\,dy=\pi\int_0^2 (2y)^2dy=\frac{4\pi}{3}y^3\Big]_0^2=\frac{32\pi}{3}$

13. $V=\pi\int_{-1}^1 (1-y^2)^2\,dy=2\pi\int_0^1 (1-2y^2+y^4)\,dy=2\pi\left[y-\frac{2}{3}y^3+\frac{1}{5}y^5\right]_0^1=\frac{16\pi}{15}$

15. $V=\pi\int_1^2 \left(\frac{1}{y}\right)^2 dy=-\frac{\pi}{y}\Big]_1^2=\frac{\pi}{2}$

17. $V=\pi\int_0^{\frac{\pi}{2}} (2\sin 2x)^2\,dx=4\pi\int_0^{\frac{\pi}{2}}\left(\frac{1}{2}-\frac{1}{2}\cos 4x\right)dx$

$=2\pi\left[x-\frac{1}{4}\sin 4x\right]_0^{\frac{\pi}{2}}=\pi^2$

19. (a) $V=\int_0^4 [(2)^2-(\sqrt{x})^2]\,dx=\pi\int_0^4 (4-x)\,dx=\pi\left[4x-\frac{1}{2}x^2\right]_0^4=8\pi$

(b) $V=\pi\int_0^4 (2-\sqrt{x})^2\,dx=\pi\int_0^4 (4-4\sqrt{x}+x)dx=\pi\left[4x-\frac{8}{3}x^{3/2}+\frac{1}{2}x^2\right]_0^4=\frac{8\pi}{3}$

21. $V = \pi\int_0^1 (1-y)^2 dx = \pi\int_0^1 (1-x^2)^2 d = \pi\int_0^1 (1-x^2+x^4)\,dx$

$= \pi\left(x - \frac{1}{3}x^3 + \frac{1}{5}x^5 \right)\Big]_0^1 = \frac{8\pi}{15}$

23. $V = 2\pi\int_0^{\pi} (\cos x + 1)^2\,dx = 2\pi\int_0^{\pi} (\cos^2 x + 2\cos x + 1)dx$

$= 2\pi\int_0^{\pi}\left[\left(\frac{1}{2} + \frac{1}{2}\cos 2x\right) + 2\cos x + 1\right] dx$

$= 2\pi\left[\frac{3}{2}x + \frac{1}{4}\sin 2x \;\; 2\sin x\right]_0^{\pi} = 3\pi^2$

25. $V = \pi\int_{-1}^{1} [4-(3x^2+1)]^2\,dx = 2\pi\int_0^1 9(1-2x^2+x^4)\,dx$

$= 18\pi\left[x - \frac{2}{3}x^3 + \frac{1}{5}x^5\right]_0^1 = \frac{144}{15}$

27. $V = \pi\int_0^{\pi} (c - \sin x)^2\,dx = \pi\int_0^{\pi} (c^2 - 2c\sin x + \sin^2 x)\,dx$

$= \pi\int_0^{\pi}\left(c^2 - 2c\sin x + \frac{1}{2} - \frac{1}{2}\cos 2x\right)dx$

$\pi\left[c^2 x + 2c\cos x + \frac{1}{2}x - \frac{1}{4}\sin 2x\right]_0^{\pi} = \pi\left(c^2\pi - 4c + \frac{\pi}{2}\right)$

If $V(c) = \pi\left(c^2\pi - 4c + \frac{\pi}{2}\right)$ then $V'(c) = \pi(2c\pi - 4) = 0 \Leftrightarrow c = \frac{2}{\pi}$.

$V''(c) = 2\pi^2 > 0 \Rightarrow c = \frac{2}{\pi}$ minimizes the volume.

29. (a) $V = \pi\int_0^h \left(\sqrt{a^2-y^2}\right)^2 dy = \frac{\pi h^2}{3}(3a - h)$

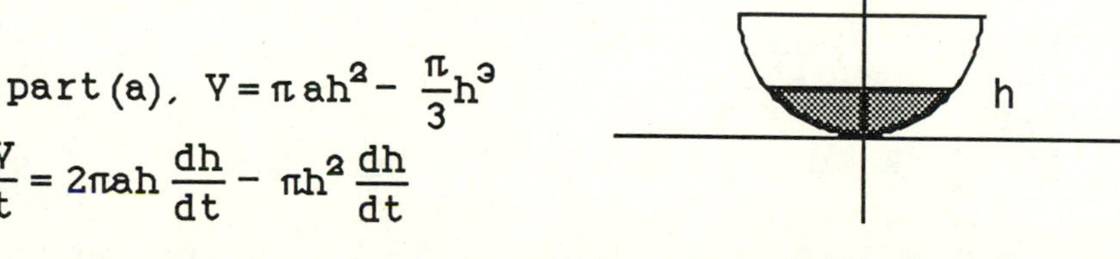

(b) From part (a), $V = \pi a h^2 - \frac{\pi}{3}h^3$

$\frac{dV}{dt} = 2\pi a h\frac{dh}{dt} - \pi h^2\frac{dh}{dt}$

$(.2) = [\,2\pi(5)(4) - \pi(16)]\,\frac{dh}{dt}$

$\frac{dh}{dt} = \frac{1}{120\pi}$ ft/s

31. The volume of the pyramid is $V = Bh$, where B = base area. We use the formula $B = \frac{1}{2}d_1d_2$, d_1 and d_2 diagonals. The diagonals of a square are equal, and $d = |y_1 - y_2| = 5x^2 - (-5x^2) = 10x^2$. Therefore,

$$V = \int_0^4 \frac{1}{2}(10x^2)(10x^2)\,dx = 10x^5\Big]_0^4 = 10,240$$

33. The base is a square with side a chord of the circle, running from a point on the top, $P\left(x, \sqrt{a^2 - x^2}\right)$, to a corresponding point on the bottom, $Q\left(x, -\sqrt{a^2 - x^2}\right)$. A side of the base is then $s = 2\sqrt{a^2 - x^2}$ and the base area is $B = s^2 = 4(a^2 - x^2)$. The volume is then:

$$V = 4\int_{-a}^{a} (a^2 - x^2)\,dx = 8\left[a^2x - \frac{1}{3}x^3\right]_0^a = \frac{16}{3}a^3$$

35. The base is a leg of an isosceles right triangle with length, as in problem 33, equal to $2\sqrt{a^2 - y^2}$. The base area is given

$B = \frac{1}{2}bh = \frac{1}{2}\left(2\sqrt{a^2 - y^2}\right)^2 = 2(a^2 - y^2)$. The volume is

$$V = \int_{-a}^{a} 2(a^2 - y^2)\,dy = \frac{8a^3}{3}.$$

37. $$V \approx \frac{50 - 0}{2(10)}[6 + 2(8.2) + 2(9.1) + 2(9.9) + 2(10.5) + 2(11.0) + 2(11.5) + 2(11.9) + 2(12.3) + 2(12.7) + 13] = 15,990$$

5.4 VOLUMES MODELED WITH WASHERS AND CYLINDRICAL SHELLS

1. We use a disk, with radius $y = 2 - x$. The limits run from the origin to the x-intercept of the line, $x = 2$.

$$V = \pi\int_0^2 (2 - x)^2\,dx = -\frac{\pi}{3}(2 - x)^3\Big]_0^2 = \frac{8\pi}{3}$$

3. We use a washer, with outer radius $R = 3x - x^2$ and inner radius $r = x$. The limits are the points where $3x - x^2 = x$, or $x = 0, 2$. The volume is $V = \pi\int(R^2 - r^2)\,dx$

$$= \int_0^2 [(3x - x^2)^2 - x^2]\,dx = \pi\int_0^2 (8x^2 - 6x^3 + x^4)\,dx$$

$$= \pi\left[\frac{8}{3}x^3 - \frac{3}{2}x^4 + \frac{1}{5}x^5\right]_0^2 = \frac{56\pi}{15}$$

5. We use a washer, with outer radius $R = 4$ and inner radius $r = x^2$. The limits are the points where $x^2 = 4$, or $x = \pm 2$. We also use the symmetry of the figure. The volume is $V = \pi \int (R^2 - r^2)\,dx$

$$= 2\pi \int_0^2 [(4)^2 - (x^2)^2]\,dx = 2\pi \int_0^2 (16 - x^4)\,dx$$

$$= 2\pi \left[16x - \frac{1}{5}x^5 \right]_0^2 = \frac{256\pi}{5}$$

7. We use a washer, with outer radius $R = x + 3$ and inner radius $r = x^2 + 1$. The limits are the points where $x^2 + 1 = x + 3$, or $x = 2, -1$. The volume is $V = \pi \int (R^2 - r^2)\,dx$

$$= \pi \int_{-1}^2 [(x+3)^2 - (x^2+1)^2]\,dx = \pi \int_{-1}^2 (-x^4 - x^2 + 6x + 8)\,dx$$

$$= \pi \left[-\frac{1}{5}x^5 - \frac{1}{3}x^3 + 3x^2 + 8x \right]_{-1}^2 = \frac{117\pi}{5}$$

9. We use a shell, with height $h = x^4$ and radius $r = x$. The volume $V = \int 2\pi rh\,dr$

$$= 2\pi \int_0^1 x^5\,dx = 2\pi \left[\frac{1}{6}x^6 \right]_0^1 = \frac{\pi}{3}$$

11. The equations of the sides of the triangle are $x = 1$, $y = 1$, and $y = x$.

(a) We use a shell, with radius $r = y$ and height $h = x - 1$.

$$V = 2\pi \int_1^2 y(1 - x)\,dy = 2\pi \int_1^2 y(1 - y)\,dy = \frac{5\pi}{3}$$

(b) We use a shell, with radius x and height $h = 2 - y$.

$$V = 2\pi \int_1^2 x(2 - y)\,dx = 2\pi \int_1^2 x(2 - x)\,dx = \frac{4\pi}{3}$$

13. (a) We use a shell, with radius $r = 1 - x = 1 - (y - y^3)$ and height $h = y$. The curve $y - y^3$ intersects the y-axis at $y = 0, 1$.

$$V = 2\pi \int_0^1 y(1 - y + y^3)\,dy = 2\pi \left[\frac{1}{2}y^2 - \frac{1}{3}y^3 + \frac{1}{5}y^5 \right]_0^1 = \frac{11\pi}{15}$$

(b) We use a washer, with outer radius $R = 1$ and inner radius $r = x = y - y^3$.

$$V = \pi\int_0^1 [(1)^2 - (y - y^3)^2]\,dy = \pi\int_0^1 (1 - y^2 + 2y^4 - y^6)\,dy$$
$$= \pi\left[y - \frac{1}{3}y^3 + \frac{2}{5}y^5 - \frac{1}{7}y^7\right]_0^1 = \frac{97\pi}{105}$$

(c) We use a disk, with radius $R = 1 - x = 1 - (y - y^3)$.

$$V = \pi\int_0^1 [(1 - y + y^3)^2]\,dy = \pi\int_0^1 (1 - 2y + y^2 + 2y^3 - 2y^4 + y^6)\,dy$$
$$= \pi\left[y - y^2 + \frac{1}{3}y^3 + \frac{1}{2}y^4 - \frac{2}{5}y^5 + \frac{1}{7}y^7\right]_0^1 = \frac{121\pi}{210}$$

(d) We use a shell, with radius $r = 1 - x = 1 - (y - y^3)$ and height $h = 1 - y$.

$$V = 2\pi\int_0^1 \Big[(1 - y)(1 - y + y^3)\,dy = 2\pi\int_0^1 (1 - 2y + y^2 + y^3 - y^4)\,dy$$
$$= 2\pi\left[y - y^2 + \frac{1}{3}y^3 + \frac{1}{4}y^4 - \frac{1}{5}y^5\right]_0^1 = \frac{23\pi}{30}$$

15. The curve $y = x^3$ intersects $y = 4x$ in the point $(2, 8)$.

(a) We use a shell, with height h = horizontal distance to the cubic − the distance to the line $= y^{1/3} - \frac{y}{4}$, and radius $r = y$. Then

$$V = 2\pi\int_0^8 y\left(y^{1/3} - \frac{y}{4}\right)dy = 2\pi\left[\frac{3}{7}y^{7/3} - \frac{1}{12}y^3\right]_0^8 = \frac{512\pi}{21}$$

(b) We use a disk, with outer radius $R = 8 - x^3$ and inner radius $r = 8 - 4x$. Then

$$V = \pi\int_0^2 [(8 - x^3)^2 - (8 - 4x)^2]\,dx = \pi\left[-4x^4 - \frac{16}{3}x^3 + 32x^2 + \frac{1}{7}x^7\right]_0^2 = \frac{832\pi}{21}$$

17. The curve $y = 2x - x^2$ intersects $y = x$ where $2x - x^2 = x$, or $x = 1, 0$.

(a) We use a shell with radius $r = x$ and height $h = x - x^2$.

$$V = 2\pi \int_0^1 x(x - x^2)\,dx = 2\pi \left[\frac{1}{3}x^3 - \frac{1}{4}x^4\right]_0^1 = \frac{\pi}{6}$$

(b) We use a shell with radius $r = 1 - x$ and height $h = x - x^2$.

$$V = 2\pi \int_0^1 (1 - x)(x - x^2)\,dx = 2\pi \int_0^1 (x - 2x^2 + x^3)\,dx$$

$$= 2\pi \left[\frac{1}{2}x^2 - \frac{2}{3}x^3 + \frac{1}{4}x^4\right]_0^1 = \frac{\pi}{6}$$

19. $$V = \pi \int_0^{\frac{\pi}{4}} (\cos^2 x - \sin^2 x)\,dx = \pi \int_0^{\frac{\pi}{4}} \cos 2x\,dx = \frac{\pi}{2}\sin 2x\Big]_0^{\frac{\pi}{4}} = \frac{\pi}{2}$$

21. We use a shell with radius $r = x$ and height $h = 2x^2 - (x^4 - 2x^2)$.

The curves intersect in $x = 0, \pm 2$.

$$V = 2\pi \int_0^2 x[2x^2 - (x^4 - 2x^2)]\,dx = 2\pi \int_0^2 (4x^3 - x^5)\,dx$$

$$= \left[x^4 - \frac{1}{6}x^6\right]_0^2 = \frac{32\pi}{3}$$

23. $$V = 2\pi \int_0^{\pi} x \sin x\,dx = 2\pi\,[\sin x - x\cos x\Big]_0^{\pi} = 2\pi^2$$

25. We use a shell of height $2\sqrt{a^2 - x^2}$ and radius $r = b - x$.

$$V = 2\pi \int_{-a}^{a} 2(b - x)\sqrt{a^2 - x^2}\,dx$$

$$= 4\pi b \int_{-a}^{a} \sqrt{a^2 - x^2}\,dx - 4\pi \int_{-a}^{a} x\sqrt{a^2 - x^2}\,dx$$

$$= 4\pi b\left(\frac{\pi a^2}{2}\right) - \frac{4\pi}{3}(a^2 - x^2)^{3/2}\Big]_{-a}^{a} = 2\pi^2 a^2 b$$

5.5 LENGTHS OF PLANE CURVES

1. $y = \frac{1}{3}(x^2+2)^{3/2}$, $0 \le x \le 3$. $\frac{dy}{dx} = x\sqrt{x^2+2}$.

$$s = \int_0^3 \sqrt{1+\left(\frac{dy}{dx}\right)^2} = \int_0^3 \sqrt{1+x^2(x^2+2)}\,dx = \int_0^3 \sqrt{(x^2+1)^2}\,dx$$

$$= \int_0^3 (x^2+1)\,dx = \frac{1}{3}x^3 + x\Big]_0^3 = 12$$

3. $9x^2 = 4y^3$ from $(0,0)$ to $(2\sqrt{3}, 3)$. $x = \frac{2}{3}y^{3/2} \Rightarrow \frac{dx}{dy} = \sqrt{y}$.

$$s = \int_0^3 \sqrt{1+\left(\frac{dx}{dy}\right)^2}\,dy = \int_0^3 \sqrt{1+y}\,dy = \frac{2}{3}(1+y^{3/2}\Big]_0^3 = \frac{14}{3}$$

5. $x = \frac{y^4}{4} + \frac{1}{8y^2}$, $1 \le y \le 2$. $\frac{dy}{dx} = y^3 - \frac{1}{4y^3}$. $\left(\frac{dy}{dx}\right)^2 = y^6 - \frac{1}{2} + \frac{1}{16y^6}$.

$$s = \int_1^2 \sqrt{1+y^6-\frac{1}{2}+\frac{1}{16y^6}}\,dy = \int_1^2 \sqrt{y^6+\frac{1}{2}+\frac{1}{16y^6}} = \int_1^2 \sqrt{\left(y^3+\frac{1}{4y^3}\right)^2}\,dy$$

$$= \int_1^2 \left(y^3+\frac{1}{4y^3}\right)dy = \frac{1}{4}y^4 - \frac{1}{8y^2}\Big]_1^2 = \frac{123}{32}$$

7. $x = a\cos^3 t \Rightarrow \frac{dx}{dt} = -3a\cos^2 t \sin t \Rightarrow \left(\frac{dx}{dt}\right)^2 = 9a^2\cos^4 t \sin^2 t$

$y = a\sin^3 t \Rightarrow \frac{dy}{dt} = 3a\sin^2 t\cos t \Rightarrow \left(\frac{dy}{dt}\right)^2 = 9a^2\sin^4 t\cos^2 t$

$$ds = \sqrt{\left(\frac{dx}{dt}\right)^2 + \left(\frac{dy}{dt}\right)^2}\,dt = \sqrt{9a^2\cos^4 t\sin^2 t + 9a^2\sin^4 t\cos^2 t\,dt}$$

$$= 3a\cos t\sin t\,dt.$$

$$s = 4\int_0^{\frac{\pi}{2}} 3a\cos t\sin t\,dt = 12a\left(\frac{\sin^2 t}{2}\right)\Big]_0^{\frac{\pi}{2}} = 6a$$

9. $$s = \int_0^\pi \sqrt{(-\sin t)^2 + (1+\cos t)^2}\,dt = \int_0^\pi \sqrt{2+2\cos t}\,dt = 2\int_0^\pi \sqrt{\frac{1+\cos t}{2}}\,dt$$

$$= 2\int_0^\pi \cos\frac{t}{2}\,dt = 4\sin\frac{t}{2}\Big]_0^\pi = 4$$

11. $x = a\cos t + at\sin t \Rightarrow \frac{dx}{dt} = -a\sin t + a\sin t + at\cos t$

$y = a\sin t - at\cos t \Rightarrow \frac{dy}{dt} = a\cos t - a\cos t + at\sin t$

$$s = \int_0^{\frac{\pi}{2}} \sqrt{a^2t^2\cos^2 t + a^2t^2\sin^2 t}\, dt = \int_0^{\frac{\pi}{2}} at\, dt = \frac{1}{2}at^2 \Big]_0^{\frac{\pi}{2}} = \frac{a\pi^2}{8}$$

13. $x = \frac{1}{3}(2t+3)^{3/2} \Rightarrow \frac{dx}{dt} = (2t+3)^{1/2}$; $y = \frac{t^2}{2} + t \Rightarrow \frac{dy}{dt} = t + 1.$

$$s = \int_0^3 \sqrt{2t + 3 + t^2 + 2t + 1}\, dt = \int_0^3 \sqrt{(t+2)^2}\, dt = \int_0^3 (t+2)\, dt = \frac{21}{2}$$

15. $5y^3 = x^2$ intersects the circle $x^2 + y^2 = 6$ where $5y^3 = 6 - y^2 \Rightarrow$ $5y^3 + y^2 - 6 = 0$, which clearly has a root at $y = 1$.

$x = \sqrt{5}y^{3/2} \Rightarrow \frac{dx}{dt} = \frac{3\sqrt{5}}{2}y^{1/2}.$

$$s = 2\int_0^1 \sqrt{1 + \frac{45}{4}y}\, dy = \frac{16}{135}\left(1 + \frac{45}{4}y\right)^{3/2}\Big]_0^1 = \frac{134}{27}$$

17. If $y = f(x)$, then Equation 4 states that

$$L = \int_a^b \sqrt{1 + [f'(x)]^2}\, dx.$$

If we regard f a parametized by $x = x$, $y = f(x)$, then Equation 7 states that

$$L = \int_a^b \sqrt{\left(\frac{dx}{dx}\right)^2 + \left(\frac{dy}{dx}\right)^2}\, dx = \int_a^b \sqrt{1 + \left(\frac{dy}{dx}\right)^2}\, dx.$$

19. $y = 25\cos\frac{\pi x}{50} \Rightarrow \frac{dy}{dx} = -\frac{\pi}{2}\sin\frac{\pi x}{50}$. Then $s = 2\int_0^{25} \sqrt{1 + \frac{\pi^2}{4}\sin^2\frac{\pi x}{50}}\, dx.$

We approximate this integral using Simpson's Rule, with $n = 10$ and $h = 2.5$ to get $s \approx 73.17466440$. The cost is then approximately $73.1747 \times 300 \times 1.75 \approx \$38,400$.

5.6 THE AREA OF A SURFACE OF REVOLUTION

1. $S = 2\pi \int_0^1 y\sqrt{1 + (3x^2)^2}\, dx = 2\pi \int_0^1 x^3 \sqrt{1 + 9x^4}\, dx$

$$= \frac{\pi}{27}(1 + 9x^4)^{3/2}\Big]_0^1 = \frac{\pi}{27}[10\sqrt{10} - 1]$$

3. $$S = 2\pi\int_1^3 (y+1)\sqrt{1+\left(x^2-\frac{1}{4}x^{-2}\right)^2}\,dx$$

$$= 2\pi\int_1^3 \left(1+\frac{x^3}{3}+\frac{1}{4x}\right)\sqrt{1+x^4-\frac{1}{2}+\frac{1}{16x^4}}\,dx$$

$$= 2\pi\int_1^3 \left(1+\frac{x^3}{3}+\frac{1}{4x}\right)\sqrt{x^4+\frac{1}{2}+\frac{1}{16x^4}}\,dx$$

$$= 2\pi\int_1^3 \left(1+\frac{x^3}{3}+\frac{1}{4x}\right)\left(x^2+\frac{1}{4x^2}\right)dx$$

$$= 2\pi\int_1^3 \left(x^2+\frac{x^5}{3}+\frac{x}{4}+\frac{1}{4}x^{-2}+\frac{x}{12}+\frac{1}{16}x^{-3}\right)dx$$

$$= 2\pi\left[\frac{x^3}{3}+\frac{x^6}{18}+\frac{x^2}{8}-\frac{1}{4x}+\frac{x^2}{24}-\frac{1}{32x^2}\right]_1^3 = \frac{1823\pi}{18}$$

5. $$ds = \sqrt{1+\left(y^3-\frac{1}{4y^3}\right)^2}\,dy = \sqrt{1+y^6-\frac{1}{2}+\frac{1}{16y^6}}\,dy$$

$$= \sqrt{y^6+\frac{1}{2}+\frac{1}{16y^6}}\,dy = \sqrt{\left(y^3+\frac{1}{4y^3}\right)^2}\,dy.$$

$$S = 2\pi\int_1^2 y\,ds = 2\pi\int_1^2 y\left(y^3+\frac{1}{4y^3}\right)dy = 2\pi\left[\frac{y^5}{5}-\frac{1}{4y}\right]_0^{2\pi} = \frac{253\pi}{20}$$

7. $y = \frac{1}{3}(x^2+2)^{3/2}$, $0 \le x \le 3$. $\frac{dy}{dx} = x\sqrt{x^2+2}$.

$$ds = \sqrt{1+\left(\frac{dy}{dx}\right)^2} = \sqrt{1+x^2(x^2+2)}\,dx = \sqrt{(x^2+1)^2}\,dx$$

$$S = \int 2\pi\rho\,ds = 2\pi\int_0^3 x(x^2+1)\,dx = 2\pi\left[\frac{x^4}{4}+\frac{x^2}{2}\right]_0^3 = \frac{99\pi}{2}$$

9. $\frac{y}{x} = \tan\theta \Rightarrow y = (\tan\theta)x \Rightarrow 1 = (\tan\theta)\frac{dx}{dy} \Rightarrow \frac{dx}{dy} = \cot\theta.$

$ds = \sqrt{1+\cot^2\theta}\,d\theta = \csc\theta$. Let L be the slant height of the frustrum.

$$S = 2\pi\int\rho\,ds = 2\pi\int_{r_1}^{r_2}\csc\theta\,y\,dy = \csc\theta\cdot\frac{1}{2}y^2\Big]_{r_1}^{r_2}$$

$$= 2\pi\frac{r_2^2-r_1^2}{2}\cdot\frac{L}{r_2-r_1} = \pi(r_2+r_1)L.$$

11. (a) $S = 2\pi\int_0^{2\pi}(1+\sin t)\sqrt{(-\sin t)^2+\cos^2 t}\,dt$

$$= 2\pi\int_0^{2\pi}(1+\sin t)\,dt = 2\pi(t-\cos t)\Big]_0^{2\pi} = 4\pi^2$$

(b) Example 4 found the general formuls to be $S = 4\pi^2 a^2$.

If $a = 1$, $S = 4\pi^2$ as found above.

13. $x = a\cos^3 t \Rightarrow \dfrac{dx}{dt} = -3a\cos^2 t\sin t \Rightarrow \left(\dfrac{dx}{dt}\right)^2 = 9a^2\cos^4 t\sin^2 t$

$y = a\sin^3 t \Rightarrow \dfrac{dy}{dt} = 3a\sin^2 t\cos t \Rightarrow \left(\dfrac{dy}{dt}\right)^2 = 9a^2\sin^4 t\cos^2 t$

$$ds = \sqrt{\left(\frac{dx}{dt}\right)^2+\left(\frac{dy}{dt}\right)^2}\,dt = \sqrt{9a^2\cos^4 t\sin^2 t + 9a^2\sin^4 t\cos^2 t}\,dt$$

$= 3a\cos t\sin t\,dt$. $S = 2\pi\int\rho\,ds$. Therefore,

$$S = 2\cdot 2\pi\int_0^{\frac{\pi}{2}}(a\sin^3 t)(3a\cos t\sin t)\,dt = 12\pi a^2\left(\frac{\sin^5 t}{5}\right)\Big]_0^{\frac{\pi}{2}} = \frac{12\pi a^2}{5}$$

15. The limits of integration will be from θ_1 to θ_2, where $\cos\theta_1 = \dfrac{a+h}{r}$ and $\cos\theta_2 = \dfrac{a}{r}$. We parametize the circle by $x = r\cos\theta$ and $y = r\sin\theta$.

Then $ds = \sqrt{(-r\sin\theta)^2+(r\cos\theta)^2}\,d\theta = r\,d\theta$.

$$S = 2\pi\int_{\theta_1}^{\theta_2} y\,ds = 2\pi\int_{\cos^{-1}\frac{a+h}{r}}^{\cos^{-1}\frac{a}{r}} r^2\sin\theta\,d\theta = 2\pi r^2[-\cos\theta\Big]_{\cos^{-1}\frac{a+h}{r}}^{\cos^{-1}\frac{a}{r}} = 2\pi r h.$$

This is independent of a, the location of h.

5.7 THE AVERAGE VALUE OF A FUNCTION

1. (a) $y_{av} = \dfrac{1}{\frac{\pi}{2}}\displaystyle\int_0^{\frac{\pi}{2}}\sin x\,dx = -\frac{2}{\pi}\cos x\Big]_0^{\frac{\pi}{2}} = \frac{2}{\pi}$

(b) $y_{av} = \dfrac{1}{2\pi}\displaystyle\int_0^{\frac{\pi}{2}}\sin x\,dx = -\frac{2}{\pi}\cos x\Big]_0^{2\pi} = 0$

3. $y_{av} = \dfrac{1}{8}\displaystyle\int_4^{12}\sqrt{2x+1}\,dx = \frac{1}{24}(2x+1)^{3/2}\Big]_4^{12} = \frac{49}{12}$

5. (a) $y_{av} = \frac{\text{area}}{b - a} = \frac{1}{2}$ (b) $y_{av} = \frac{2}{4} = \frac{1}{2}$

7. If $v = V \sin \omega t$ then $v^2 = V^2 \sin^2 \omega t$. Then

$$(v^2)_{av} = \frac{\omega}{2\pi} \int_0^{2\pi} V^2 \sin^2 \omega t \, dt = \frac{V^2 \omega}{2\pi} \int_0^{2\pi} \left(\frac{1}{2} - \frac{1}{2} \cos 2\omega t \right) dt$$

$$= \frac{V^2 \omega}{2\pi} \left[\frac{1}{2} t - \frac{1}{4} \cos 2\omega t \right]_0^{2\pi} = \frac{V^2}{2}.$$ Therefore,

$$V_{rms} = \sqrt{\frac{V^2}{2}} = \frac{V}{\sqrt{2}}$$

9. Average daily inventory $= \frac{1}{30} \int_0^{30} \left(450 - \frac{x^2}{2} \right) dx = 300.$

Average daily holding cost $= 300 \times .02 = \$6.00.$

11. $f_{av} = \frac{1}{365} \int_0^{365} 37 \sin \left[\frac{2\pi}{365} (x - 101) + 25 \right] dx$

$$= \frac{1}{365} \left[25x - \frac{37 \cdot 365}{2\pi} \cos \left(\frac{2\pi}{365} (x - 101) \right) \right]_0^{365} = 25$$

13. (b) $v_{av} = \frac{1}{2} \int_0^2 32 \, t \, dt = 8t^2 \Big]_0^2 = 32$

(c) $v_{av} = \frac{1}{64} \int_0^{64} 8\sqrt{s} \, ds = \frac{1}{12} s^{3/2} \Big]_0^{64} = \frac{128}{3}$

15. $f'_{av} = \frac{1}{b - a} \int_a^b f'(x) \, dx = \frac{1}{b - a} f(x) \Big]_a^b = \frac{f(b) - f(a)}{b - a}$

5.8 MOMENTS AND CENTERS OF MASS

1. $100x = 80(5) \Rightarrow x = 4$ feet from the fulcrum.

3. $M_0 = \int_0^L x \, dx = \frac{1}{2} x^2 \Big]_0^L = \frac{L^2}{2}$; $M = \int_0^L dx = L$; $\bar{x} = \frac{M_0}{M} = \frac{L}{2}$.

5. $M_0 = \int_0^L \left(1 + \frac{x}{L} \right)^2 x \, dx = \int_0^L \left(x + \frac{2x^2}{L} + \frac{x^3}{L} \right) dx = \frac{1}{2} x^2 + \frac{2}{3L} x^3 + \frac{1}{4L} x^4 \Big]_0^L = \frac{17}{12} L^2$

$$M = \int_0^L \left(1 + \frac{x}{L} \right)^2 dx = \int_0^L \left(1 + \frac{2x}{L} + \frac{x^2}{L^2} \right) dx = x + \frac{x^2}{L} + \frac{x^3}{3L^2} \Big]_0^L = \frac{7L}{3}$$

$$\bar{x} = \frac{M_0}{M} = \frac{17 L^2}{12} \cdot \frac{3}{7L} = \frac{17 L}{28}$$

7. $M_x = \int_0^1 6x^2\,dx = 2x^3\Big]_0^1 = 2;\ \bar{y} = \frac{2}{3}$

9. $M = \int_0^1 (y - y^3)\,dy = \frac{1}{2}y^2 - \frac{1}{4}y^4\Big]_0^1 = \frac{1}{4}$

$$M_x = \int_0^1 y\,dM = \int_0^1 y\,(y - y^3)\,dy = \frac{1}{3}y^3 - \frac{1}{5}y^5\Big]_0^1 = \frac{2}{15}.\quad \bar{y} = \frac{2}{15}\cdot\frac{4}{1} = \frac{8}{15}$$

$$M_y = \int_0^1 \frac{1}{2}x\,dM = \int_0^1 \frac{1}{2}(y - y^3)^2\,dy = \frac{1}{2}\int_0^1 (y^2 - 2y^4 + y^6)\,dy$$

$$= \frac{1}{2}\left[\frac{1}{3}y^3 - \frac{2}{5}y^5 + \frac{1}{7}y^7\right]_0^1 = \frac{4}{105}.\quad \bar{x} = \frac{4}{105}\cdot\frac{4}{1} = \frac{16}{105}$$

$$(\bar{x}, \bar{y}) = \left(\frac{16}{105}, \frac{8}{15}\right)$$

11. $M = \int_0^2 [(x - x^2) - (-x)]\,dx = x^2 - \frac{1}{3}x^3\Big]_0^2 = \frac{4}{3}.$

$$M_x = \int_0^2 \frac{1}{2}(x - x^2 + (-x))(2x - x^2)\,dx = -\frac{1}{2}\int_0^2 x^2(2x - x^2)\,dx$$

$$= -\frac{1}{2}\left[\frac{1}{2}x^4 - \frac{1}{5}x^5\right]_0^2 = -\frac{4}{5};\ \bar{y} = \left(-\frac{4}{5}\right)\left(\frac{3}{4}\right) = -\frac{3}{5}$$

$$M_y = \int_0^2 x(2x - x^2)\,dx = \frac{2}{3}x^3 - \frac{1}{4}x^4\Big]_0^2 = \frac{4}{3}.\quad \bar{x} = \frac{4}{3}\cdot\frac{3}{4} = 1$$

$$(\bar{x}, \bar{y}) = \left(1, -\frac{3}{5}\right)$$

13. $M = \frac{\pi a^2}{4};\ M_x = \frac{1}{2}\int_0^a \left(\sqrt{a^2 - x^2}\right)^2 dx = \frac{1}{2}\left[a^2x - \frac{x^3}{3}\right]_0^a = \frac{a^3}{3}.$

$\bar{y} = \frac{a^3}{3}\cdot\frac{4}{\pi a^2} = \frac{4a}{3\pi}.\quad \bar{x} = \bar{y}$ by symmetry.

15. $M = a^2 - \frac{\pi a^2}{4};\ M_x = \int_0^a \frac{a + \sqrt{a^2 - x^2}}{2}\left(a - \sqrt{a^2 - x^2}\right)dx$

$$= \frac{1}{2}\int_0^a x^2\,dx = \frac{1}{6}x^3\Big]_0^a = \frac{a^3}{6}.\quad \bar{y} = \frac{a^3}{6}\cdot\frac{4}{a^2 - \pi a^2} = \frac{2a}{3(4 - \pi)}.$$

By symmetry, $\bar{x} = \bar{y}$.

17. $M = \int_0^2 (2y - y^2)\,dy = y^2 - \frac{1}{3}y^3\Big]_0^2 = \frac{4}{3}$

$M_y = \frac{1}{2}\int_0^2 (2y - y^2)^2\,dy = \frac{2}{3}y^3 - \frac{1}{2}y^4 + \frac{1}{10}y^5\Big]_0^2 = \frac{8}{15}$

$\bar{x} = \frac{8}{15}\cdot\frac{3}{4} = \frac{2}{5}$. $\bar{y} = 1$ by symmetry.

19. $M = \int_0^2 [(2x - x^2) - (2x^2 - 4x)]\,dx = \int_0^2 (6x - 3x^2)\,dx = 3x^2 - x^3\Big]_0^2 = 4$

$M_y = \int_0^2 x(6x - 3x^2)\,dx = 2x^3 - \frac{3}{4}x^4\Big]_0^2 = 4 \quad \therefore \bar{x} = \frac{4}{4} = 1$

$M_x = \int_0^2 \frac{1}{2}2x^2 - 4x + 2x - x^2)(6x - 3x^2)\,dx = -\frac{3}{2}\int_0^2 (x^2 - 2x)^2\,dx$

$= -\frac{3}{2}\left[\frac{x^5}{5} - x^4 + \frac{4x^3}{3}\right]_0^2 = -\frac{8}{5}$. $\bar{y} = -\frac{8}{5}\cdot\frac{1}{4}. = -\frac{2}{5}$

21. A median lies along the y-axis, and the centroid lies $\frac{2}{3}$ of the distance from (0,3) to (0,0) at the point (0,1).

23. The median from the vertex at the origin is $\frac{3\sqrt{2}}{2}$ units long. The centroid is located $\frac{2}{3}$ of this length away from the origin, or $\sqrt{2}$ units along the median. The coordinates of this point are (1,1).

25. $M_x = \int_0^2 \sqrt{x}\left(\sqrt{1 + \frac{1}{4x}}\right)dx = \int_0^2 \sqrt{x + \frac{1}{4}}\,dx = \left(x + \frac{1}{4}\right)^{3/2}\Big]_0^2 = \frac{13}{6}$
This differs from the answer in Prob. 1, Art. 5.6 by a factor of 2π.

27. $M = \int_0^h k\sqrt{y}\left(b - \frac{b}{h}\right)y\,dy$

$= \frac{2k\,(bh - b)}{5h}\,y^{5/2}\Big]_0^h = \frac{2k(bh - b)h^{3/2}}{5}$

$M_x = \int_0^h (k\sqrt{y})(y)\left(b - \frac{b}{h}\right)y\,dy$

$= \frac{2k(b - h)}{7h}\,y^{7/2}\Big]_0^h = \frac{2k\,(bh - b)h^{5/2}}{7}.$

$\bar{y} = \frac{2k\,(bh - b)h^{5/2}}{7} \cdot \frac{5}{2k(bh - b)h^{3/2}} = \frac{5h}{7}$

29. $M = \int_0^1 12x\,(x - x^2)\;dx = 12\left[\frac{x^3}{3} - \frac{x^4}{4}\right]_0^1 = 1$

$M_x = \int_0^1 (12x)\left(\frac{x + x^2}{2}\right)(x - x^2)\,dx = 6\int_0^1 (x^3 - x^5)\,dx = \frac{1}{2}$

$M_y = \int_0^1 12x\,(x - x^2)\,dx = \frac{3}{5}. \quad \therefore (\bar{x}, \bar{y}) = \left(\frac{3}{5}, \frac{1}{2}\right)$

31. $M = ka\int_0^\pi a\sin\theta\,d\theta = -ak\cos\theta\Big]_0^\pi = 2ak$

$M_y = ka\int_0^\pi \sin\theta\,(a\cos\theta)\,d\theta = ka^2\left(\frac{1}{2}\sin^2\theta\right]_0^\pi = 0$

$M_x = ka^2\int_0^\pi \sin^2\theta = ka^2\left[\frac{1}{2}\theta - \frac{1}{4}\sin 2\theta\right]_0^\pi = \frac{a^2k\pi}{2}.$

$\bar{x} = 0;\ \bar{y} = \frac{a^2k\pi}{2}\frac{1}{2ak} = \frac{\pi a}{4}$

5.9 WORK

1. $6 = (0.4)k \Rightarrow k = 15$

$W = \int_0^{0.2} 15x\,dx = \frac{15}{2}x^2\Big]_0^{0.2} = 7.5\,(.04) = 0.3$ newtons

3. (a) $800 = 4k \Rightarrow k = 200$

(b) $\int_0^2 200x\,dx = 400$ in–lbs

(c) $200\,x = 1600 \Rightarrow x = 8$ in

5. $5k = 1 \Rightarrow k = \frac{1}{5}$; $W = \int_0^5 \frac{1}{5}x\,dx = \frac{x^2}{10}\Big]_0^5 = \frac{5}{2}$ ft-lbs

$\frac{1}{5}x = 2 \Rightarrow x = 10$. Total length is 12 ft.

7. (a) $W = \int_{-1}^{0} \frac{k}{(1-x)^2}dx = \frac{k}{1-x}\Big]_{-1}^{0} = \frac{k}{2}$

(b) $W = \int_3^5 \left[\frac{k}{(x+1)^2} + \frac{k}{(x-1)^2}\right]dx = \frac{-k}{x+1} + \frac{-k}{x-1}\Big]_3^5 = \frac{k}{3}$

9. $18\frac{dw}{dx} = 72 \Rightarrow \frac{dw}{dx} = 4$ lbs/ft. $\therefore$ $w = 4x + C$. $x = 0 \Rightarrow w = 144$ so $C = 144$.

$$W = \int_0^{18} (144 - 4x)\,dx = 144x - 2x^2\Big]_0^{18} = 1{,}944 \text{ ft-lbs}$$

11. $W = 48\int_0^h \pi x^2 y(\pi + h - y)\,dy$

$= 48\int_0^h \pi\left(\frac{ry}{h}\right)^2(\pi + h - y)\,dy$

$= \frac{48r^2\pi}{h^2}\int_0^h (\pi y^2 + hy^2 - y^3)\,dy$

$= 4\pi r^2 h(4\pi + h)$

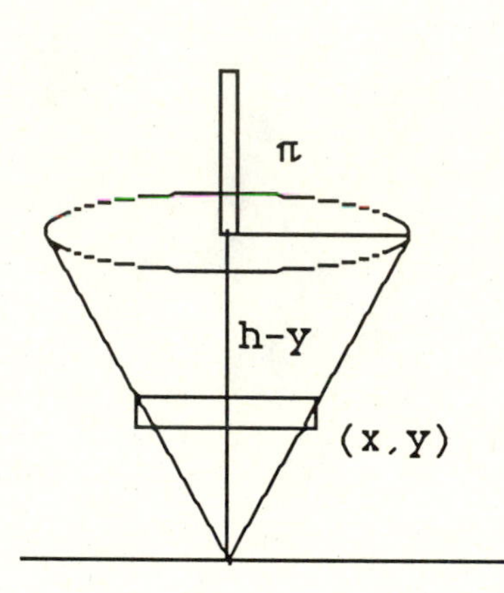

Note: If you know that the centroid of a right circular cone is located $\frac{3}{4}$ the distance from the vertex to the base, then you can compute

$$W = (48)\left(\frac{\pi r^2 h}{3}\right)\left(\frac{1}{4}h + \pi\right) = 4\pi r^2 h\,(4\pi + h)$$

13. $V = 9800\int_0^{16} \pi y(16 - y)\,dy = 9800\left[8y^2 - \frac{1}{3}y^3\right]_0^{16} = \frac{20{,}070{,}400\pi}{3}$ Newtons

15. The center of mass is 2 feet below the surface.

$$W = (62.5)\left(\frac{100\pi}{3}\right)(8)(2 + 6) = \frac{400{,}000\pi}{3} \text{ ft-lbs} = \frac{200\pi}{3} \text{ ft-tons}$$

17. $\frac{dw}{dh} = \frac{400 \text{ gal}}{4750 \text{ ft}} \cdot \frac{8 \text{ lbx}}{\text{gal}} = \frac{64\text{lbs}}{95 \text{ ft}}$.

$w = \frac{64}{95}h + C$. $C = 800 \cdot 8 = 6400$ when $h = 0$.

$$W = \int_0^{4750} \left(6400 - \frac{64}{75}h\right)dh = 22{,}800{,}000 \text{ ft-lbs} = 11.400 \text{ ft-tons}$$

19. Work to fill the pipe: $62.5 \int_0^{360} \left(\frac{1}{3}\right)^2 \pi \, dx = 450,000\pi$ ft–lbs

Work to fill tank: $62.5 \int_0^{25} (100\pi)(360 + x)\, dx = 58,203,125\,\pi$ ft–lbs

$$\text{Time} = \frac{(450,000\pi + 58,203,125\pi) \text{ ft–lbs}}{1650 \text{ ft–lbs/sec}} = 31.02 \text{ hrs.}$$

21. $W = \frac{1}{2}mv_2^2 - \frac{1}{2}mv_1^2 = \frac{0.3125 \text{ lb}\cdot(132)^2 \text{ ft}^2/\text{s}^2}{32 \text{ ft/s}^2} = 85.1$ ft–lb

5.10 HYDROSTATIC FORCE

1. $F = \int_0^3 w(2y)\, h\, dh$

$$= 62.5 \int_0^3 \frac{4}{3} h(3 - h)\, dh \qquad \frac{y}{3-h} = \frac{2}{3}$$

$$= (62.5)\frac{4}{3}\left[\frac{3h^2}{2} - \frac{h^3}{3}\right]_0^3 = 375 \text{ lbs}$$

3. Using similar triangles, $\frac{L}{4} = \frac{5 - (h - 1)}{5} \Rightarrow L = 4\left(\frac{6}{5} - \frac{h}{5}\right)$. Then

$$p = \frac{4}{5}(62.5)\int_1^6 h(6-h)\, dh = \frac{4}{5}(62.5)\left[3h^2 - \frac{h^3}{3}\right]_1^6 = 1,666\frac{2}{3} \text{ lb.}$$

5. $p = 62.5 \int_0^1 2h\sqrt{1 - h^2}\, dh = 62.5\left[-\frac{2}{3}(1 - h^2)^{3/2}\right]_0^1 = 41\frac{2}{3}$ lb

7. If $d =$ the depth of the water, then the area $A = \frac{1}{2}bh$

$= \left(\frac{1}{2}d\right)(2)\left(\frac{2d}{5}\right) = \frac{2d^2}{5}$. Then, using the fact that the centroid is $\frac{d}{3}$ units below the base, we must have

$62.5\left(\frac{2d^2}{5}\right)\left(\frac{d}{3}\right) = 6667 \Rightarrow d \approx 9.3$ ft. Then $V = 30\left(\frac{2}{5}\right)(9.3)^2 \approx 1,033.4 \text{ ft}^3$

9. $F = \int_0^2 (62.5)\left(2\sqrt{4 - h^2}\right) dy = 62.5\left(-\frac{2}{3}(4 - h^2)^{3/2}\right]_0^2 = \frac{1000}{3}$ lbs.

$100x = 333.3 \Rightarrow x = 3.33$ ft. Yes, the trough will overflow because the spring will have moved only 3.33 ft by the time the the tank if full.

5.M MISCELLANEOUS PROBLEMS

1. $\text{T.D.} = \int_0^3 |3t^2 - 15t + 18|\,dt = 3\int_0^2 (t^2 - 5t + 6)\,dt - 3\int_2^3 (t^2 - 5t + 6)\,dt$

$$= 3\left[\frac{1}{3}t^3 - \frac{5}{2}t^2 + 6t\right]_0^2 - 3\left[\frac{1}{3}t^3 - \frac{5}{2}t^2 + 6t\right]_2^3 = \frac{29}{2}$$

$$\text{N.D.} = 3\int_0^3 (t^2 - 5t + 6)\,dt = 3\left[\frac{1}{3}t^3 - \frac{5}{2}t^2 + 6t\right]_0^3 = \frac{27}{2}$$

3. $y = 2 - x^2$ intersects $y = -x$ in the points $x = 2, -1$.

$$A = \int_{-1}^{2} (2 - x^2 + x)\,dx = 2x - \frac{1}{3}x^3 + \frac{1}{2}x^2\Big]_{-1}^{2} = \frac{9}{2}$$

5. $y = x$ intersects $y = \frac{1}{\sqrt{x}}$ at the point $(1, 1)$.

$$A = \int_1^2 \left(x - \frac{1}{\sqrt{x}}\right)dx = \frac{1}{2}x^2 - 2\sqrt{x}\Big]_1^2 = \frac{7}{2} - 2\sqrt{2}$$

7. $y = 2x^2$ intersects $y = x^2 + 2x + 3$ where $2x^2 = x^2 + 2x + 3$, or where $x = 3, -1$.

$$A = \int_{-1}^{3} [(x^2 + 2x + 3) - 2x^2]\,dx = \int_{-1}^{3} (-x^2 + 2x + 3)\,dx = -\frac{1}{3}x^3 + x^2 + 3x\,\Big]_{-1}^{3} = \frac{32}{3}$$

9. $4x = y^2 - 4$ intersects $4x = y + 16$ where $y^2 - 4 = y + 16$, or where $y = 5, -4$.

$$A = \int_{-4}^{5} \left(\frac{y}{4} + 4 - \frac{y^2}{4} + 1\right)dy = \frac{1}{8}y^2 + 5y - \frac{1}{12}y^3\Big]_{-4}^{5} = \frac{243}{8}$$

11. $y = \sin x$ intersects $y = \frac{\sqrt{2}}{2}x$ at the origin. To find the other point of intersection, we use Newton's Method on $f(x) = \sin x - \frac{\sqrt{2}}{2}x$ to find $x \approx 1.39$. Then

$$A = \int_0^{1.39} \left(\sin x - \frac{\sqrt{2}}{2}x\right)dx = -\cos x - \frac{\sqrt{2}}{4}x^2\Big]_0^{1.39} \approx 0.137$$

13. $y^2 = 9x$ intersects $y = \frac{3x^2}{8}$ where $9x = \left(\frac{3x^2}{8}\right)^2$, or $x = 0, 4$.

$$A = \int_0^4 \left(3\sqrt{x} - \frac{3}{8}x^2\right)dx = 2x^{3/2} - \frac{x^3}{8}\Big]_0^4 = 8$$

15. $y^2 = 4x$ intersects $y = 4x - 2$ where $y = y^2 - 2$, or $y = 2, -1$.

$$A = \int_{-1}^{2}\left(\frac{y}{4} + \frac{1}{2} - \frac{y^2}{4}\right)dy = \frac{y^2}{8} + \frac{y}{2} - \frac{y^3}{12}\Big]_{-1}^{2} = \frac{9}{8}$$

17. $$A = \int_0^{\pi}(1 - |\cos x|)\,dx = \int_0^{\frac{\pi}{2}}(1 - \cos x)\,dx + \int_{\frac{\pi}{2}}^{\pi}(1 + \cos x)\,dx$$

$$= x - \sin x\Big]_0^{\frac{\pi}{2}} + x + \sin x\Big]_{\frac{\pi}{2}}^{\pi} = \pi - 2$$

19. $y = x^3 - 3x^2 \Rightarrow y' = 3x^2 - 6x = 0 \Leftrightarrow x = 0, 2$. $y' < 0$ for $x < 0$, $y' > 0$ for $x > 0$ $\Rightarrow$ there is maximum at the point $(0,0)$. $y' < 0$ for $x < 2$ and $y' > 0$ for $x > 2 \Rightarrow$ there is minimum at the point $(2,-4)$.

$$A = \int_0^3 (3x^2 - x^3)\,dx = x^3 - \frac{1}{4}x^4\Big]_0^3 = \frac{27}{4}$$

21. $$V = 2\pi\int_0^2 x(2x - x^2)\,dx = 2\pi\left[\frac{2}{3}x^3 - \frac{1}{4}x^4\right]_0^2 = \frac{8\pi}{3}$$

23. We write the volume as a funtion

$$V(x) = \int_a^x \pi[f(t)]^2\,dt = x^2 - ax.\ \text{Then } V'(x) = \pi\,[f(x)]^2 = 2x - a.$$

$$\text{Then } [f(x)]^2 = \frac{2x - a}{\pi} \quad \text{or } f(x) = \pm\sqrt{\frac{2x - a}{\pi}}.$$

25. $$V = 2\pi\int_0^4 y\left(y - \frac{y^2}{4}\right)dy = 2\pi\left[\frac{y^3}{3} - \frac{y^4}{16}\right]_0^4 = \frac{32\pi}{3}$$

27. We use a shell with radius $r = x$ and height $h = y = \dfrac{x}{\sqrt{x^3 + 8}}$.

$$V = 2\pi\int_0^2 x\left(\frac{x}{\sqrt{x^3 + 8}}\right)dx = 2\pi\int_0^2 \frac{x^2\,dx}{\sqrt{x^3 + 8}} = \frac{4\pi}{3}\sqrt{x^3 + 8}\Big]_0^2 = \frac{4\pi}{3}[4 - 2\sqrt{2}]$$

29. We use a shell with radius $r = 2a - x$, and height $h = 2\,(2\sqrt{ax})$.

$$V = 2\pi\int_0^a (2a - x)(4\sqrt{ax})\,dx = 8\pi\sqrt{a}\int_0^a (2a\sqrt{x} - x^{3/2})\,dx$$

$$= 16\pi\sqrt{a}\left(\frac{2}{3}x^{3/2}\right) - \frac{16\pi\sqrt{a}}{5}x^{5/2}\Big]_0^a = \frac{112\pi a^3}{15}$$

31. $$V = \pi\int_0^{\frac{\pi}{2}} \sin^2 2x\,dx = \frac{\pi}{2}\int_0^{\frac{\pi}{2}}(1 - \cos 4x)\,dx = \frac{\pi}{2}\left[x - \frac{1}{4}\sin 4x\right]_0^{\frac{\pi}{2}} = \frac{\pi^2}{4}$$

33. The curves $x^2 = 4y$ and $y^2 = 4x$ intersect where $x^4 = 16y^2 = 16(4x) \Rightarrow$ $x^4 - 64x = 0$ or $x = 0, 4$. A diameter in the base runs from $y = 2\sqrt{x}$ to $y = \frac{x^2}{4}$ so that a radius $r = \frac{1}{2}\left(2\sqrt{x} - \frac{x^2}{4}\right)$ and the base area is $B = \pi\left(\sqrt{x} - \frac{x^2}{8}\right)^2$. The volume of the solid is then

$$V = \int_0^4 \pi\left(\sqrt{x} - \frac{x^2}{8}\right)^2 = \pi\int_0^4 \left(x - \frac{x^{5/2}}{4} + \frac{x^4}{64}\right)dx = \pi\left[\frac{x^2}{2} - \frac{x^{7/2}}{14} + \frac{x^5}{320}\right]_0^4 = \frac{72\pi}{35}$$

35. If $V(a) = \int_0^a \pi\,[f(x)]^2\,dx = a^2 + a$, then $V'(a) = 2a + 1$ and

$V'(x) = \pi\,[f(x)]^2$ also. Thus $\pi\,[f(x)]^2 = 2x + 1 \Rightarrow f(x) = \pm\sqrt{\frac{2x+1}{\pi}}$.

37. (a) If $y = 2\sqrt{7}x^{3/2} - 1$, then $\frac{dy}{dx} = 3\sqrt{7x}$ and $\left(\frac{dy}{dx}\right)^2 = 63x$. Then

$$s = \int_0^1 \sqrt{1 + 63x}\,dx = \frac{2}{189}(1 + 63x)^{3/2}\Big]_0^1 = \frac{1022}{189}$$

(b) If $y = \frac{2}{3}x^{3/2} - \frac{1}{2}x^{1/2}$, then $\frac{dy}{dx} = x^{1/2} - \frac{1}{4}x^{-1/2}$ and

$$\left(\frac{dy}{dx}\right)^2 = x - \frac{1}{2} + \frac{1}{16}x^{-1}, \text{ so}$$

$$1 + \left(\frac{dy}{dx}\right)^2 = x + \frac{1}{2} + \frac{1}{16}x^{-1} = \left(x^{1/2} + \frac{1}{4}x^{-1/2}\right)^2. \text{ Then}$$

$$s = \int_0^4 \left(x^{1/2} + \frac{1}{4}x^{-1/2}\right)dx = \frac{2}{3}x^{3/2} + \frac{1}{2}x^{1/2}\Big]_0^4 = \frac{19}{3}$$

(c) If $x = \frac{3}{5}y^{5/3} - \frac{3}{4}y^{1/3}$, then $\frac{dx}{dy} = y^{2/3} - \frac{1}{4}y^{-2/3}$ and

$$\left(\frac{dx}{dy}\right)^2 = y^{4/3} - \frac{1}{2} + \frac{1}{16}y^{-4/3}, \text{ so}$$

$$1 + \left(\frac{dx}{dy}\right)^2 = y^{4/3} + \frac{1}{2} + \frac{1}{16}y^{-4/3} = \left(y^{2/3} + \frac{1}{4}y^{-2/3}\right)^2. \text{ Then}$$

$$s = \int_0^1 \left(y^{2/3} + \frac{1}{4}y^{-2/3}\right)dy = \frac{3}{5}y^{5/3} + \frac{3}{4}y^{1/3}\Big]_0^1 = \frac{27}{20}$$

39. (a) $S = 2\pi\int_0^4 x\left(x^{1/2} + \frac{1}{4}x^{-1/2}\right)dx = 2\pi\int_0^4\left(x^{3/2} + \frac{1}{4}x^{1/2}\right)dx$

$= 2\pi\left[\frac{2}{5}x^{5/2} + \frac{1}{6}x^{3/2}\right]_0^4 = \frac{424\pi}{15}$

(b) $S = 2\pi\int_0^1 (y+1)\left(y^{2/3} + \frac{1}{4}y^{-2/3}\right)dy$

$= 2\pi\int_0^1\left(y^{5/3} + \frac{1}{4}y^{1/3} + y^{2/3} + \frac{1}{4}y^{-2/3}\right)dy$

$= 2\pi\left[\frac{3}{8}y^{8/3} + \frac{3}{16}y^{4/3} + \frac{3}{5}y^{5/3} + \frac{3}{4}y^{1/3}\right]_0^1 = \frac{153\pi}{40}$

41. Every continuous function over a closed interval must have both a maximum and a minimum value on that interval. Let $u \in [a,b]$ and $v \in [a,b]$ be such that $f(u)$ is the minimum value and $f(v)$ is the maximum value of f on $[a,b]$. Then

$$f(u)(b-a) \le \int_a^b f(x)\,dx \le f(v)(b-a).$$

By the Intermediate Value Theorem, there must exist a range value of f, say $f(c)$, where $c \in [a,b]$, such that

$$f(c)(b-a) = \int_a^b f(x)\,dx, \text{ or } f(c) = \frac{1}{b-a}\int_a^b f(x)\,dx.$$

43. Average length $= \frac{1}{2a}\int_{-a}^{a} 2\sqrt{a^2 - x^2}\,dx = \frac{1}{a}\cdot\frac{\pi a^2}{2} = \frac{\pi a}{2}$

45. Average length $= \frac{1}{2a}\int_{-a}^{a}\left(2\sqrt{a^2 - x^2}\right)^2 dx = \frac{2}{a}\left[a^2x - \frac{x^3}{3}\right]_{-a}^{a} = \frac{8a^2}{3}$

47. $s = 120t - 16t^2 \Rightarrow v = 120 - 32t$

$v_{av} = \frac{1}{3}\int_0^3 (120 - 32t)\,dt = \frac{1}{3}\left[120t - 16t^2\right]_0^3 = 72$

$s(3) - s(0) = 216$, $ds = (120 - 32t)\,dt$

$v_{av} = \frac{1}{216}\int_0^{216} v(s)\,ds = \frac{1}{216}\int_0^3 v(t)(120 - 32t)\,dt = \frac{1}{216}\int_0^3 (120 - 32t)^2\,dt$

$= \frac{1}{216}\cdot\frac{-1}{96}(120 - 32t)^3\Big]_0^3 = 82\frac{2}{3}$

49. $A = \int_0^2 (2\sqrt{2}x^{1/2} - x^2)\,dx = \frac{4\sqrt{2}}{3}x^{3/2} - \frac{1}{3}x^3\Big]_0^2 = \frac{8}{3}$

$M_x = \int_0^2 \frac{1}{2}(2\sqrt{2}x^{1/2} + x^2)(2\sqrt{2}x^{1/2} - x^2)\,dx = \frac{1}{2}\int_0^2 (8x - x^4)\,dx$

$= \frac{1}{2}\left[4x^2 - \frac{1}{5}x^5\right]_0^2 = \frac{24}{5}.\quad \therefore \bar{y} = \frac{24}{5}\cdot\frac{3}{8} = \frac{9}{5}$

$M_y = \int_0^2 x\,(2\sqrt{2}x^{1/2} - x^2)\,dx = \int_0^2 (2\sqrt{2}x^{3/2} - x^3)\,dx$

$= \frac{4\sqrt{2}}{5}x^{5/2} - \frac{1}{4}x^4\Big]_0^2 = \frac{12}{5}.\quad \therefore \bar{x} = \frac{12}{5}\cdot\frac{3}{8} = \frac{9}{10}$

51. $y^2 = x$ intersects $x = 2y$ in the points $(4,2)$ and $(0,0)$.

$A = \int_0^4 \left(\sqrt{x} - \frac{x}{2}\right)dx = \frac{2}{3}x^{3/2} - \frac{x^2}{4}\Big]_0^4 = \frac{4}{3}$

$M_x = \int_0^4 \frac{1}{2}\left(\sqrt{x} + \frac{x}{2}\right)\left(\sqrt{x} - \frac{x}{2}\right)dx = \frac{1}{2}\int_0^4 \left(x - \frac{x^2}{4}\right)dx$

$= \frac{1}{2}\left[\frac{x^2}{2} - \frac{x^3}{12}\right]_0^4 = \frac{4}{3}.\quad \therefore \bar{y} = \frac{4}{3}\cdot\frac{3}{4} = 1$

$M_y = \int_0^4 x\left(\sqrt{x} - \frac{x}{2}\right)dx = \int_0^4 \left(x^{3/2} - \frac{x^2}{2}\right)dx$

$= \frac{2}{5}x^{5/2} - \frac{1}{6}x^3\Big]_0^4 = \frac{32}{15}.\quad \therefore \bar{x} = \frac{32}{15}\cdot\frac{3}{4} = \frac{8}{5}$

53. $A = \int_0^1 x^2\,dx = \frac{1}{3}x^3\Big]_0^1 = \frac{1}{3}$

$M_x = \int_0^1 x^4\,dx = \frac{1}{10}x^5\Big]_0^1 = \frac{1}{10};\ \bar{y} = \frac{1}{10}\cdot\frac{3}{1} = \frac{3}{10}$

$M_y = \int_0^1 x^3\,dx = \frac{1}{4}x^4\Big]_0^1 = \frac{1}{4};\quad \bar{x} = \frac{1}{4}\cdot\frac{3}{1} = \frac{3}{4}$

55. (a) $M = 2\int_0^a 2\,\rho\sqrt{ax}\,dx = 4\sqrt{a}\int_0^a kx\sqrt{x}\,dx = \frac{8k\sqrt{a}}{5}x^{5/2}\Big]_0^a = \frac{8ka^3}{5}$

$M_y = 4k\sqrt{a}\int_0^a x^{5/2}\,dx = \frac{8k\sqrt{a}}{7}x^{7/2}\Big]_0^a = \frac{8ka^4}{7}$

$\bar{x} = \frac{8ka^4}{7}\cdot\frac{5}{8ka^3} = \frac{5a}{7}.\quad \bar{y} = 0$ by symmetry.

(b) $M = \int_{-2a}^{2a} |y| \left(a - \frac{y^2}{4a}\right) dy = 2\int_0^{2a} \left(ay - \frac{y^3}{4a}\right) dy = 2\left[\frac{ay^2}{2} - \frac{y^4}{16a}\right]_0^{2a} = 2a^3$

$$M_y = \frac{1}{2}\int_{-2a}^{2a}\left(a + \frac{y^2}{4a}\right)|y|\left(a - \frac{y^2}{4a}\right)dy = \frac{1}{2}\int_{-2a}^{2a}|y|\left(a^2 - \frac{y^4}{16a^2}\right)dy$$

$$= \int_0^{2a}\left(a^2y - \frac{y^5}{16a^2}\right)dy = \frac{a^2y^2}{2} - \frac{y^6}{96a^2}\Big]_0^{2a} = \frac{4a^4}{3}.$$

$\bar{x} = \frac{4a^4}{3}\cdot\frac{1}{2a^3} = \frac{2a}{3}$. $\bar{y} = 0$ by symmetry.

59. $W = 60\int_0^{10}\pi x^2(10 - y)\,dy = 60\pi\int_0^{10}\left(3y^2 - \frac{1}{4}y^3\right)dy$

$$= 60\pi\left[y^3 - \frac{1}{16}y^4\right]_0^{10} = 22,500\pi \text{ ft-lbs}.$$

22,500π ft-lbs) / 275 ft-lbs/s ≈ 257 s or 4.3 minutes.

61. $W = \int_a^b \frac{k}{x^2}dx = -\frac{k}{x}\Big]_a^b = \frac{k(b-a)}{ab}$

63. $W = \int_0^4 57(10 + x)(20)2\sqrt{16 - x^2}\,dx$

$$= 22800\int_0^4\sqrt{16 - x^2}\,dx + 2280\int_0^4 x\sqrt{16 - x^2}\,dx$$

$$= 22800\left(\text{area of } \frac{1}{4}\text{circle with } r = 4\right) + 2280\left[\frac{1}{3}(16 - x^2)^{3/2}\right]_0^4$$

$$= 22800(4\pi) + 2280\left(\frac{64}{3}\right) \approx 335,153.3 \text{ ft-lbs} \approx 167.6 \text{ ft-tons}$$

65. $A = 2\int_0^1(4 - 4x^2)\,dx = 2\left\{4x - \frac{4}{3}x^3\right]_0^1 = \frac{16}{3}$

$$M_x = \frac{1}{2}\int_{-1}^1(4 + 4x^2)(4 - 4x^2)\,dx = \frac{1}{2}\left[16x - \frac{16}{5}x^5\right]_{-1}^1 = \frac{64}{5}$$

$\bar{y} = \frac{64}{5}\cdot\frac{3}{16} = \frac{12}{5}$. The depth to centroid is $\frac{33}{5}$ ft.

$F = \left(\frac{33}{5}\text{ft}\right)\left(62.5\,\frac{\text{lb}}{\text{ft}^3}\right)\left(\frac{16}{3}\text{ ft}^2\right) = 2200$ lbs

67. By similar triangle, $\frac{L}{4} = \frac{h-2}{4}$ or $L = h - 2$.

$$F = 62.5\int_2^6 h(h-2)\,dh = 62.5\left[\frac{h^3}{3} - h^2\right]_2^6 = 2,333\tfrac{1}{3} \text{ lbs}.$$

69. (a) $M_x = \int_0^h y\, f(y)\, dy = \int_0^h w\, b\, y^2\, dy = \frac{wh^3}{3}$. The total force is

$$F = \int_0^h wby\, dy = \frac{wbh^2}{2}. \quad \therefore \left(\frac{wbh^2}{2}\right)\bar{y} = \frac{wbh^3}{3} \text{ or } \bar{y} = \frac{2h}{3}.$$

(b) $M_x = \int_a^{a+h} y\, f(y)\, dy = \int_a^{a+h} y\, (wy)\, 1\, dy = \int_a^{a+h} wy^2 \left(\frac{b}{h}(y-a)\right) dy$

and $F = \int_a^{a+h} w\, y \left(\frac{b}{h}(y-a)\right) dy$. Since $\bar{y}\, F = M_x$, we have

$$\int_a^{a+h} (y^3 - ay^2)\, dy = \left(\int_a^{a+h} (y^2 - ay)\, dy\right)\bar{y}$$

$$\bar{y}\left[\frac{1}{3}y^3 - \frac{a}{2}y^2\right]_a^{a+h} = \left[\frac{1}{4}y^4 - \frac{a}{3}y^3\right]_a^{a+h}$$

$$\bar{y} = \frac{3\,(a+h)^4 - 4a\,(a+h)^3 - 3a^4 + 4a^4}{4(a+h)^3 - 6a\,(a+h)^2 - 4a^3 + 6a^3} = \frac{6a^2 + 8ah + 3h^2}{2\,(3a + 2h)}$$

71. $V = 2\pi\,(2)(8) = 32\pi$; $S = 2\pi\,(2)(8\sqrt{2}) = 32\sqrt{2}\,\pi$

73. $V = 2\pi\,(2)\pi = 4\pi^2$

75. $A = 4\pi a^2$. $\therefore 4\pi a^2 = 2\pi\,(\pi a)\,\bar{y} \Rightarrow \bar{y} = \frac{2a}{\pi}$. $\bar{x} = 0$ by symmetry.

77. $V = \frac{4}{3}\pi a^3$. $\therefore \frac{4}{3}\pi a^3 = 2\pi\left(\frac{\pi a^2}{2}\right)\bar{y} \Rightarrow \bar{y} = \frac{4a}{3\pi}$. $\bar{x} = 0$.

79. $$y = \frac{4a\sqrt{2}}{3\pi} + \frac{a - \frac{4a}{3\pi}}{\sqrt{2}} = \frac{2\sqrt{2}\,a}{3\pi} + \frac{a\sqrt{2}}{2}.$$

$$V = 2\pi\left[\frac{2\sqrt{2}\,a}{3\pi} + \frac{a\sqrt{2}}{2}\right]\frac{\pi a^2}{2} = \frac{2\sqrt{2}\,a^3\pi}{3} + \frac{\pi^2\sqrt{2}\,a^3}{2} = \frac{\pi a^3\,(4 + 3\pi)}{3\sqrt{2}}$$

CHAPTER 6

TRANSCENDENTAL FUNCTIONS

6.1 INVERSE FUNCTIONS

1. $f(x)=2x+3$, and $f(1)=-1$. Interchanging and solving:

$x=2y+3 \Rightarrow y=\frac{1}{2}x-\frac{3}{2}=g(x)$.

$f'(x)=2$ and $g'(x)=\frac{1}{2}$. $\therefore f'(-1)=\frac{1}{g'(1)}$.

3. $f(x)=\frac{1}{5}x+7$ and $f(-1)=\frac{34}{5}$. Interchanging and solving:

$x=\frac{1}{5}y+7 \Rightarrow y=5x+35=g(x)$.

$f'(x)=\frac{1}{5}$ and $g'(x)=5$ $\therefore f'(-1)=\frac{1}{g'\left(\frac{34}{5}\right)}$.

5. $f(x)=x^2+1$, $x\geq 0$ and $f(5)=26$. Interchanging and solving:

$x=y^2+1 \Rightarrow y=\sqrt{x-1}=g(x)$.

$f'(x)=2x$ and $g'(x)=\frac{1}{2\sqrt{x-1}}$. $\therefore f'(5)=10=\frac{1}{g'(26)}=\frac{1}{\frac{1}{10}}$.

7. $f(x)=x^3-1$, and $f(2)=7$. Interchanging and solving:

$x=y^3-1 \Rightarrow y^3=x+1 \Rightarrow y=(x+1)^{1/3}=g(x)$.

$f'(x)=3x^2$ and $g'(x)=\frac{1}{3(x+1)^{2/3}}$. $\therefore f'(2)=12$ and $g'(7)=\frac{1}{12}$.

9. $f(x)=x^5$. Interchanging and solving: $x=y^5 \Rightarrow y=x^{\frac{1}{5}}$.

$f(f^{-1}(x))=(x^{\frac{1}{5}})^5=x$ and $f^{-1}(f(x))=(x^5)^{\frac{1}{5}}=x$.

11. $f(x)=x^{\frac{2}{3}}$, $x\geq 0$. Interchanging and solving: $x=y^{\frac{2}{3}} \Rightarrow y=x^{\frac{3}{2}}$.

$f(f^{-1}(x))=(x^{\frac{3}{2}})^{\frac{2}{3}}=x$ and $f^{-1}(f(x))=(x^{\frac{2}{3}})^{\frac{3}{2}}=x$.

13. $f(x) = (x-1)^2$, $x \ge 1$. Interchanging and solving: $x = (y-1)^2$

$\Rightarrow y = \sqrt{x} + 1$, $x \ge 0$. $f(f^{-1}(x)) = (\sqrt{x} + 1 - 1)^2 = x$

and $f^{-1}(f(x)) = \sqrt{(x-1)^2} + 1 = x$.

15. $f(x) = x^{-2}$, $x > 0$. Interchanging and solving: $x = y^{-2} \Rightarrow y = x^{-\frac{1}{2}}$.

$f(f^{-1}(x)) = (x^{-\frac{1}{2}})^{-2} = x$ and $f^{-1}(f(x)) = (x^{-2})^{-\frac{1}{2}} = x$.

17.

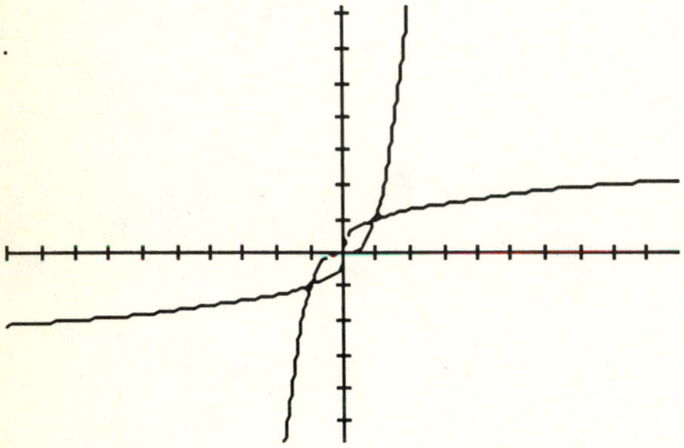

19. $f(x) = x^2 - 4x - 3$, $x > 2$. $f(x) = 2 \Leftrightarrow x^2 - 4x - 3 = 2 \Rightarrow$

$(x-5)(x+1) = 0 \Rightarrow x = 5, -1$. $f'(x) = 2x - 4 \Rightarrow f'(5) = 6$.

$\therefore g'(2) = \dfrac{1}{f'(5)} = \dfrac{1}{6}$.

21. New volume $= \dfrac{4\pi}{3}(10)^3 + 1$. Since $r = \left(\dfrac{3V}{4\pi}\right)^{\frac{1}{3}}$, this change

in volume would be produced by a radius of $r = \left(1000 + \dfrac{3}{4\pi}\right)^{\frac{1}{3}} =$

10.0008 cm. This is a change of .0008 cm.

.1 mm/hr = .01 cm/hr. $\therefore$ $.01t = .0008 \Rightarrow t = .08$ hr or 4.8 min.

6.2 THE INVERSE TRIGONOMETRIC FUNCTIONS

1. (a) $\frac{\pi}{4}$ (b) $\frac{\pi}{3}$ (c) $\frac{\pi}{6}$

3. (a) $\frac{\pi}{6}$ (b) $\frac{\pi}{4}$ (c) $\frac{\pi}{3}$

5. (a) $\frac{\pi}{3}$ (b) $\frac{\pi}{4}$ (c) $\frac{\pi}{6}$

7. (a) $\frac{\pi}{4}$ (b) $\frac{\pi}{6}$ (c) $\frac{\pi}{3}$

9. (a) $\frac{\pi}{4}$ (b) $\frac{\pi}{3}$ (c) $\frac{\pi}{6}$

11. (a) $\frac{\pi}{4}$ (b) $\frac{\pi}{6}$ (c) $\frac{\pi}{3}$

13. $\alpha = \sin^{-1}\frac{1}{2} \Rightarrow \cos\alpha = \frac{\sqrt{3}}{2}$, $\tan\alpha = \frac{1}{\sqrt{3}}$, $\sec\alpha = \frac{2}{\sqrt{3}}$, $\csc\alpha = 2$

15. $\sin\left(\cos^{-1}\frac{\sqrt{2}}{2}\right) = \frac{\sqrt{2}}{2}$

17. $\sec\left(\cos^{-1}\frac{1}{2}\right) = 2$

19. $\csc(\sec^{-1} 2) = \frac{2}{\sqrt{3}}$

21. $\cos(\cot^{-1} 1) = \frac{\sqrt{2}}{2}$

23. $\cot(\cos^{-1} 0) = 0$

25. $\tan(\sec^{-1} 1) = 0$

27. $\sin^{-1} 1 - \sin^{-1}(-1) = \pi$

29. $\sec^{-1} 2 - \sec^{-1}(-2) = -\frac{\pi}{3}$

31. $\cos(\sin^{-1} 0.8) = 0.6$ (a 3–4–5 right triangle)

33. $\cos^{-1}\left(-\sin\frac{\pi}{6}\right) = \frac{2\pi}{3}$

35 $\lim_{x \to 1^-} \sin^{-1} x = \sin^{-1} 1 = \frac{\pi}{2}$.

37. $\lim_{x \to \infty} \tan^{-1} x = \frac{\pi}{2}$.

39. $\lim_{x \to \infty} \sec^{-1} x = \lim_{x \to \infty} \cos^{-1}\left(\frac{1}{x}\right) = \cos^{-1} 0 = \frac{\pi}{2}$.

41. $\lim_{x \to \infty} \csc^{-1} x = \lim_{x \to \infty} \sin^{-1}\left(\frac{1}{x}\right) = \sin^{-1} 0 = 0$.

43. $65^\circ = \alpha + \beta$ and $\beta = \tan^{-1}\frac{21}{50}$. $\therefore \alpha = 65^\circ - \tan^{-1}\frac{21}{50} = 42.2^\circ$.

45. $\frac{h}{3} = \cos\alpha$ and $\frac{r}{3} = \sin\alpha$. $V = \frac{1}{3}\pi r^2 h = \frac{\pi}{3}(9\sin^2\alpha)(3\cos\alpha)$

$V' = 9\pi(-\sin^3\alpha + 2\sin\alpha\cos^2\alpha)$. $V' = 0 \Leftrightarrow 9\pi\sin\alpha(2\cos^2\alpha - \sin^2\alpha) = 0$

$\Rightarrow 2\cos^2\alpha = \sin^2\alpha \Rightarrow \tan\alpha = \sqrt{2}$. $\alpha = \tan^{-1}\sqrt{2} = 54.7^\circ$.

$V'' = 9\pi(-3\sin^2\alpha\cos\alpha + 2\cos^3\alpha - 4\sin^2\alpha\cos\alpha) =$

$= 9\pi\cos\alpha(-7\sin^2\alpha + 2\cos^2\alpha) = 9\pi\left(\frac{1}{\sqrt{3}}\right)\left[-7\left(\frac{2}{3}\right) + 2\left(\frac{1}{3}\right)\right] < 0$

$\Rightarrow$ this angle gives maximum volume.

47. $\sec^{-1}(-x) = \cos^{-1}\left(-\frac{1}{x}\right) = \pi - \cos^{-1}\left(\frac{1}{x}\right) = \pi - \sec^{-1}x$

49.

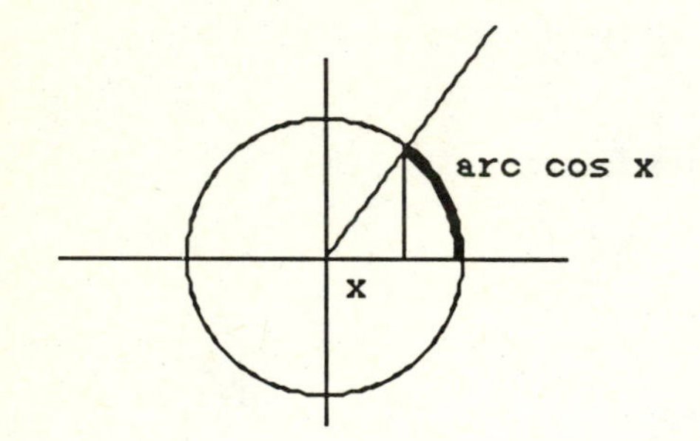

6.3 DERIVATIVES OF INVERSE TRIGONOMETRIC FUNCTIONS

1. $y = \cos^{-1}x^2 \Rightarrow \frac{dy}{dx} = \frac{-1}{\sqrt{1-(x^2)^2}}\frac{d}{dx}(x^2) = \frac{-2x}{\sqrt{1-x^4}}$.

3. $y = 5\tan^{-1}3x \Rightarrow \frac{dy}{dx} = 5\left(\frac{1}{1+(3x)^2}\right)\frac{d}{dx}(3x) = \frac{15}{1+9x^2}$.

5. $y = \sin^{-1}\frac{x}{2} \Rightarrow \frac{dy}{dx} = \frac{1}{\sqrt{1-\left(\frac{x}{2}\right)^2}}\frac{d}{dx}\left(\frac{x}{2}\right) = \frac{1}{\sqrt{4-x^2}}$.

7. $y = \sec^{-1}5x \Rightarrow \frac{dy}{dx} = \frac{1}{|5x|\sqrt{(5x)^2-1}}\frac{d}{dx}(5x) = \frac{1}{|x|\sqrt{25x^2-1}}$.

9. $y = \csc^{-1}(x^2+1) \Rightarrow \frac{dy}{dx} = \frac{-1}{|x^2+1|\sqrt{(x^2+1)^2-1}}\frac{d}{dx}(x^2+1) = \frac{-2x}{(x^2+1)\sqrt{x^4+2x^2}}$.

11. $y = \csc^{-1}\sqrt{x} + \sec^{-1}\sqrt{x} = \frac{\pi}{2}$ since the secant and cosecant are cofunctions. $\therefore \frac{dy}{dx} = 0$.

13. $y = \cot^{-1}\sqrt{x-1} \Rightarrow \frac{dy}{dx} = \frac{-1}{1+(\sqrt{x-1})^2}\frac{d}{dx}(\sqrt{x-1}) = \frac{-1}{2x\sqrt{x-1}}$.

15. $y = \sqrt{x^2-4} - 2\sec^{-1}\frac{x}{2} \Rightarrow \frac{dy}{dx} = \frac{1}{2}(x^2-4)^{-\frac{1}{2}}(2x) - 2\left(\frac{\frac{1}{2}}{|\frac{x}{2}|\sqrt{\left(\frac{x}{2}\right)^2-1}}\right)$

$= \frac{x}{\sqrt{x^2-4}} - \frac{1}{|\frac{x}{2}|\sqrt{\frac{x^2-4}{4}}} = \frac{x|x|-4}{|x|\sqrt{x^2-4}}$.

If $x > 2$, $|x| = x$ and $\frac{dy}{dx} = \frac{\sqrt{x^2-4}}{x}$. If $x < -2$, $|x| = -x$ and $\frac{dy}{dx} = \frac{x^2+4}{x\sqrt{x^2-4}}$.

17. $y = \tan^{-1}\frac{x-1}{x+1} \Rightarrow \frac{dy}{dx} = \frac{1}{1+\left(\frac{x-1}{x+1}\right)^2}\frac{d}{dx}\left(\frac{x-1}{x+1}\right)$

$= \left(\frac{(x+1)^2}{(x+1)^2+(x-1)^2}\right)\left(\frac{x+1-x+1}{(x+1)^2}\right) = \frac{1}{x^2+1}$.

19. $y = x(\sin^{-1}x)^2 - 2x + 2\sqrt{1-x^2}\sin^{-1}x$

$\frac{dy}{dx} = (\sin^{-1}x)^2 + x(2\sin^{-1}x)\left(\frac{1}{\sqrt{1-x^2}}\right) - 2 + 2\left[\frac{1}{2}(1-x^2)^{-\frac{1}{2}}(-2x)\sin^{-1}x\right.$

$\left. + 2\sqrt{1-x^2}\left(\frac{1}{\sqrt{1-x^2}}\right)\right] =$

$(\sin^{-1}x)^2 + \frac{2x\sin^{-1}x}{\sqrt{1-x^2}} + \frac{-2x\sin^{-1}x}{\sqrt{1-x^2}} = (\sin^{-1}x)^2$.

21. $\displaystyle\int_0^{\frac{1}{2}} \frac{dx}{\sqrt{1-x^2}} = \sin^{-1} x \Big|_0^{\frac{1}{2}} = \sin^{-1}\frac{1}{2} - \sin^{-1} 0 = \frac{\pi}{6}$

23. $\displaystyle\int_{\sqrt{2}}^{2} \frac{dx}{x\sqrt{x^2-1}} = \sec^{-1}|x| \Big]_{\sqrt{2}}^{2} = \sec^{-1} 2 - \sec^{-1}\sqrt{2} = \frac{\pi}{3} - \frac{\pi}{4} = \frac{\pi}{12}$

25. $\displaystyle\int_{-1}^{0} \frac{4dx}{1+x^2} = 4\tan^{-1} x \Big]_{-1}^{0} = 0 - 4\tan^{-1}(-1) = \pi$

27. $\displaystyle\int_0^{\frac{\sqrt{2}}{2}} \frac{xdx}{\sqrt{1-x^4}} = \int_0^{\frac{1}{2}} \frac{\frac{1}{2}du}{\sqrt{1-u^2}} = \frac{1}{2}\sin^{-1} u \Big]_0^{\frac{1}{2}} = \frac{\pi}{12}$

Let $u = x^2 \Rightarrow du = 2xdx$. Then $x = \frac{\sqrt{2}}{2} \Rightarrow u = \frac{1}{2}$; $x = 0 \Rightarrow u = 0$.

29. $\displaystyle\int_{\frac{1}{\sqrt{3}}}^{1} \frac{dx}{x\sqrt{4x^2-1}} = \int_{\frac{2}{\sqrt{3}}}^{2} \frac{\frac{1}{2}du}{\frac{1}{2}u\sqrt{u^2-1}} = \sec^{-1} u \Big]_{\frac{2}{\sqrt{3}}}^{2} = \frac{\pi}{3} - \frac{\pi}{6} = \frac{\pi}{6}$

Let $u = 2x \Rightarrow u^2 = 4x^2$ and $du = 2dx$. $x = \frac{1}{\sqrt{3}} \Rightarrow u = \frac{2}{\sqrt{3}}$ and $x = 1 \Rightarrow u = 2$

31. $\displaystyle\int_0^{\sqrt{2}} \frac{4\,dx}{\sqrt{4-x^2}} = \int_0^{\frac{\pi}{4}} \frac{4\,(2\cos u\,du)}{\sqrt{4-4\sin^2 u}} = 4\int_0^{\frac{\pi}{4}} du = 4u\Big]_0^{\frac{\pi}{4}} = \pi$

Let $x = 2\sin u \Rightarrow dx = 2\cos u\,du$. $x = 0 \Rightarrow u = 0$; $x = \sqrt{2} \Rightarrow u = \frac{\pi}{4}$

33. $\displaystyle\int_{\sqrt{2}}^{\sqrt[4]{2}} \frac{xdx}{x^2\sqrt{x^4-1}} = \int_2^{\sqrt{2}} \frac{\frac{1}{2}du}{u\sqrt{u^2-1}} = \frac{1}{2}\sec^{-1}|u| \Big]_2^{\sqrt{2}} = -\frac{\pi}{24}$

Let $u = x^2 \Rightarrow du = 2xdx$; $x = \sqrt{2} \Rightarrow u = 2$ and $x = \sqrt[4]{2} \Rightarrow u = \sqrt{2}$.

35. $\displaystyle\int_0^2 \frac{dx}{1+(x-1)^2} = \tan^{-1}(x-1)\Big]_0^2 = \tan^{-1}1 - \tan^{-1}(-1) = \frac{\pi}{4} - \left(-\frac{\pi}{4}\right) = \frac{\pi}{2}$

37. $\displaystyle\int_{\frac{1}{2}}^{\frac{3}{4}} \frac{dx}{\sqrt{x}\sqrt{1-x}} = \int_{\frac{1}{\sqrt{2}}}^{\frac{\sqrt{3}}{2}} \frac{2u\,du}{u\sqrt{1-u^2}} = 2\sin^{-1}u\Big]_{\frac{1}{\sqrt{2}}}^{\frac{\sqrt{3}}{2}} = \frac{\pi}{6}$

Let $u = \sqrt{x} \Rightarrow u^2 = x$ and $2u\,du = dx$. $x = \frac{1}{2} \Rightarrow u = \frac{1}{\sqrt{2}}$ and $x = \frac{3}{4} \Rightarrow u = \frac{\sqrt{3}}{2}$

39. $\displaystyle\int_{-\frac{2}{3}}^{-\frac{\sqrt{2}}{3}} \frac{dx}{x\sqrt{9x^2-1}} = \int_{-2}^{-\sqrt{2}} \frac{\frac{1}{3}du}{\frac{1}{3}u\sqrt{u^2-1}} = \sec^{-1}|u|\Big]_{-2}^{-\sqrt{2}} = -\frac{\pi}{12}$

Let $u = 3x \Rightarrow u^2 = 9x^2$ and $du = 3dx$. $x = -\frac{2}{3} \Rightarrow u = -2$; $x = -\frac{\sqrt{2}}{3} \Rightarrow u = -\sqrt{2}$

41. $\displaystyle\lim_{x\to 0} \frac{\sin^{-1}2x}{x} = \lim_{x\to 0} \frac{\frac{2}{\sqrt{1-4x^2}}}{1} = 2$

43. $\displaystyle\lim_{x\to 0} x^{-3}(\sin^{-1}x - x) = \lim_{x\to 0} \frac{\frac{1}{\sqrt{1-x^2}} - 1}{3x^2} = \lim_{x\to 0} \frac{\frac{x}{\sqrt{1-x^2}}}{6x} = \frac{1}{6}$

45. $f(x) = \sin^{-1}x + \cos^{-1}x \equiv \frac{\pi}{2}$. $\therefore f'(x) = 0$ and $f(.32) = \frac{\pi}{2}$.

47. $\tan\theta = x \Rightarrow \sec^2\theta \frac{d\theta}{dt} = \frac{dx}{dt}$. $x = 2 \Rightarrow \sec\theta = \sqrt{5}$.

$(\sqrt{5})^2 \frac{d\theta}{dt} = 4 \Rightarrow \frac{d\theta}{dt} = \frac{4}{5}$ rad/hr

49. Yes, $\sin^{-1}x = \frac{\pi}{2} - \cos^{-1}x \Rightarrow$ these functions differ by at most a constant.

51. Let $u = 1 - x^2$, $du = -2x\,dx$. Then $\int \frac{x\,dx}{\sqrt{1 - x^2}} = -\frac{1}{2}\int u^{-\frac{1}{2}}du =$

$-\sqrt{u} + C = -\sqrt{1 - x^2} + C$.

53. $(x^2 + 1)\frac{dy}{dx} = -y^2$, $y = 1$ when $x = 0$

$$-\frac{dy}{y^2} = \frac{dx}{x^2 + 1} \Rightarrow \frac{1}{y} = \tan^{-1}x + C.\ 1 = \tan^{-1}0 + C \Rightarrow C = 1$$

$$\therefore\ y = \frac{1}{\tan^{-1}x + 1}$$

55. $x\sqrt{x^2 - 1}\,\frac{dy}{dx} = \sqrt{1 - y^2}$, $y = -\frac{1}{2}$ when $x = 2$.

$$\frac{dy}{\sqrt{1 - y^2}} = \frac{dx}{x\sqrt{x^2 - 1}} \Rightarrow \sin^{-1}y = \sec^{-1}|x| + C$$

$$\sin^{-1}\left(-\frac{1}{2}\right) = \sec^{-1}2 + C \Rightarrow C = -\frac{\pi}{6} - \frac{\pi}{3} = -\frac{\pi}{2}$$

$$\therefore\ \sin^{-1}y = \sec^{-1}|x| - \frac{\pi}{2}$$

57. (a) Let $y = \cos^{-1}u \Rightarrow \cos y = u \Rightarrow -\sin y\,\frac{dy}{du} = 1 \Rightarrow$

$$\frac{dy}{du} = -\frac{1}{\sin y} = -\frac{1}{\sqrt{1 - u^2}}.\ \text{Then } \frac{dy}{dx} = \frac{dy}{du}\frac{du}{dx} = -\frac{1}{\sqrt{1 - u^2}}\frac{du}{dx}$$

(b) Let $y = \tan^{-1}u \Rightarrow \tan y = u \Rightarrow \sec^2 y\,\frac{dy}{du} = 1 \Rightarrow$

$$\frac{dy}{du} = \frac{1}{\sec^2 y} = \cos^2 y = \frac{1}{1 + u^2}.\ \text{Then } \frac{dy}{dx} = \frac{1}{1 + u^2}\frac{du}{dx}.$$

(c) $\cot^{-1}u = \frac{\pi}{2} - \tan^{-1}u \Rightarrow \frac{d}{dx}(\cot^{-1}u) = -\frac{d}{dx}(\tan^{-1}u) = \frac{-1}{1 + u^2}\frac{du}{dx}$

(d) $\frac{d}{dx}(\csc^{-1}u) = \frac{d}{dx}\left(\sin^{-1}\frac{1}{u}\right) = \frac{1}{\sqrt{1 - \left(\frac{1}{u}\right)^2}}\left(\frac{-1}{u^2}\right)\frac{du}{dx} =$

$$\frac{|u|}{\sqrt{u^2 - 1}}\left(\frac{-1}{u^2}\right)\frac{du}{dx} = \frac{-1}{|u|\sqrt{u^2 - 1}}\frac{du}{dx}.$$

6.4 THE NATURAL LOGARITHM AND ITS DERIVATIVE

1. $y = \ln 2x \Rightarrow \frac{dy}{dx} = \frac{1}{2x}\frac{d}{dx}(2x) = \frac{2}{2x} = \frac{1}{x}$

3. $y = \ln kx \Rightarrow \frac{dy}{dx} = \frac{1}{kx}\frac{d}{dx}(kx) = \frac{k}{kx} = \frac{1}{x}$

5. $y = \ln\left(\frac{10}{x}\right) = \ln(10x^{-1}) = -\ln 10x \Rightarrow \frac{dy}{dx} = -\frac{1}{x}$

7. $y = (\ln x)^3 \Rightarrow \frac{dy}{dx} = 3(\ln x)^2\frac{1}{x} = \frac{3\ln^2 x}{x}$

9. $y = \ln(\sec x + \tan x) \Rightarrow \frac{dy}{dx} = \frac{1}{\sec x + \tan x}(\sec x\tan x + \sec^2 x)$

$= \frac{\sec x(\tan x + \sec x)}{\sec x + \tan x} = \sec x$

11. $y = x^3 \ln 2x \Rightarrow \frac{dy}{dx} = x^3\left(\frac{2}{2x}\right) + 3x^2 \ln 2x = x^2(1 + 3\ln 2x)$

13. $y = \tan^{-1}(\ln x) \Rightarrow \frac{dy}{dx} = \frac{1}{1 + \ln^2 x}\left(\frac{1}{x}\right) = \frac{1}{x(1 + \ln^2 x)}$

15. $y = x^2 \ln(x^2) \Rightarrow \frac{dy}{dx} = x^2\left(\frac{2x}{x^2}\right) + 2x\ln(x^2) = 2x[1 + \ln(x^2)]$

17. $y = \ln x - \frac{1}{2}\ln(1 + x^2) - \frac{\tan^{-1} x}{x}$

$\frac{dy}{dx} = \frac{1}{x} - \frac{1}{2}\left(\frac{2x}{1 + x^2}\right) - \frac{1}{x}\left(\frac{1}{1 + x^2}\right) - (\tan^{-1} x)\left(-\frac{1}{x^2}\right)$

$= \frac{1}{x(1 + x^2)} - \frac{1}{x(1 + x^2)} + \frac{\tan^{-1} x}{x} = \frac{\tan^{-1} x}{x}$

19. $y = x[\sin(\ln x) + \cos(\ln x)]$

$\frac{dy}{dx} = [\sin(\ln x) + \cos(\ln x)] + x\left[\cos(\ln x)\left(\frac{1}{x}\right) - \sin(\ln x)\left(\frac{1}{x}\right)\right]$

$= 2\cos(\ln x)$

21. $\int \frac{dx}{x} = \ln|x| + C$

23. $\int \frac{dx}{2x} = \frac{1}{2}\ln|x| + C$

25. $\int_0^1 \frac{dx}{x+1} = \ln|x+1|\Big]_0^1 = \ln 2 - \ln 1 = \ln 2$

27. $\int_{-1}^0 \frac{dx}{2x+3} = \frac{1}{2}\ln|2x+3|\Big]_{-1}^0 = \frac{1}{2}[\ln 3 - \ln 1] = \frac{1}{2}\ln 3$

29. $\int_0^1 \frac{x\,dx}{4x^2+1} = \frac{1}{8}\ln(4x^2+1)\Big]_0^1 = \frac{1}{8}\ln 5$

31. $\int \tan 3x\,dx = \int \frac{\sin 3x}{\cos 3x}dx = -\frac{1}{3}\ln|\cos 3x| + C$

33. $\int \frac{x^2\,dx}{4-x^3} = -\frac{1}{3}\ln|4-x^3| + C$

35. $\int \frac{dx}{x\ln x} = \int \frac{du}{u} = \ln|u| + C = \ln|\ln x| + C$

Let $u = \ln x \Rightarrow du = \frac{1}{x}dx$

37. $\int_1^2 \frac{(\ln x)^2}{x}dx = \int_0^{\ln 2} u^2\,du = \frac{1}{3}u^3\Big]_0^{\ln 2} = \frac{1}{3}\ln^3 2$

Let $u = \ln x \Rightarrow du = \frac{1}{x}dx$; $x = 1 \Rightarrow u = \ln 1 = 0$; $x = 2 \Rightarrow u = \ln 2$

39. $\int \frac{\sec^2 x + \sec x \tan x}{\sec x + \tan x}dx = \int u^{-1}du = \ln|u| + C = \ln|\sec x + \tan x| + C$

Let $u = \sec x + \tan x \Rightarrow du = \sec x \tan x + \sec^2 x\,dx$

41. $\lim_{x\to\infty} \frac{\ln x}{x} = \lim_{x\to\infty}\left(\frac{\frac{1}{x}}{1}\right) = 0$

43. $\lim_{t\to 0} \frac{\ln(1+2t) - 2t}{t^2} = \lim_{t\to 0}\left(\frac{\frac{2}{1+2t} - 2}{2t}\right) = \lim_{t\to 0}\left(\frac{\frac{-4}{(1+2t)^2}}{2}\right) = -2$

45. $A = \int_1^2 \left(\frac{2}{x} - 1\right)dx = 2\ln|x| - x\Big]_1^2 = 2\ln 2 - 1$

47. $A = \int_0^1 \frac{2}{1+x^2}\,dx = 2\tan^{-1}x]_0^1 = \frac{\pi}{2}$

$\bar{x} = \frac{2}{\pi}\int_0^1 x\left(\frac{2}{1+x^2}\right)dx = \frac{2}{\pi}\ln|1+x^2|\Big]_0^1 = \frac{2\ln 2}{\pi}$

$\bar{y} = 0$ by symmetry. $\therefore\ (\bar{x},\bar{y}) = \left(\frac{2\ln 2}{\pi}, 0\right)$.

49. $x = \ln(\sec t + \tan t) - \sin t \Rightarrow \frac{dx}{dt} = \frac{\sec t\tan t + \sec^2 t}{\sec t + \tan t} - \cos t$

$\left(\frac{dx}{dt}\right)^2 = (\sec t - \cos t)^2 = \sec^2 t - 2 + \cos^2 t$.

$y = \cos t \Rightarrow \frac{dy}{dt} = -\sin t \Rightarrow \left(\frac{dy}{dt}\right)^2 = \sin^2 t$

$s = \int_0^{\frac{\pi}{3}} \sqrt{\left(\frac{dx}{dt}\right)^2 + \left(\frac{dy}{dt}\right)^2}\,dt = \int_0^{\frac{\pi}{3}} \sqrt{\sec^2 t - 2 + \cos^2 t + \sin^2 t}\,dt$

$\int_0^{\frac{\pi}{3}} \sqrt{\sec^2 t - 1}\,dt = \int_0^{\frac{\pi}{3}} \tan t\,dt = \int_0^{\frac{\pi}{3}} \frac{\sin t}{\cos t}\,dt =$

$-\ln|\cos t|\Big]_0^{\frac{\pi}{3}} = -\left[\ln\frac{1}{2} - \ln 1\right] = -\ln\frac{1}{2} = \ln 2$

6.5 PROPERTIES OF THE NATURAL LOGARITHM

1. $\ln 16 = \ln 2^4 = 4\ln 2$

3. $\ln 2\sqrt{2} = \ln 2^{\frac{3}{2}} = \frac{3}{2}\ln 2$

5. $\ln\frac{4}{9} = \ln 4 - \ln 9 = 2\ln 2 - 2\ln 3$

7. $\ln\frac{9}{8} = \ln 9 - \ln 8 = 2\ln 3 - 3\ln 2$

9. $\ln 4.5 = \ln\frac{9}{2} = \ln 9 - \ln 2 = 2\ln 3 - \ln 2$

11. $y = \ln\sqrt{x^2+5} = \frac{1}{2}\ln(x^2+5) \quad \Rightarrow \quad \frac{dy}{dx} = \frac{1}{2}\left(\frac{2x}{x^2+5}\right) = \frac{x}{x^2+5}$

13. $y = \ln\frac{1}{x\sqrt{x+1}} = -\ln x\sqrt{x+1} = -(\ln x + \frac{1}{2}\ln(x^2+5)$

$$\frac{dy}{dx} = -\left[\frac{1}{x} + \frac{1}{2(x+1)}\right]$$

15. $y = \ln(\sin x \sin 2x) \Rightarrow \frac{dy}{dx} = \frac{1}{\sin x \sin 2x} \cdot \frac{d}{dx}(\sin x \sin 2x)$

$$= \frac{\sin x(2\cos x) + (\sin 2x)\cos x}{\sin x \sin 2x}$$

$$= \frac{2\cos 2x + 2\cos^2 x}{\sin 2x}$$

17. $y = \ln(3x\sqrt{x+2}) = \ln 3x + \frac{1}{2}\ln(x+2)$

$$\frac{dy}{dx} = \frac{1}{x} + \frac{1}{2(x+2)}$$

19. $y = \frac{1}{3}\ln\left(\frac{x^3}{1+x^3}\right)$

$$= \frac{1}{3}\left[\ln x^3 - \ln(1+x^3)\right]$$

$$\frac{dy}{dx} = \frac{1}{3}\left[\frac{3x^2}{x^3} - \frac{3x^2}{1+x^3}\right]$$

21. $y = \ln\frac{(x^2+1)^5}{\sqrt{1-x}} = 5\ln(x^2+1) - \frac{1}{2}\ln(1-x)$

$$\frac{dy}{dx} = \frac{10x}{x^2+1} + \frac{1}{2(1-x)}$$

23. $y^2 = x(x+1),\ x > 0$

$2\ln y = \ln x + \ln(x+1)$

$$\frac{2}{y}\frac{dy}{dx} = \frac{1}{x} + \frac{1}{x+1} \Rightarrow \frac{dy}{dx} = \frac{y}{2}\left(\frac{1}{x} + \frac{1}{x+1}\right)$$

25. $y = \sqrt{x+2}\sin x\cos x,\ 0 < x < \frac{\pi}{2} \Rightarrow \ln y = \frac{1}{2}\ln(x+2) + \ln\sin x + \ln\cos x$

$$\frac{1}{y}\frac{dy}{dx} = \frac{1}{2(x+2)} + \frac{\cos x}{\sin x} - \frac{\sin x}{\cos x}$$

$$\frac{dy}{dx} = y\left[\frac{1}{2(x+2)} + \cot x - \tan x\right]$$

27. $y = \sqrt[3]{\dfrac{x(x-2)}{x^2+1}}$, $x > 2 \Rightarrow \ln y = \dfrac{1}{3}[\ln x + \ln(x-2) - \ln(x^2+1)]$

$$\frac{1}{y}\frac{dy}{dx} = \frac{1}{3}\left[\frac{1}{x} + \frac{1}{x-2} - \frac{2x}{x^2+1}\right] \Rightarrow \frac{dy}{dx} = \frac{y}{3}\left[\frac{1}{x} + \frac{1}{x-2} - \frac{2x}{x^2+1}\right]$$

29. $y = \sqrt{\dfrac{x(x+1)(x-2)}{(x^2+1)(2x+3)}}$, $x > 2$

$$\ln y = \frac{1}{3}[\ln x + \ln(x+1) + \ln(x-2) - \ln(x^2+1) - \ln(2x+3)$$

$$\frac{1}{y}\frac{dy}{dx} = \frac{1}{3}\left[\frac{1}{x} + \frac{1}{x+1} + \frac{1}{x-2} - \frac{2x}{x^2+1} - \frac{2}{2x+3}\right]$$

$$\frac{dy}{dx} = \frac{y}{3}\left[\frac{1}{x} + \frac{1}{x+1} + \frac{1}{x-2} - \frac{2x}{x^2+1} - \frac{2}{2x+3}\right]$$

31. $\sqrt{y} = \dfrac{x^5 \tan^{-1} x}{(3-2x)\sqrt[3]{x}} \Rightarrow$

$$\frac{1}{2}y = 5\ln x + \ln \tan^{-1} x - \ln(3-2x) - \frac{1}{3}\ln x$$

$$\frac{1}{2y}\frac{dy}{dx} = \frac{5}{x} + \frac{1}{\tan^{-1} x} \cdot \frac{1}{1+x^2} + \frac{2}{3-2x} - \frac{1}{3x}$$

$$\frac{dy}{dx} = 2y\left[\frac{14}{3x} + \frac{2}{3-2x} + \frac{1}{(1+x^2)\tan^{-1} x}\right]$$

33. $\displaystyle\int_{-1}^{1} \frac{dx}{x+3} = \ln|x+3|\Big]_{-1}^{1} = \ln 4 - \ln 2 = \ln 2$

35. $\displaystyle\int_{\frac{\pi}{4}}^{\frac{\pi}{2}} \cot x\,dx = \int_{\frac{\pi}{4}}^{\frac{\pi}{2}} \frac{\cos x}{\sin x}\,dx = \ln|\sin x|\Big]_{\frac{\pi}{4}}^{\frac{\pi}{2}} = \ln\sqrt{2}$

37. $\displaystyle\int_{2}^{4} \frac{2x-5}{x}\,dx = \int_{2}^{4}\left(2 - \frac{5}{x}\right)dx = 2x - 5\ln|x|\Big]_{2}^{4} = 4 - 5\ln 2$

39. (a) $\displaystyle\int_{0}^{\frac{3}{5}} \frac{x\,dx}{1-x^2} = -\frac{1}{2}\int_{1}^{\frac{16}{25}} \frac{du}{u} = -\frac{1}{2}\ln|u|\Big]_{1}^{\frac{16}{25}} = \ln\left(\frac{16}{25}\right)^{-1/2} = \ln\frac{5}{4}$

Let $u = 1 - x^2 \Rightarrow du = -2x\,dx \Rightarrow x\,dx = -\dfrac{1}{2}u$;

$x = 0 \Rightarrow u = 1$; $x = \dfrac{3}{5} \Rightarrow u = \dfrac{16}{25}$

(b) $\displaystyle\int_0^{\frac{3}{5}} \frac{dx}{\sqrt{1-x^2}} = \sin^{-1} x\Big]_0^{\frac{3}{5}} = \sin^{-1}\frac{3}{5}$

(c) $\displaystyle\int_0^{\frac{3}{5}} \frac{x\,dx}{\sqrt{1-x^2}} = -\sqrt{1-x^2}\,\Big]_0^{\frac{3}{5}} = \frac{1}{5}$

41. (a) $y = \log|x|$

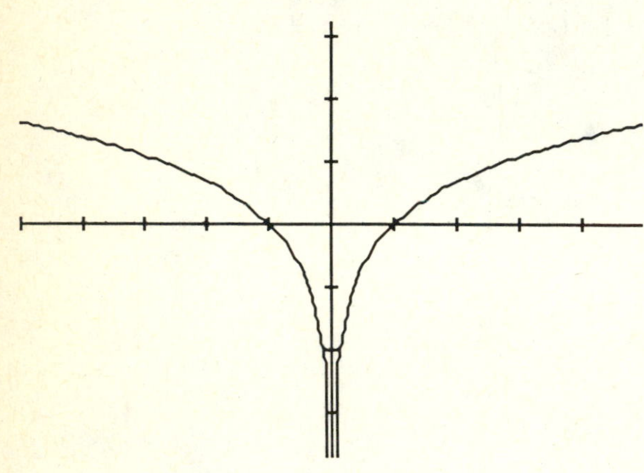

(b) $y = |\ln x|$

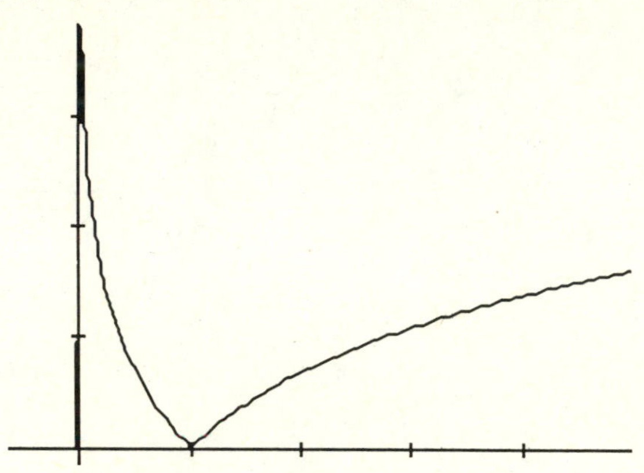

43. $\displaystyle\lim_{x\to\infty}\int_x^{2x} \frac{1}{t}\,dt = \lim_{x\to\infty}[\ln t\Big]_x^{2x} = \lim_{x\to\infty}[\ln 2x - \ln x] = \ln 2$

45. The region is bounded by $y = \frac{4}{x}$ and $y = (x-3)^2$. These curves intersect where $\frac{4}{x} = (x-3)^2 \Leftrightarrow x^3 - 6x^2 + 9x - 4 = 0 \Leftrightarrow (x-1)^2(x-4) = 0$ or $x = 1, 4$. Then:

$$A = \int_1^4 \left[\frac{4}{x} - (x-3)^2\right]dx = 4\ln|x| - \frac{1}{3}(x-3)^2\Big]_1^4 = 4\ln 4 - 3$$

$$V = \pi\int_1^4 \left[\left(\frac{4}{x}\right)^2 - (x-3)^4\right]dx = \pi\left[-\frac{16}{x} - \frac{1}{5}(x-3)^5\right]_0^4 = \frac{27\pi}{5}$$

47. (a) If $f(x) = \ln(1+x)$, then $f'(x) = \frac{1}{1+x}$, $f''(x) = -\frac{1}{(1+x)^2}$ and $f'''(x) = \frac{2}{(1+x)^3}$. Then $L(x) \approx f(0) + f'(0)x = \ln 1 + (1)x = x$.

$$Q(x) \approx f(0) + f'(0)x + \frac{f''(0)}{2}x^2 = \ln 1 + (1)x - \frac{1}{2}x^2 = x - \frac{x^2}{2}.$$

(b) $|e_1(x)| \le \frac{1}{2}M_1x^2$, where $M_1 = \max\{|f''(x)| : 0 \le x \le .1\} = 1$

$\therefore\ |e_1(x)| \le \frac{1}{2}(1)(.1)^2 = 0.005$

$|e_2(x)| \le \frac{1}{6}M_2\ |x|^3$, where $M_2 = \max\{|f'''(x)| : 0 \le x \le .1\} = 2.$

$\therefore\ |e_2(x)| \le \frac{1}{6}(2)(.1)^3 = 0.000\overline{3}$

49. (a) $\ln 1.2 = \ln(1 + 0.2) \approx 0.2$ by (16)

$$\approx 0.2 - \frac{(.2)^2}{2} = 0.1800 \text{ by } (17)$$

$$\ln 1.2 \approx \frac{.1}{3}\left[1 + 4\left(\frac{1}{1.1}\right) + \frac{1}{1.2}\right] \approx 0.18232 \text{ by Simpson's Rule.}$$

(b) $\ln .8 = \ln(1 - 0.2) \approx -0.2$ by (16)

$$\approx -0.2 - \frac{(-.2)^2}{2} = -0.2200 \text{ by } (17)$$

$$\ln .8 \approx -\frac{.1}{3}\left[\frac{1}{.8} + 4\left(\frac{1}{.9}\right) + 1\right] \approx -0.22315 \text{ by Simpson's Rule.}$$

6.6 THE EXPONENTIAL FUNCTION e^x

1. $e^{\ln x} = x$

3. $e^{-\ln(x^2)} = e^{\ln(x^{-2})} = x^{-2}$

5. $\ln e^{\frac{1}{x}} = \frac{1}{x}$

7. $e^{\ln 2 + \ln x} = e^{\ln 2x} = 2x$

9. $\ln e^{(x - x^2)} = x - x^2$

11. $e^{(x + \ln x)} = e^x e^{\ln x} = xe^x$

13. $e^{\sqrt{y}} = x^2 \Rightarrow \sqrt{y}\ln e = \ln x^2 \Rightarrow \sqrt{y} = 2\ln x \Rightarrow y = 4\ln^2 x$

15. $e^{(x^2)}e^{2x+1} = e^y \Leftrightarrow y = x^2 + 2x + 1$

17. $\ln(y - 2) = \ln(\sin x) - x \Rightarrow \ln\frac{y - 2}{\sin x} = -x \Rightarrow \frac{y - 2}{\sin x} = e^{-x}$

$y = e^{-x}\sin x + 2.$

19. $y = e^{3x} \Rightarrow \frac{dy}{dx} = 3e^{3x}$

21. $y = e^{5-7x} \Rightarrow \frac{dy}{dx} = -7e^{5-7x}$

23. $y = x^2 e^x = \frac{dy}{dx} = x^2 e^x + 2xe^x = xe^x (x+2)$

25. $y = e^{\sin x} \Rightarrow \frac{dy}{dx} = (\cos x)e^{\sin x}$

27. $y = \ln (3xe^{-x}) = \ln 3 + \ln x - x \Rightarrow \frac{dy}{dx} = \frac{1}{x} - 1 = \frac{1-x}{x}$

29. $y = e^{\sin^{-1} x} \Rightarrow \frac{dy}{dx} = e^{\sin^{-1} x}\left(\frac{1}{\sqrt{1-x^2}}\right)$

31. $y = (9x^2 - 6x + 2)e^{3x} \Rightarrow \frac{dy}{dx} = 3(9x^2 - 6x + 2)e^{3x} + (18x - 6)e^{3x} = 27x^2 e^{3x}$

33. $y = x^2 e^{-(x^2)} \Rightarrow \frac{dy}{dx} = x^2(-2xe^{-(x^2)}) + 2xe^{-(x^2)} = 2xe^{-(x^2)}(1 - x^2)$

35. $y = \tan^{-1}(e^x) \Rightarrow \frac{dy}{dx} = \frac{e^x}{1 + e^{2x}}$

37. $y = x^3 e^{-2x} \cos 5x \Rightarrow \ln y = 3 \ln x - 2x + \ln (\cos 5x)$

$$\frac{1}{y}\frac{dy}{dx} = \frac{3}{x} - 2 - \frac{5 \sin 5x}{\cos 5x} = \frac{3}{x} - 2 - 5 \tan 5x$$

$$\frac{dy}{dx} = y\left(\frac{3}{x} - 2 - 5 \tan 5x\right)$$

39. $\ln y = x \sin x \Rightarrow \frac{1}{y}\frac{dy}{dx} = \sin x + x \cos x \Rightarrow \frac{dy}{dx} = y(\sin x + x \cos x)$

41. $e^{2x} = \sin (x + 3y) \Rightarrow 2e^{2x} = \cos (x + 3y)\left[1 + 3\frac{dy}{dx}\right]$

$$2e^{2x} = \cos (x + 3y) + 3\cos (x + 3y)\frac{dy}{dx}$$

$$\frac{dy}{dx} = \frac{2e^{2x} - \cos (x + 3y)}{3 \cos(x + 3y)}$$

43. $\int_{\ln 3}^{\ln 5} e^{2x}\, dx = \frac{1}{2} e^{2x}\Big]_{\ln 3}^{\ln 5} = \frac{1}{2}[e^{2 \ln 5} - e^{2 \ln 3}] = \frac{1}{2}(25 - 9) = 8$

45. $\int_0^{\pi} e^{\sin x} \cos x\, dx = e^{\sin x}\Big]_0^{\pi} = 0$

47. $\int_{-\ln(a+1)}^{0} e^{-x}\, dx = -e^{-x}\Big]_{-\ln(a+1)}^{0} = -[e^0 - e^{\ln(a+1)}] = a$

49. $\int_0^1 e^{\ln \sqrt{x}}\, dx = \int_0^1 \sqrt{x}\, dx = \frac{2}{3}x^{3/2}\Big]_0^1 = \frac{2}{3}$

51. $\displaystyle\int_0^{\ln 2} \frac{24\,dx}{e^{3x}} = -8\,e^{-3x}\Big]_0^{\ln 2} = -8\,(e^{-3\ln 2} - e^0) = -8\left(\frac{1}{8} - 1\right) = 7$

53. $\displaystyle\int_0^{\ln 13} \frac{e^x\,dx}{1 + 2e^x} = \frac{1}{2}\ln(1 + 2e^x)\Big]_0^{\ln 13} = \frac{1}{2}(\ln 27 - \ln 3) = \ln 3$

55. $\displaystyle\int_0^{\ln 2} \frac{e^x}{1 + e^{2x}}\,dx = \tan^{-1} e^x\Big]_0^{\ln 2} = \tan^{-1} 2 - \frac{\pi}{4}$

57. $\displaystyle\lim_{h\to 0}\frac{e^h - (1 + h)}{h^2} = \lim_{h\to 0}\frac{e^h - 1}{2h} = \lim_{h\to 0}\frac{e^h}{2} = \frac{1}{2}$

59. $\displaystyle\lim_{x\to\infty}\frac{x^2 + e^x}{x + e^x} = \lim_{x\to\infty}\frac{2x + e^x}{1 + e^x} = \lim_{x\to\infty}\frac{2 + e^x}{e^x} = \lim_{x\to\infty}\frac{e^x}{e^x} = 1$

61. (a) $y = e^{\sin x}$, $-\pi \le x \le 2\pi$. $y'(x) = (\cos x)e^{\sin x} = 0 \Leftrightarrow \cos x = 0$.

$\therefore x = -\frac{\pi}{2}, \frac{\pi}{2}, \frac{3\pi}{2}$. $y\left(\frac{\pi}{2}\right) = e^{\sin\left(\frac{\pi}{2}\right)} = e$. $y\left(-\frac{\pi}{2}\right) = y\left(\frac{3\pi}{2}\right) = \frac{1}{e}$.

Testing endpoints: $y(\pi) = e^{\sin\pi} = 1$ and $y(2\pi) = 1$.

The absolute maximum $= e$ at $x = \frac{\pi}{2}$ and the

absolute minimum $= \frac{1}{e}$ at $x = -\frac{\pi}{2}$ or $\frac{3\pi}{2}$.

(b)

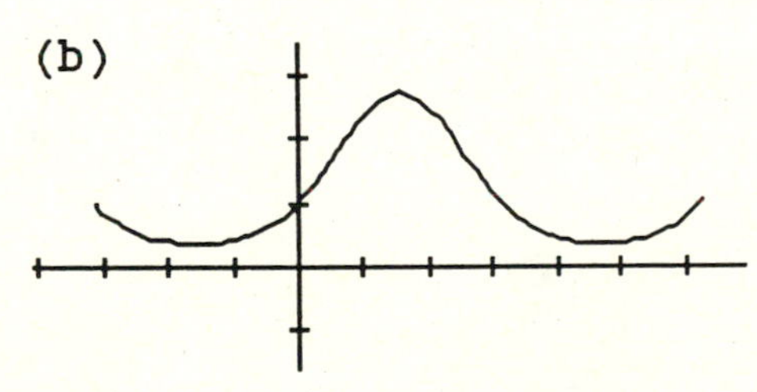

63. $f(x) = x^2 \ln\frac{1}{x} = -x^2 \ln x \Rightarrow f'(x) = x^2\left(\frac{1}{x}\right) - 2x\ln x = 0$ if

$-x(1 + 2\ln x) = 0 \Leftrightarrow x = 0$ or $x = e^{-1/2}$. $f''(x) = -3 - 2\ln x$

and $f''(e^{-1/2}) = -2 \Rightarrow f(e^{-1/2}) = \frac{1}{2e}$ is a maximum.

65. (a) $f(x) = f'(x) = f''(x) = f'''(x) = e^x$

Linear: $e^x \approx f(0) + f'(0)x = 1 + x$

Quadratic: $e^x \approx f(0) + f'(0)x + \frac{f''(0)}{2}x^2 = 1 + x + \frac{1}{2}x^2$

(b) $|e_1(x)| \le \frac{1}{2}Mx^2 = \frac{1}{2}(1)(0.1)^2 = 0.005$

$|e_2(x)| \le \frac{1}{6}Mx^3 = \frac{1}{6}(1)(0.1)^3 \approx 0.00017$

(c)

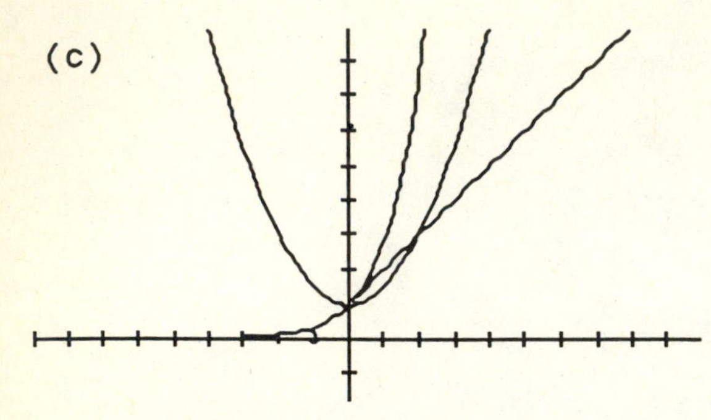

67. $A(t) = \int_0^t e^{-x}\,dx = -e^{-x}\Big]_0^t = -e^{-t} + 1$

$V(t) = \pi\int_0^t (e^{-x})^2\,dx = \pi\left[-\frac{1}{2}e^{-2x}\right]_0^t = \frac{\pi}{2}(1 - e^{-2t})$

(a) $\lim_{t\to\infty} A(t) = \lim_{t\to\infty} (-e^{-t} + 1) = 1$

(b) $\lim_{t\to\infty} \frac{V(t)}{A(t)} = \lim_{t\to\infty} \frac{\pi}{2}\left(\frac{1 - e^{-2t}}{-e^{-t} + 1}\right) = \frac{\pi}{2}$

(c) $\lim_{t\to 0^+} \frac{V(t)}{A(t)} = \lim_{t\to 0^+} \frac{\pi}{2}\left(\frac{1 - e^{-2t}}{-e^{-t} + 1}\right) = \lim_{t\to 0^+} \frac{\pi}{2}\left(\frac{2e^{-2t}}{e^{-t}}\right) = \pi$

69. (a) $y = Ce^{ax} \Rightarrow \frac{dy}{dx} = aCe^{ax} = ay$

(b) $\frac{dy}{dt} = -2y \Rightarrow y = Ce^{-2t}$. $3 = Ce^0 \Rightarrow C = 3$. $\therefore y = 3e^{-2t}$

71. $\frac{dy}{dx} = e^{-x}$, $y = 0$ when $x = 4 \Rightarrow y = -e^{-x} + C$.

$0 = -e^{-4} + C \Rightarrow C = e^{-4}$. $\therefore\ y = -e^{-x} + e^{-4}$

73. $\frac{1}{y+1}\frac{dy}{dx} = \frac{1}{2x}$, $x > 0$, $y = 1$ when $x = 2$

$\frac{dy}{y+1} = \frac{dx}{2x} \Rightarrow \ln|y+1| = \frac{1}{2}\ln|x| + C$

$C = \ln 2 - \ln\sqrt{2} = \ln\frac{2}{\sqrt{2}} = \ln\sqrt{2}$

$\ln|1+y| = \ln\sqrt{x} + \ln\sqrt{2} = \ln\sqrt{2x} \Rightarrow y + 1 = \sqrt{2x}$ or $(y+1)^2 = 2x$

75. $\frac{d}{dx}(\cosh x) = \frac{d}{dx}\left[\frac{1}{2}(e^x + e^{-x})\right] = \frac{1}{2}(e^x - e^{-x}) = \sinh x$

77. $\cosh(-x) = \frac{1}{2}(e^{-x} + e^x) = \frac{1}{2}(e^x - e^{-x}) = \cosh x$

$\sinh(-x) = \frac{1}{2}(e^{-x} - e^x) = -\frac{1}{2}(e^x - e^{-x}) = -\sinh x$

6.7 THE FUNCTIONS a^x and a^u

1. If $y = 2^x = e^{\ln 2^x} = e^{x \ln 2}$, then $\frac{dy}{dx} = (\ln 2)\, e^{x \ln 2} = (\ln 2)2^x$

   or $y = 2^x \Rightarrow \ln y = \ln 2^x = x \ln 2$, so $\frac{1}{y}\frac{dy}{dx} = \ln 2$, $\frac{dy}{dx} = y \ln 2 = (\ln 2)2^x$

3. $y = 8^x = e^{\ln 8^x} = e^{x \ln 8} \Rightarrow \frac{dy}{dx} = (\ln 8)\, e^{x \ln 8} = (\ln 8)8^x$

5. $y = 9^x = e^{\ln 9^x} = e^{x \ln 9} \Rightarrow \frac{dy}{dx} = (\ln 9)\, e^{x \ln 9} = (\ln 9)9^x$

7. $y = (2^x)^2 = (2^2)^x = 4^x$. $\frac{dy}{dx} = (\ln 4)4^x$

9. $y = (\sin x)^{\tan x}$, $\sin x > 0 \Rightarrow \ln y = (\tan x)\ln(\sin x)$

   $\frac{1}{y}\frac{dy}{dx} = (\tan x)\left(\frac{\cos x}{\sin x}\right) + [\ln(\sin x)](\sec^2 x)$

   $\frac{dy}{dx} = y\,[1 + (\sec^2 x)\ln(\sin x)]$

11. $y = x^{\ln x}$, $x > 0 \Rightarrow \ln y = (\ln x)(\ln x) = \ln^2 x$

    $\frac{1}{y}\frac{dy}{dx} = 2 \ln x\left(\frac{1}{x}\right) \Rightarrow \frac{dy}{dx} = 2y\left(\frac{\ln x}{x}\right) = \frac{2 \ln x\,(x^{\ln x})}{x}$

13. $y = (1 - x)^x$, $x < 1 \Rightarrow \ln y = x \ln(1 - x)$

    $\frac{1}{y}\frac{dy}{dx} = \frac{-x}{1 - x} + \ln(1 - x) \Rightarrow \frac{dy}{dx} = y\left(\frac{x}{x - 1} + \ln(1 - x)\right)$ or

    $\frac{dy}{dx} = (1 - x)^x\left(\frac{x}{x - 1} + \ln(1 - x)\right)$

15. $y = 2^x \ln x \Rightarrow \frac{dy}{dx} = 2^x\left(\frac{1}{x}\right) + \ln x\,(\ln 2)2^x = 2^x\left[\frac{1}{x} + \ln 2\,(\ln x)\right]$

17. $\int_0^1 5^x\,dx = \frac{1}{\ln 5}5^x\Big]_0^1 = \frac{1}{\ln 5}(5 - 1) = \frac{4}{\ln 5}$

19. $\int_0^1 \frac{1}{2^x}dx = \int_0^1 2^{-x}dx = -\frac{1}{\ln 2}2^{-x}\Big]_0^1 = -\frac{1}{\ln 2}\left[\frac{1}{2} - 1\right] = \frac{1}{2 \ln 2}$

21. $\int_0^1 3^{2x}\,dx = \frac{1}{2\ln 3}3^{2x}\Big]_0^1 = \frac{1}{2 \ln 3}[3^2 - 1] = \frac{4}{\ln 3}$

23. $\int_{-1}^0 4^{-x} \ln 2\,dx = -\frac{\ln 2}{\ln 4}\,4^{-x}\Big]_{-1}^0 = -\frac{\ln 2}{\ln 4}[1 - 4] = \frac{3 \ln 2}{\ln 4}$

25. $\int_1^2 5^{(2x-2)}\,dx = \frac{1}{2 \ln 5}\,5^{(2x-2)}\Big]_1^2 = \frac{1}{2 \ln 5}\,[5^2 - 5^0] = \frac{24}{\ln 25}$

27. $\int_0^{\frac{\pi}{2}} 2^{\cos x} \sin x\,dx = \frac{-1}{\ln 2}\,2^{\cos x}\Big]_0^{\frac{\pi}{2}} = -\frac{1}{\ln 2}[2^0 - 2] = \frac{1}{\ln 2}$

29. (b), because $\int_0^1 2^{3x}\,dx = \int_0^1 8^x\,dx$, and $\int_0^1 3^{2x}\,dx = \int_0^1 9^x\,dx$

31. $\lim_{x\to\infty} 2^{-x} = 0$

33. $\lim_{x\to 0} \dfrac{3^{\sin x} - 1}{x} = \lim_{x\to 0} \dfrac{3^{\sin x}(\cos x)(\ln 3)}{1} = \ln 3$

35. Let $y = (e^x + x)^{\frac{1}{x}}$ so that $\ln y = \frac{1}{x}\ln(e^x + x)$. We first find

$$\lim_{x\to 0} \frac{\ln(e^x + x)}{x} = \lim_{x\to 0} \frac{e^x + 1}{e^x + x} = 2.$$ Thus $\lim_{x\to 0} \ln y = 2$, so

$$\lim_{x\to 0} e^{\ln y} = \lim_{x\to 0}(e^x + x)^{\frac{1}{x}} = e^2.$$

37. (a) $\lim_{x\to\infty} \dfrac{3^x - 5}{4(3^x + 2)} = \lim_{x\to\infty} \dfrac{1 - \frac{5}{3^x}}{4\left(1 + \frac{2}{3^x}\right)} = \dfrac{1}{4}$

(b) $\lim_{x\to -\infty} \dfrac{3^x - 5}{4(3^x + 2)} = \dfrac{0 - 5}{4(0 + 2)} = -\dfrac{5}{8}$

39. Let $f(x) = 2^x - x^2$, so that $f'(x) = (\ln 2)\,2^x - 2x$. Let $x_0 = -0.5$.

Using the iterative formula $x_n = x_{n-1} - \dfrac{f'(x_{n-1})}{f(x_{n-1})}$, we find

$x = -0.766664696$, $y = 0.587774756$.

41. Let $y = x^{(1/x^n)}$ and consider $\ln y = \dfrac{1}{x^n}\ln x$.

$$\lim_{x\to\infty} \ln y = \lim_{x\to\infty} \frac{\ln x}{x^n} = \lim_{n\to\infty} \frac{\frac{1}{x}}{nx^{n-1}} = \lim_{n\to\infty} \frac{1}{nx^n} = 0.$$ Hence

$$\lim_{x\to\infty} e^{\ln y} = \lim_{x\to\infty} x^{(1/x^n)} = e^0 = 1$$

43. Equation 2 implies line (a), commutativity of multiplication implies line (b), Equaion 4 implies line (c), and Equation 2 implies line (d).

6.8 THE FUNCTIONS $Y = \log_a u$. RATES OF GROWTH

1. $\log_4 16 = \dfrac{\ln 16}{\ln 4} = \dfrac{2\ln 4}{\ln 4} = 2$

3. $\log_5 0.04 = \dfrac{\ln 4 - \ln 100}{\ln 5} = \dfrac{\ln 4 - \ln 25 - \ln 4}{\ln 5} = -\dfrac{2\ln 5}{\ln 5} = -2$

5. $\log_2 4 = \dfrac{\ln 4}{\ln 2} = \dfrac{2\ln 2}{\ln 2} = 2$

7. $\log_8 16 = \dfrac{\ln 16}{\ln 8} = \dfrac{4\ln 2}{3\ln 2} = \dfrac{4}{3}$

9. $3^{\log_3 7} + 2^{\log_2 5} = 5^{\log_5 x} \Rightarrow x = 7 + 5 = 12$

11. $\lim\limits_{x\to\infty} \dfrac{\log_2 x}{\log_3 x} = \lim\limits_{x\to\infty} \dfrac{\dfrac{\ln x}{\ln 2}}{\dfrac{\ln x}{\ln 3}} = \dfrac{\ln 3}{\ln 2}$

13. $\lim\limits_{x\to\infty} \dfrac{\log_9 x}{\log_3 x} = \lim\limits_{x\to\infty} \dfrac{\dfrac{\ln x}{\ln 9}}{\dfrac{\ln x}{\ln 3}} = \dfrac{\ln 3}{2\ln 3} = \dfrac{1}{2}$

15. $y = \log_4 x = \dfrac{\ln x}{\ln 4} \quad \Rightarrow \quad \dfrac{dy}{dx} = \dfrac{1}{x\ln 4}$

17. $y = \log_{10} e^x = x\log_{10} e \quad \Rightarrow \quad \dfrac{dy}{dx} = \log_{10} e = \dfrac{\ln e}{\ln 10} = \dfrac{1}{\ln 10}$

19. $y = (\ln 2)\log_2 x = \dfrac{(\ln 2)(\ln x)}{\ln 2} = \ln x. \quad \therefore \quad \dfrac{dy}{dx} = \dfrac{1}{x}$

21. $y = \log_{10}\sqrt{x+1} = \dfrac{1}{2}\log_{10}(x+1). \quad \therefore \quad \dfrac{dy}{dx} = \dfrac{1}{2\ln 10\,(x+1)}$

23. $y = \dfrac{1}{\log_2 x} = (\log_2 x)^{-1} \quad \Rightarrow \quad \dfrac{dy}{dx} = -(\log_2 x)^{-2}\dfrac{1}{x\ln 2} = -\dfrac{1}{x\log_2^2 x\ln 2}$

25. $y = \log_5 (x+1)^2 = \dfrac{\ln (x+1)^2}{\ln 5} = \dfrac{2\ln(x+1)}{\ln 5}. \quad \dfrac{dy}{dx} = \dfrac{2}{(x+1)\ln 5}$

27. $y = \log_7(\sin x) = \dfrac{\ln(\sin x)}{\ln 7}. \quad \dfrac{dy}{dx} = \dfrac{\cos x}{(\ln 7)\sin x} = \dfrac{\cot x}{\ln 7}$

29. If $y = \ln x$, then $\frac{dy}{dx} = \frac{1}{x}\Big|_{x=10} = \frac{1}{10}$. If $y = \log_2 x$, then $\frac{dy}{dx} = \frac{1}{x \ln 2}\Big|_{x=10} = \frac{1}{10 \ln 2}$. Since $\frac{1}{10} < \frac{1}{10 \ln 2}$, $\log_2 x$ is changing faster.

31. $$\int_1^{10} \frac{\log_{10} x}{x} dx = \int_0^{10} \frac{\ln x}{(\ln 10)\, x} dx = \frac{1}{\ln 10}\left[\frac{1}{2} \ln^2 x\right]_1^{10} = \frac{\ln 10}{2}$$

33. $$\int_1^8 \frac{\log_4 (x^2)}{x} dx = \int_1^8 \frac{\ln x^2}{(\ln 4)\, x} dx = \frac{2}{\ln 4} \int_1^8 \frac{\ln x}{x} dx = \frac{2}{\ln 4}\left[\frac{\ln^2 x}{x}\right]_1^8 = \frac{\ln^2 8}{\ln 4}$$

35. $$\int_1^{125} \frac{(\log_5 x)^2}{x} dx = \int_1^{125} \frac{\ln^2 x}{(\ln^2 5)x} dx = \frac{1}{\ln^2 5}\left[\frac{\ln^3 x}{3}\right]_1^{125} = \frac{\ln^3 125}{3 \ln^2 5} = \frac{3^3 \ln^3 5}{3 \ln^2 5} = 9 \ln 5$$

37. $$\int_{\sqrt{e}}^{e} \frac{dx}{x \log_{10} x} = \int_{\sqrt{e}}^{e} \frac{\ln 10}{x \ln x} dx = \ln 10\, [\ln (\ln x)]_{\sqrt{e}}^{e}$$
$$= \ln 10\, [\ln (\ln e) - \ln (\ln \sqrt{e}\,)] = \ln 10 \left[\ln 1 - \ln\left(\frac{1}{2}\right)\right] = (\ln 10)(\ln 2)$$

39. (a) $\lim\limits_{x\to\infty} \frac{x + 3}{e^x} = 0$; slower

(b) $\lim\limits_{x\to\infty} \frac{x^3 - 3x + 1}{e^x} = 0$; slower

(c) $\lim\limits_{x\to\infty} \frac{\sqrt{x}}{e^x} = \lim\limits_{x\to\infty} \frac{1}{2\sqrt{x} e^x} = 0$; slower

(d) $\lim\limits_{x\to\infty} \frac{4^x}{e^x} = \lim\limits_{x\to\infty} \left(\frac{4}{e}\right)^x = \infty$; faster

(e) $\lim\limits_{x\to\infty} \frac{(5/2)^x}{e^x} = \lim\limits_{x\to\infty} \left(\frac{5}{2e}\right)^x = 0$; slower $\left(\frac{5}{2e} < 1\right)$

(f) $\lim\limits_{x\to\infty} \frac{\ln x}{e^x} = \lim\limits_{x\to\infty} \frac{1}{xe^x} = 0$; slower

41. (a) $\lim_{x\to\infty} \frac{\log_3 x}{\ln x} = \lim_{x\to\infty} \frac{\ln x}{(\ln 3)\ln x} = \frac{1}{\ln 3}$; same rate

(b) $\lim_{x\to\infty} \frac{\log_2 x^2}{\ln x} = \lim_{x\to\infty} \frac{2 \ln x}{(\ln 2)\ln x} = \frac{2}{\ln 2}$; same rate

(c) $\lim_{x\to\infty} \frac{\log_{10} \sqrt{x}}{\ln x} = \lim_{x\to\infty} \frac{\ln x}{(2 \ln 10) \ln x} = \frac{1}{2 \ln 10}$; same rate

(d) $\lim_{x\to\infty} \frac{\frac{1}{x}}{\ln x} = \lim_{x\to\infty} \frac{1}{x \ln x} = 0$; slower

(e) $\lim_{x\to\infty} \frac{\frac{1}{\sqrt{x}}}{\ln x} = \lim_{x\to\infty} \frac{1}{\sqrt{x} \ln x} = 0$; slower

(f) $\lim_{x\to\infty} \frac{e^{-x}}{\ln x} = 0$; slower

(g) $\lim_{x\to\infty} \frac{x}{\ln x} = \lim_{x\to\infty} \frac{1}{\frac{1}{x}} = \infty$; faster

(h) $\lim_{x\to\infty} \frac{5 \ln x}{\ln x} = 5$; same rate

(i) $\lim_{x\to\infty} \frac{2}{\ln x} = 0$; slower

(j) $\lim_{x\to\infty} \frac{\sin x}{\ln x} = 0$; slower

43. $\ln x$ grows faster because $\lim_{x\to\infty} \frac{\ln x}{\ln(\ln x)} = \lim_{x\to\infty} \frac{\frac{1}{x}}{\frac{1}{x \ln x}} = \lim_{x\to\infty} (\ln x) = \infty$

45. Let $p_1(x)$ and $p_2(x)$ be polynomials with degrees n_1 and n_2. Then

$n_1 > n_2 \Rightarrow p_1(x)$ grows faster than $p_2(x)$;

$n_1 < n_2 \Rightarrow p_1(x)$ grows slower than $p_2(x)$;

$n_1 = n_2 \Rightarrow p_1(x)$ grows at the same rate as $p_2(x)$.

47. Let $x = [H_3O^+]$ amd $S - x = [OH^-]$. Then $x(S - x) = 10^{-14} \Rightarrow S - x = \dfrac{10^{-14}}{x}$

(a) If $f(x) = x + \dfrac{10^{-14}}{x}$, then $f'(x) = 1 - \dfrac{10^{-14}}{x^2} = 0$ if $x = 10^{-7}$.

$f''(x) = \dfrac{10^{-14}}{x^3} > 0 \Rightarrow$ a minimum value.

(b) $pH = -\log_{10}[10^{-7}] = 7$

(c) $\dfrac{[OH^-]}{[H_3O^+]} = 1$

49. (a) $\log_a uv = \dfrac{\ln uv}{\ln a} = \dfrac{\ln u}{\ln a} + \dfrac{\ln v}{\ln a} = \log_a u + \log_a u$

(b) $\log_a \dfrac{u}{v} = \dfrac{\ln(u/v)}{\ln a} = \dfrac{\ln u}{\ln a} - \dfrac{\ln v}{\ln a} = \log_a u - \log_a v$

6.9 APPLICATIONS OF EXPONENTIAL AND LOGARITHMIC FUNCTIONS

1. $y = y_0 e^{kt}$; when $t = 0$, $y_0 = 1 \Rightarrow y = e^{kt}$.

when $t = 30$ min $= \dfrac{1}{2}$hr, $2 = e^{.5k} \Rightarrow .5k = \ln 2 \Rightarrow k = 2\ln 2$.

After 24 hrs, $y = e^{24(2\ln 2)} = 2^{48} \approx 2.81 \times 10^{14}$.

3. $y = y_0 e^{kt}$; $t = 0 \Rightarrow y_0 = 10{,}000$ so $y = 10{,}000\, e^{kt}$. When $t = 1$, there are 10% fewer cases or 9000 remaining, so $9000 = 10000\, e^{k} \Rightarrow$
$k = \ln \dfrac{9}{10}$. Then $1000 = 10000 e^{(\ln .9)t} \Rightarrow (0.9)^t = 0.1 \Rightarrow$
$t = \dfrac{\ln .1}{\ln .9} \approx 21.9$ years.

5. $\dfrac{dQ}{dt} = kQ \Rightarrow \dfrac{dQ}{Q} = kt \Rightarrow \ln Q = kt + C$. When $t = 0$, $C = \ln Q_0$ so

$\ln Q = kt + \ln Q_0 \Rightarrow \ln \dfrac{Q}{Q_0} = kt \Rightarrow \dfrac{Q}{Q_0} = e^{kt} \Rightarrow Q = Q_0 e^{kt}$.

7. $A = A_0 e^{rt}$, where $A_0 = 1000$ lbs. In 100 years, $A = 90{,}000$.
Therefore, $90000 = 1000\, e^{100r} \Rightarrow 100r = \ln 90$ or $r = 4.5$ %

9. To double, we need $A = 2A_0$ in the formula $A = A_0 e^{rt}$. Thus $2A_0 = A_0 e^{rt} \Rightarrow 2 = e^{rt} \Rightarrow rt = \ln 2$. If the value $\ln 2 \approx 0.7$ is used, then $rt = 0.70$ or in percent, $rt = 70$.

11. (a) To have a half-life of 5700 years means that $y = \frac{1}{2} y_0$ $\left(\text{the amount deteriorates by } \frac{1}{2}\right)$ in 5700 years, or $\frac{1}{2} y_0 = y_0 e^{5700k} \Rightarrow 5700k = \ln \frac{1}{2} \Rightarrow k = \frac{-\ln 2}{5700}$

(b) 90% decayed means 10% remaining, so

$$.1 = e^{\left(\frac{-\ln 2}{5700}\right)t} \Rightarrow t = \frac{(\ln 0.1)5700}{(-\ln 2)} \approx 18,935 \text{ years}$$

(c) $t = \dfrac{(\ln 0.445)\,5700}{(-\ln 2)} \approx 6,658$ years

13. (a) To have a half-life of 140 days means that $y = \frac{1}{2} y_0$ $\left(\text{the amount deteriorates by } \frac{1}{2}\right)$ in 140 days, or $\frac{1}{2} y_0 = y_0 e^{140k} \Rightarrow 140\,k = \ln \frac{1}{2} \Rightarrow k = \frac{-\ln 2}{140}$.

(b) 90% decayed means 10% remaining, so

$$.1 = e^{\left(\frac{-\ln 2}{140}\right)t} \Rightarrow t = \frac{(\ln 0.1)140}{(-\ln 2)} \approx 465 \text{ days}$$

15. In the formula $T - T_s = (T_0 - T_s)e^{kt}$, $T_s = 30^\circ$ is given.

When $t = 10$ minutes, $T = 0^\circ \Rightarrow -30 = (T_0 - 30)e^{10k}$.

When $t = 20$ minutes, $T = 15^\circ \Rightarrow -15 = (T_0 - 30)e^{20k}$.

$$\therefore (T_0 - 30)\left(\frac{-30}{T_0 - 30}\right)^2 = -15 \Rightarrow \frac{900}{T_0 - 30} = -15 \Rightarrow T_0 = -30^\circ.$$

17. $T_0 - T_s = 60$, so in the formula $T - T_s = (T_0 - T_s)e^{kt}$,

$T - T_s = 60\,e^{kt}$. When $t = -20$ minutes, $T - T_s = 70$.

$\therefore\ 70 = 60\,e^{-20k} \Rightarrow k = -\frac{1}{20}\ln\frac{7}{6}$.

(a) In 15 minutes, $T - T_s = 60\,e^{\left(-\frac{1}{20}\ln\frac{7}{6}\right)15} = 60\left(\frac{7}{6}\right)^{-3/4} \approx 53.4^\circ$

(b) After 2 hours = 120 minutes, $T - T_s = 60\left(\frac{7}{6}\right)^{-6} \approx 23.8^\circ$

(c) $60\left(\frac{7}{6}\right)^{-t/20} = 10 \Rightarrow -\frac{t}{20}\ln\frac{7}{6} = \ln\frac{1}{6} \Rightarrow t = \dfrac{-20\ln\frac{1}{6}}{\ln\frac{7}{6}} = 232.5 \text{ min} = 3.9 \text{ hrs}$

19. $i = \frac{V}{R}\left(1 - e^{-\frac{R}{L}t}\right)$ and $t = \frac{L}{R} \Rightarrow i = \frac{V}{R}\left(1 - \frac{1}{e}\right)$

21. $F = -kv \Rightarrow m\frac{dv}{dt} = -kv \Rightarrow \frac{dv}{v} = -\frac{k}{m}\,dt \Rightarrow \ln v = -\frac{k}{m}t + C_1$.

$C_1 = \ln v_0 \Rightarrow \ln\frac{v}{v_0} = -\frac{k}{m}t \Rightarrow v = \frac{dx}{dt} = v_0\,e^{-\frac{k}{m}t}$.

$x = -\frac{m}{k}v_0 e^{-\frac{k}{m}t} + C_2$. If $x_0 = 0$, then $C_2 = \frac{m}{k}v_0$ and $x = \frac{m}{k}v_0\left(1 - e^{-\frac{k}{m}t}\right)$.

23. $800 = 1000e^{10k} \Rightarrow 10k = \ln\frac{4}{5} \Rightarrow k = \frac{1}{10}\ln\frac{4}{5}$. Another 14 hours

means $t = 24$, so $y = 1000\,e^{(0.1\ln 0.8)24} = 1000\,(0.8)^{2.4} = 585.4$ kg

6.M MISCELLANEOUS EXERCISES

1. $\tan^{-1}x - \cot^{-1}x = \frac{\pi}{4} \Rightarrow \tan(\tan^{-1}x - \cot^{-1}x) = \tan\frac{\pi}{4} = 1$.

Using the formula $\tan(x - y) = \dfrac{\tan x - \tan y}{1 - \tan x \tan y}$ on the left,

$\dfrac{\tan(\tan^{-1}x) - \tan(\cot^{-1}x)}{1 - \tan(\tan^{-1}x)\tan(\cot^{-1}x)} = 1$, or $\dfrac{x - \frac{1}{x}}{2} = 1$. Thus,

$x^2 - 2x - 1 = 0$, or $x = 1 \pm \sqrt{2}$.

3. $\lim\limits_{b\to 1^-}\int_0^b \dfrac{dx}{\sqrt{1 - x^2}} = \lim\limits_{b\to 1^-}(\sin^{-1}x)\Big]_0^b\ \lim\limits_{b\to 1^-}\sin^{-1}1 = \frac{\pi}{2}$

5. $V = \pi \int_0^{\frac{\pi}{3}} \sec^2 y \, dy = \pi \tan y \Big]_0^{\frac{\pi}{3}} = \pi \sqrt{3}$

7. (a) $L = k\left(\frac{a - b\cot\theta}{R^4} + \frac{b\csc\theta}{r^4}\right) \Rightarrow \frac{dL}{d\theta} = k\left[\frac{-b}{R^4}(-\csc^2\theta) - \frac{b}{r^4}(\csc\theta\cot\theta)\right] = 0$

$\Leftrightarrow b\csc\theta\left(\frac{\csc\theta}{R^4} - \frac{\cot\theta}{r^4}\right) = 0 \Leftrightarrow \frac{r^4}{R^4} = \frac{\cot\theta}{\csc\theta} = \cos\theta \Leftrightarrow \theta = \cos^{-1}\frac{r^4}{R^4}.$

(b) $\frac{r}{R} = \frac{5}{6} \Rightarrow \frac{r^4}{r^4} = \left(\frac{5}{6}\right)^4 \Rightarrow \theta = \cos^{-1}\left(\frac{5}{6}\right)^4 \approx 61°.$

9. $\int_0^1 \frac{x^4(1-x)^4}{1+x^2}dx = \int_0^1 \frac{x^8 - 4x^7 + 6x^6 - 4x^5 + x^4}{x^2+1}dx =$

$\int_0^1 \left(x^6 - 4x^5 + 5x^4 - 4x^2 - 4 - \frac{4}{x^2+1}\right)dx =$

$\frac{1}{7}x^7 - \frac{2}{3}x^6 + x^5 - \frac{4}{3}x^3 - 4x - 4\tan^{-1}x\Big]_0^1 = \frac{22}{7} - \pi$

11. $y = x\ln x \Rightarrow \frac{dy}{dx} = x\left(\frac{1}{x}\right) + \ln x = 1 + \ln x$

13. $y = \frac{\ln x}{x} \Rightarrow \frac{dy}{dx} = \frac{x\left(\frac{1}{x}\right) - \ln x}{x^2} = \frac{1 - \ln x}{x^2}$

15. $y = \frac{x}{\ln x} \Rightarrow \frac{dy}{dx} = \frac{\ln x - x\left(\frac{1}{x}\right)}{\ln^2 x} = \frac{\ln x - 1}{\ln^2 x}$

17. $y = x^3\ln x \Rightarrow \frac{dy}{dx} = x^3\left(\frac{1}{x}\right) + 3x^2\ln x = x^2(1 + 3\ln x)$

19. $y = \ln(3x^2) \Rightarrow \frac{dy}{dx} = \frac{6x}{3x^2} = \frac{2}{x}$

21. $y = \frac{1}{2}\ln\frac{1+x}{1-x} = \frac{1}{2}[\ln(1+x) - \ln(1-x)] \Rightarrow \frac{dy}{dx} = \frac{1}{2}\left[\frac{1}{1+x} + \frac{1}{1-x}\right] = \frac{1}{1-x^2}$

23. $y = \ln \frac{x}{\sqrt{x^2+1}} = \ln x - \frac{1}{2}\ln (x^2+1) \Rightarrow \frac{dy}{dx} = \frac{1}{x} - \frac{x}{x^2+1} = \frac{1}{x(x^2+1)}$

25. $y = x \sec^{-1} x - \ln\left(x + \sqrt{x^2-1}\right)$, $x > 1$

$$\frac{dy}{dx} = x\left(\frac{1}{x\sqrt{x^2-1}}\right) + \sec^{-1} x - \left(\frac{1}{x+\sqrt{x^2-1}}\right)\left(1 + \frac{x}{\sqrt{x^2-1}}\right)$$

$$= \sec^{-1} x + \frac{1}{\sqrt{x^2-1}} - \frac{\sqrt{x^2-1}+x}{\sqrt{x^2-1}\left(\sqrt{x^2-1}+x\right)} = \sec^{-1} x$$

27. $y = \frac{x(x-2)}{(x^2+1)^{1/3}} \Rightarrow \ln y = \ln x + \ln (x-2) - \frac{1}{3}\ln (x^2+1)$

$$\frac{1}{y}\frac{dy}{dx} = \frac{1}{x} + \frac{1}{x-2} - \frac{2x}{3(x^2+1)} \Rightarrow \frac{dy}{dx} = \frac{x(x-2)}{(x^2+1)^{1/3}}\left[\frac{2x-2}{x(x-2)} - \frac{2x}{3(x^2+1)}\right]$$

29. $\int_0^{\frac{8}{3}} \frac{dx}{4-3x} = -\frac{1}{3} \ln|4-3x|\Big]_0^{\frac{8}{3}} = -\frac{1}{3}[\ln 12 - \ln 4] = -\frac{1}{3}\ln 3$

31. $\int_0^2 \frac{x\,dx}{x^2+2} = \frac{1}{2}\ln (x^2+2)\Big]_0^2 = \frac{1}{2}[\ln 6 - \ln 2] = \ln\sqrt{3}$

33. $\int_1^2 \frac{5\,dx}{x-3} = 5 \ln|x-3|\Big]_1^2 = -5 \ln 2 = -\ln 32$

35. $\int \frac{x^3\,dx}{x^4+1} = \frac{1}{4} \ln (x^4+1) + C$

37. $\int x^2 \cot (2+x^3)\,dx = \int \frac{x^2 \cos (2+x^3)}{\sin (2+x^3)} = \frac{1}{3} \ln (\sin (2+x^3)) + C$

39. $\int \frac{dx}{\sqrt{1-x^2}\,(3+\sin^{-1} x)} = \ln (3+\sin^{-1} x) + C$

41. (a) $\lim_{h\to 0^+} h \ln h = \lim_{h\to 0^+} \frac{\ln h}{\frac{1}{h}} = \lim_{h\to 0^+} \frac{\frac{1}{h}}{-\frac{1}{h^2}} = 0$

(b) $\lim_{x\to 0^+} x^p \ln x.\ p > 0 = \lim_{x\to 0^+} \frac{\ln x}{x^{-p}} = \lim_{x\to 0^+} \frac{\frac{1}{x}}{\frac{-p}{x^{p+1}}} = \lim_{x\to 0^+}\left(-\frac{x^p}{p}\right) = 0$

43. $\lim_{n\to\infty} \frac{1}{n}\left(\frac{n}{n+1} + \frac{n}{n+2} + \ldots + \frac{n}{p\cdot n}\right) = \lim_{n\to\infty} \frac{1}{n}\left[\frac{1}{1+\frac{1}{n}} + \frac{1}{1+\frac{2}{n}} + \ldots + \frac{1}{1+p-1}\right]$

$= \int_0^{p-1} \frac{dx}{1+x} = \ln|1+x|\Big]_0^{p-1} = \ln p$, $p \geq 2$, p an integer.

45. (a) $f(x) = \ln(\sec x + \tan x)$; $f'(x) = \sec x$; $f''(x) = \sec x \tan x$.

$Q(x) = f(0) + f'(0)x + \frac{1}{2}f''(0)x^2 = 0 + x + 0 = x$.

(b) $f(x) = \ln x$; $b'(x) = \frac{1}{x}$; $f''(x) = -\frac{1}{x^2}$.

$Q(x) = f(1) + f'(1)(x-1) + \frac{1}{2}f''(1)(x-1)^2 = x - 1 + \frac{1}{2}(x-1)^2 = -\frac{3}{2} + 2x - \frac{x^2}{2}$

47. $x = a\left(\cos t + \ln\tan\frac{t}{2}\right) \Rightarrow \frac{dx}{dt} = a\left(-\sin t + \frac{\frac{1}{2}\sec^2\frac{t}{2}}{\tan\frac{t}{2}}\right)$

$= a\left(-\sin t + \frac{1}{2\cos\frac{t}{2}\sin\frac{t}{2}}\right) = a(-\sin t + \csc t)$

$y = a\sin t \Rightarrow \frac{dy}{dt} = a\cos t$. $y = a = a\sin t \Rightarrow t = \frac{\pi}{2}$; $y_1 = a\sin t \Rightarrow$

$\sin t = \frac{y_1}{a} \Rightarrow t = \sin^{-1}\frac{y_1}{a}$. $\therefore$ $s = \int_{\frac{\pi}{2}}^{\sin^{-1}\frac{y_1}{a}} \sqrt{\left(\frac{dx}{dt}\right)^2 + \left(\frac{dy}{dt}\right)^2}\, dt =$

$\int_{\frac{\pi}{2}}^{\sin^{-1}\frac{y_1}{a}} \sqrt{a^2 - 2a^2 + a^2\csc^2 t}\, dt = \int_{\frac{\pi}{2}}^{\sin^{-1}\frac{y_1}{a}} a\sqrt{\cot^2 t}\, dt =$

$a\ln|\sin t|\Big]_{\frac{\pi}{2}}^{\sin^{-1}\frac{y_1}{a}} = a\ln\left|\sin\left(\sin^{-1}\frac{y_1}{a}\right)\right| = a\ln\left|\frac{y_1}{a}\right|$

49. $\frac{dy}{dx} = \frac{1+\frac{1}{x}}{1+\frac{1}{y}}$, $x > 0$, $y > 0$, $x = 1$ when $y = 1$

$\left(1+\frac{1}{y}\right)dy = \left(1+\frac{1}{x}\right)dx \Rightarrow y + \ln y = x + \ln x + C$. $1 + \ln 1 = 1 + \ln 1 + C \Rightarrow C = 0$.

$\therefore$ $\ln y - \ln x = x - y \Rightarrow \ln\frac{y}{x} = x - y \Rightarrow \frac{y}{x} = e^{x-y}$ or $y = xe^{x-y}$.

51. $A = \int_0^x f(t)\,dt = \frac{1}{3}xy \Rightarrow f(x) = \frac{d}{dx}\left(\frac{1}{3}xy\right)$ or $y = \frac{1}{3}\left(y + x\frac{dy}{dx}\right)$.

$\frac{2}{3}y = \frac{1}{3}x\frac{dy}{dx} \Rightarrow \frac{2\,dx}{x} = \frac{dy}{y} \Rightarrow \ln x^2 = \ln y + C$. (a,b) belongs to

the curve $\Rightarrow C = \ln\frac{a^2}{b}$. $\therefore\ \ln x^2 = \ln y + \ln\frac{a^2}{b}$, or $y = \frac{b}{a^2}x^2$.

53. The slope of the tangent line between (x, y) and $(-x, 0)$ is $m = \frac{y-0}{x-(-x)} = \frac{y}{2x}$.

Therefore, $\frac{dy}{dx} = \frac{y}{2x} \Rightarrow \ln y = \frac{1}{2}\ln x + C$. $(1, 2)$ belongs to the curve $\Rightarrow$

$\ln 2 = \frac{1}{2}\ln 1 + C \Rightarrow C = \ln 2$. Thus, $\ln y = \frac{1}{2}\ln x + \ln 2 \Rightarrow y = 2\sqrt{x}$ or $y^2 = 4x$.

55. $y = e^{1/x} \Rightarrow \frac{dy}{dx} = -\frac{1}{x^2}e^{1/x}$

57. $y = x e^{-x} \Rightarrow \frac{dy}{dx} = e^{-x} - xe^{-x} = e^{-x}(1-x)$

59. $y = \frac{\ln x}{e^x} \Rightarrow \frac{dy}{dx} = \frac{e^x\left(\frac{1}{x}\right) - e^x\ln x}{e^{2x}} = e^{-x}\left(\frac{1}{x} - \ln x\right)$

61. $y = \ln(2xe^{2x}) = \ln 2x + 2x \Rightarrow \frac{dy}{dx} = \frac{1}{x} + 2$

63. $y = e^{\sin^2 x} \Rightarrow \frac{dy}{dx} = e^{\sin^2 x}(2\sin x\cos x) = (\sin 2x)e^{\sin^2 x}$

65. $y = \ln\left(\frac{e^x}{1+e^x}\right) = \ln e^x - \ln(1+e^x) \Rightarrow \frac{dy}{dx} = 1 - \frac{e^x}{1+e^x} = \frac{1}{1+e^x}$

67. (a) $y = 2x - \frac{1}{2}e^{2x} \Rightarrow \frac{dy}{dx} = 2 - e^{2x}$

(b) $y = e^{\left(2x - \frac{1}{2}e^{2x}\right)} \Rightarrow \frac{dy}{dx} = (2 - e^{2x})e^{\left(2x - \frac{1}{2}e^{2x}\right)}$

69. $y = \frac{e^{2x} - e^{-2x}}{e^{2x} + e^{-2x}} \Rightarrow \frac{dy}{dx} = \frac{(e^{2x} + e^{-2x})(2e^{2x} + 2e^{-2x}) - (e^{2x} - e^{-2x})(2e^{2x} - 2e^{-2x})}{(e^{2x} + e^{-2x})^2}$

$= \frac{2[(e^{2x} + e^{-2x})^2 - (e^{2x} - e^{-2x})^2]}{(e^{2x} + e^{-2x})^2} = \frac{8}{(e^{2x} + e^{-2x})^2}$

71. $y = \sin^{-1}(x^2) - xe^{(x^2)} \Rightarrow \frac{dy}{dx} = \frac{2x}{\sqrt{1-x^4}} - e^{(x^2)} - 2x^2e^{(x^2)}$

73. $y = e^{2x}(\sin 3x + \cos 3x)$

$\frac{dy}{dx} = e^{2x}(3\cos 3x - 3\sin 3x) + 2e^{2x}(\sin 3x + \cos 3x) = e^{2x}(5\cos 3x - \sin 3x)$

75. $\int_1^e \frac{x+1}{x}dx = \int_1^e \left(1 + \frac{1}{x}\right)dx = x + \ln x\Big]_1^e = e$

77. $\int_0^{\ln 2} e^{-2x}\,dx = -\frac{1}{2}e^{-2x}\Big]_0^{\ln 2} = \frac{3}{8}$

79. $\int_{-1}^{1} \frac{e^x - e^{-x}}{e^x + e^{-x}}dx = \ln(e^x + e^{-x})\Big]_{-1}^{1} = 0$

81. $\lim_{x\to\infty}\frac{x^5}{e^x} = \lim_{x\to\infty}\frac{5x^4}{e^x} = \lim_{x\to\infty}\frac{5\cdot 4x^3}{e^x} = \ldots = \lim_{x\to\infty}\frac{5!}{e^x} = 0$

83. $\lim_{x\to 4}\frac{e^{x-4} + 4 - x}{\cos^2(\pi x)} = \frac{e^0}{\cos^2(4\pi)} = 1$

85. $f(x) = x + e^{4x}$ at $x = 0$; $f'(x) = 1 + 4e^{4x}$; $L(x) = f(0) + f'(0)x = 1 + 5x$

87. (a) $y = \frac{\ln x}{\sqrt{x}}$; intercept $(1, 0)$

$y' = \frac{2 - \ln x}{2x^{3/2}} = 0$ if $x = e^2$

$\left(e^2, \frac{2}{e}\right)$ is maximum value

$y'' = \frac{-8 + 3\ln x}{x^{5/2}} = 0$ if $x = e^{8/3}$

concave down if $0 < x < e^{8/3}$;

concave up if $x > e^{8/3}$.

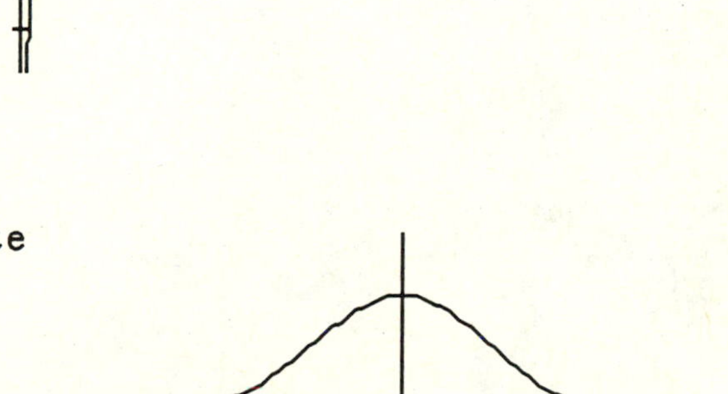

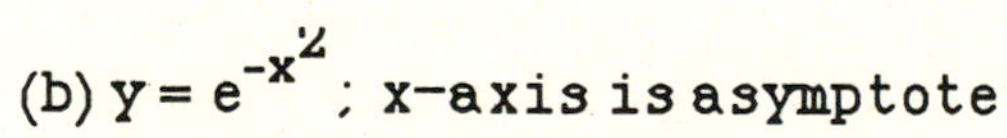

(b) $y = e^{-x^2}$; x-axis is asymptote

$y' = -2xe^{-x^2} = 0$ if $x = 0$

$(0, 1)$ is maximum value

$y'' = 2e^{-x^2}(2x^2 - 1) = 0$ if $x = \pm\frac{1}{\sqrt{2}}$

$\left(\pm\frac{1}{\sqrt{2}}, e^{-1/2}\right)$ inflection points

symmetry to y-axis

(c) $y = (1+x)e^{-x}$; x-axis is asymptote

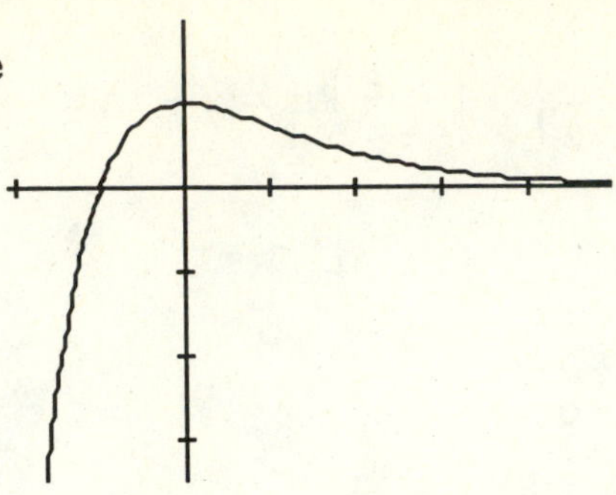

$y' = -xe^{-x} = 0$ if $x = 0$
$(0, 1)$ is maximum value

$y'' = e^{-x}(x-1) = 0$ if $x = 1$

$\left(1, \frac{2}{e}\right)$ inflection point

89. $S = \int_0^x \sqrt{1+\left(\frac{dy}{dx}\right)^2}\, dx = e^x + y - 1 \Rightarrow e^x + \frac{dy}{dx} = \sqrt{1+\left(\frac{dy}{dx}\right)^2}$

$\therefore e^{2x} + 2e^x\frac{dy}{dx} + \left(\frac{dy}{dx}\right)^2 = 1 + \left(\frac{dy}{dx}\right)^2 \Rightarrow 2e^x\frac{dy}{dx} = 1 - e^{2x}.\ \frac{dy}{dx} = \frac{1}{2}e^{-x} - \frac{1}{2}e^x \Rightarrow$

$y = -\frac{1}{2}e^{-x} - \frac{1}{2}e^x + C$. $(0, 0)$ is on curve so $C = 1$ and $y = -\frac{1}{2}e^{-x} - \frac{1}{2}e^x + 1$.

91. $y = \frac{1}{2}(e^x + e^{-x}) \Rightarrow \frac{dy}{dx} = \frac{1}{2}(e^x - e^{-x}) \Rightarrow \left(\frac{dy}{dx}\right)^2 = \frac{1}{4}(e^{2x} - 2 + e^{-2x})$;

$ds = \sqrt{1 + \frac{1}{4}(e^{2x} - 2 + e^{-2x})}\, dx = \sqrt{\left(\frac{e^x + e^{-x}}{2}\right)^2}\, dx$.

$s = \int_0^{\frac{1}{2}\ln 2} 2\pi y\, ds = 2\pi\int_0^{\frac{1}{2}\ln 2} \frac{1}{2}(e^x + e^{-x})\frac{1}{2}(e^x + e^{-x})\, dx$

$= \frac{\pi}{2}\int_0^{\frac{1}{2}\ln 2}(e^{2x} + 2 + e^{-2x})\, dx = \frac{\pi}{2}\left[\frac{1}{2}e^{2x}n - \frac{1}{2}e^{-2x} + 2x\right]_0^{\frac{1}{2}\ln 2} = \frac{\pi}{2}\left(\frac{3}{4} + \ln 2\right)$

93. $= \int_{-a}^{a} \frac{a}{2}(e^{x/a} + e^{-x/a})\, dx = \frac{a}{2}\left[a(e^{x/a} - ae^{-x/a}\right]_{-a}^{a} = a^2\left(e - \frac{1}{e}\right)$

95. $y = e^{-(2x^2)} \Rightarrow y' = -4x\, e^{-(2x^2)}$

$y'' = 16x^3\, e^{-(2x^2)} - 4\, e^{-(2x^2)}$

$4\, e^{-(2x^2)}(4x^2 - 1) = 0 \Leftrightarrow x = \pm\frac{1}{2}$

$x = \frac{1}{2} \Rightarrow y = \frac{1}{\sqrt{e}}$. The line $y = \frac{2}{\sqrt{e}}x$ intersects $y = 1$ when

$x = \frac{\sqrt{e}}{2}$. $\therefore A = \frac{1}{2}\left(\frac{\sqrt{e}}{2}\right)(1) = \frac{\sqrt{e}}{4}$

97. The tangent line to the graph of $y = e^x$ through $P(x_1, e^{x_1})$ is:

$y - e^{x_1} = e^{x_1}(x - x_1)$. The x-intercept of this line is: $e^{x_1}(x_0 - x_1) = -e^{x_1}$

or $x_0 = \dfrac{x_1 e^{x_1} + e^{x_1}}{e^{x_1}} = x_1 + 1$. Observe that $|x_1 + 1 - x_1| = 1$.

99. $\dfrac{dy}{dx} = y^2 e^{-x}$, $y = 2$ when $x = 0$. $y^{-2}dy = e^{-x}\,dx \Rightarrow -y^{-1} = -e^{-x} + C$.

$-\dfrac{1}{2} = -e^0 + C \Rightarrow C = \dfrac{1}{2}$. $\therefore -\dfrac{1}{y} = -e^{-x} + \dfrac{1}{2}$ or $y = \dfrac{2e^x}{2 - e^x}$.

101. $x = ae^{\omega t} + be^{-\omega t} \Rightarrow v = \dfrac{dx}{dt} = a\omega e^{\omega t} - b\omega e^{-\omega t}$ and

$a = \dfrac{d^2x}{dt^2} = a\omega^2 e^{\omega t} + b\omega^2 e^{-\omega t} = \omega^2(ae^{\omega t} + be^{-\omega t}) = \omega^2 x$. $\therefore F = ma = m\omega^2 x$.

103. Show that $\dfrac{d^n}{dx^n}(xe^x) = (x + n)e^x$. If $n = 1$, then $\dfrac{d}{dx}(xe^x) = e^x + xe^x = (x + 1)e^x$.

Assume that $\dfrac{d^k}{dx^k}(xe^x) = (x + k)e^x$. Then $\dfrac{d^{k+1}}{dx^{k+1}}(xe^x) = \dfrac{d}{dx}\left(\dfrac{d^k}{dx^k}(xe^x)\right) =$

$\dfrac{d}{dx}\left((x + k)e^x\right) = (x + k)e^x + e^x = e^x(x + k + 1) = e^x[x + (k + 1)]$.

105. (a) $\lim\limits_{h\to 0}\dfrac{e^h - 1}{h} = \lim\limits_{h\to 0}\dfrac{e^h - e^0}{h} = f'(0) = 1$, where $f(x) = e^x$

(b) Show that $\lim\limits_{n\to\infty} n(\sqrt[n]{x} - 1) = \ln x$, for any $x > 0$.

Let $n = \dfrac{\ln x}{h}$, so that $\ln x = nh$ or $x = e^{nh}$. Observe that

$n \to \infty \Leftrightarrow h \to 0$. Then $\lim\limits_{n\to\infty} n(\sqrt[n]{x} - 1) = \lim\limits_{h\to 0}\left(\dfrac{\ln x}{h}(e^h - 1)\right) = \ln x$.

107. (a) $y = x^{\tan 3x} \Rightarrow \ln y = \tan 3x \ln x \Rightarrow \dfrac{1}{y}\dfrac{dy}{dx} = \tan 3x\left(\dfrac{1}{x}\right) + \ln x\,(3\sec^2 3x)$

or $\dfrac{dy}{dx} = x^{\tan 3x}\left(\dfrac{\tan 3x + 3x\ln x\sec^2 3x}{x}\right)$

(b) $x^{\ln y} = 2 \Rightarrow \ln y \ln x = \ln 2 \Rightarrow \left(\dfrac{1}{y}\ln x + \dfrac{1}{x}\ln y\right)\dfrac{dy}{dx} = 0$ so $\dfrac{dy}{dx} = -\dfrac{y\ln y}{x\ln x}$

(c) $y = (x^2 + 2)^{2-x} \Rightarrow \ln y = (2 - x)\ln(x^2 + 2) \Rightarrow \dfrac{1}{y}\dfrac{dy}{dx} = \dfrac{2x(2 - x)}{x^2 + 2} - \ln(x^2 + 2)$

$\dfrac{dy}{dx} = \dfrac{1}{y}\left(\dfrac{4x - 2x^2 - (x^2 + 2)\ln(x^2 + 2)}{x^2 + 2}\right) = (x^2 + 2)^{1-x}[4x - 2x^2 - (x^2 + 2)\ln(x^2 +$

(d) $y = x^{1/x} \Rightarrow \ln y = \frac{1}{x}\ln x \Rightarrow \frac{1}{y}\frac{dy}{dx} = \frac{1}{x^2} - \frac{\ln x}{x^2}$ so $\frac{dy}{dx} = x^{1/x}\left(\frac{1}{x^2} - \frac{\ln x}{x^2}\right)$

109. (a) $\int_0^{\frac{\pi}{6}} (\cos x)4^{-\sin x}\,dx = -\frac{1}{\ln 4}4^{-\sin x}\Big]_0^{\frac{\pi}{6}} = \frac{1}{2\ln 4}$

(b) $\int_{\ln(\log_4(\ln 4))}^{\ln(\log_4(\ln 16))} e^x\, 4^{(e^x)}\,dx = \frac{1}{\ln 4}4^{(e^x)}\Big]_{\ln(\log_4(\ln 4))}^{\ln(\log_4(\ln 16))}$

$= \frac{1}{\ln 4}[\,4^{e^{\ln(\log_4(\ln 16))}} - 4^{e^{\ln(\log_4(\ln 4))}}\,] = \frac{1}{\ln 4}[\ln 16 - \ln 4] = 1$

111. (a) If $a^b = b^a$, then $\ln a^b = \ln b^a$, $b\ln a = a\ln b$, or $\frac{\ln a}{a} = \frac{\ln b}{b}$.

(b) $f(x) = \frac{\ln x}{x} \Rightarrow f'(x) = \frac{1-\ln x}{x^2}$

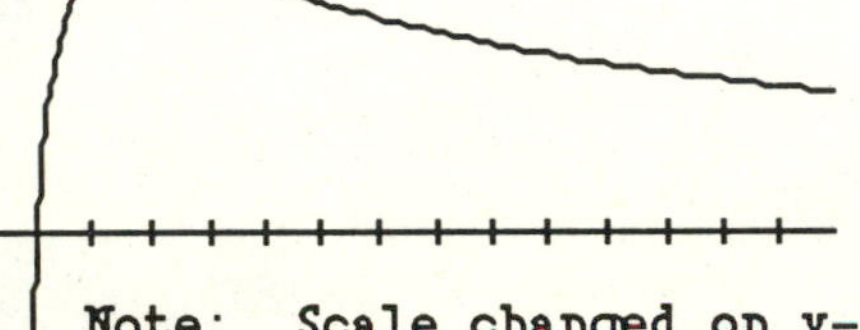

Note: Scale changed on y-axis to emphasize behavior at the maximum.

$1 - \ln x = 0$ if $x = e$. $f'(x) > 0$ if $x < e$ and $f'(x) < 0$ if $x < e \Rightarrow$ maximum at $x = e$. $f''(x) = \frac{2\ln x - 3}{x^3}$.

$2\ln x - 3 = 0$ if $x = e^{3/2}$. $f''(x) < 0$ if $x < e^{3/2}$ and $f''(x) > 0$ if $x > e^{3/2} \Rightarrow$ inflection point at $x = e^{3/2}$.

$\lim_{x\to\infty}\frac{\ln x}{x} = \lim_{x\to\infty}\frac{1}{x} = 0$; $\lim_{x\to 0^+}\frac{\ln x}{x} = \lim_{t\to\infty}\frac{\ln\frac{1}{t}}{\frac{1}{t}} = \lim_{t\to\infty}(-t\ln t) = -\infty$.

(c) $f(x) = \frac{\ln x}{x}$ is increasing on $0 < x \le 1$; $\frac{\ln a}{a} = \frac{\ln b}{b} \Rightarrow \ln a = \ln b \Rightarrow a = b$.

$f(x) = \frac{\ln x}{x}$ increases on $(1, e)$, decreases on (e, ∞), reaches absolute maximum at $x = e$, so for each $0 < y < \frac{1}{e}$ such that $\frac{\ln a}{a} = y$, there exists exactly one b for which $\frac{\ln b}{b} = y$.

113. $\frac{dN}{dt} = 0.02\,N \Rightarrow \ln N = 0.02t + C$. If $t = 0 \Rightarrow C = N_0$ and $N = N_0 e^{0.02t}$.

To double, $N = 2N_0$ and $N_0 e^{0.02t} = 2N_0 \Leftrightarrow 0.02t = \ln 2$ or $t = 34.7$ years.

115. $\frac{dx}{dt} = kx \Rightarrow x = x_0 e^{kt}$. If $t = 0 \Rightarrow x_0 = 2$ and $x = 2e^{kt}$. When $t = 10$,

then $x = 4$ so $4 = 2e^{10k} \Rightarrow k = \frac{1}{10}\ln 2$. If $t = 5$, then $x = 2e^{\left(\frac{1}{10}\ln 2\right)5} = 2\sqrt{2}$.

(b) $f(x) = \dfrac{\ln x}{x} \Rightarrow f'(x) = \dfrac{1 - \ln x}{x^2}$

$1 - \ln x = 0$ if $x = e$. $f'(x) > 0$ if $x < e$ and $f'(x) < 0$ if $x < e \Rightarrow$

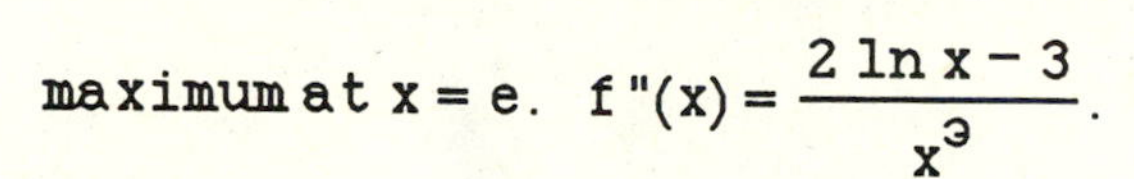

maximum at $x = e$. $f''(x) = \dfrac{2\ln x - 3}{x^3}$.

$2\ln x - 3 = 0$ if $x = e^{3/2}$. $f''(x) < 0$ if $x < e^{3/2}$ and $f''(x) > 0$ if $x > e^{3/2} \Rightarrow$ inflection point at $x = e^{3/2}$.

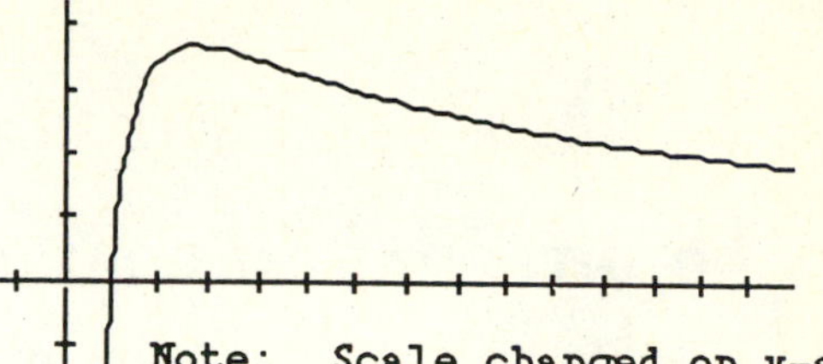

Note: Scale changed on y-axis to emphasize behavior at the maximum.

$$\lim_{x\to\infty} \frac{\ln x}{x} = \lim_{x\to\infty} \frac{1}{x} = 0;\ \lim_{x\to 0^+} \frac{\ln x}{x} = \lim_{t\to\infty} \frac{\ln \frac{1}{t}}{\frac{1}{t}} = \lim_{t\to\infty} (-t \ln t) = -\infty.$$

(c) (i) $f(x) = \dfrac{\ln x}{x}$ is increasing on $0 < x \le 1$, so that if $\dfrac{\ln a}{a} = \dfrac{\ln b}{b}$ then $\ln a = \ln b$ so that $a = b$.

(ii) $f(x) = \dfrac{\ln x}{x}$ increases on $(1, e)$, decreases on (e, ∞) and attains an absolute maximum at $x = e$ of $\dfrac{1}{e}$.

Therefore, for each y such that $0 < y < \dfrac{1}{e}$ such that $\dfrac{\ln a}{a} = y$, there exists exactly one b for which $\dfrac{\ln b}{b} = y$.

113. $\dfrac{dN}{dt} = 0.02\,N \Rightarrow \ln N = 0.02t + C$. If $t = 0 \Rightarrow N = N_0$, then $C = N_0$ and $N = N_0 e^{0.02t}$. For the population to double, $N = 2N_0$ and $N_0 e^{0.02t} = 2N_0 \Leftrightarrow 0.02t = \ln 2$ or $t = 34.7$ years.

115. $\dfrac{dx}{dt} = kx \Rightarrow x = x_0 e^{kt}$. If $t = 0$, then $x = 2$ so $x_0 = 2$ and $x = 2e^{kt}$. When $t = 10$, then $x = 4$ so $4 = 2e^{10k} \Rightarrow k = \dfrac{1}{10}\ln 2$.

Therefore, if $t = 5$, then $x = 2e^{\left(\frac{1}{10}\ln 2\right)5} = 2e^{\ln\sqrt{2}} = 2\sqrt{2}$.

CHAPTER 7

INTEGRATION METHODS

7.1 BASIC INTEGRATION FORMULAS

1. Let $u = 3x^2 + 5 \Rightarrow du = 6x\,dx$

$$\int 6x\sqrt{3x^2+5}\,dx = \int \sqrt{u}\,du = \frac{2}{3}u^{3/2} + C = \frac{2}{3}(3x^2+5)^{3/2} + C$$

3. Let $u = x^2 \Rightarrow du = 2x\,dx \Rightarrow x\,dx = \frac{1}{2}du$

$$\int xe^{(x^2)}\,dx = \int \frac{1}{2}e^u\,du = \frac{1}{2}e^u + C$$

$$\int_0^{\sqrt{\ln 2}} xe^{(x^2)}\,dx = \frac{1}{2}e^{(x^2)}\Big]_0^{\sqrt{\ln 2}} = \frac{1}{2}e^{\ln 2} - \frac{1}{2} = \frac{1}{2}$$

5. Let $u = 8x^2 + 1 \Rightarrow du = 16x\,dx \Rightarrow x\,dx = \frac{1}{16}du$

$$\int \frac{x\,dx}{\sqrt{8x^2+1}} = \int \frac{1}{16}u^{-1/2}\,du = \frac{1}{8}\sqrt{u} + C = \frac{1}{8}\sqrt{8x^2+1} + C$$

$$\int_0^1 \frac{x\,dx}{\sqrt{8x^2+1}} = \frac{1}{8}\sqrt{8x^2+1}\Big]_0^1 = \frac{1}{4}$$

7. Let $u = e^x \Rightarrow du = e^x\,dx$

$$\int e^x\sec^2(e^x)\,dx = \int \sec^2 u\,du = \tan u + C = \tan(e^x) + C$$

9. Let $u = 3 + 4e^x \Rightarrow du = 4e^x\,dx \Rightarrow e^x\,dx = \frac{1}{4}du$

$$\int e^x\sqrt{3+4e^x}\,dx = \int \frac{1}{4}\sqrt{u}\,du = \frac{1}{6}u^{3/2} + C = \frac{1}{6}(3+4e^x)^{3/2} + C$$

11. Let $u = \ln x \Rightarrow du = \frac{1}{x}\,dx$. $\int \frac{dx}{x\ln x} = \ln|\ln x| + C$

13. Let $u = \sqrt{x} \Rightarrow du = \frac{1}{2\sqrt{x}}\,dx \Rightarrow dx = 2\sqrt{x}\,du = 2u\,du$

$$\int_0^1 e^{\sqrt{x}}\,dx = \int_0^1 2ue^u\,du = 2(ue^u - e^u)\Big]_0^1 = 2$$

Note: Although it is not difficult to guess that $\frac{d}{du}(ue^u - e^u) = ue^u$, it can be done by the technique of integration by parts taught in the next section.

15. Let $u = \tan x \Rightarrow du = \sec^2 x\,dx$.

$$\int \tan x\sec^2 x\,dx = \int u\,du = \frac{1}{2}u^2 + C = \frac{1}{2}\tan^2 x + C$$

17. Let $u = \sin x \Rightarrow du = \cos x\,dx$

$$\int \cos^3 x\,dx = \int \cos^2 x \cos x\,dx = \int (1 - \sin^2 x)\cos x\,dx =$$

$$\int (1 - u^2)\,du = u - \frac{1}{3}u^3 + C = \sin x - \frac{1}{3}\sin^3 x + C$$

19. Let $u = \sec x \Rightarrow du = \sec x \tan x\,dx$. $\int \tan^3 x \sec x\,dx =$

$$\int (\sec^2 x - 1)\sec x \tan x\,dx = \int (u^2 - 1)\,du = \int \frac{1}{3}u^3 - u + C = \frac{1}{3}\sec^3 x - \sec x + C$$

21. Let $u = 2x + 3 \Rightarrow du = 2\,dx \Rightarrow dx = \frac{1}{2}\,du$, $x = \frac{1}{2} \Rightarrow u = 3$, $x = 3 \Rightarrow u = 9$.

$$\int_{\frac{1}{2}}^{3} \sqrt{2x + 3}\,dx = \int_4^9 \frac{1}{2}\sqrt{u}\,du = \frac{1}{2}u^{3/2}\Big]_4^9 = \frac{19}{3}$$

23. Let $u = 2x - 7 \Rightarrow du = 2\,dx \Rightarrow dx = \frac{1}{2}\,du$.

$$\int \frac{dx}{(2x - 7)^2} = \int \frac{1}{2}u^{-2}\,du = -\frac{1}{2}u^{-1} + C = -\frac{1}{2}(2x - 7)^{-1} + C$$

25. Let $u = \cos x \Rightarrow -du = \sin x\,dx$, $x = -\pi \Rightarrow u = -1$, $x = 0 \Rightarrow u = 1$

$$\int_{-\pi}^{0} \frac{\sin x\,dx}{2 + \cos x} = \int_{-1}^{1} \frac{-du}{2 + u} = -\ln|2 + u|\Big]_{-1}^{1} = -\ln 3$$

27. Let $u = \sin x \Rightarrow du = \cos x\,dx$.

$$\int \sqrt{\sin x}\cos x\,dx = \int \sqrt{u}\,du = \frac{2}{3}u^{3/2} + C = \frac{2}{3}(\sin x)^{2/3} + C$$

29. Let $u = \sin x \Rightarrow du = \cos x\,dx$.

$$\int \frac{\cos x\,dx}{(1 + \sin x)^2} = \int (1 + u)^{-2}\,du = -(1 + u)^{-1} + C = \frac{-1}{1 + \sin x} + C$$

31. $$\int_{-\frac{\pi}{2}}^{\frac{\pi}{2}} \sin^2 2x \cos^3 2x\,dx = \int_{-\frac{\pi}{2}}^{\frac{\pi}{2}} \sin^2 2x\,(1 - \sin^2 2x)\cos 2x\,dx =$$

$$\int_{-\frac{\pi}{2}}^{\frac{\pi}{2}} \sin^2 2x \cos 2x\,dx - \int_{-\frac{\pi}{2}}^{\frac{\pi}{2}} \sin^4 2x \cos 2x\,dx = \frac{1}{6}\sin^3 2x - \frac{1}{10}\sin^5 2x\Big]_{-\frac{\pi}{2}}^{\frac{\pi}{2}} = 0$$

33. Let $u = 1 - 4x^2 \Rightarrow du = -8x\,dx \Rightarrow x\,dx = -\frac{1}{8}\,du$

$$\int \frac{x\,dx}{\sqrt{1 - 4x^2}} = -\frac{1}{8}\int \frac{du}{\sqrt{u}} = -\frac{1}{4}\sqrt{u} + C = -\frac{1}{4}\sqrt{1 - 4x^2} + C$$

35. $$\int_0^{\frac{\sqrt{2}}{2}} \frac{dx}{\sqrt{1 - x^2}} = \sin^{-1} x\Big]_0^{\frac{\sqrt{2}}{2}} = \frac{\pi}{4}$$

37. $\displaystyle\int_0^1 \frac{3x}{1+x^2}\,dx = \frac{3}{2}\int_0^1 \frac{2x}{1+x^2}\,dx = \frac{3}{2}\ln(1+x^2)\Big]_0^1 = \frac{3}{2}\ln 2$

39. $\displaystyle\int_0^{\frac{\pi}{4}} \frac{\sin^2 2x\,dx}{1+\cos 2x} = \int_0^{\frac{\pi}{4}} \frac{1-\cos^2 2x}{1+\cos 2x}\,dx = \int_0^{\frac{\pi}{4}}(1-\cos 2x)\,dx =$

$$x - \frac{1}{2}\sin 2x\Big]_0^{\frac{\pi}{4}} = \frac{\pi-2}{4}$$

41. Let $u = 2x \Rightarrow du = 2\,dx$, $x = 0 \Rightarrow u = 0$, $x = \frac{1}{2} \Rightarrow u = 1$

$$\int_0^{\frac{1}{2}} \frac{2\,dx}{\sqrt{1-4x^2}} = \int_0^1 \frac{du}{\sqrt{1-u^2}} = \sin^{-1} u\Big]_0^1 = \frac{\pi}{2}$$

43. Let $u = 3x^2 + 4 \Rightarrow du = 6x\,dx \Rightarrow x\,dx = \frac{1}{6}du$

$$\int \frac{x\,dx}{(3x^2+4)^3} = \frac{1}{6}\int u^{-3}\,du = -\frac{1}{12}u^{-2} + C = -\frac{1}{12}(3x^2+4)^{-2} + C$$

45. Let $u = x^3 + 5 \Rightarrow du = 3x^2\,dx \Rightarrow x^2\,dx = \frac{1}{3}\,du$

$$\int \frac{x^2\,dx}{\sqrt{x^3+5}} = \frac{1}{3}\int \frac{1}{\sqrt{u}}\,du = \frac{2}{3}\sqrt{u} + C = \frac{2}{3}\sqrt{x^3+5} + C$$

47. $\displaystyle\int_0^{\ln 2} e^{2x}\,dx = \frac{1}{2}e^{2x}\Big]_0^{\ln 2} = \frac{3}{2}$

49. $\displaystyle\int \frac{dx}{e^{3x}} = \int e^{-3x}\,dx = -\frac{1}{3}e^{-3x} + C$

51. Let $u = e^x \Rightarrow du = e^x\,dx$. $\displaystyle\int \frac{e^x}{1+e^{2x}}\,dx = \int \frac{du}{1+u^2} = \tan^{-1}u + C = \tan^{-1}(e^x) + C$

53. Let $u = \cos x \Rightarrow du = -\sin x\,dx$. $\displaystyle\int \cos^2 x \sin x\,dx =$

$$\int -u^2\,du = -\frac{1}{3}u^3 + C = -\frac{1}{3}\cos^3 x + C$$

55. $\displaystyle\int_{\frac{\pi}{6}}^{\frac{\pi}{2}} \cot^3 x \csc^2 x\,dx = -\frac{1}{4}\cot^4 x\Big]_{\frac{\pi}{6}}^{\frac{\pi}{2}} = \frac{9}{4}$

57. Let $u = e^{2x} - e^{-2x} \Rightarrow du = 2e^{2x} - 2e^{-2x} \Rightarrow \frac{1}{2}du = e^{2x} - e^{-2x}$.

$$\int \frac{e^{2x}+e^{-2x}}{e^{2x}-e^{-2x}}\,dx = \frac{1}{2}\int \frac{du}{u} = \frac{1}{2}\ln|u| + C = \frac{1}{2}\ln(e^{2x}-e^{-2x}) + C$$

59. $\displaystyle\int_{-\frac{\pi}{2}}^{\frac{\pi}{2}} (1+\cos\theta)^3 \sin\theta\,d\theta = -\frac{1}{4}(1+\cos\theta)^4\Big]_{-\frac{\pi}{2}}^{\frac{\pi}{2}} = 0$

61. Let $u = 2t \Rightarrow dt = \frac{1}{2}\,du$ and $t = \frac{1}{2}u$.

$$\int \frac{dt}{t\sqrt{4t^2-1}} = \int \frac{\frac{1}{2}\,du}{\frac{1}{2}u\sqrt{u^2-1}} = \sec^{-1}|u| + C = \sec^{-1}|2t| + C$$

63. $\displaystyle\int \frac{\cos x\,dx}{\sin x} = \ln|\sin x| + C$

65. $\displaystyle\int \sec^3 x \tan x\,dx = \int \sec^2 x(\sec x \tan x)\,dx = \frac{1}{3}\sec^3 x + C$

67. Let $u = \tan 3x \Rightarrow du = 3\sec^2 3x\,dx$

$$\int e^{\tan 3x}\sec^2 3x\,dx = \int \frac{1}{3}e^u\,du = \frac{1}{3}e^u + C = \frac{1}{3}e^{\tan 3x} + C$$

69. $\displaystyle\int_{\frac{\pi}{6}}^{\frac{\pi}{4}} \frac{1+\cos 2x}{\sin^2 2x}\,dx = \int_{\frac{\pi}{6}}^{\frac{\pi}{4}} \csc^2 2x\,dx + \int_{\frac{\pi}{6}}^{\frac{\pi}{4}} \cot 2x \csc 2x\,dx =$

$$-\frac{1}{2}\cot 2x - \frac{1}{2}\csc 2x\Big]_{\frac{\pi}{6}}^{\frac{\pi}{4}} = \frac{\sqrt{3-1}}{2}$$

71. Let $u = \cot 2t \Rightarrow du = = -2\csc^2 2t\,dt$

$$\int \frac{\csc^2 2t}{\sqrt{1+\cot 2t}}\,dt = \int \frac{-\frac{1}{2}du}{\sqrt{1+u}} = = -\sqrt{1+u} + C = -\sqrt{1+\cot 2t} + C$$

73. Let $u = \tan^{-1} 2t \Rightarrow du = \frac{2\,dt}{1+4t^2}$, $t = 0 \Rightarrow u = 0$, $t = \frac{1}{2} \Rightarrow u = \frac{\pi}{4}$

$$\int_0^{\frac{1}{2}} \frac{e^{\tan^{-1}2t}}{1+4t^2}\,dt = \int_0^{\frac{\pi}{4}} \frac{1}{2}e^u\,du = \frac{1}{2}e^u\Big]_0^{\frac{\pi}{4}} = \frac{1}{2}[e^{\frac{\pi}{4}} - 1]$$

75. Let $u = \sqrt{x} \Rightarrow du = \frac{1}{2\sqrt{x}}\,dx$, $x = 1 \Rightarrow u = 1$, $x = 4 \Rightarrow u = 2$

$$\int_1^4 \frac{2^{\sqrt{x}}}{2\sqrt{x}}\,dx = \int_1^2 2^u\,du = \frac{1}{\ln 2}(2^u)\Big]_1^2 = \frac{2}{\ln 2}$$

77. Let $u = \ln x \Rightarrow du = \frac{1}{x}\,dx$. $\displaystyle\int \frac{\ln x}{x}dx = \int u\,du = \frac{1}{2}u^2 + C = \frac{1}{2}\ln^2 x + C$

79. Let $u = \sqrt{x} \Rightarrow du = \frac{1}{2\sqrt{x}}\,dx \Rightarrow 2\,du = \frac{1}{\sqrt{x}}dx$, $x = 1 \Rightarrow u = 1$, $x = 9 \Rightarrow u = 3$.

$$\int_1^9 \frac{dx}{\sqrt{x}(1+\sqrt{x})} = \int_1^3 \frac{2\,du}{1+u} = 2\ln|1+u|\Big]_1^3 = 2\ln 2$$

81. (a) Let $u = x^2 \Rightarrow du = 2x\,dx$. $\displaystyle\int xe^{(x^2)}\,dx = \int \frac{1}{2}e^u\,du = \frac{1}{2}e^u + C = \frac{1}{2}e^{(x^2)} + C$

(b) Let $u = e^{(x^2)} \Rightarrow du = 2xe^{(x^2)}\,dx$. $\displaystyle\int e^{(x^2)}\,dx = \int u\,du = \frac{1}{2}u^2 + C = \frac{1}{2}e^{(x^2)} + C$

7.2 INTEGRATION BY PARTS

1. Let $u = x$ and $dv = \sin x\,dx \Rightarrow du = dx$ and $v = -\cos x$

$$\int x \sin x\,dx = -x\cos x - \int (-\cos x)\,dx = -x\cos x + \sin x + C$$

3. $\int x^2 \sin x\,dx = -x^2 \cos x + 2x \sin x + 2\cos x + C$

x^2	+	$\sin x$
$2x$	−	$-\cos x$
2	+	$-\sin x$
0		$\cos x$

5. Let $u = \ln x$ and $dv = x\,dx \Rightarrow du = \frac{1}{x}\,dx$ and $v = \frac{1}{2}x^2$

$$\int_1^2 \ln x\,dx = \frac{1}{2}x^2 \ln x\Big]_1^2 - \int_1^2 \frac{1}{2}x\,dx = \frac{1}{2}x^2 \ln x - \frac{1}{4}x^2\Big]_1^2 = 2\ln 2 - \frac{3}{4}.$$

7. Let $u = \ln x$ and $dv = x^3\,dx \Rightarrow du = \frac{1}{x}\,dx$ and $v = \frac{1}{4}x^4$

$$\int x^3 \ln x\,dx = \frac{1}{4}x^4 \ln x - \int \frac{1}{4}x^3\,dx = \frac{1}{4}x^4 \ln x - \frac{1}{16}x^4 + C$$

9. Let $u = \tan^{-1} x$ and $dv = dx \Rightarrow du = \frac{1}{1+x^2}\,dx$ and $v = x$

$$\int \tan^{-1} x\,dx = x\tan^{-1} x - \int \frac{x\,dx}{1+x^2} = x\tan^{-1} x - \frac{1}{2}\ln(1+x^2) + C$$

11. Let $u = \sin^{-1} x$ and $dv = dx \Rightarrow du = \frac{dx}{\sqrt{1-x^2}}$ and $v = x$

$$\int \sin^{-1} x\,dx = x\sin^{-1} x - \int \frac{x\,dx}{\sqrt{1-x^2}} = x\sin^{-1} x + \sqrt{1-x^2} + C$$

13. Let $u = x$ and $dv = \sec^2 x\,dx \Rightarrow du = dx$ and $v = \tan x$

$$\int x\sec^2 x\,dx = x\tan x - \int \tan x\,dx = x\tan x + \ln|\cos x| + C$$

15. $\int x^3 e^x\,dx = x^3 e^x - 3x^2 e^x + 6xe^x - 6e^x + C = e^x(x^3 - 3x^2 + 6x - 6) + C$

x^3	+	e^x
$3x^2$	−	e^x
$6x$	+	e^x
6	−	e^x
0		e^x

17. $\int (x^2 - 5x)e^x dx = (x^2 - 5x)e^x - (2x - 5)e^x + 2e^x - 2e^x + C$

$= e^x(x^2 - 7x + 7) + C$

	sign	
$x^2 - 5x$	$+$	e^x
$2x - 5$	$-$	e^x
2	$+$	e^x
0		e^x

19. $\int x^5 e^x dx = x^5 e^x - 5x^4 e^x + 20x^3 e^x - 60x^2 e^x + 120xe^x - 120e^x + C$

$e^x(x^5 - 5x^4 + 20x^3 - 60x^2 + 120x - 120) + C$

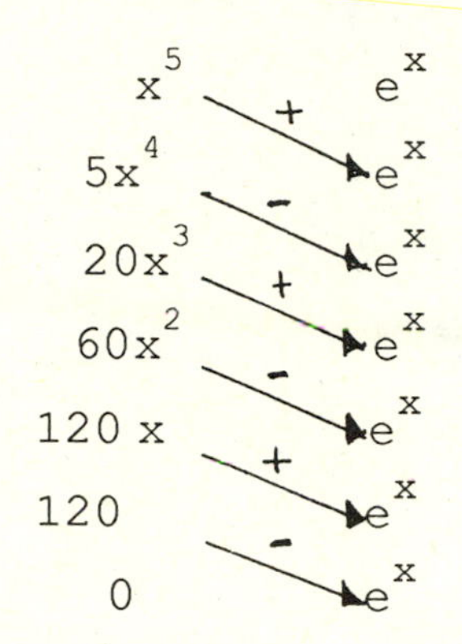

21. $\int_0^{\frac{\pi}{2}} x^2 \sin 2x\, dx = -\frac{1}{2}x^2 \cos 2x + \frac{1}{2}x \sin 2x + \frac{1}{4}\cos 2x \Big]_0^{\frac{\pi}{2}} = \frac{\pi^2}{8} - \frac{1}{2}$

	sign	
x^2	$+$	$\sin 2x$
$2x$	$-$	$-\frac{1}{2}\cos 2x$
2	$+$	$-\frac{1}{4}\sin 2x$
0		$\frac{1}{8}\cos 2x$

23. $\int x^4 \cos x\, dx = x^4 \sin x + 4x^3 \cos x - 12x^2 \sin x - 24x \cos x + 24 \sin x + C$

	sign	
x^4	$+$	$\cos x$
$4x^3$	$-$	$\sin x$
$12x^2$	$+$	$-\cos x$
$24x$	$-$	$-\sin x$
24	$+$	$\cos x$
0		$\sin x$

25. $\int x^2 \cos ax\,dx = \frac{x^2}{a}\sin ax + \frac{2x}{a^2}\cos ax - \frac{2}{a^3}\sin ax + C$

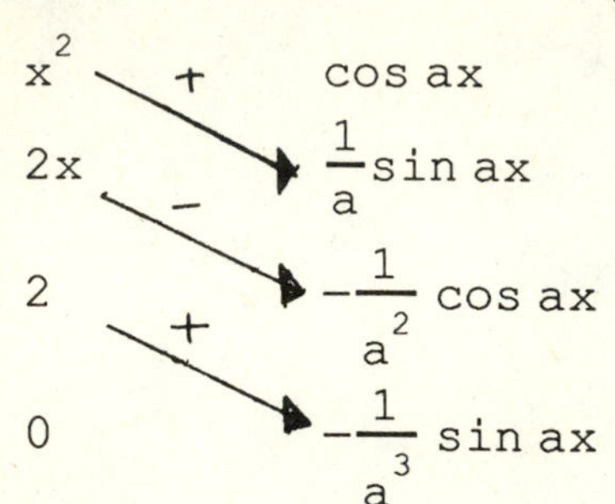

27. Let $u = \sec^{-1} x$ and $dv = x\,dx \Rightarrow du = \frac{dx}{x\sqrt{x^2-1}}$ and $v = \frac{1}{2}x^2$

$$\int_1^2 x\sec^{-1} x\,dx = \frac{1}{2}x^2\sec^{-1} x\Big]_1^2 - \frac{1}{2}\int_1^2 \frac{x}{\sqrt{x^2-1}}dx =$$

$$\frac{1}{2}x^2\sec^{-1} x - \frac{1}{2}\sqrt{x^2-1}\Big]_1^2 = \frac{2\pi}{3} - \frac{\sqrt{3}}{2}$$

29. $\int_1^e \frac{\ln x}{x}dx = \frac{1}{2}\ln^2 x\Big]_1^e = \frac{1}{2}$

31. Let $u = \sin^{-1}\left(\frac{1}{x}\right)$ and $dv = x\,dx \Rightarrow du = \frac{-dx}{x\sqrt{x^2-1}}$ and $v = \frac{1}{2}x^2$

$$\int x\sin^{-1}\left(\frac{1}{x}\right)dx = \frac{1}{2}x^2\sin^{-1}\left(\frac{1}{x}\right) + \int\frac{x\,dx}{\sqrt{x^2-1}} = \frac{1}{2}x^2\sin^{-1}\left(\frac{1}{x}\right) + \frac{1}{2}\sqrt{x^2-1} + C$$

33. Let $u = e^x$ and $dv = \sin x\,dx \Rightarrow du = e^x\,dx$ and $v = -\cos x$.

$$\int e^x \sin x\,dx = -e^x\cos x + \int e^x\cos x\,dx$$

Now let $u = e^x$ and $dv = \cos x\,dx \Rightarrow du = e^x\,dx$ and $v = \sin x$.

$$\int e^x\sin x\,dx = -e^x\cos x + e^x\sin x - \int e^x\sin x\,dx$$

$$2\int e^x\sin x\,dx = e^x(\sin x - \cos x)$$

$$\int e^x\sin x\,dx = \frac{1}{2}e^x(\sin x - \cos x) + C$$

35. Let $u = e^{2x}$ and $dv = \cos 3x\,dx \Rightarrow du = 2e^{2x}\,dx$ and $v = \frac{1}{3}\sin 3x$

$$\int e^{2x}\cos 3x\,dx = \frac{1}{3}e^{2x}\sin 3x - \frac{2}{3}\int e^{2x}\sin 3x\,dx$$

Now let $u = e^{2x}$ and $dv = \sin 3x\,dx \Rightarrow du = 2e^{2x}\,dx$ and $v = -\frac{1}{3}\cos 3x$.

$$\int e^{2x}\cos 3x\,dx = \frac{1}{3}e^{2x}\sin 3x + \frac{2}{9}e^{2x}\cos 3x - \frac{4}{9}\int e^{2x}\cos 3x\,dx$$

$$\frac{13}{9}\int e^{2x}\cos 3x\,dx = \frac{1}{9}e^{2x}(3\sin 3x + 2\cos 3x)$$

$$\int e^{2x}\cos 3x\,dx = \frac{1}{13}e^{2x}(3\sin 3x + 2\cos 3x) + C$$

37. Let $u = \sin(\ln x)$ and $dv = dx \Rightarrow du = \cos(\ln x)\frac{1}{x}dx$ and $v = x$

$$\int \sin(\ln x)dx = x\sin(\ln x) - \int \cos(\ln x)$$

Now let $u = \cos(\ln x)$ and $dv = dx \Rightarrow du = -\sin(\ln x)\frac{1}{x}dx$ and $v = x$

$$= x\sin(\ln x) - x\cos(\ln x) - \int \sin(\ln x)dx$$

$$2\int \sin(\ln x)dx = x\sin(\ln x) - x\cos(\ln x)$$

$$\int \sin(\ln x)dx = \frac{x}{2}(\sin(\ln x) - \cos(\ln x)) + C$$

39. (a) $\int_0^{\pi} x\sin x\,dx = -x\cos x + \sin x\Big]_0^{\pi} = \pi$ (See problem 1)

(b) $\int_{\pi}^{2\pi} |x\sin x|\,dx = \int_{\pi}^{2\pi} (-x\sin x)\,dx = x\cos x - \sin x\Big]_{\pi}^{2\pi} = 3\pi$

41. $V = 2\pi\int_0^1 xe^{-x}\,dx = 2\pi(-xe^{-x} - e^{-x})\Big]_0^1 = 2\pi\left(\frac{e-2}{2}\right)$

43. $A = \int_0^1 x^2e^x\,dx = x^2e^x - 2xe^x + 2e^x\Big]_0^1 = e - 2$

$$\bar{x} = \frac{1}{e-2}\int_0^1 x^3e^x\,dx = \frac{1}{e-2}\left[e^x(x^3 - 3x^2\ \ 6x - 6)\right]_0^1 = \frac{6-2e}{2-e}$$

$$\bar{y} = \frac{1}{e-2}\int_0^1 \frac{1}{2}x^4e^{2x}\,dx = \frac{1}{2(e-2)}\left[e^{2x}\left(\frac{x^4}{2} - x^3 + \frac{3x^2}{2} - \frac{3x}{2} + \frac{3}{4}\right)\right]_0^1$$

$$= \frac{e^2-3}{8(e-2)}$$

45. (a) $V = \pi\int_0^{\pi} x^2 \sin^2 x\, dx = \frac{x^3}{2} - \frac{x^2 \sin 2x}{4}\Big]_0^{\pi} - \int_0^{\pi}\left(x^2 - \frac{x \sin 2x}{2}\right)dx$

Let $u = x^2$ $\quad dv = \sin^2 x\, dx$ $\qquad$ Let $u = \frac{x}{2}$ $\quad dv = \sin 2x\, dx$

$du = 2x\, dx$ $\quad v = \frac{x}{2} - \frac{\sin 2x}{4}$ $\qquad du = \frac{1}{2}dx$ $\quad v = -\frac{1}{2}\cos 2x\, dx$

$$= \frac{x^3}{2} - \frac{x^2 \sin 2x}{4} - \frac{x^3}{3} - \frac{x \cos 2x}{4} + \frac{\sin 2x}{8}\Big]_0^{\pi} = \frac{\pi^4}{6} - \frac{\pi^2}{4}$$

(b) $V = 2\pi\int_0^{\pi} (\pi - x)\, x \sin x\, dx$

$$= 2\pi^2\, [-x \cos x + \sin x\Big]_0^{\pi} - 2\pi\, [-x^2 \cos x + 2x \sin x + 2 \cos x\Big]_0^{\pi} = 8\pi$$

47. $M_y = x\,\delta\, dA = \int_0^{\pi} x(1 + x) \sin x\, dx$

$$= -(x + x^2) \cos x + (1 + 2x)\sin x + 2 \cos x\Big]_0^{\pi} = \pi^2 + \pi - 4$$

49. $x = e^t \sin t \Rightarrow \frac{dx}{dt} = e^t(\sin t + \cos t)$.

$y = e^t \cos t \Rightarrow \frac{dy}{dt} = e^t(-\sin t + \cos t)$. $ds = \sqrt{\left(\frac{dx}{dt}\right)^2 + \left(\frac{dy}{dt}\right)^2}\, dt$

$= e^t\sqrt{2}\, dt$. $S = \int 2\pi\, r\, ds \; = 2\pi\sqrt{2}\int_0^{\pi} e^{2t}\cos t\, dt$

$$= \frac{2\pi\sqrt{2}}{5} e^{2t}(\sin t + 2 \cos t)\Big]_0^{\pi} = \frac{2\pi\sqrt{2}\,(e^{\pi} - 2)}{5}$$

51. Let $u = \tan^{-1} x \Rightarrow du = \frac{dx}{1 + x^2}$ and $dv = x\, dx \Rightarrow v = \frac{x^2}{2} + C$. Take $C = \frac{1}{2}$.

$$\int x \tan^{-1} x\, dx = \frac{x^2 + 1}{2}\tan^{-1} x - \int \frac{x^2 + 1}{2}\cdot\frac{dx}{1 + x^2} = \frac{x^2 + 1}{2}\tan^{-1} x - \frac{x}{2} + C$$

7.3 PRODUCTS AND POWERS OF TRIGONOMETRIC FUNCTIONS

1. $\int \sin^5 x\, dx = \int (\sin^2 x)^2 \sin x\, dx = \int (1 - \cos^2 x)^2 \sin x\, dx$

$$= \int (1 - 2\cos^2 x + \cos^4 x) \sin x\, dx = -\cos x + \frac{2}{3}\cos^3 x - \frac{1}{5}\cos^5 x + C$$

3. $\int_{-\frac{\pi}{2}}^{\frac{\pi}{2}} \cos^3 x\, dx = \int_{-\frac{\pi}{2}}^{\frac{\pi}{2}} (1 - \sin^2 x)\cos x\, dx = \sin x - \frac{1}{3}\sin^3 x\Big]_{-\frac{\pi}{2}}^{\frac{\pi}{2}} = \frac{4}{3}$

5. $\int \sin^7 x\,dx = \int (\sin^2 x)^3 \sin x\,dx = \int (1-\cos^2 x)^3 \sin x\,dx$

$= \int (1 - 3\cos^2 x + 3\cos^4 x - \cos^6 x) \sin x\,dx$

$= -\cos x + \cos^3 x - \frac{3}{5}\cos^5 x + \frac{1}{7}\cos^7 x + C$

7. $\int \cos^{\frac{2}{3}} x \sin^5 x\,dx = \int \cos^{\frac{2}{3}} x (1-\cos^2 x)^2 \sin x\,dx$

$= \int (\cos^{\frac{2}{3}} x - 2\cos^{\frac{8}{3}} x + \cos^{\frac{14}{3}} x) \sin x\,dx$

$= -\frac{3}{5}\cos^{\frac{5}{3}} x + \frac{6}{11}\cos^{\frac{11}{3}} x - \frac{3}{17}\cos^{\frac{17}{3}} x + C$

9. $\int_0^{\frac{\pi}{3}} \sec x\,dx = \ln|\sec x + \tan x|\Big]_0^{\frac{\pi}{3}} = \ln(2+\sqrt{3})$

11. $\int_{-\frac{\pi}{3}}^{0} \sec^3 x\,dx = \frac{1}{2}\sec x \tan x + \frac{1}{2}\ln|\sec x + \tan x|\Big]_{-\frac{\pi}{3}}^{0}$

$= \sqrt{3} - \frac{1}{2}\ln(2-\sqrt{3})$ (See Example 3 for integration)

13. $\int e^x \sec^3 e^x\,dx = \int \sec^3 u\,du$, where $u = e^x$, $du = e^x dx$. By Example 3,

$\int e^x \sec^3 e^x\,dx = \frac{1}{2}\sec e^x \tan e^x + \frac{1}{2}\ln|\sec e^x + \tan e^x| + C$

15. $\int_{\frac{\pi}{4}}^{\frac{\pi}{2}} \frac{dx}{\sin^4 x} = \int_{\frac{\pi}{4}}^{\frac{\pi}{2}} \csc^4 x\,dx = \int_{\frac{\pi}{4}}^{\frac{\pi}{2}} (1+\cot^2 x)\csc^2 x\,dx = -\cot x - \frac{1}{3}\cot^3 x\Big]_{\frac{\pi}{4}}^{\frac{\pi}{2}} = \frac{4}{3}$

17. $\int_0^{\frac{\pi}{4}} \sec^8 x\,dx = \int_0^{\frac{\pi}{4}} (\sec^2 x)^3 \sec^2 x\,dx = \int_0^{\frac{\pi}{4}} (1+\tan^2 x)^3 \sec^2 x\,dx$

$= \int_0^{\frac{\pi}{4}} (1 + 3\tan^2 x + 3\tan^4 x + \tan^6 x)\sec^2 x\,dx$

$= \tan x + \tan^3 x + \frac{3}{5}\tan^5 x + \frac{1}{7}\tan^7 x\Big]_0^{\frac{\pi}{4}} = \frac{96}{35}$

19. $\int_0^{\frac{\pi}{4}} \tan^3 dx = \int_0^{\frac{\pi}{4}} \tan^2 x \tan x\, dx = \int_0^{\frac{\pi}{4}} (\sec^2 x - 1)\tan x\, dx$

$= \int_0^{\frac{\pi}{4}} \tan x \sec^2 x\, dx - \int_0^{\frac{\pi}{4}} \tan x\, dx = \frac{1}{2}\tan^2 x - \ln|\cos x|\Big]_0^{\frac{\pi}{4}} = \frac{1}{2} - \ln\frac{\sqrt{2}}{2}$

21. $\int \tan^6 x\, dx = \int (\tan^4 x)(\tan^2 x)\, dx = \int \tan^4 x(\sec^2 x - 1)\, dx$

$= \frac{1}{5}\tan^5 x - \int \tan^4 x\, dx = \frac{1}{5}\tan^5 x - \frac{1}{3}\tan^3 x + \tan x - x + C$

(See Example 4 for integration of $\int \tan^4 x\, dx$)

23. $\int \csc x\, dx = \int \frac{\csc x(\csc x + \cot x)}{\csc x + \cot x}\, dx = -\ln|\csc x + \cot x| + C$

25. (a) $\int_0^{\frac{\pi}{4}} \frac{dx}{\sqrt{1-\sin^2 x}} = \int_0^{\frac{\pi}{4}} \sec x\, dx = \ln|\sec x + \tan x|\Big]_0^{\frac{\pi}{4}} = \ln(\sqrt{2}+1)$

(b) $\int_{\frac{\pi}{3}}^{\frac{\pi}{2}} \frac{dx}{\sqrt{1-\cos^2 x}} = \int_{\frac{\pi}{3}}^{\frac{\pi}{2}} \csc x\, dx = -\ln|\csc x + \cot x|\Big]_{\frac{\pi}{3}}^{\frac{\pi}{2}} = \ln\sqrt{3}$

27. $\int_{-\pi}^{0} \sin 3x \cos 2x\, dx = \frac{1}{2}\int_{-\pi}^{0} (\sin x + \sin 5x)\, dx = -\frac{1}{2}\cos x - \frac{1}{10}\cos 5x\Big]_{-\pi}^{0} = -\frac{6}{5}$

29. $\int_{-\pi}^{\pi} \sin^2 3x\, dx = \int_{-\pi}^{\pi} \left(\frac{1}{2} - \frac{1}{2}\cos 6x\right)dx = \frac{1}{2}x - \frac{1}{12}\sin 6x\Big]_{-\pi}^{\pi} = \pi$

31. $\int_0^{\pi} \cos 3x \cos 4x\, dx = \frac{1}{2}\int_0^{\pi} (\cos x + \cos 7x)\, dx = \frac{1}{2}\sin x + \frac{1}{7}\sin 7x\Big]_0^{\pi} = 0$

33. $\int \frac{\sec 2x \csc 2x}{2}\, dx = \int \frac{dx}{2\cos 2x \sin 2x} = \int \frac{dx}{\sin 4x}$

$= \int \csc 4x\, dx = -\frac{1}{4}\ln|\csc 4x + \cot 4x| + C$

35. $A = 2\int_0^{\frac{\pi}{4}} (2\cos x - \sec x)\, dx = 4\sin x - 2\ln|\sec x + \tan x|\Big]_0^{\frac{\pi}{4}} = 2\sqrt{2} - 2\ln(\sqrt{2}+1)$

37. $s = \int_0^{\frac{\pi}{4}} \sqrt{1 + \left(\frac{dy}{dx}\right)^2}\, dx = \int_0^{\frac{\pi}{4}} \sqrt{1+\tan^2 x}\, dx = \ln|\sec x + \tan x|\Big]_0^{\frac{\pi}{4}} = \ln(\sqrt{2}+1)$

39. (a) $d = 25\int_{\frac{\pi}{6}}^{\frac{\pi}{4}} \sec x\, dx = 25\ln|\sec x + \tan x|\Big]_{\frac{\pi}{6}}^{\frac{\pi}{4}} \approx 8.3$ cm

(b) $d = 25\int_{\frac{\pi}{4}}^{\frac{\pi}{3}} \sec x\, dx = 25\ln|\sec x + \tan x|\Big]_{\frac{\pi}{4}}^{\frac{\pi}{3}} \approx 10.9$ cm

41. (a) $\int_k^{k+2\pi} \sin mx \sin nx\, dx = \frac{1}{2}\int_k^{k+2\pi} [\cos(m-n)x - \cos(m+n)x]\, dx$

$$= \frac{\sin(m-n)x}{2(m-n)} - \frac{\sin(m+n)x}{2(m+n)},\ m^2 \neq n^2 \Bigg]_k^{k+2\pi} = 0$$

since since $\sin k = \sin(k+2\pi)$

(b) $\int_k^{k+2\pi} \cos mx \cos nx\, dx = \frac{1}{2}\int_k^{k+2\pi} [\cos(m-n)x + \cos(m+n)x]\, dx$

$$= \frac{\sin(m-n)x}{2(m-n)} - \frac{\sin(m+n)x}{2(m+n)},\ m^2 \neq n^2 \Bigg]_k^{k+2\pi} = 0 \text{ as in part (a).}$$

7.4 EVEN POWERS OF SINES AND COSINES

1. $\int_{-\pi}^{\pi} \sin^2 x\, dx = \int_{-\pi}^{\pi}\left(\frac{1}{2} - \frac{1}{2}\cos 2x\right) dx = \frac{1}{2}x - \frac{1}{4}\sin 2x\Big]_{-\pi}^{\pi} = \pi$

3. $\int \sin^2 2t\, dt = \int\left(\frac{1}{2} - \frac{1}{2}\cos 4t\right) dt = \frac{1}{2}t - \frac{1}{8}\sin 4t + C$

5. $\int_{-\frac{\pi}{4}}^{\frac{\pi}{4}} \sin^2 x \cos^2 x\, dx = \frac{1}{4}\int_{-\frac{\pi}{4}}^{\frac{\pi}{4}} (2\sin x\cos x)^2\, dx = \frac{1}{4}\int_{-\frac{\pi}{4}}^{\frac{\pi}{4}} \sin^2 2x\, dx$

$$= \frac{1}{4}\int_{-\frac{\pi}{4}}^{\frac{\pi}{4}}\left(\frac{1}{2} - \frac{1}{2}\cos 4x\right) dx = \frac{1}{8}x - \frac{1}{32}\sin 4x\Big]_{\frac{-\pi}{4}}^{\frac{\pi}{4}} = \frac{\pi}{16}$$

7. $\int_0^{\frac{\pi}{a}} \sin^4 ax\, dx = \int_0^{\frac{\pi}{a}} (\sin^2 ax)^2\, dx = \int_0^{\frac{\pi}{a}}\left(\frac{1}{2} - \frac{1}{2}\cos 2ax\right)^2 dx$

$$= \int_0^{\frac{\pi}{a}}\left(\frac{1}{4} - \frac{1}{2}\cos 2ax + \frac{1}{4}\cos^2 2ax\right) dx$$

$$= \int_0^{\frac{\pi}{a}}\left[\left(\frac{1}{4} - \frac{1}{2}\cos 2ax + \frac{1}{4}\left(\frac{1}{2} + \frac{1}{2}\cos 4ax\right)\right)\right] dx$$

$$= \frac{3}{8}x - \frac{1}{4a}\sin 2ax + \frac{1}{32a}\sin 4ax\Big]_0^{\frac{\pi}{a}} = \frac{3\pi}{8a}$$

9. $\int \frac{\sin^4 x}{\cos^2 x}\, dx = \int \frac{\sin^2 x\,(1-\cos^2 x)}{\cos^2 x}\, dx = \int (\tan^2 x - \sin^2 x)\, dx$

$$= \int\left(\sec^2 x - 1 - \left(\frac{1}{2} - \frac{1}{2}\cos 2x\right)\right) dx = \tan x - \frac{3}{2}x + \frac{1}{4}\sin 2x + C$$

11. $\int_0^{\pi} \sin^4 y \cos^2 y\, dy = \int_0^{\pi} \sin^4 y (1 - \sin^2 y)\, dy = I_1 - I_2$, where

$$I_1 = \int_0^{\pi} \sin^4 y\, dy = \int_0^{\pi} \left(\frac{1}{2} - \frac{1}{2}\cos 2y\right)^2 dy$$

$$= \int_0^{\pi} \left(\frac{1}{4} - \frac{1}{2}\cos 2y + \frac{1}{4}\cos^2 2y\right) dy$$

$$= \int_0^{\pi} \left(\frac{1}{4} - \frac{1}{2}\cos 2y + \frac{1}{4}\left(\frac{1}{2} + \frac{1}{2}\cos 4y\right)\right) dy$$

$$= \frac{3}{8}y - \frac{1}{4}\sin 2y + \frac{1}{32}\sin 4y \Big]_0^{\pi} = \frac{3\pi}{8}$$

and $I_2 = \int_0^{\pi} \sin^6 y\, dy = \int_0^{\pi} \left(\frac{1}{2} - \frac{1}{2}\cos 2y\right)^3 dy$

$$= \frac{1}{8}\int_0^{\pi} (1 - 3\cos 2y + 3\cos^2 2y - \cos^3 2y)\, dy$$

$$= \frac{1}{8}\int_0^{\pi} \left[1 - 3\cos 2y + 3\left(\frac{1}{2} + \frac{1}{2}\cos 4y\right) - (1 - \sin^2 2y)\cos 2y\right] dy$$

$$= \frac{1}{8}\left[\frac{5}{2}y - 2\sin 2y + \frac{3}{8}\sin 4y + \frac{1}{6}\sin^3 y\right]_0^{\pi} = \frac{5\pi}{16}$$

Combining, $\frac{3\pi}{8} - \frac{5\pi}{16} = \frac{\pi}{16}$

13. $\int \frac{\sin^6\theta}{\cos^2\theta}\, d\theta = \int \frac{\sin^2\theta (1 - \cos^2\theta)^2}{\cos^2\theta}\, d\theta = \int \tan^2\theta (1 - 2\cos^2\theta + \cos^4\theta)\, d\theta$

$$= \int (\sec^2\theta - 1)\, d\theta - 2\int \left(\frac{1}{2} - \frac{1}{2}\cos 2\theta\right) d\theta + \frac{1}{4}\int (2\sin\theta\cos\theta)^2\, d\theta$$

$$= \tan\theta - 2\theta + \frac{1}{2}\sin 2\theta + \frac{1}{4}\int \left(\frac{1}{2} - \frac{1}{2}\cos 4\theta\right) d\theta$$

$$= \tan\theta + \frac{1}{2}\sin 2\theta - \frac{1}{32}\sin 4\theta - \frac{15}{8}\theta + C$$

15. $\int_0^{2\pi} \sqrt{\frac{1 - \cos t}{2}}\, dt = \int_0^{2\pi} \left|\sin\frac{t}{2}\right| dt = \int_0^{2\pi} \sin\frac{t}{2}\, dt = -2\cos\frac{t}{2}\Big]_0^{2\pi} = 4$

Note: $0 \le t \le 2\pi \Rightarrow 0 \le \frac{t}{2} \le \pi \Rightarrow \sin\frac{t}{2} \ge 0$ on this interval.

17. $\int_0^{\frac{\pi}{10}} \sqrt{1 + \cos 5\theta}\, d\theta = \sqrt{2}\int_0^{\frac{\pi}{10}} \sqrt{\frac{1 + \cos 5\theta}{2}}\, d\theta = \sqrt{2}\int_0^{\frac{\pi}{10}} \cos\frac{5\theta}{2}\, d\theta$

$$= \frac{2\sqrt{2}}{5}\sin\frac{5\theta}{2}\Big]_0^{\frac{\pi}{10}} = \frac{2}{5}$$

19. $\int_0^{\frac{\pi}{2}} \theta\sqrt{1-\cos\theta}\, d\theta = \sqrt{2}\int_0^{\frac{\pi}{2}} \theta \sin\frac{\theta}{2}\, d\theta = \sqrt{2}\left[-2\theta\cos\frac{\theta}{2} + \int_0^{\frac{\pi}{2}} 2\cos\frac{\theta}{2}\, d\theta\right]$

$= \sqrt{2}\left(-2\theta\cos\frac{\theta}{2} + 4\sin\frac{\theta}{2}\Big]_0^{\frac{\pi}{2}}\right) = 4-\pi$

21. $\int_{-\frac{\pi}{4}}^{\frac{\pi}{4}} \sqrt{1+\tan^2 x}\, dx = \int_{-\frac{\pi}{4}}^{\frac{\pi}{4}} |\sec x|\, dx = \int_{-\frac{\pi}{4}}^{\frac{\pi}{4}} \sec x\, dx$

$= 2\ln|\sec x + \tan x|\Big]_{-\frac{\pi}{4}}^{\frac{\pi}{4}} = 2\ln(\sqrt{2}+1)$

23. $\int_0^{\pi} \sqrt{1-\cos^2\theta}\, d\theta = \int_0^{\pi} |\sin\theta|\, d\theta = -\cos\theta\Big]_0^{\pi} = 2$

25. $\int_{\frac{\pi}{4}}^{\frac{3\pi}{4}} \sqrt{\cot^2\theta + 1}\, d\theta = \int_{\frac{\pi}{4}}^{\frac{3\pi}{4}} |\csc\theta|\, d\theta = \int_{\frac{\pi}{4}}^{\frac{3\pi}{4}} \csc\theta\, d\theta$

$= -\ln|\csc\theta - \cot\theta|\Big]_{\frac{\pi}{4}}^{\frac{3\pi}{4}} = \ln\frac{\sqrt{2}+1}{\sqrt{2}-1} = \ln(3+2\sqrt{2})$

27. $\int_{-\pi}^{\pi} \sqrt{1-\cos^2 x}\sin x\, dx = 0$ because it the integral of an odd function over a symmetric interval about 0.

29. $\int \frac{1}{\sin^4 x}\, dx = \int \csc^4 dx = \int (\cot^2 x + 1)\csc^2 x\, dx = -\frac{1}{3}\cot^3 x - \cot x + C$

31. $\int_{-\frac{\pi}{4}}^{\frac{\pi}{4}} \frac{\sin^2 x}{\cos^2 x}\, dx = \int_{-\frac{\pi}{4}}^{\frac{\pi}{4}} \tan^2 x\, dx = \int_{-\frac{\pi}{4}}^{\frac{\pi}{4}} (\sec^2 x - 1)\, dx = \tan x - x\Big]_{-\frac{\pi}{4}}^{\frac{\pi}{4}} = 2 - \frac{\pi}{2}$

33. $\int \frac{\sin^2 t}{\cos t}\, dt = \int \frac{1-\cos^2 t}{\cos t}\, dt = \int (\sec t - \cos t)\, dt = \ln|\sec t + \tan t| - \sin t + C$

35. $\int_0^{\pi} \sqrt{1+\cos 4x}\, dx = \sqrt{2}\int_0^{\pi} |\cos 2x|\, dx = 4\sqrt{2}\int_0^{\frac{\pi}{4}} \cos 2x\, dx = 2\sqrt{2}\sin 2x\Big]_0^{\frac{\pi}{4}} = 2\sqrt{2}$

37. $V = \pi\int_0^{\pi} x^2\sin^2 x\, dx = \pi\int_0^{\pi} x^2\left(\frac{1}{2} - \frac{1}{2}\cos 2x\right) dx = \frac{\pi}{2}\int_0^{\pi} (x^2 - x^2\cos 2x)\, dx$

$= \frac{\pi}{2}\left[\frac{x^3}{3} - \frac{x^2}{2}\sin 2x + \frac{x}{2}\cos 2x - \frac{1}{4}\sin 2x\right]_0^{\pi} = \frac{\pi^4}{6} - \frac{\pi^2}{4}$

7.5 TRIGONOMETRIC SUBSTITUTIONS

1. Let $x = 2\tan u \Rightarrow dx = 2\sec^2 u\,du$

$$\int_{-2}^{2} \frac{dx}{4 + x^2} = \int_{-\frac{\pi}{4}}^{\frac{\pi}{4}} \frac{2\sec^2 u\,du}{4 + 4\tan^2 u} = \int_{-\frac{\pi}{4}}^{\frac{\pi}{4}} \frac{1}{2}\,du = \frac{1}{2}u\Big]_{-\frac{\pi}{4}}^{\frac{\pi}{4}} = \frac{\pi}{4}$$

3. $\displaystyle\int \frac{dx}{1 + 4x^2} = \frac{1}{2}\int \frac{2\,dx}{1 + (2x)^2} = \frac{1}{2}\tan^{-1} 2x + C$

5. $\displaystyle\int_0^{\frac{\pi}{8}} \frac{2\,dx}{\sqrt{1 - 4x^2}} = \sin^{-1}(2x)\Big]_0^{\frac{\pi}{8}} = \frac{\sqrt{2}}{2}$

7. Let $5\tan u = y \Rightarrow 5\sec^2 u\,du = dy$. $\displaystyle\int \frac{dy}{\sqrt{25 + y^2}} = \int \frac{5\sec^2 u\,du}{\sqrt{25 + 25\tan^2 u}}$

$$= \int \sec u\,du = \ln|\sec u + \tan u| = \ln\left|\frac{\sqrt{25 + y^2}}{5} + \frac{y}{5}\right| + C$$

9. Let $5\tan u = 3y \Rightarrow 5\sec^2 u\,du = 3\,dy$. $\displaystyle\int \frac{dy}{\sqrt{25 + 9y^2}} = \int \frac{\frac{5}{3}\sec^2 u\,du}{\sqrt{25 + 25\tan^2 u}}$

$$= \frac{1}{3}\int \sec u\,du = \frac{1}{3}\ln|\sec u + \tan u| = \frac{1}{3}\ln\left|\frac{\sqrt{25 + 9y^2}}{5} + \frac{3y}{5}\right| + C$$

(Note: the denominator may be absorbed into the constant)

11. Let $3z = \sec u \Rightarrow 3\,dz = \sec u\tan u\,du$. Then $\displaystyle\int \frac{3\,dz}{\sqrt{9z^2 - 1}} = \int \frac{\sec u\tan u\,du}{\sqrt{\sec^2 u - 1}}$

$$= \int \sec u\,du = \ln|\sec u + \tan u| = \ln|3z + \sqrt{9z^2 - 1}| + C$$

13. Let $x = 4\sec u \Rightarrow dx = 4\sec u\tan u\,du$. Then $\displaystyle\int_{\frac{8}{\sqrt{3}}}^{8} \frac{dx}{x\sqrt{x^2 - 16}}$

$$= \int_{\frac{\pi}{6}}^{\frac{\pi}{3}} \frac{4\sec u\tan u\,du}{4\sec u\sqrt{16\sec^2 u - 16}} = \frac{1}{4}\int_{\frac{\pi}{6}}^{\frac{\pi}{3}} du = \frac{1}{4}u\Big]_{\frac{\pi}{6}}^{\frac{\pi}{3}} = \frac{\pi}{24}$$

15. Let $x = \frac{1}{2}\sec u \Rightarrow dx = \frac{1}{2}\sec u \tan u\, du$. Then $\int_{\frac{\sqrt{2}}{2}}^{1} \frac{dx}{2x\sqrt{4x^2-1}}$

$$= \int_{\frac{\pi}{4}}^{\frac{\pi}{3}} \frac{\frac{1}{2}\sec u \tan u\, du}{\sec u\sqrt{\sec^2 u - 1}} = \frac{1}{2}\int_{\frac{\pi}{4}}^{\frac{\pi}{3}} du = \frac{1}{2}u\Big]_{\frac{\pi}{4}}^{\frac{\pi}{3}} = \frac{\pi}{24}$$

17. Let $x = 2\sec w \Rightarrow dx = 2\sec w \tan w dx$, $x = 4 \Rightarrow w = \frac{\pi}{3}$, $x = 2 \Rightarrow w = 0$

$$\int_2^4 \sqrt{x^2-4}\, dx = 2\int_0^{\frac{\pi}{3}} \sqrt{4\sec^2 w - 4}\, \sec w \tan w\, dw = 4\int_0^{\frac{\pi}{3}} \tan^2 w \sec w\, dw$$

$$= 4\int_0^{\frac{\pi}{3}} (\sec^2 w - 1)\sec w\, dw = 4\int_0^{\frac{\pi}{3}} (\sec^3 w - \sec w)\, dw$$

$$= 4\left(\frac{1}{2}\sec w \tan w + \frac{1}{2}\ln|\sec w + \tan w| - \ln|\sec w + \tan w|\right)\Big]_0^{\frac{\pi}{3}}$$

$$= 2\sec w \tan w - 2\ln|\sec w + \tan w|\Big]_0^{\frac{\pi}{3}} = 4\sqrt{3} - 2\ln(2+\sqrt{3})$$

(For integration of $\int \sec^3 w\, dw$, see Sect. 7.3, Ex. 3)

19. Let $x = 3\sin u \Rightarrow dx = 3\cos u\, du$. Then $\int \frac{dx}{x^2\sqrt{9-x^2}} =$

$$\int \frac{3\cos u\, du}{9\sin^2 u\sqrt{9 - 9\sin^2 u}} = \frac{1}{9}\int \csc^2 u\, du = -\frac{1}{9}\left(\frac{\sqrt{9-x^2}}{x}\right) + C$$

21. Let $x = \csc u \Rightarrow dx = -\csc u \cot u\, du$. Then $\int_{\frac{5}{4}}^{\frac{5}{3}} \frac{dx}{x^2\sqrt{x^2-1}} =$

$$\int_{x=\frac{5}{4}}^{x=\frac{5}{3}} \frac{-\csc u \cot u\, du}{\csc^2 u\sqrt{\csc^2 u - 1}} = \int_{x=\frac{5}{4}}^{x=\frac{5}{3}} (-\sin u)\, du = \frac{\sqrt{x^2-1}}{x}\Big]_{\frac{5}{4}}^{\frac{5}{3}} = \frac{1}{5}$$

23. Let $x = 5\sin u \Rightarrow dx = 5\cos u\, du$, $x = 0 \Rightarrow u = 0$, $x = 5 \Rightarrow u = \frac{\pi}{2}$.

$$\int_0^5 \sqrt{25-x^2}\, dx = \int_0^{\frac{\pi}{2}} \sqrt{25 - 25\sin^2 u}\cdot 5\cos u\, du = 25\int_0^{\frac{\pi}{2}} \cos^2 u\, du$$

$$= 25\int_0^{\frac{\pi}{2}} \left(\frac{1}{2} + \frac{1}{2}\cos 2u\right) du = 25\left(\frac{1}{2}u + \frac{1}{4}\sin 2u\right)\Big]_0^{\frac{\pi}{2}} = \frac{25\pi}{4}$$

25. Let $x-1=2\sin u \Rightarrow dx = 2\cos u\,du$, $x=2 \Rightarrow u=\frac{\pi}{6}$, $x=1 \Rightarrow u=0$.

$$\int_1^2 \frac{dx}{\sqrt{4-(x-1)^2}} = \int_0^{\frac{\pi}{6}} \frac{2\cos u\,du}{\sqrt{4-4\sin^2 u}} = \int_0^{\frac{\pi}{6}} du = u\Big]_0^{\frac{\pi}{6}} = \frac{\pi}{6}$$

27. Let $x=2\sin u \Rightarrow dx = 2\cos u\,du$, $x=1 \Rightarrow u=\frac{\pi}{6}$, $x=0 \Rightarrow u=0$.

$$\int_0^1 \frac{12\,dx}{\sqrt{4-x^2}} = \int_0^{\frac{\pi}{6}} \frac{24\cos u\,du}{\sqrt{4-4\sin^2 u}} = 12\int_0^{\frac{\pi}{6}} du = 12\,u\Big]_0^{\frac{\pi}{6}} = 2\pi$$

29. Let $x=\tan u \Rightarrow dx = \sec^2 u\,du$, $x=0 \Rightarrow u=0$, $x=1 \Rightarrow u=\frac{\pi}{4}$.

$$\int_0^1 \frac{x^3\,dx}{\sqrt{x^2+1}} = \int_0^{\frac{\pi}{4}} \frac{\tan^3 u\sec^2 u\,du}{\sqrt{\tan^2 u+1}} = \int_0^{\frac{\pi}{4}} \tan^3 u\sec u\,du =$$

$$\frac{1}{3}\sec^3 u - \sec u\Big]_0^{\frac{\pi}{4}} = \frac{2-\sqrt{3}}{3}$$

31. Let $x=\frac{1}{2}\sec u \Rightarrow dx = \frac{1}{2}\sec u\tan u\,du$. Then

$$\int \frac{dx}{x\sqrt{x^2-\frac{1}{4}}} = \int \frac{\frac{1}{2}\sec u\tan u\,du}{\frac{1}{2}\sec u\sqrt{\frac{1}{4}\sec^2 u-\frac{1}{4}}} = \int 2\,du = 2\sec^{-1}2x + C$$

33. Let $x=\sin u \Rightarrow dx = \cos u\,du$. Then $\displaystyle\int \frac{\sqrt{1-x^2}}{x^2}\,dx = \int \frac{\sqrt{1-\sin^2 u}}{\sin^2 u}\cos u\,du$

$$= \int \cot^2 u\,du = -\cot u - u + C = -\frac{\sqrt{1-x^2}}{x} - \sin^{-1} x + C$$

35. $$\int_{\frac{3}{4}}^{\frac{4}{5}} \frac{dx}{x^2\sqrt{1-x^2}} = \int_{x=\frac{3}{4}}^{x=\frac{4}{5}} \frac{\cos u\,du}{\sin^2 u\sqrt{1-\sin^2 u}} = \int_{x=\frac{3}{4}}^{x=\frac{4}{5}} \csc^2 u\,du$$

$$= -\cot u\Big]_{x=\frac{3}{4}}^{x=\frac{4}{5}} = -\frac{\sqrt{1-x^2}}{x}\Big]_{x=\frac{3}{4}}^{x=\frac{4}{5}} = -\left(\frac{3}{4}-\frac{\sqrt{7}}{3}\right)$$

37. $$\int \frac{dx}{(a^2-x^2)^{\frac{3}{2}}} = \int \frac{a\cos u\,du}{\left(\sqrt{a^2-a^2\sin^2 u}\right)^3} = \frac{1}{a^2}\int \sec^2 u\,du = \frac{1}{a^2}\left(\frac{x}{\sqrt{a^2-x^2}}\right) + C$$

39. (a) Let $u^2 = 16 - y^2 \Rightarrow u\,du = -y\,dy$

$$\int \frac{y\,dy}{\sqrt{16-y^2}} = -\int du = -u + C = -\sqrt{16-y^2} + C$$

(b) Let $y = 4\sin u \Rightarrow dy = 4\cos u\,du$

$$\int \frac{y\,dy}{\sqrt{16-y^2}} = \int \frac{16\sin u\cos u\,du}{\sqrt{16-16\sin^2 u}} = -4\cos u + C = -\sqrt{16-y^2} + C$$

41. (a) Let $x^2 - 1 = u^2 \Rightarrow x\,dx = u\,du$. Then $\int \frac{x}{(x^2-1)^{\frac{3}{2}}}dx =$

$$\int u^{-2} + C = -u^{-1} + C = -\frac{1}{\sqrt{x^2-1}} + C$$

(b) Let $x = \sec u \Rightarrow dx = \sec u\tan u\,du$. Then $\int \frac{x}{(x^2-1)^{\frac{3}{2}}}dx =$

$$\int \frac{\sec u \cdot \sec u\tan u\,du}{(\sec^2 u - 1)^{\frac{3}{2}}} = \int \frac{\sec^2 u\,du}{\tan^2 u} = -\frac{1}{\tan u} + C = -\frac{1}{\sqrt{x^2-1}} + C$$

43. (a) $\displaystyle\int \frac{du}{u^2+a^2} = \int \frac{a\,dz}{a^2z^2+a^2} = \frac{1}{a}\int \frac{dz}{z^2+1} = \frac{1}{a}\tan^{-1}z + C = \frac{1}{a}\tan^{-1}\frac{u}{a} + C$

(b) $\displaystyle\int \frac{du}{\sqrt{a^2-u^2}} = \int \frac{a\,dz}{\sqrt{a^2-a^2z^2}} = \int \frac{dz}{\sqrt{1-z^2}} = \sin^{-1}z + C = \sin^{-1}\frac{u}{a} + C$

45. (a) $\sin\left(\tan^{-1}\frac{u}{a}\right) = \dfrac{u}{\sqrt{a^2+u^2}}$ (b) $\cos\left(\sec^{-1}\frac{u}{a}\right) = \dfrac{a}{u}$

47. $S = 2\pi y\,ds = 2\pi\displaystyle\int_0^{\pi} \sin t\sqrt{(-2\sin t)^2 + \cos^2 t}\,dt = 2\pi\int_0^{\pi}\sqrt{4-3\cos^2 t}\,\sin t\,dt$

Let $\frac{2}{\sqrt{3}}\sin\theta = \cos t \Rightarrow -\frac{2}{\sqrt{3}}\cos\theta\,d\theta = \sin t\,dt$

$$= 2\pi\int_{\frac{\pi}{3}}^{-\frac{\pi}{3}}\left(-\frac{2}{\sqrt{3}}\cos\theta\right)(2\cos\theta)\,d\theta = -\frac{8\pi}{\sqrt{3}}\int_{\frac{\pi}{3}}^{-\frac{\pi}{3}}\cos^2\theta\,d\theta =$$

$$= -\frac{4\pi}{\sqrt{3}}\int_{\frac{\pi}{3}}^{-\frac{\pi}{3}}(1+\cos 2\theta)\,d\theta = -\frac{4\pi}{\sqrt{3}}\left(\theta + \frac{1}{2}\sin 2\theta\right)\Bigg]_{\frac{\pi}{3}}^{-\frac{\pi}{3}} = \frac{3\pi^2 + 6\sqrt{3}}{3\sqrt{3}}$$

49. $\frac{dy}{dx} = y^2\left(1 - \frac{1}{\sqrt{4-x^2}}\right) \Rightarrow \frac{dy}{y^2} = dx - \frac{dx}{\sqrt{4-x^2}} \Rightarrow$

$-\frac{1}{y} = x - \sin^{-1}\frac{x}{2} + C$. $-\frac{\pi}{4} = \sqrt{2} - \sin^{-1}\frac{1}{\sqrt{2}} \Rightarrow C = -\sqrt{2}$. Therefore,

$$y = \frac{-1}{x - \sin^{-1}\frac{x}{2} - \sqrt{2}}$$

51. $M_x = \delta\int y\,ds = \int_0^{\ln\sqrt{3}} e^x\sqrt{1+e^{2x}}\,dx = \int_{\frac{\pi}{4}}^{\frac{\pi}{3}} \sqrt{1+\tan^2 u}\,\sec^2 u\,du$

$$= \int_{\frac{\pi}{4}}^{\frac{\pi}{3}} \sec^3 u\,du = \frac{1}{2}\sec u\tan u + \frac{1}{2}\ln|\sec u + \tan u|\Big]_{\frac{\pi}{4}}^{\frac{\pi}{3}}$$

$$= \frac{1}{2}\left(2\sqrt{3} - \sqrt{2} + \ln\frac{2+\sqrt{3}}{1+\sqrt{2}}\right)$$

7.6 INTEGRALS INVOLVING $ax^2 + bx + c$

1. $\int_1^3 \frac{dx}{x^2-2x+5} = \int_1^3 \frac{dx}{(x-1)^2+4} = \frac{1}{2}\tan^{-1}\left(\frac{x-1}{2}\right)\Big]_1^3 = \frac{\pi}{8}$

3. $\int_1^3 \frac{x\,dx}{x^2-2x+5} = \int_1^3 \frac{x\,dx}{(x-1)^2+4} = \int_1^3 \frac{(x-1)\,dx}{(x-1)^2+4} + \int_1^3 \frac{dx}{(x-1)^2+4} =$

$$\frac{1}{2}\ln|(x-1)^2+4| + \frac{1}{2}\tan^{-1}\frac{x-1}{2}\Big]_1^3 = \frac{\pi}{8} - \ln\frac{\sqrt{2}}{2}$$

5. $\int_1^2 \frac{dx}{x^2-2x+4} = \int_1^2 \frac{dx}{(x-1)^2+3} = \frac{1}{\sqrt{3}}\tan^{-1}\frac{x-1}{\sqrt{3}}\Big]_1^2 = \frac{\pi}{6\sqrt{3}}$

7. $\int \frac{dx}{\sqrt{9x^2-6x+5}} = \int \frac{dx}{\sqrt{(3x-1)^2+4}} = \int \frac{\frac{2}{3}\sec^2 u\,du}{\sqrt{4\tan^2 u+4}} = \int \frac{1}{3}\sec u\,du$

Let $3x - 1 = 2\tan u \Rightarrow dx = \frac{2}{3}\sec^2 u\,du$

$$= \frac{1}{3}\ln|\sec u + \tan u| + C = \frac{1}{3}\ln\left|\frac{\sqrt{9x^2-6x+5}}{2} + \frac{3x-1}{2}\right| + C$$

9. $$\int \frac{x\,dx}{\sqrt{9x^2 - 6x + 5}} = \int \frac{x dx}{\sqrt{(3x-1)^2 + 4}} = \frac{1}{3}\int \frac{3(x - 1 + 1)\,dx}{\sqrt{(3x-1)^2 + 4}} =$$

$$\frac{1}{3}\int \frac{(3x-1)\,dx}{\sqrt{(3x-1)^2+4}} + \frac{1}{3}\int \frac{dx}{\sqrt{(3x-1)^2+4}} =$$

$$\frac{1}{9}\sqrt{(3x-1)^2+4} + \frac{1}{9}\ln\left|\frac{\sqrt{(3x-1)^2+4}}{2} + \frac{3x-1}{2}\right| + C$$

11. $$\int \frac{dx}{\sqrt{x^2+2x}} = \int \frac{dx}{\sqrt{(x+1)^2 - 1}} = \int \frac{\sec u \tan u\,du}{\sqrt{\sec^2 u - 1}} =$$

$$\ln|\sec u + \tan u| = \ln|x + 1 + \sqrt{x^2+2x}| + C$$

13. $$\int_{-2}^{2} \frac{x\,dx}{\sqrt{x^2+4x+13}} = \int_{-2}^{2} \frac{x\,dx}{\sqrt{(x+2)^2+9}} = \int_{-2}^{2} \frac{(x+2)\,dx}{\sqrt{(x+2)^2+9}} - \int_{-2}^{2} \frac{2\,dx}{\sqrt{(x+2)^2+9}}$$

$$= \sqrt{(x+2)^2+9} - 2\ln\left|\sqrt{(x+2)^2+9}\right|\Big]_{-2}^{2} = 2 - \ln 9$$

15. $$\int \frac{dx}{\sqrt{x^2+4x+13}} = \int \frac{dx}{\sqrt{(x-1)^2-4}} = \ln\left|x - 1 + \sqrt{(x-1)^2-4}\right| + C$$

17. $$\int \frac{(x+1)\,dx}{\sqrt{2x-x^2}} = \int \frac{(x+1)^2\,dx}{\sqrt{1-(x-1)^2}} = -\sqrt{2x-x^2} + 2\sin^{-1}(x-1) + C$$

19. $$\int \frac{x\,dx}{\sqrt{5+4x-x^2}} = \int \frac{x\,dx}{\sqrt{9-(x-2)^2}} = \int \frac{(x-2)\,dx}{\sqrt{9-(x-2)^2}} + \int \frac{2\,dx}{\sqrt{9-(x-2)^2}} =$$

$$-\sqrt{9-(x-2)^2} + 2\sin^{-1}\frac{x-2}{3} + C$$

21. $$\int_0^1 \frac{(1-x)\,dx}{\sqrt{8+2x-x^2}} = \int_0^1 \frac{(1-x)\,dx}{\sqrt{9-(1-x)^2}} = \sqrt{9-(1-x)^2}\Big]_0^1 = 3 - 2\sqrt{2}$$

23. $$\int_{-2}^{-1} \frac{x\,dx}{x^2+4x+5} = \int_{-2}^{-1} \frac{x\,dx}{(x+2)^2+1} = \int_0^{\frac{\pi}{4}} \frac{(\tan u - 2)\sec^2 u\,du}{\sec^2 u} =$$

$$-\ln|\cos u| - 2u\Big]_0^{\frac{\pi}{4}} = \ln\sqrt{2} - \frac{\pi}{2}$$

25. $$V = \pi\int_{-2}^{11}\left(\frac{20}{\sqrt{x^2-2x+17}}\right)^2 dx = 400\pi\int_{-2}^{11} \frac{dx}{(x-1)^2+16} =$$

$$100\pi\tan^{-1}\left(\frac{x-1}{4}\right)\Big]_{-2}^{11} = 100\pi\left[\tan^{-1}\frac{5}{2} - \tan^{-1}\left(-\frac{3}{4}\right)\right] \approx 183.3\pi$$

27. $S = 2\pi\int_{-1}^{0}\sqrt{x^2+2x+3}\sqrt{\dfrac{2x^2+4x+4}{x^2+2x+3}}\,dx = 2\sqrt{2}\,\pi\int_{-1}^{0}\sqrt{(x+1)^2+1}\,dx$

$= 2\sqrt{2}\,\pi\int_{1}^{\frac{\pi}{4}}\sec^3 u\,du = 2\pi + \sqrt{2}\,\pi\ln(\sqrt{2}+1)$

29. $A = \int_{1}^{5}\dfrac{2}{\sqrt{x^2-2x+10}}\,dx = \int_{1}^{5}\dfrac{2\,dx}{\sqrt{(x-1)^2+9}} = \ln 9$

$\bar{x} = \dfrac{1}{\ln 9}\int_{1}^{5}\dfrac{2x\,dx}{\sqrt{(x-1)^2+9}} = \dfrac{4+2\ln 3}{\ln 9}$

$\bar{y} = \dfrac{1}{\ln 9}\int_{1}^{5}\dfrac{2\,dx}{\left(\sqrt{(x-1)^2+9}\right)^2} = \dfrac{2}{\ln 9}\int_{1}^{5}\dfrac{dx}{(x-1)^2+9} = \dfrac{2}{3\ln 9}\tan^{-1}\dfrac{4}{3}$

31. $\lim_{a\to -5^+}\int_{a}^{-4}\dfrac{dx}{\sqrt{-x^2-8x-15}} = \lim_{a\to -5^+}\int_{a}^{-4}\dfrac{dx}{\sqrt{1-(x+4)^2}} = \lim_{a\to -5^+}\sin^{-1}(x+4)\Big]_a^{-4} = \dfrac{\pi}{2}$

7.7 THE INTEGRATION OF RATIONAL FUNCTIONS

1. $\dfrac{5x-13}{(x-3)(x-2)} = \dfrac{A}{x-3} + \dfrac{B}{x-2} = \dfrac{A(x-2)+B(x-3)}{(x-3)(x-2)} \Rightarrow$

$5x-13 = A(x-2)+B(x-3)$. Let $x=2 \Rightarrow -3=-B \Rightarrow B=3$.

Let $x=3 \Rightarrow 2=A$. $\therefore \dfrac{5x-13}{(x-3)(x-2)} = \dfrac{2}{x-3} + \dfrac{3}{x-2}$

3. $\dfrac{x+4}{(x+1)^2} = \dfrac{A}{x+1} + \dfrac{B}{(x+1)^2} \Rightarrow x+4 = A(x+1)+B$. $x=-1 \Rightarrow B=3$.

Differentiating both sides $\Rightarrow A=1$. $\therefore \dfrac{x+4}{(x+1)^2} = \dfrac{1}{x+1} + \dfrac{3}{(x+1)^2}$

5. $\dfrac{x+1}{x^2(x-1)} = \dfrac{A}{x} + \dfrac{B}{x^2} + \dfrac{C}{x-1} \Rightarrow x+1 = Ax(x-1)+B(x-1)+Cx^2$

$x=0 \Rightarrow B=-1$. $x=1 \Rightarrow C=2$. If we let $x=0$, then $A=-2$.

$\therefore \dfrac{x+1}{x^2(x-1)} = -\dfrac{2}{x} - \dfrac{1}{x^2} + \dfrac{2}{x-1}$

7. By long division, $\dfrac{x^2+8}{x^2-5x+6} = 1 + \dfrac{5x+2}{x^2-5x+6}$. Then

$\dfrac{5x+2}{x^2-5x+6} = \dfrac{A}{x-2} + \dfrac{B}{x-3} \Rightarrow 5x+2 = A(x-3) + B(x-2)$. If $x=3$, $B=17$

and if $x=2$, $A=-12$. $\therefore \dfrac{x^2+8}{x^2-5x+6} = 1 + \dfrac{17}{x-3} - \dfrac{12}{x-2}$

9. $\dfrac{3}{x^2(x^2+9)} = \dfrac{A}{x} + \dfrac{B}{x^2} + \dfrac{Cx+D}{x^2+9} \Rightarrow Ax(x^2+9) + B(x^2+9) + (Cx+D)x^2 = 3$

If $x=0$, then $B = \dfrac{1}{3}$. Equating coefficients, we have

$(A+C)x^3 + (B+D)x^2 + 9Ax + 9B = 3$. $9A = 0 \Rightarrow A = 0$. $A+C=0 \Rightarrow C=0$.

$B+D=0 \Rightarrow D = -\dfrac{1}{3}$. $\therefore \dfrac{3}{x^2(x^2+9)} = \dfrac{1}{3x^2} - \dfrac{1}{3(x^2+9)}$.

11. $\displaystyle\int_0^{\frac{1}{2}} \frac{dx}{1-x^2} = \int_0^{\frac{1}{2}} \frac{\frac{1}{2}\,dx}{1-x} + \int_0^{\frac{1}{2}} \frac{\frac{1}{2}\,dx}{1+x} = -\frac{1}{2}\ln|1-x| + \frac{1}{2}\ln|1+x|\Big]_0^{\frac{1}{2}} = \ln\sqrt{3}$

13. $\displaystyle\int_0^{2\sqrt{2}} \frac{x^3}{x^2+1}\,dx = \int_0^{2\sqrt{2}} \left(x - \frac{x}{x^2+1}\right)dx = \frac{x^2}{2} - \frac{1}{2}\ln(x^2+1)\Big]_0^{2\sqrt{2}} = 4 - \ln 3$

15. $\displaystyle\int_{\frac{1}{4}}^{\frac{3}{4}} \frac{dx}{x(1-x)} = \int_{\frac{1}{4}}^{\frac{3}{4}} \left(\frac{1}{x} + \frac{1}{1-x}\right)dx = \ln x - \ln|1-x|\Big]_{\frac{1}{4}}^{\frac{3}{4}} = 2\ln 3$

17. $\displaystyle\int \frac{(x+4)\,dx}{(x+6)(x-1)} = \frac{2}{7}\int \frac{dx}{x+6} + \frac{5}{7}\int \frac{dx}{x-1} = \frac{2}{7}\ln|x+6| + \frac{5}{7}\ln|x-1| + C$

19. $\displaystyle\int_0^1 \frac{3x^2\,dx}{x^2+2x+1} = \int_0^1 \left(3 + \frac{-6x+3}{(x+1)^2}\right)dx = \int_0^1 \left(3 - \frac{6}{x+1} + \frac{9}{(x+1)^2}\right)dx =$

$\displaystyle 3x - 6\ln|x+1| - \frac{3}{x+1}\Big]_0^1 = \frac{9}{2} - 6\ln 2$

21. $\displaystyle\int \frac{x\,dx}{(x+5)(x-1)} = \frac{5}{6}\int \frac{dx}{x+5} + \frac{1}{6}\int \frac{dx}{x-1} = \frac{5}{6}\ln|x+5| + \frac{1}{6}\ln|x-1| + C$

23. $\displaystyle\int \frac{(x+1)\,dx}{(x+5)(x-1)} = \frac{2}{3}\int \frac{dx}{x+5} + \frac{1}{3}\int \frac{dx}{x-1} = \frac{2}{3}\ln|x+5| + \frac{1}{3}\ln|x-1| + C$

25. $\displaystyle\int_1^3 \frac{dx}{x(x+1)^2} = \int_1^3 \left(\frac{1}{x} - \frac{1}{x+1} - \frac{1}{(x+1)^2}\right)dx = \ln x - \ln|x+1| + \frac{1}{x+1}\Big]_1^3 = \ln\frac{3}{2} - \frac{1}{4}$

27. $\displaystyle\int \frac{(x+3)\,dx}{2x(x-2)(x+2)} = -\frac{3}{4}\int\frac{dx}{2x} + \frac{5}{16}\int\frac{dx}{x-2} + \frac{1}{16}\int\frac{dx}{x+2} =$

$\displaystyle -\frac{3}{8}\ln|x| + \frac{5}{16}\ln|x-2| + \frac{1}{16}\ln|x+2| + C$

29. $\displaystyle\int_0^{\sqrt{3}} \frac{5x^2}{x^2+1} = \int_0^{\sqrt{3}}\left(5 - \frac{5}{x^2+1}\right)dx = 5x + 5\tan^{-1}x\Big]_0^{\sqrt{3}} = 5\sqrt{3} - \frac{5\pi}{3}$

31. $\displaystyle\int_{-1}^{1} \frac{x^3+x}{x^2+1}\,dx = \int_{-1}^{1} x\,dx = 0$

33. $\displaystyle\int \frac{3x^2+x+4}{x(x^2+1)}dx = \int\left(\frac{4}{x} + \frac{-x+1}{x^2+1}\right)dx = 4\ln|x| - \frac{1}{2}\ln(x^2+1) + \tan^{-1}x + C$

35. $\displaystyle\int \frac{x^3+4x^2}{x^2+4x+3}dx = \int\left(x - \frac{3x}{x^2+4x+3}\right)dx = \int\left(x + \frac{3}{2(x+1)} - \frac{9}{2(x+3)}\right)dx$

$\displaystyle \frac{x^2}{2} + \frac{3}{2}\ln|x+1| - \frac{9}{2}\ln|x+3| + C$

37. $\displaystyle\int_0^1 \frac{x^2+2x+1}{(x^2+1)^2}\,dx \int_0^1\left(\frac{1}{x^2+1} + \frac{2x}{(x^2+1)^2}\right)dx = \tan^{-1}x - \frac{1}{x^2+1}\Big]_0^1 = \frac{\pi}{4} + \frac{1}{2}$

39. $\displaystyle\int_{-1}^{0} \frac{2x\,dx}{(x^2+1)(x-1)^2} = \int_{-1}^{0}\left(\frac{-1}{x^2+1} + \frac{1}{(x-1)^2}\right)dx = -\tan^{-1}x - \frac{1}{x-1}\Big]_{-1}^{0} = \frac{1}{2} - \frac{\pi}{4}$

41. Let $u = e^t \Rightarrow du = e^t\,dt$, $t = 0 \Rightarrow u = 1$, $t = \ln 2 \Rightarrow u = 2$

$\displaystyle\int_0^{\ln 2} \frac{e^t\,dt}{e^{2t} + 3e^t + 2}dt = \int_1^2 \frac{du}{u^2+3u+2} = \int_1^2\left(\frac{1}{u+1} - \frac{1}{u+2}\right)du$

$\displaystyle \ln|u+1| - \ln|u+2|\Big]_1^2 = 2\ln 3 - 3\ln 2$

43. $\int_0^1 \frac{x^4\,dx}{(x^2+1)^2} = \int_0^1 \left(1 - \frac{2x^2+1}{(x^2+1)^2}\right) dx = x\Big]_0^1 - \int_0^1 \frac{2x^2+1}{(x^2+1)^2}\,dx$

Let $x = \tan u$, $dx = \sec^2 u\,du$, $x = 0 \Rightarrow u = 0$, $x = 1 \Rightarrow u = \frac{\pi}{4}$.

$$1 - \int_0^1 \frac{2x^2+1}{(x^2+1)^2}\,dx = 1 - \int_0^{\frac{\pi}{4}} \frac{(2\tan^2 u + 1)\sec^2 u\,du}{\sec^4 u} =$$

$$1 - \frac{3}{2}u + \frac{1}{4}\sin 2u \Big]_0^{\frac{\pi}{4}} = 1 - \frac{3\pi}{8} + \frac{1}{4} = \frac{5}{4} - \frac{3\pi}{8}$$

45. Let $u = \sqrt{x} \Rightarrow u^2 = x$ and $2u\,du = dx$. Then $\int \frac{1-\sqrt{x}}{1+\sqrt{x}}\,dx = \int \frac{2u - 2u^2}{1+u}\,du$

$$= \int \left(-2u + 4 - \frac{4}{u+1}\right) du = -u^2 + 4u - 4\ln|u+1| = -x + 4\sqrt{x} - 4\ln(\sqrt{x}+1) + C$$

47. Let $x = u^6 \Rightarrow dx = 6u^5\,du$. Then $\int \frac{dx}{\sqrt{x}+\sqrt[3]{x}} = \int \frac{6u^5\,du}{u^3+u^2} = \int \frac{6u^3\,du}{u+1} =$

$$\int \left(6u^2 - 6u + 6 - \frac{6}{y+1}\right) dy = 2u^3 - 3u^2 + 6u - 6\ln|u+1| + C =$$

$$2x^{\frac{1}{2}} - 3x^{\frac{1}{3}} + 6x^{\frac{1}{6}} - 6\ln(x^{\frac{1}{6}} + 1) + C$$

49. $\int_0^1 \ln(x^2+1)\,dx = x\ln(x^2+1) - \int \frac{2x^2\,dx}{x^2+1} = x\ln(x^2+1) - \int \left(2 - \frac{2}{x^2+1}\right) dx =$

$$x\ln(x^2+1) - 2x + 2\tan^{-1}x\Big]_0^1 = \ln 2 - 2 + \frac{\pi}{2}$$

51. $V = \pi \int_{\frac{1}{2}}^{\frac{5}{2}} \frac{9}{3x - x^2}\,dx = \pi \int_{\frac{1}{2}}^{\frac{5}{2}} \left(\frac{3}{x} + \frac{3}{3-x}\right) dx = 3\pi(\ln x - \ln|3-x|)\Big]_{\frac{1}{2}}^{\frac{5}{2}} = 6\pi\ln 5$

53. $A = \int_3^5 \frac{4x^2 + 13x - 9}{x^3 + 2x^2 - 3x}\,dx = \int_3^5 \left(\frac{3}{x} - \frac{1}{x+3} + \frac{2}{x-1}\right) dx = \ln\frac{125}{9}$

$$M_y = \int_3^5 \frac{x(4x^2 + 13x - 9)}{x^3 + 2x^2 - 3x}\,dx = \int_3^5 \left(4 + \frac{1}{x-1} - \frac{1}{x+3}\right) dx = 8 + 8\ln 2 - 3\ln 3$$

$$\bar{y} = \frac{8 + 8\ln 2 - 3\ln 3}{3\ln 5 - 2\ln 3}.$$

55. $\frac{dx}{dt} = .004\,x(1000-x) \Rightarrow \int \frac{dx}{x(1000-x)} = .004\,dt \Rightarrow$

$\int\left(\frac{1}{x} + \frac{1}{1000-x}\right)dx = 4\,dt \Rightarrow \ln \frac{x}{1000-x} = 4t + C \Rightarrow \frac{x}{1000-x} = Ce^{4t}.$

$C = \frac{2}{1000-2} = \frac{1}{499}.$

(a) $\frac{x}{1000-x} = \frac{1}{499}e^{4t}$

(b) $\frac{500}{500} = \frac{1}{499}e^{4t} \Leftrightarrow t = \frac{\ln 499}{4} \approx 1.55$ days

57. $\frac{dx}{dt} = kx(a-x) \Rightarrow \frac{dx}{x(a-x)} = k\,dt$

$\int\left(\frac{1}{ax} + \frac{1}{a(a-x)}\right)dx = \int k\,dt$

$\frac{1}{a}\ln|x| - \frac{1}{a}\ln|a-x| = kt + C$

$\ln \frac{x}{a-x} = akt + C$ or $\frac{x}{a-x} = Ce^{akt}$. Since $C = \frac{x_0}{a-x_0}$,

$\frac{x}{a-x} = \left(\frac{x_0}{a-x_0}\right)e^{akt}$

7.8 IMPROPER INTEGRALS

1. $\int_0^\infty \frac{dx}{x^2+1}$ converges, since $\lim_{b\to\infty} \int_0^b \frac{dx}{x^2+1} = \lim_{b\to\infty} \left[\tan^{-1} x\right]_0^b = \frac{\pi}{2}.$

3. $\int_{-1}^1 \frac{dx}{x^{2/3}} = \lim_{b\to 0^-} \int_{-1}^b \frac{dx}{x^{2/3}} + \lim_{b\to 0+} \int_b^1 \frac{dx}{x^{2/3}} = \lim_{b\to 0^-} 3x^{1/3}\Big]_{-1}^b + \lim_{b\to 0^+} 3x^{1/3}\Big]_b^1 = 6$

Hence, $\int_{-1}^1 \frac{dx}{x^{2/3}}$ converges.

5. $\int_0^4 \frac{dx}{\sqrt{4-x}}$ converges, since $\lim_{b\to 4^-} \int_0^b \frac{dx}{\sqrt{4-x}} = \lim_{b\to 4^-} \left[-2\sqrt{4-x}\right]_0^b = 4$

7. $\int_0^1 \frac{dx}{x^{.999}}$ converges, since $\lim_{b\to 0^+} \int_b^1 \frac{dx}{x^{.999}} = \lim_{b\to 0^+} 1000x^{.001}\Big]_b^1 = 1000.$

9. $\int_2^\infty \frac{dx}{x(x-1)}$ converges, since $\lim_{b\to\infty}\int_2^b \left(-\frac{1}{x}+\frac{1}{x-1}\right)dx$

$= \lim_{b\to\infty}\left[-\ln|x| + \ln|x-1|\right]_2^b = \ln 2$

11. $\int_1^\infty \frac{dx}{\sqrt{x}} = \lim_{b\to\infty}\int_1^b \frac{dx}{\sqrt{x}} = \lim_{b\to\infty} 2\sqrt{x}\Big]_1^b$ diverges

13. $\int_1^\infty \frac{dx}{x^3+1}$ converges, since $\frac{1}{x^3+1} \le \frac{1}{x^3}$, $x \ge 1$, and $\lim_{b\to\infty}\int_1^b \frac{dx}{x^3} = \lim_{b\to\infty} -x^{-2}\Big]_1^b = \frac{1}{2}$.

15. $\int_0^\infty \frac{dx}{x^{\frac{3}{2}}+1} = \int_0^1 \frac{dx}{x^{\frac{3}{2}}+1} + \int_1^\infty \frac{dx}{x^{\frac{3}{2}}+1}$. The first integral is finite, and the converges because $x^{\frac{3}{2}} + 1 \ge x^{\frac{3}{2}} \Rightarrow \frac{1}{x^{3/2}+1} \le \frac{1}{x^{3/2}}$ and $\int_1^\infty \frac{dx}{x^{3/2}}$ converges.

17. $\int_0^{\frac{\pi}{2}} \tan x\, dx$ diverges, since $\lim_{b\to\frac{\pi}{2}} (\ln|\sec x|)\Big]_0^b = \lim_{b\to\frac{\pi}{2}} (\ln|\sec b|) = \infty$.

19. $\int_{-1}^1 \frac{dx}{x^{2/5}} = \lim_{b\to 0^-}\int_{-1}^b \frac{dx}{x^{2/5}} + \lim_{b\to 0^+}\int_b^1 \frac{dx}{x^{2/5}} = \lim_{b\to 0^-}\frac{5}{3}x^{3/5}\Big]_{-1}^b + \lim_{b\to 0^+}\frac{5}{3}x^{3/5}\Big]_b^1 = \frac{10}{3}$

Hence, $\int_{-1}^1 \frac{dx}{x^{2/5}}$ converges.

21. $\int_2^\infty \frac{dx}{\sqrt{x-1}} = \lim_{b\to\infty}\int_2^b \frac{dx}{\sqrt{x-1}} = \lim_{b\to\infty} 2\sqrt{x-1}\Big]_2^b = \lim_{b\to\infty} 2\sqrt{b-1} - 2$, which diverges.

23. $\int_0^2 \frac{dx}{1-x^2} = \lim_{b\to 1^+}\int_0^b \frac{dx}{1-x^2} + \lim_{b\to 1^-}\int_b^2 \frac{dx}{1-x^2}$. Consider $\lim_{b\to 1^+}\int_0^b \frac{dx}{1-x^2}$

$= \lim_{b\to 1^+}\frac{1}{2}\left(\int_0^b \frac{dx}{1-x} + \int_0^b \frac{dx}{1+x}\right)$. Now $\lim_{b\to 1^+}\int_0^b \frac{dx}{1-x}$ diverges, since

$\lim_{b\to 1^+}\int_0^b \frac{dx}{1-x} = \lim_{b\to 1^+}(-\ln|1-x|)\Big]_0^b = -\lim_{b\to 1^+}\ln(1-b|)$. Thus $\int_0^2 \frac{dx}{1-x^2}$ diverges.

25. $\int_0^\infty \frac{dx}{\sqrt{x^6+1}}$ converges since $\int_0^\infty \sqrt{\frac{1}{x^6}}\,dx = \int_0^\infty \frac{1}{x^3}\,dx$ converges, and

$\lim_{x\to\infty}\sqrt{\frac{1}{x^6+1}}\cdot\sqrt{\frac{x^6}{1}} = 1$.

27. $\int_0^\infty x^2e^{-x}\,dx = \lim_{b\to\infty}\int_0^b x^2e^{-x}\,dx = \lim_{b\to\infty}\left[-e^{-x}(x^2+2x+2)\right]_0^b = 2$, so converges.

29. Since $2+\cos x \geq 1$, and $\int_\pi^\infty \frac{dx}{x}$ diverges, $\int_\pi^\infty \frac{2+\cos x}{x}dx$ diverges.

31. $\int_6^\infty \frac{1}{\sqrt{x+5}}dx = \lim_{b\to\infty}\int_6^b \frac{1}{\sqrt{x+5}}dx = \lim_{b\to\infty} 2\sqrt{x+5}\Big]_6^b$ which diverges.

33. $\int_2^\infty \frac{2\,dx}{x^2-1}$ converges, since $\lim_{b\to\infty}\int_2^b \frac{2\,dx}{x^2-1} = \lim_{b\to\infty}\int_2^b\left(\frac{1}{x-1}-\frac{1}{x+1}\right)dx$

$$= \lim_{b\to\infty}\left[\ln|x-1| - \ln|x+1|\right]_2^b = \lim_{b\to\infty}\ln\left|\frac{b-1}{b+1}\right| + \ln 3 = \ln 3.$$

35. $\int_2^\infty \frac{dx}{\ln x}$ diverges, since $\ln x < x$ for $x \geq 2 \Rightarrow \frac{1}{\ln x} > \frac{1}{x}$ and $\int_2^\infty \frac{dx}{x}$ diverges.

37. $\int_1^\infty \frac{1}{e^x - 2^x}\,dx$ converges, by limit comparison test, since

$$\lim_{x\to\infty}\frac{e^x}{e^x-2^x} = \lim_{x\to\infty}\frac{1}{1-\left(\frac{2}{e}\right)^x} = 1 \neq 0 \text{ and } \int_1^\infty \frac{dx}{e^x} \text{ converges.}$$

39. $\int_0^\infty \frac{dx}{\sqrt{x+x^4}} = \lim_{b\to0^+}\int_b^1 \frac{dx}{\sqrt{x+x^4}} + \lim_{b\to\infty}\int_1^b \frac{dx}{\sqrt{x+x^4}}.$

$\lim_{b\to0^+}\int_b^1 \frac{dx}{\sqrt{x+x^4}}$ is finite, since $\frac{1}{\sqrt{x+x^4}} \leq \frac{1}{\sqrt{x}}$ and $\int_0^1 \frac{dx}{\sqrt{x}} = 2$.

$\lim_{b\to\infty}\int_1^b \frac{dx}{\sqrt{x+x^4}}$ is finite, since $\lim_{x\to\infty}\frac{\sqrt{x+x^4}}{x^2} = \lim_{x\to\infty}\sqrt{1+\frac{1}{x^3}} = 1$

and $\int_1^\infty \frac{dx}{x^2} = 1$. Therefore, $\int_0^\infty \frac{dx}{\sqrt{x+x^4}}$ converges.

41. $\int_3^\infty e^{-3x}dx = \lim_{b\to\infty}\int_3^b e^{-3x}dx = \lim_{b\to\infty}\left[-\frac{1}{3}e^{-3x}dx\right]_3^b = \frac{1}{3}e^{-9} \approx 0.0000041.$

For $x \geq 3$, $x^2 \geq 3x \Rightarrow -x^2 \leq -3x \Rightarrow e^{(-x^2)} \leq e^{-3x}$.

So $\int_3^\infty e^{(-x^2)}dx \leq \int_3^\infty e^{-3x}dx < 0.000042$. Simpson's Rule with

$n = 6$ gives $\int_0^3 e^{(-x^2)}dx \approx 0.8862$.

43. $\int_1^\infty \frac{dx}{x^p} = \lim_{b\to\infty}\int_1^b \frac{dx}{x^p} = \lim_{b\to\infty}\left[\frac{x^{-p+1}}{1-p}\right]_1^b = \frac{1}{p-1} + \lim_{b\to\infty}\frac{b^{1-p}}{1-p}.$

If $p > 1$, $1 - p < 0$ and $b^{1-p} \to 0$ as $b \to \infty$. If $p < 1$, $1 - p > 0$ and $b^{1-p} \to \infty$ as $b \to \infty$.

45. $A = \int_0^\infty e^{-x}\,dx = \lim_{b\to\infty}\int_0^b e^{-x}\,dx = \lim_{b\to\infty}[e^{-x}\,dx]_0^b = 1$

47. $V = 2\pi\int_0^\infty xe^{-x}\,dx = \lim_{b\to\infty} 2\pi\int_0^b xe^{-x}\,dx = \lim_{b\to\infty} 2\pi\,[xe^{-x} - e^{-x}]_0^b = 2\pi$

49. $A = 2\int_0^1 \frac{dx}{\sqrt{1-x^2}} = \lim_{b\to1^-} 2\int_0^b \frac{dx}{\sqrt{1-x^2}} = 2\lim_{b\to1^-}[\sin^{-1}x]_0^b = 2\left(\frac{\pi}{2}\right) = \pi.$

$\bar{y} = 0$ by symmetry. $\bar{x} = \frac{1}{\pi}\int_0^1 \frac{2x\,dx}{\sqrt{1-x^2}} = \frac{1}{\pi}\lim_{b\to1^-}\int_0^b \frac{2x\,dx}{\sqrt{1-x^2}} =$

$\frac{1}{\pi}\lim_{b\to1^-}\left[-2\sqrt{1-x^2}\right]_0^b = \frac{2}{\pi}$. $(\bar{x}, \bar{y}) = \left(\frac{2}{\pi}, 0\right)$.

51. $A = 2\int_1^\infty \frac{1}{1+x^2}dx = 2\lim_{b\to\infty}\int_1^b \frac{1}{1+x^2}dx = 2\lim_{b\to\infty}\tan^{-1}x\Big]_1^b = 2\left(\frac{\pi}{2}\right) = \pi,$

which is the area of a disk with radius $= 1$.

7.9 USING INTEGRAL TABLES

1. $\int_0^\infty e^{(-x^2)}\,dx = \frac{\sqrt{\pi}}{2}$ (Formula 140)

3. $\displaystyle\int_6^9 \frac{dx}{x\sqrt{x-3}} = \frac{2}{\sqrt{3}}\left[\tan^{-1}\sqrt{\frac{x-3}{3}}\right]_6^9 = \frac{2}{\sqrt{3}}\left(\tan^{-1}\sqrt{2} - \frac{\pi}{4}\right)$ (Formula 13a)

5. $\displaystyle\int \frac{dx}{(9-x^2)^2} = \frac{x}{18(9-x^2)} + \frac{1}{18}\int \frac{dx}{9-x^2}$ (Formula 19)

$\displaystyle = \frac{x}{18(9-x^2)} + \frac{1}{108}\ln\left|\frac{x+3}{x-3}\right| + C$ (Formula 18)

7. $\displaystyle\int_3^{11} \frac{dx}{x^2\sqrt{7+x^2}} = -\frac{\sqrt{7+x^2}}{7x}\Bigg]_3^{11} = \frac{44-24\sqrt{2}}{231}$ (Formula 27)

9. $\displaystyle\int_{-2}^{-\sqrt{2}} \frac{\sqrt{x^2-2}}{x}\,dx = \sqrt{x^2-2} - \sqrt{2}\sec^{-1}\left|\frac{x}{\sqrt{2}}\right|\Bigg]_{-2}^{-\sqrt{2}} = \sqrt{2}\left(\frac{\pi}{4} - 1\right)$ (Formula 42)

11. $\displaystyle\int \frac{dx}{4+5\sin 2x} = -\frac{1}{6}\ln\left|\frac{5+4\sin 2x+3\cos 2x}{4+5\sin 2x}\right| + C$ (Formula 71)

13. $\displaystyle\int x\sqrt{2x-3}\,dx = \frac{(2x-3)^{3/2}}{4}\left[\frac{2}{5}(2x-3) + \frac{2}{9}\right] + C = \frac{(2x-3)^{3/2}(x+1)}{5} + C$

15. $\displaystyle\int_0^{\infty} x^{10}e^{-x}\,dx = \Gamma(11) = 10!$ (Formula 139)

17. $\displaystyle\int_0^1 \sin^{-1}\sqrt{x}\,dx = 2\int_0^1 u\sin^{-1}u\,du = 2\left(\frac{u^2}{2}\sin^{-1}u - \frac{1}{2}\int_0^1 \frac{u^2\,du}{\sqrt{1-u^2}}\right)$ (Formula 99)

$\displaystyle = \left[u^2\sin^{-1}u - \left(\left[\frac{1}{2}\sin^{-1}u - \frac{1}{2}u\sqrt{1-u^2}\right]\right)\right]_0^1 = \frac{\pi}{4}$ (Formula 33)

19. $\displaystyle\int_0^{\frac{1}{2}} \frac{\sqrt{x}}{\sqrt{1-x}}\,dx = \int_0^{\frac{1}{\sqrt{2}}} \frac{2u^2\,du}{\sqrt{1-u^2}} = 2\left[\frac{1}{2}\sin^{-1}u - \frac{1}{2}u\sqrt{1-u^2}\right]_0^{\frac{1}{\sqrt{2}}} = \frac{\pi}{4} - \frac{1}{2}$

21. $\displaystyle A = \int_0^3 \frac{dx}{\sqrt{x+1}} = 2\sqrt{x+1}\Bigg]_0^3 = 2$

$\displaystyle \bar{x} = \frac{1}{2}\int_0^3 \frac{x}{\sqrt{x+1}}\,dx = \frac{1}{2}\left(\sqrt{x+1}\left(\frac{2}{3}(x+1) - 2\right)\Bigg]_0^3\right) = \frac{4}{3}$

$\displaystyle \bar{y} = \frac{1}{2}\int_0^3 \frac{1}{2\sqrt{x+1}}\cdot\frac{1}{\sqrt{x+1}}\,dx = \frac{1}{4}\int_0^3 \frac{dx}{x+1} = \frac{1}{4}\ln(x+1)\Bigg]_0^3 = \frac{1}{4}\ln 4 = \frac{1}{2}\ln 2$

23. $$\frac{d}{dx}\left(-\frac{1}{a}\sqrt{\frac{2a-x}{x}}\right) = -\frac{1}{a}\left[\frac{1}{2}\left(\frac{2a-x}{x}\right)^{-1/2}\right]\cdot\frac{x(-1)-(2a-x)}{x^2} =$$

$$-\frac{1}{2a}\sqrt{\frac{x}{2a-x}}\left(\frac{-2a}{x^2}\right) = \sqrt{\frac{x}{x^4(2a-x)}} = \frac{1}{x\sqrt{2ax-x^2}}.$$

Therefore, $\displaystyle\int \frac{1}{x\sqrt{2ax-x^2}}\,dx = -\frac{1}{a}\sqrt{\frac{2a-x}{x}} + C.$

25. $$\int x(ax+b)^{-2}\,dx = \int\left(\frac{u-b}{a}\cdot\frac{1}{u^2}\cdot\frac{1}{a}\right)du = \frac{1}{a^2}\int\frac{u-b}{u^2}du$$

$$= \frac{1}{a^2}(\ln|u|\ \ bu^{-1}) + C = \frac{1}{a^2}\left[\ln|ax+b| + \frac{b}{ax+b}\right] + C$$

7.10 REDUCTION FORMULAS

1. $$\int_{-\pi}^{\pi}\cos^4 x\,dx = \frac{\cos^3 x\sin x}{4} + \frac{3}{4}\left(\frac{\cos x\sin x}{2} + \frac{x}{2}\right)\Bigg]_{-\pi}^{\pi} = \frac{3\pi}{4}$$

3. $$\int\cos^6 x\,dx = \frac{\cos^5 x\sin x}{6} + \frac{5}{6}\int\cos^4 x\,dx =$$

$$\frac{\cos^5 x\sin x}{6} + \frac{5}{6}\left[\frac{\cos^3 x\sin x}{4} + \frac{3}{4}\left(\frac{\cos x\sin x}{2} + \frac{x}{2}\right)\right] + C =$$

$$\frac{1}{6}\cos^5 x\sin x + \frac{5}{24}\cos^3 x\sin x + \frac{5}{16}(\cos x\sin x + x) + C$$

5. $$\int_0^{\pi}\sin^4 x\,dx = -\frac{\sin^3 x\cos x}{4} + \frac{3}{4}\int_0^{\pi}\sin^2 x\,dx =$$

$$-\frac{\sin^3 x\cos x}{4} + \frac{3}{4}\left[-\frac{\sin x\cos x}{2} + \frac{x}{2}\right]_0^{\pi} = \frac{3\pi}{8}$$

7. $$\int_0^{\frac{\pi}{2}}\sin^5 x\,dx = -\frac{\sin^4 x\cos x}{5} + \frac{4}{5}\int_0^{\frac{\pi}{2}}\sin^3 x\,dx =$$

$$-\frac{\sin^4 x\cos x}{5} + \frac{4}{5}\left[-\frac{\sin^2 x\cos x}{3} + \frac{2}{3}\int_0^{\frac{\pi}{2}}\sin x\,dx\right] =$$

$$-\frac{\sin^4 x\cos x}{5} + \frac{4}{5}\left[-\frac{\sin^2 x\cos x}{3} - \frac{2}{3}\cos x\right]_0^{\frac{\pi}{2}} = \frac{8}{15}$$

9. $$\int\tan^3 2x\,dx = \frac{\tan^2 2x}{4} - \int\tan 2x\,dx = \frac{\tan^2 2x}{4} + \frac{1}{2}\ln|\cos 2x| + C$$

11. $$\int \tan^5 x\,dx = \frac{\tan^4 x}{4} - \int \tan^3 x\,dx = \frac{\tan^4 x}{4} - \left[\frac{\tan^2 x}{2} - \int \tan x\,dx\right]$$

$$= \frac{\tan^4 x}{4} - \frac{\tan^2 x}{2} + \ln|\cos x| + C$$

13. $$\int \cot^3 x\,dx = -\frac{\cot^2 x}{2} - \int \cot x\,dx = -\frac{\cot^2 x}{2} - \ln|\sin x| + C$$

15. $$\int_{\frac{\pi}{4}}^{\frac{3\pi}{4}} \cot^4 x\,dx = -\frac{\cot^3 x}{3} - \int_{\frac{\pi}{4}}^{\frac{3\pi}{4}} \cot^2 x\,dx = -\frac{\cot^3 x}{3} + \cot x + x\Big]_{\frac{\pi}{4}}^{\frac{3\pi}{4}} = \frac{\pi}{2} - \frac{4}{3}$$

17. $$\int_{-\frac{\pi}{3}}^{\frac{\pi}{3}} \sec^4 x\,dx = \frac{\sec^2 x \tan x}{3} + \frac{2}{3}[\tan x\Big]_{-\frac{\pi}{3}}^{\frac{\pi}{3}} = 4\sqrt{3}$$

19. $$\int \sec^5 x\,dx = \frac{\sec^3 x \tan x}{4} + \frac{3}{4}\int \sec^3 x\,dx =$$

$$\frac{\sec^3 x \tan x}{4} + \frac{3}{4}\left[\frac{\sec x \tan x}{2} + \frac{1}{2}\int \sec x\,dx\right] =$$

$$\frac{\sec^3 x \tan x}{4} + \frac{3}{8}\sec x \tan x + \frac{3}{8}\ln|\sec x + \tan x| + C$$

21. $$\int_{\frac{\pi}{4}}^{\frac{3\pi}{4}} \csc^3 x\,dx = -\frac{1}{2}\csc x \cot x - \frac{1}{2}\ln|\csc x + \cot x|\Big]_{\frac{\pi}{4}}^{\frac{3\pi}{4}} = \sqrt{2} + \frac{1}{2}\ln\frac{\sqrt{2}+1}{\sqrt{2}-1}$$

23. $$\int \csc^4 x\,dx = -\frac{\csc^2 x \cot x}{3} - \frac{2}{3}\cot x + C$$

25. $$\int_0^1 (x^2+1)^{-3/2}\,dx = \int_0^{\frac{\pi}{4}} (\tan^2 u + 1)^{-3/2} \sec^2 u\,du = \sin u\Big]_0^{\frac{\pi}{4}} = \frac{1}{\sqrt{2}}$$

27. $$\int_0^{\frac{3}{5}} \frac{dx}{(1-x^2)^3} = \int_0^{\sin^{-1}\frac{3}{5}} \frac{\cos u\,du}{(1-\sin^2 u)^3} = \int_0^{\sin^{-1}\frac{3}{5}} \sec^5 u\,du =$$

$$\frac{1}{4}\sec^3 u \tan u + \frac{3}{8}\sec u \tan u + \frac{3}{8}\ln|\sec u + \tan u|\Big]_0^{\sin^{-1}\frac{3}{5}} = \frac{735}{1024} + \frac{3}{8}\ln 2$$

29. $$\frac{d}{dx}\left(-\frac{\sin^{n-1}x\cos x}{n}\right) = -\frac{1}{n}\sin^{n-1}x\,(-\sin x) - \left(\frac{1}{n}\cos x\right)(n-1)(\sin^{n-2}x)(\cos x)$$

$$= \frac{1}{n}\sin^n x - \frac{n-1}{n}\sin^{n-2}x\cos^2 x$$

$$\frac{d}{dx}\left(\frac{n-1}{n}\int \sin^{n-2}x\,dx\right) = \frac{n-1}{n}\sin^{n-2}x$$

$$\frac{1}{n}\sin^n x - \frac{n-1}{n}\sin^{n-2}x\cos^2 x + \frac{n-1}{n}\sin^{n-2}x =$$

$$\frac{1}{n}\sin^n x + \frac{n-1}{n}\sin^n x\left(\frac{1-\cos^2 x}{\sin^2 x}\right) = \sin^n x\left(\frac{1}{n} + \frac{n-1}{n}\right) = \sin^n x$$

31. (a) Let $u = x^n \Rightarrow du = nx^{n-1}\,dx$, $dv = \sin x\,dx \Rightarrow v = -\cos x$

$$\int x^n \sin x\,dx = -x^n\cos x - \int nx^{n-1}(-\cos x)\,dx$$

$$= -x^n\cos x + n\int x^{n-1}\cos x\,dx$$

(b) $$\int x^n \sin ax\,dx = -\frac{x^n}{a}\cos ax + \frac{n}{a}\int x^{n-1}\cos ax\,dx$$

33. (a) Let $dv = x^m\,dx \Rightarrow v = \frac{1}{m+1}x^{m+1}$, $v = (\ln x)^n \Rightarrow dv = \frac{n(\ln x)^{n-1}}{x}dx$

$$\int x^m(\ln x)^n\,dx = \frac{1}{m+1}x^{m+1}(\ln x)^n - \int \frac{1}{m+1}x^{m+1}\left(\frac{n(\ln x)^{n-1}}{x}\right)dx$$

$$= \frac{x^{m+1}(\ln x)^n}{m+1} - \frac{n}{m+1}\int x^m(\ln x)^{n-1}\,dx$$

(b) $$\int_1^e x^3(\ln x)^2\,dx = \frac{x^4(\ln x)^2}{4} - \frac{1}{2}\int x^3\ln x\,dx$$

$$= \frac{x^4(\ln x)^2}{4} - \frac{1}{2}\left[\frac{x^4\ln x}{4} - \frac{1}{4}\int x^3 dx\right]$$

$$= \frac{x^4(\ln x)^2}{4} - \frac{x^4\ln x}{8} + \frac{x^4}{32}\Big]_1^e = \frac{e^4-1}{16}$$

7.M MISCELLANEOUS PROBLEMS

1. $$\int \frac{\cos x\,dx}{\sqrt{1+\sin x}}\,dx = 2\sqrt{1+\sin x} + C$$

3. $$\int \tan x\sec^2 x\,dx = \frac{1}{2}\tan^2 x + C$$

5. $\int_0^9 e^{\ln\sqrt{x}}\,dx = \int_0^9 \sqrt{x}\,dx = \frac{2}{3}x^{3/2}\Big]_0^9 = 18$

7. Let $\tan u = x + 1 \Rightarrow \sec^2 u\,du = dx$. Then $\int \frac{dx}{\sqrt{x^2 + 2x + 2}} = \int \frac{dx}{\sqrt{(x+1)^2 + 1}} =$

$$\int \frac{\sec^2 u\,du}{\sqrt{\tan^2 u + 1}} = \int \sec u\,du = \ln|\sec u + \tan u| + C =$$

$$\ln|\sqrt{x^2 + 2x + 2} + x + 1| + C$$

9. $\int x^2 e^x\,dx = e^x(x^2 - 2x + 2) + C$

$$\begin{array}{lcl} x^2 & + & e^x \\ 2x & - & e^x \\ 2 & + & e^x \\ 0 & & e^x \end{array}$$

11. Let $u = e^t \Rightarrow du = e^t\,dt$. Then $\int \frac{e^t\,dt}{1 + e^{2t}} = \int \frac{du}{1 + u^2} = \tan^{-1} u = \tan^{-1}(e^t) + C$

13. Let $u^2 = x \Rightarrow 2u\,du = dx$.

Then $\int \frac{dx}{(x+1)\sqrt{x}} = \int \frac{2u\,du}{u(u^2 + 1)} = 2\tan^{-1} u = 2\tan^{-1}\sqrt{x} + C$.

15. Let $u = t^{\frac{5}{3}} + 1 \Rightarrow du = \frac{5}{3}t^{\frac{2}{3}}\,dt \Rightarrow t^{\frac{2}{3}}\,dt = \frac{3}{5}\,du$

$$\int t^{\frac{2}{3}}(t^{\frac{5}{3}} + 1)^{\frac{2}{3}}\,dt = \frac{3}{5}\int u^{\frac{2}{3}}\,du = \frac{3}{5}\left(\frac{3}{5}u^{\frac{5}{3}}\right) + C = \frac{9}{25}(t^{\frac{5}{3}} + 1)^{\frac{5}{3}} + C$$

17. Let $u^2 = e^t + 1 \Rightarrow 2u\,du = e^t\,dt \Rightarrow dt = \frac{2u\,du}{e^t} = \frac{2u\,du}{u^2 - 1}$.

$$\int \frac{dt}{\sqrt{e^t + 1}} = \int \frac{2u\,du}{u(u^2 - 1)} = \int\left(\frac{1}{u-1} - \frac{1}{u+1}\right)du = \ln|u - 1| - \ln|u + 1| + C$$

$$= \ln|\sqrt{e^t + 1} - 1| - \ln|\sqrt{e^t + 1} + 1| + C = \ln\left|\frac{\sqrt{e^t + 1} - 1}{\sqrt{e^t + 1} + 1}\right| + C$$

19. $\int_0^{\frac{\pi}{2}} \frac{\cos x\,dx}{1 + \sin^2 x} = \tan^{-1}(\sin x)\Big]_0^{\frac{\pi}{2}} = \frac{\pi}{4}$

21. $\displaystyle\int_0^{\frac{\pi}{2}} \frac{\sin x\,dx}{1+\cos^2 x} = -\tan^{-1}(\cos x)\Big]_0^{\frac{\pi}{2}} = \frac{\pi}{4}$

23. $\displaystyle\int_0^{\frac{\pi}{3}} \frac{dx}{\sin x \cos x} = \int_0^{\frac{\pi}{3}} \frac{2\,dx}{\sin 2x} = \int_0^{\frac{\pi}{3}} 2\csc 2x = -\ln|\csc 2x + \cot 2x|\Big]_0^{\frac{\pi}{3}} = \ln\sqrt{3}$

25. $\displaystyle\int_{-\frac{\pi}{2}}^{0} \sqrt{1-\sin x}\,dx = \int_{-\frac{\pi}{2}}^{0} \frac{\sqrt{1-\sin x}}{1}\cdot\frac{\sqrt{1+\sin x}}{\sqrt{1+\sin x}} = \int_{-\frac{\pi}{2}}^{0} \frac{\cos x\,dx}{\sqrt{1+\sin x}}$

$$= 2\sqrt{1+\sin x}\,\Big]_{-\frac{\pi}{2}}^{0} = 2$$

27. Let $u = \sqrt[3]{1+e^x} \Rightarrow e^x = u^3 - 1 \Rightarrow e^x\,dx = 3u^2\,du$. Then

$$\int \frac{e^{2x}\,dx}{\sqrt[3]{1+e^x}} = \int \frac{(u^3-1)\,3u^2\,du}{u} = \int (3u^4 - 3u)\,du = \frac{3}{5}u^5 - \frac{3}{2}u^2 + C$$

$$= \frac{3}{5}(1+e^x)^{5/3} - \frac{3}{2}(1+e^x)^{2/3} + C$$

29. Let $x = \sec u \Rightarrow dx = \sec u \tan u\,du$

$$\int_{\frac{5}{4}}^{\frac{5}{3}} \frac{dx}{(x^2-1)^{3/2}} = \int_{x=\frac{5}{4}}^{x=\frac{5}{3}} \frac{\sec u \tan u\,du}{(\tan^2 u)^{3/2}} = \int_{x=\frac{5}{4}}^{x=\frac{5}{3}} \sin^{-2} u \cos u\,du$$

$$= -\csc u = -\frac{x}{\sqrt{x^2-1}}\Bigg]_{\frac{5}{4}}^{\frac{5}{3}} = \frac{5}{12}$$

31. Let $u = 2 + \ln x \Rightarrow du = \dfrac{dx}{x}$. Then

$$\int \frac{dx}{x(2+\ln x)} = \int \frac{du}{u} = \ln|u| + C = \ln|2+\ln x| + C$$

33. Let $u = \sqrt[4]{e^x+1} \Rightarrow u^4 = e^x + 1 \Rightarrow 4u^3\,du = e^x\,dx$,

$x = 0 \Rightarrow u = \sqrt[4]{2}$, $x = \ln 2 \Rightarrow u = \sqrt[4]{3}$.

$$\int_0^{\ln 2} \frac{e^{2x}\,dx}{\sqrt[4]{e^x+1}} = \int_{\sqrt[4]{2}}^{\sqrt[4]{3}} \frac{(u^4-1)\,4u^3\,du}{u} = \int_{\sqrt[4]{2}}^{\sqrt[4]{3}} (4u^6 - 4u^2)\,du$$

$$\frac{4}{7}u^7 - \frac{4}{3}u^3\Big]_{\sqrt[4]{2}}^{\sqrt[4]{3}} = \frac{4}{21}(\sqrt[4]{8} - 2\sqrt[4]{27})$$

35. $\int_0^3 (16+x^2)^{-3/2}\,dx = \int_0^{\tan^{-1}\frac{3}{4}} (16+16\tan^2 u)^{-3/2}\ 4\sec^2 u\,du =$

$= \frac{1}{16}\int_0^{\tan^{-1}\frac{3}{4}} \cos u\,du = \frac{1}{16}\sin u\Big]_0^{\tan^{-1}\frac{3}{4}} = \frac{3}{80}$

37. Let $u^2 = x+1 \Rightarrow 2u\,du = dx$. Then $\int \sin\sqrt{x+1}\,dx = \int 2u\sin u\,du$

$= -2u\cos u + 2\sin u + C = -2\sqrt{x+1}\cos\sqrt{x+1} + 2\sin\sqrt{x+1} + C$

39. $\int_0^1 \frac{dx}{4-x^2} = \frac{1}{4}\int_0^1 \frac{dx}{2+x} + \frac{1}{4}\int_0^1 \frac{dx}{2-x} = \frac{1}{4}\ln|2+x| - \frac{1}{4}\ln|2-x| = \frac{1}{4}\ln 3$

41. $\frac{1}{y(2y^3+1)^2} = \frac{A}{y} + \frac{By^2+Cy+D}{2y^3+1} + \frac{Ey^2+Fy+G}{(2y^3+1)^2}$

$A(2y^3+1)^2 + (By^2+Cy+D)(2y^3+1)y + (Ey^2+Fy+G)y = 1$

$(4A+2B)y^6 + 2Cy^5 + 2Dy^4 + (4A+B+E)y^3 + (C+F)y^2 + (D+G)y + A = 1$

$A = 1 \Rightarrow B = -2 \Rightarrow E = -2$. $C = D = F = G = 0$.

$\int \frac{dy}{y(2y^3+1)^2} = \int\left(\frac{1}{y} - \frac{2y^2}{2y^3+1} - \frac{y^2}{(2y^3+1)^2}\right)dy$

$= \ln|y| - \frac{1}{3}\ln|2y^3+1| + \frac{1}{3}(2y^3+1)^{-1} + C$

43. $\int \frac{dx}{x(x^2+1)^2} = \int \frac{\sec^2 u\,du}{\tan u(\sec^2 u)^2} = \int \frac{\cos^3 u}{\sin u}\,du$

$= \int \frac{\cos u\,du}{\sin u} - \int \sin u\cos u\,du = \ln|\sin u| + \frac{1}{2}\cos^2 u + C$

$= \ln\left|\frac{x}{\sqrt{x^2+1}}\right| + \frac{1}{2}(x^2+1)^{-1} + C$

45. Let $u = e^x \Rightarrow du = e^x\,dx \Rightarrow dx = \frac{du}{e^x} = \frac{du}{u}$. Then $\int \frac{dx}{e^x-1} = \int \frac{du}{u(u-1)} =$

$\int\left(-\frac{1}{u} + \frac{1}{u-1}\right)du = -\ln|u| + \ln|u-1| + C = \ln|e^x-1| - x + C$

47. $\int \frac{x\,dx}{x^2+4x+3} = \frac{3}{2}\int \frac{dx}{x+3} - \frac{1}{2}\int \frac{dx}{x+1} = \frac{3}{2}\ln|x+3| - \frac{1}{2}\ln|x+1| + C$

49. $\int \frac{4\,dx}{x^3+4x} = \int \frac{1}{x}\,dx - \int \frac{x}{x^2+4}\,dx = \ln|x| - \frac{1}{2}\ln(x^2+4) + C = \ln\left|\frac{x}{\sqrt{x^2+4}}\right| + C$

51. $\displaystyle\int \frac{\sqrt{x^2-1}}{x}\,dx = \int \frac{\sqrt{\sec^2 u - 1}\ \sec u \tan u\, du}{\sec u} = \int (\sec^2 u - 1)\,du$

$\displaystyle = \tan u - u + C = \sqrt{x^2-1} - \tan^{-1}\sqrt{x^2-1} + C$

53. Let $u = \sqrt{x} \Rightarrow u^2 = x \Rightarrow 2u\,du = dx$. Then $\displaystyle\int \frac{dx}{x(3\sqrt{x}+1)} =$

$\displaystyle\int \frac{2u\,du}{u^2(3u+1)} = 2\int \frac{du}{u} - 6\int \frac{du}{3u+1} = 2\ln\sqrt{x} - 2\ln(3\sqrt{x}+1) + C$

55. Let $\tan u = \sin\theta \Rightarrow \sec^2 u\,du = \cos\theta\,d\theta$, $\theta = \frac{\pi}{6} \Rightarrow u = \tan^{-1}\frac{1}{2}$, $\theta = \frac{\pi}{2} \Rightarrow u = \frac{\pi}{4}$

$\displaystyle\int_{\pi/6}^{\pi/2} \frac{\cot\theta\,d\theta}{1+\sin^2\theta} = \int_{\pi/6}^{\pi/2} \frac{\cos\theta\,d\theta}{\sin\theta(1+\sin^2\theta)} = \int_{\tan^{-1}\frac{1}{2}}^{\pi/4} \frac{\sec^2 u\,du}{\tan u \sec^2 u}$

$\displaystyle = \ln|\sin u|\Big]_{\tan^{-1}\frac{1}{2}}^{\pi/4} = \ln\frac{\sqrt{5}}{2} - \ln\frac{\sqrt{2}}{2} = \frac{1}{2}\ln\frac{5}{2}$

57. Let $u^3 = 1 + e^{2t} \Rightarrow 3u^2\,du = 2e^{2t}\,dt$

$\displaystyle\int \frac{e^{4t}\,dt}{(1+e^{2t})^{2/3}} = \frac{3}{2}\int \frac{(u^3-1)u^2\,du}{(u^3)^{2/3}} = \frac{3}{2}\int (u^3-1)\,du = \frac{3}{8}u^4 - \frac{3}{2}u + C$

$\displaystyle = \frac{3}{8}(1+e^{2t})^{4/3} - \frac{3}{2}(1+e^{2t})^{1/3} + C$

$\displaystyle = \frac{3}{8}(1+e^{2t})^{1/3}(e^{2t}-3) + C$

59. $\displaystyle\int \frac{(x^3+x^2)\,dx}{x^2+x-2} = \int\left(x + \frac{2x}{x^2+x-2}\right)dx = \int x\,dx + \frac{4}{3}\int\frac{dx}{x+2} + \frac{2}{3}\int\frac{dx}{x-1}$

$\displaystyle = \frac{1}{2}x^2 + \frac{4}{3}\ln|x+2| + \frac{2}{3}\ln|x-1| + C$

61. $\displaystyle\int \frac{x\,dx}{(x-1)^2} = \int\frac{dx}{x-1} + \int\frac{dx}{(x-1)^2} = \ln|x-1| - \frac{1}{x-1} + C$

63. Let $\frac{1}{2}\sec u = y + \frac{1}{2} \Rightarrow \sec u = 2y+1$ and $\frac{1}{2}\sec u \tan u\,du = dy$. Then

$\displaystyle\int \frac{dy}{(2y+1)\sqrt{y^2+y}} = \int \frac{dy}{(2y+1)\sqrt{\left(y+\frac{1}{2}\right)^2 - \frac{1}{4}}} = \int \frac{\frac{1}{2}\sec u \tan u\,du}{\frac{1}{2}\sec u \tan u}$

$\displaystyle = u + C = \sec^{-1}(2y+1) + C \text{ or } \tan^{-1}\left(2\sqrt{y^2+y}\right) + C$

65. Let $x = \sin u \Rightarrow dx = \cos u\, du$. Then $\int (1 - x^2)^{3/2} = \int \cos^4 u\, du$

$$= \frac{1}{4}\int (1 + \cos 2u)^2\, du = \frac{1}{4}\int \left(1 + 2\cos 2u + \frac{1}{2} + \frac{1}{2}\cos 4u\right) du$$

$$= \frac{3}{8}u + \frac{1}{4}\sin 2u + \frac{1}{32}\sin 4u$$

$$= \frac{3}{8}u + \frac{1}{2}\sin u \cos u + \frac{1}{8}\sin u \cos u\,(\cos^2 u - \sin^2 u)$$

$$= \frac{3}{8}\sin^{-1} x + \frac{1}{2}x\sqrt{1 - x^2} + \frac{1}{8}x\sqrt{1 - x^2}\,(1 - 2x^2) + C$$

67. $\int x \tan^2 x\, dx = \int x \sec^2 x\, dx - \int x\, dx = x \tan x + \ln|\cos x| - \frac{1}{2}x^2 + C$

$x \xrightarrow{+} \sec^2 x$
$1 \xrightarrow{-} \tan x$
$0 \quad -\ln|\cos x|$

69. $\int_0^{\pi} x^2 \sin x\, dx = -x^2 \cos x + 2x \sin x + 2\cos x \Big]_0^{\pi} = \pi^2 - 4$

$x^2 \xrightarrow{+} \sin x$
$2x \xrightarrow{-} -\cos x$
$2 \xrightarrow{+} -\sin x$
$0 \quad \cos x$

71. $\int_0^1 \frac{dt}{t^4 + 4t^2 + 3} = \frac{1}{2}\int_0^1 \frac{dt}{t^2 + 1} - \frac{1}{2}\int_0^1 \frac{dt}{t^2 + 3} = \frac{1}{2}\tan^{-1} t - \frac{\sqrt{3}}{6}\tan^{-1}\frac{t}{\sqrt{3}}\Big]_0^1 = \frac{(9 - 2\sqrt{3})\pi}{72}$

$$\frac{1}{(t^2 + 1)(t^2 + 3)} = \frac{At + B}{t^2 + 1} + \frac{Ct + D}{t^2 + 3}$$

$$At^3 + Bt^2 + 3At + 3B + Ct^3 + Dt^2 + Ct + D = 1$$

$$(A + C)t^3 + (B + D)t^2 + (3A + C)t + (3B + D) = 1$$

$$A = C = 0,\ B = -\frac{1}{2},\ D = \frac{1}{2}$$

73. Let $u = \ln(x + 2) \Rightarrow du = \frac{dx}{x + 2}$, $dv = x\, dx \Rightarrow v = \frac{1}{2}x^2$. Then

$$\int x \ln\sqrt{x + 2}\, dx = \frac{1}{2}\left(\frac{1}{2}x^2 \ln(x + 2) - \frac{1}{2}\int_0^2 \frac{x^2\, dx}{x + 2}\right)$$

$$= \frac{1}{4}x^2 \ln(x + 2) - \frac{1}{4}\int \left(x - 2 + \frac{4}{x + 2}\right) dx$$

$$= \frac{1}{4}x^2 \ln(x + 2) - \frac{1}{8}x^2 + \frac{1}{2}x - \ln|x + 2| + C.$$ Therefore

$$\int_0^2 x \ln\sqrt{x + 2}\, dx = \frac{1}{4}x^2 \ln(x + 2) - \frac{1}{8}x^2 + \frac{1}{2}x - \ln|x + 2|\Big]_0^2 = \frac{1}{2} + \ln 2$$

75. Let $u = \sec^{-1} x \Rightarrow du = \dfrac{dx}{x\sqrt{x^2 - 1}}$, $dv = dx \Rightarrow v = x$. Then

$$\int \sec^{-1} x\, dx = x \sec^{-1} x - \int \frac{dx}{\sqrt{x^2-1}} = x\sec^{-1} x - \int \frac{\sec u \tan u\, du}{\tan u}$$

$$= x\sec^{-1} x - \ln|x + \sqrt{x^2 - 1}| + C$$

77. $$\int \frac{x\, dx}{x^4 - 16} = \frac{1}{16}\ln|x-2| + \frac{1}{16}\ln|x+2| - \frac{1}{16}\ln(x^2+4) + C = \frac{1}{16}\ln\left|\frac{x^2-4}{x^2+4}\right| + C$$

$$\frac{x}{x^4 - 16} = \frac{A}{x-2} + \frac{B}{x+2} + \frac{Cx + D}{x^2+4}$$

$$A(x+2)(x^2+2) + B(x-2)(x^2+2) + (Cx+D)(x^2-4) = x$$

$$(A+B+C)x^3 + (2A - 2B + D)x^2 + (4A + 4B - 4C)x + (8A - 8B - 4D) = x$$

$$A = B = \frac{1}{16},\ C = -\frac{1}{8},\ D = 0$$

79. $$\int \frac{\cos x\, dx}{\sin^3 x - \sin x} = \int \frac{\cos x\, dx}{\sin x(-\cos^2 x)} = \int \frac{-2\, dx}{\sin 2x} = 2\ln|\csc 2x + \cot 2x| + C$$

81. Let $u^2 = x \Rightarrow 2u\, du = dx$, $x = 0 \Rightarrow u = 0$, $x = 1 \Rightarrow u = 1$

$$\int_0^1 \frac{x\, dx}{1\ \sqrt{x} + x} = \int_0^1 \frac{2u^3\, du}{u^2 + u + 1} = 2\int_0^1 \left(u - 1 + \frac{1}{u^2+u+1}\right)dt$$

$$= u^2 - 2u\Big]_0^1 + \int_0^1 \frac{du}{\left(u + \frac{1}{2}\right)^2 + \frac{3}{4}}$$

$$= u^2 - 2u + \frac{4}{\sqrt{3}}\tan^{-1}\left(\frac{2u+1}{\sqrt{3}}\right)\Bigg]_0^1 = \frac{2\pi\sqrt{3} - 9}{9}$$

83. Let $u = \tan t \Rightarrow du = \sec^2 t\, dt$. Then

$$\int \frac{dt}{\sec^2 t + \tan^2 t} = \int \frac{dt}{2\tan^2 t + 1} = \int \frac{dt}{2\tan^2 t + 1}\cdot\frac{\sec^2 t}{1 + \tan^2 t}$$

$$= \int \frac{du}{(2u^2+1)(u^2+1)} = 2\int \frac{du}{2u^2+1} - \int \frac{du}{u^2+1}$$

$$= \sqrt{2}\tan^{-1}(\sqrt{2}u) - \tan^{-1} u = \sqrt{2}\tan^{-1}(\sqrt{2}\tan t) - t + C$$

85. Let $u = e^t \Rightarrow du = e^t dt$ and $dv = e^t \cos(e^t)\, dt \Rightarrow v = \sin(e^t)$. Then

$$\int e^{2t}\cos(e^t)\, dt = e^t \sin(e^t) - \int e^t \sin(e^t)\, dt$$

$$= e^t \sin(e^t) + \cos(e^t) + C$$

87. Let $u = \ln(x^3 + x) \Rightarrow du = \dfrac{3x^2+1}{x^3+x}\,dx \quad dv = x\,dx \Rightarrow v = \dfrac{1}{2}x^2$. Then

$$\int x\ln(x^3+x)\,dx = \frac{1}{2}x^2\ln(x^3+x) - \int \frac{1}{2}x^2\cdot\frac{3x^2+1}{x^3+x}\,dx$$
$$= \frac{1}{2}x^2\ln(x^3+x) - \frac{1}{2}\int\left(3x - \frac{2x}{x^2+1}\right)dx$$
$$= \frac{1}{2}x^2\ln(x^3+x) - \frac{3}{4}x^2 + \frac{1}{2}\ln(x^2+1) + C$$

89. Let $\sin x = \sqrt{3}\tan u \Rightarrow \cos x\,dx = \sqrt{3}\sec^2 u\,du$

$$\int \frac{\cos x\,dx}{\sqrt{4-\cos^2 x}} = \int \frac{\cos x\,dx}{\sqrt{3+\sin^2 x}} = \int \frac{\sec^2 u\,du}{\sqrt{3+3\tan^2 u}} = \int \sec u\,du$$

$$= \ln|\sec u + \tan u| + C = \ln|\sqrt{3+\sin^2 x} + \sin x| + C$$

91. $\displaystyle\int x^2\sin(1-x)\,dx = x^2\cos(1-x) + 2x\sin(1-x) - 2\cos(1-x) + C$

x^2	+	$\sin(1-x)$
$2x$	−	$\cos(1-x)$
2	+	$-\sin(1-x)$
0		$-\cos(1-x)$

93. $$\int \frac{dx}{\cot^3 x} = \int \tan^3 x\,dx = \int \tan x(\sec^2 x - 1)\,dx$$
$$= \frac{1}{2}\tan^2 x + \ln|\cos x| + C$$

95. $$\int \frac{x^3\,dx}{(x^2+1)^2} = \int \frac{\tan^3 u\sec^2 u\,du}{(\tan^2 u+1)^2} = \int \frac{\sin^3 u\,du}{\cos u} = \int \frac{(1-\cos^2 u)\sin u\,du}{\cos u}$$
$$= \int \tan u\,du - \int \cos u\sin u\,du = \ln|\sec u| + \frac{1}{2}\cos^2 u + C$$
$$= \ln\sqrt{x^2+1} + \frac{1}{2(x^2+1)} + C$$

97. Let $u \Rightarrow u^2 = 2x+1 \Rightarrow 2u\,du = 2\,dx$, $x = 0 \Rightarrow u = 1$, $x = 4 \Rightarrow u = 3$. Then

$$\int_0^4 x\sqrt{2x+1}\,dx = \int_1^3 \frac{u^2-1}{2}\cdot u^2\,du = \frac{1}{2}\int_1^3 (u^4 - u^2)\,du = \frac{1}{10}u^5 - \frac{1}{6}u^3\Big]_1^3 = \frac{298}{15}$$

99. Let $u=\ln\left(x-\sqrt{x^2-1}\right) \Rightarrow du=-\dfrac{dx}{\sqrt{x^2-1}}$, $dv=dx \Rightarrow v=x$. Then

$$\int \ln\left(x-\sqrt{x^2-1}\right)dx = x\ln\left(x-\sqrt{x^2-1}\right)-\int\frac{x\,dx}{\sqrt{x^2-1}}$$

$$=\ln\left(x-\sqrt{x^2-1}\right)+\sqrt{x^2-1}+C$$

101. Let $u=\ln(x+\sqrt{x}) \Rightarrow du=\dfrac{2\sqrt{x}+1}{2x\sqrt{x}+2x}dx$, $dv=dx \Rightarrow v=x$. Then

$\displaystyle\int \ln(x+\sqrt{x})\,dx = x\ln|x+\sqrt{x}|-\int\frac{2\sqrt{x}+1}{2x\sqrt{x}+2x}\cdot x\,dx$. Now

$\displaystyle\int\frac{2\sqrt{x}+1}{2x\sqrt{x}+2x}x\,dx=-\frac{1}{2}\int\frac{\sqrt{x}+\sqrt{x}+1}{\sqrt{x}+1}dx=-\frac{1}{2}\int dx-\frac{1}{2}\int\frac{\sqrt{x}\,dx}{\sqrt{x}+1}$.

Finally let $u^2=x \Rightarrow 2u\,du=dx$. Then

$\displaystyle-\frac{1}{2}\int\frac{\sqrt{x}\,dx}{\sqrt{x}+1}=-\frac{1}{2}\int\frac{2u^2\,du}{u+1}=-\int\left(u-1+\frac{1}{u+1}\right)du=-\frac{1}{2}u^2+u-\ln|u+1|=$

$-\frac{1}{2}x+\sqrt{x}-\ln(\sqrt{x}+1)$. Thus, putting the pieces together,

$$\int\ln(x+\sqrt{x})\,dx = x\ln|x+\sqrt{x}|-x+\sqrt{x}-\ln|\sqrt{x}+1|+C.$$

103. $\displaystyle\int_1^2\ln(x^2+x)\,dx=\int_1^2\ln[x(x+1)]dx=\int_1^2\ln x\,dx+\int_1^2\ln(x+1)\,dx$

$$=x\ln x-x+(x-1)\ln(x+1)-x=3\ln 3-2$$

105. Let $u^2=x \Rightarrow 2u\,du=dx$, $x=0\Rightarrow u=0$, $x=\dfrac{\pi^2}{4}\Rightarrow u=\dfrac{\pi}{2}$. Then

$$\int_0^{\frac{\pi^2}{4}}\sin\sqrt{x}\,dx=\int_0^{\frac{\pi}{2}}2u\sin u\,du=-2u\cos u+2\sin u\Big]_0^{\frac{\pi}{2}}=2$$

$2u$	$+$	$\sin u$
2	$-$	$-\cos u$
0		$-\sin u$

107. Let $u = \sin^{-1}x \Rightarrow du = \dfrac{dx}{\sqrt{1-x^2}}$

$dv = \sqrt{1-x^2}\,dx \Rightarrow v = \int\sqrt{1-x^2}\,dx = \frac{1}{2}\sin^{-1}x + \frac{1}{2}x\sqrt{1-x^2}$. Then

$$\int \sin^{-1}x\sqrt{1-x^2}\,dx = \sin^{-1}x\left(\frac{1}{2}\sin^{-1}x + \frac{1}{2}x\sqrt{1-x^2}\right) - \int\left(\frac{1}{2}\sin^{-1}x + \frac{1}{2}x\sqrt{1-x^2}\right)\frac{dx}{\sqrt{1-x^2}}$$

$$= \frac{1}{2}(\sin^{-1}x)^2 + \frac{1}{2}x\sin^{-1}x\sqrt{1-x^2} - \frac{1}{4}x^2 - \frac{1}{4}(\sin^{-1}x)^2$$

$$= \frac{1}{4}(\sin^{-1}x)^2 + \frac{1}{2}x\sin^{-1}x\sqrt{1-x^2} - \frac{1}{4}x^2 + C$$

109. $$\int \frac{\tan x\,dx}{\tan x + \sec x}\cdot\frac{\tan x - \sec x}{\tan x - \sec x} = \int(-\tan^2 x + \tan x\sec x)\,dx$$

$$= \int(1 - \sec^2 x + \tan x\sec x)\,dx = x - \tan x + \sec x + C$$

111. $$\int \frac{dx}{(\cos^2 x + 4\sin x - 5)\cos x} = \int\frac{-\cos x\,dx}{(\sin^2 x - 4\sin x + 4)\cos^2 x}$$

$$= \int\frac{-\cos x\,dx}{(\sin x - 2)^2(1 - \sin^2 x)}$$

Let $u = \sin x \Rightarrow du = \cos x\,dx$. Then

$$\int\frac{-\cos x\,dx}{(\sin x - 2)^2(1-\sin^2 x)} = -\int\frac{du}{(u-2)^2(1-u^2)}$$

$$= \frac{1}{3}\int\frac{du}{(u-2)^2} - \frac{4}{9}\int\frac{du}{u-2} + \frac{1}{2}\int\frac{du}{u-1} - \frac{1}{18}\int\frac{du}{u+1}$$

$$= -\frac{1}{3}(u-2)^{-1} - \frac{4}{9}\ln|u-2| + \frac{1}{2}\ln|u-1| - \frac{1}{18}\ln|u+1| + C$$

$$= -\frac{1}{3}(\sin x - 2)^{-1} - \frac{4}{9}\ln|\sin x - 2| + \frac{1}{2}\ln|\sin x - 1| - \frac{1}{18}\ln|\sin x + 1| + C$$

113. $$\int\sqrt{\frac{1-\cos x}{\cos\alpha-\cos x}}\,dx=\int\sqrt{\frac{2\sin^2(x/2)}{2\cos^2(\alpha/2)-2\cos^2(x/2)}}\,dx$$

$$=\frac{\sqrt{\sin^2(x/2)/\cos^2(\alpha/2)}}{\sqrt{1-\cos^2(x/2)/\cos^2(\alpha/2)}}dx$$

Let $u=\frac{1}{\cos(\alpha/2)}\cos\frac{x}{2}\Rightarrow du=\left(\frac{1}{\cos(\alpha/2)}\right)\left(-\frac{1}{2}\sin\frac{x}{2}\right)$. Then

$$\frac{\sqrt{\sin^2(x/2)/\cos^2(\alpha/2)}}{\sqrt{1-\cos^2(x/2)/\cos^2(\alpha/2)}}dx=\int\frac{-2\,du}{\sqrt{1-u^2}}=-2\sin^{-1}u+C$$

$$=-2\sin^{-1}\left(\frac{\cos(x/2)}{\cos(\alpha/2)}\right)+C$$

115. Let $u=\ln(2x^2+4)\Rightarrow du=\frac{2x\,dx}{x^2+2}$ and $dv=dx\Rightarrow v=x$. Then

$$\int_0^1\ln(2x^2+4)=x\ln(2x^2+4)\Big]_0^1-\int_0^1\frac{2x^2\,dx}{x^2+2}$$

$$=x\ln(2x^2+4)\Big]_0^1-\int_0^1\left(2-\frac{4}{x^2+2}\right)dx$$

$$=x\ln(2x^2+4)-2x+2\sqrt{2}\tan^{-1}\frac{x}{\sqrt{2}}\Big]_0^1$$

$$=\ln 6-2+2\sqrt{2}\tan^{-1}\left(\frac{1}{\sqrt{2}}\right)$$

117. Let $u^2=1-e^{-t}\Rightarrow 2u\,du=e^{-t}dt\Rightarrow dt=\frac{2u\,du}{e^{-t}}=\frac{2u\,du}{1-u^2}$. Then

$$\int\frac{dt}{\sqrt{1-e^{-t}}}=\int\frac{2u\,du}{u(1-u^2)}=\int\left(\frac{1}{1-u}+\frac{1}{1+u}\right)du$$

$$=-\ln|1-u|+\ln|1+u|+C$$

$$=\ln\left|\frac{1+\sqrt{1-e^{-t}}}{1-\sqrt{1-e^{-t}}}\right|+C$$

119. Let $u = x(\sin^{-1}x)^2 \Rightarrow du = \dfrac{2\sin^{-1}x\,dx}{\sqrt{1-x^2}}$, $dv = dx \Rightarrow v = x$. Then

$$\int(\sin^{-1}x)^2\,dx = x\,(\sin^{-1}x)^2 - \int\frac{2x\sin^{-1}x\,dx}{\sqrt{1-x^2}} + C$$

Let $u = \sin^{-1}x \Rightarrow du = \dfrac{dx}{\sqrt{1-x^2}}$, $dv = \dfrac{-2x\,dx}{\sqrt{1-x^2}} \Rightarrow v = 2\sqrt{1-x^2}$. Then

$$-\int\frac{2x\sin^{-1}x\,dx}{\sqrt{1-x^2}} = 2\sin^{-1}x\sqrt{1-x^2} - 2x + C. \text{ Therefore,}$$

$$\int(\sin^{-1}x)^2\,dx = x\,(\sin^{-1}x)^2 + 2\sin^{-1}x\sqrt{1-x^2} - 2x + C$$

121. Let $u = \sin^{-1}x \Rightarrow du = \dfrac{dx}{\sqrt{1-x^2}}$, $dv = x\,dx \Rightarrow v = \dfrac{1}{2}x^2$. Then

$$\int x\sin^{-1}x\,dx = \frac{1}{2}x^2\sin^{-1}x - \frac{1}{2}\int\frac{x^2\,dx}{\sqrt{1-x^2}}.$$

Let $x = \sin v \Rightarrow dx = \cos v\,dv$. Then $-\dfrac{1}{2}\displaystyle\int\frac{x^2\,dx}{\sqrt{1-x^2}} = -\frac{1}{2}\int\sin^2 v\,dv$

$$= -\frac{1}{2}\int\left(\frac{1}{2} - \frac{1}{2}\cos 2v\right)dv = -\frac{1}{4}v + \frac{1}{4}\sin v\cos v. \text{ Therefore}$$

$$\int x\sin^{-1}x\,dx = \frac{1}{2}x^2\sin^{-1}x - \frac{1}{4}\sin^{-1}x + \frac{1}{4}x\sqrt{1-x^2} + C$$

123. $$\int\frac{d\theta}{1-\tan^2\theta} = \int\frac{\cos^2\theta\,d\theta}{\cos^2\theta - \sin^2\theta} = \frac{1}{2}\int\frac{1+\cos 2\theta\,d\theta}{\cos 2\theta}$$

$$= \frac{1}{2}\int\sec 2\theta\,d\theta + \frac{1}{2}\int d\theta$$

$$= \frac{1}{4}\ln|\sec 2\theta + \tan 2\theta| + \frac{1}{2}\theta + C$$

125. Let $t = \sin x \Rightarrow dt = \cos x\,dx$. Then $\displaystyle\int\frac{dt}{t-\sqrt{1-t^2}}$

$$= \int\frac{\cos x\,dx}{\sin x - \cos x}\cdot\frac{\sin x + \cos x}{\sin x + \cos x} = \int\frac{\frac{1}{2}\sin 2x + \frac{1}{2}\cos 2x + \frac{1}{2}}{-\cos 2x}\,dx$$

$$= -\frac{1}{2}\int\tan 2x\,dx - \frac{1}{2}\int\sec 2x\,dx - \frac{1}{2}\int dx$$

$$= \frac{1}{4}\ln|\cos 2x| - \frac{1}{4}\ln|\sec 2x + \tan 2x| - \frac{1}{2}x + C$$

$$= \frac{1}{4}\ln|1-2t^2| - \frac{1}{4}\ln\left|\frac{1+2t\sqrt{1-t^2}}{1-2t^2}\right| - \frac{1}{2}\sin^{-1}t + C$$

127. $$\int \frac{dx}{x^4+4} = \int \frac{dx}{(x^2+2)^2-4x^2} = \int \frac{dx}{(x^2+2x+2)(x^2-2x+2)}$$

$$= \int \frac{\frac{1}{8}x+\frac{1}{4}}{x^2+2x+2}\,dx + \int \frac{-\frac{1}{8}x+\frac{1}{4}}{x^2-2x+2}\,dx$$

$$= \frac{1}{16}\int \frac{2x+4}{x^2+2x+2}\,dx - \frac{1}{16}\int \frac{2x-4}{x^2-2x+2}\,dx$$

$$= \frac{1}{16}\left(\int \frac{2x+2}{x^2+2x+2}\,dx + \int \frac{2\,dx}{(x+1)^2+1} - \int \frac{2x-2}{x^2-2x+2}dx + \int \frac{2\,dx}{(x-1)^2+1}\right)$$

$$= \frac{1}{16}\ln\left|\frac{x^2+2x+2}{x^2-2x+2}\right| + \frac{1}{8}[\tan^{-1}(x+1)+\tan^{-1}(x-1)] + C$$

129. $$\lim_{n\to\infty}\left(\frac{n}{n^2+0^2}+\frac{n}{n^2+1^2}+\frac{n}{n^2+2^2}+\ldots+\frac{n}{n^2+(n+1)^2}\right) =$$

$$\lim_{n\to\infty}\sum_{k=0}^{n+1}\frac{n}{n^2+k^2} = \lim_{n\to\infty}\left(\frac{1}{n}\sum_{k=0}^{n+1}\frac{n^2}{n^2+k^2}\right) = \lim_{n\to\infty}\left(\frac{1}{n}\sum_{k=0}^{n+1}\frac{1}{1+\left(\frac{k}{n}\right)^2}\right)$$

$$= \int_0^1 \frac{dx}{1+x^2} = \tan^{-1}x\Big]_0^1 = \frac{\pi}{4}$$

131. $$\lim_{n\to\infty}\sum_{k=0}^{n-1}\frac{1}{\sqrt{n^2-k^2}} = \lim_{n\to\infty}\left(\frac{1}{n}\sum_{k=0}^{n-1}\frac{n}{\sqrt{n^2-k^2}}\right)$$

$$= \lim_{n\to\infty}\left(\frac{1}{n}\sum_{k=0}^{n-1}\frac{1}{\sqrt{1-\left(\frac{k}{n}\right)^2}}\right) = \int_0^1 \frac{dx}{\sqrt{1-x^2}} = \sin^{-1}x\Big]_0^1 = \frac{\pi}{2}$$

133. $$\int_0^1 \ln x\,dx = \lim_{b\to 0^+}\int_b^1 \ln x\,dx = \lim_{b\to 0^+}(x\ln x - x)\Big]_b^1$$

$$= \lim_{b\to 0^+}(-1 - b\ln b + b) = -1$$

Note: $\lim_{b\to 0^+} b\ln b = \lim_{b\to 0^+}\frac{\ln b}{\frac{1}{b}} = \lim_{b\to 0^+}\frac{\frac{1}{b}}{\frac{-1}{b^2}} = \lim_{b\to 0^+}(-b) = 0$

135. The line through (a, 0) and (0, a) is: $y = -x + a$.

$$A = \int_0^a \left(\sqrt{a^2 - x^2} - (a - x)\right) dx = \frac{a^2}{2}\left(\frac{\pi}{2} - 1\right)$$

$$M_y = \int_0^a x\left(\sqrt{a^2 - x^2} - (a - x)\right) dx = \frac{a^3}{6}$$

$$\bar{x} = \left(\frac{a^3}{6}\right)\left(\frac{4}{a^2(\pi - 2)}\right) = \frac{2a}{3(\pi - 2)}. \qquad \bar{y} = \bar{x} \text{ by symmetry.}$$

137. $$A = \int_0^{2\pi r} K\, s\, ds = \int_0^{2\pi} K(a\theta)\, a d\theta \text{ (since } s = a\theta\text{)} = Ka^2 \left.\frac{\theta^2}{2}\right]_0^{2\pi} = 2K\pi^2 a^2.$$

Or, since if the figure is unwrapped, it forms a right triangle, just

$$A = \frac{1}{2}bh = \frac{1}{2}(2\pi a)(2\pi K a) = 2K\pi^2 a^2$$

139. $$s = \int_1^e \sqrt{1 + \frac{1}{x^2}}\, dx = \int_1^e \frac{\sqrt{x^2 + 1}}{x}\, dx = \int_{\frac{\pi}{4}}^{\tan^{-1} e} \frac{\sec^3 u\, du}{\tan u}$$

$$= \sec u - \ln|\csc u + \cot u| \Big]_{\frac{\pi}{4}}^{\tan^{-1} e}$$

$$= \sqrt{1 + e^2} - \ln\left|\frac{\sqrt{1 + e^2}}{e} + \frac{1}{e}\right| - \sqrt{2} + \ln(1 + \sqrt{2})$$

141. $$S = 2\pi \int_0^2 e^x \sqrt{1 + e^{2x}}\, dx = 2\pi \int_{\frac{\pi}{4}}^{\tan^{-1} e^2} \tan u \sqrt{1 + \tan^2 u}\, \sec^2 u\, du$$

$$= 2\pi \int_{\frac{\pi}{4}}^{\tan^{-1} e^2} \sec^3 u\, du$$

$$= 2\pi \left(\frac{\sec u \tan u}{2} + \frac{\ln|\sec u + \tan u|}{2}\right)\Bigg]_{\frac{\pi}{4}}^{\tan^{-1} e^2}$$

$$= \pi\left[e^2\sqrt{e^4 + 1} + \ln\left|\sqrt{e^4 + 1} + e^2\right| - \sqrt{2} - \ln(\sqrt{2} + 1)\right]$$

143. $$\int_0^1 \pi(-\ln x)^2\, dx = \lim_{b \to 0^-} \int_b^1 \pi \ln^2 x\, dx$$

$$= \lim_{b \to 0^-} (x \ln^2 x - 2x \ln x + 2x)\Big]_b^1 = 2\pi$$

145. $V = 2\pi\int_0^{\frac{\pi}{2}} y^2\,dx = 2\pi\int_0^{\frac{\pi}{2}} y^2\left(\frac{dx}{dt}\right)dt$

$$= 2\pi\int_0^{\frac{\pi}{2}} (a\cos t)^2\,(a\sec t - a\cos t)\,dt$$

$$= 2\pi a^3\int_0^{\frac{\pi}{2}} (\cos t - \cos^3 t)\,dt = \frac{2\pi a^3}{3}\sin^3 t\,\Big]_0^{\frac{\pi}{2}} = \frac{2\pi a^3}{3}$$

147. $\int \frac{\ln x\,dx}{x^2} = -\frac{1}{x}\ln x + \int \frac{1}{x^2}dx = -\frac{1}{x}\ln x - \frac{1}{x} + C$

$$\therefore \int_1^{\infty}\frac{\ln x\,dx}{x^2} = \lim_{b\to\infty}\int_1^b \frac{\ln x\,dx}{x^2} = \lim_{b\to\infty}\left[-\frac{1}{x}\ln x - \frac{1}{x}\right]_1^b = 1$$

149. Let $u = \frac{1}{1+y} \Rightarrow du = -\frac{dy}{(1+y)^2}$, $dv = n\,y^{n-1}\,dy \Rightarrow v = y^n$. Then

$$\lim_{n\to\infty}\int_0^1 \frac{n\,y^{n-1}}{1+y}\,dy = \lim_{n\to\infty}\frac{y^n}{1+y}\Big]_0^1 + \lim_{n\to\infty}\int_0^1 \frac{y^n}{(1+y)^2}\,dy$$

$$= \frac{1}{2} + \lim_{n\to\infty}\int_0^1 \frac{y^n}{(1+y)^2}\,dy\,.$$

Since, for $0 \le y \le 1$, $\frac{y^n}{(1+y)^2} \le y^n$, and $\lim_{n\to\infty}\int_0^1 y^n\,dy = \lim_{n\to\infty}\frac{y^{n+1}}{n+1} = 0$,

$$\lim_{n\to\infty}\int_0^1 \frac{y^n}{(1+y)^2}\,dy = 0 \quad\text{and}\quad \lim_{n\to\infty}\int_0^1 \frac{n\,y^{n-1}}{1+y}\,dy = \frac{1}{2}.$$

151. $\int_0^{\pi}\frac{dx}{1+\sin x} = \int_0^{\infty}\frac{2\,dz}{1+z^2}\cdot\frac{1}{1+\frac{2z}{1+z^2}} = \int_0^{\infty}\frac{2\,dz}{(1+z)^2} = -\frac{2}{1+z}\Big]_0^{\infty} = 2$

153. $\int\frac{dx}{1-\sin x} = \int\frac{2\,dz}{1+z^2}\cdot\frac{1}{1-\frac{2z}{1+z^2}} = \int\frac{2\,dz}{(1-z)^2} = \frac{2}{1-z} = \frac{2}{1-\tan\frac{x}{2}} + C$

155. $\int\frac{\cos x\,dx}{1-\cos x} = \int\frac{1-z^2}{1+z^2}\cdot\frac{1}{1-\frac{1-z^2}{1+z^2}}\cdot\frac{2\,dz}{1+z^2} = \int\frac{1-z^2}{z^2(1+z^2)}dz$

$$= \int\left(\frac{1}{z^2} - \frac{2}{1+z^2}\right)dz = -\frac{1}{z} - 2\tan^{-1}z + C = -\cot\frac{x}{2} - x + C$$

157. $$\int \frac{dx}{\sin x - \cos x} = \int \frac{1}{\dfrac{2z}{1+z^2} - \dfrac{1-z^2}{1+z^2}} \cdot \frac{2\,dz}{1+z^2} = \int \frac{2}{z^2 + 2z - 1} dz$$

Let $u = z + 1$, $du = dz$. Then $\displaystyle\int \frac{2\,dz}{(z+1)^2 - 2} = \int \frac{2\,du}{u^2 - 1}$

$$= \frac{1}{\sqrt{2}} \int \left(\frac{1}{u - \sqrt{2}} - \frac{1}{u + \sqrt{2}} \right) du = \frac{1}{\sqrt{2}} \ln \left| \frac{u - \sqrt{2}}{u \ \sqrt{2}} \right| + C$$

$$= \frac{1}{\sqrt{2}} \ln \left| \frac{\tan\frac{x}{2} + 1 - \sqrt{2}}{\tan\frac{x}{2} + 1 + \sqrt{2}} \right| + C$$

CHAPTER 8

CONIC SECTIONS AND OTHER PLANE CURVES

8.1 EQUATIONS FROM THE DISTANCE FORMULA

1. $\sqrt{x^2+y^2} = |y-(-4)| \Rightarrow x^2+y^2 = (y+4)^2 \Rightarrow x^2 = 8y+16$

3. $\sqrt{(x+2)^2+(y-1)^2} = \sqrt{(x-2)^2+(y+3)^2}$

   $x^2+4x+4+y^2-2y+1 = x^2-4x+4+y^2+6y+9$

   $8x-8y=8$ or $x-y=1$

5. $\left(\sqrt{(x+2)^2+y^2}\right)\left(\sqrt{(x-2)^2+y^2}\right) = 4$

   $(x^2+4x+4+y^2)(x^2-4x+4+y^2) = 16$

   $[(x^2+4+y^2)+4x][(x^2+4+y^2)-4x] = 16$

   $(x^2+y^2+4)^2 - 16x^2 = 16$

   $(x^2+y^2)^2 + 8(y^2-x^2) = 0$

7. $|x-(-2)| = 2\sqrt{(x-2)^2+y^2} \Rightarrow x^2+4x+4 = 4(x^2-4x+4+y^2)$

   $3x^2-20x+4y^2+12=0$

9. $\sqrt{(x+3)^2+y^2} = \sqrt{(x-3)^2+y^2} + 4$

   $(x+3)^2+y^2 = (x-3)^2+y^2+8\sqrt{(x-3)^2+y^2}+16$

   $x^2+6x+9+y^2 = x^2-6x+9+8\sqrt{(x-3)^2+y^2}+16$

   $3x-4 = 2\sqrt{(x-3)^2+y^2}$

   $9x^2-24x+16 = 4(x^2-6x+9+y^2)$

   $5x^2-4y^2=20$

11. $\sqrt{(x-2)^2+(y-3)^2} = 3 \Rightarrow x^2-4x+4+y^2-6y+9=9$

    $x^2+y^2-4x-6y+4=0$

13. Let $P(x,y)$ be equidistant from each of $A(0,1)$, $B(1,0)$ and $C(4,3)$.

Then $\sqrt{(x-0)^2+(y-1)^2}=\sqrt{(x-1)^2+(y-0)^2}$, and

$\sqrt{(x-0)^2+(y-1)^2}=\sqrt{(x-4)^2+(y-3)^2}$. Then

$x^2+y^2-2y+1=x^2-2x+1+y^2$, and

$x^2+y^2-2y+1=x^2-8x+16+y^2-6y+9$. Simplifying,

we get $y=x$ and $2x+y=6$. Hence the point is $P(2,2)$.

The radius is $r=\sqrt{2^2+1^2}=\sqrt{5}$.

8.2 CIRCLES

1. $x^2+(y-2)^2=4$ or $x^2+y^2-4y=0$

3. $(x-3)^2+(y+4)^2=25$ or $x^2+y^2-6x+8y=0$

5. $(x+2)^2+(y+1)^2<6$ or $x^2+y^2+4x+2y<1$

7. $x^2+y2=16$ has $C(0,0)$ and radius $r=4$.

9. $x^2+y^2-2y=3 \Rightarrow x^2+y^2-2y+1=3+1 \Rightarrow x^2+(y-1)^2=4$

$C(0,1)$, $r=2$

11. $x^2+4x+y^2=12 \Rightarrow x^2+4x+4+y^2=12+4$

$(x+2)^2+y^2=16$ $C(-2,0)$, $r=4$

13. $x^2+2x+y^2+2y=-1 \Rightarrow x^2+2x+1+y^2+2y+1=-1+1+1$

$(x+1)^2+(y+1)^2=1$ $C(-1,-1)$, $r=1$

15. $2x^2+x+2y^2+y=0 \Rightarrow x^2+\frac{1}{2}x+y^2+\frac{1}{2}y=0$

$x^2+\frac{1}{2}x+\frac{1}{16}+y^2+\frac{1}{2}y+\frac{1}{16}=\frac{1}{8} \Rightarrow \left(x+\frac{1}{4}\right)^2+\left(y+\frac{1}{4}\right)^2=\frac{1}{8}$

$C\left(-\frac{1}{4},-\frac{1}{4}\right)$, $r=\sqrt{\frac{1}{8}}=\frac{\sqrt{2}}{4}$

17. $x^2+y^2=2x-4y+5\le 0 \Rightarrow x^2-2x+1+y^2+4y+4\le -5+1+4$

$(x+1)^2+(y-2)^2\le 0$. This is satisfied only by the point $(-1,2)$.

19. The circle with center $(2,2)$ has the form $(x-2)^2+(y-2)^2=r^2$. The point $(4,5)$ belongs to the circle, and hence must satisfy this equation. Therefore, $(4-2)^2+(5-2)^2=r^2$, or $r^2=13$. The equation is then $(x-2)^2+(y-2)^2 = 13$ or $x^2+y^2-4x-4y=5$.

21. The distance from the center to the line equals the radius. Hence, $r=\dfrac{|-1+2-4|}{\sqrt{5}}=\dfrac{3}{\sqrt{5}}$. The equation is $(x+1)^2+(y-1)^2=\dfrac{9}{5}$ or $5x^2+5y^2+10x-10y+1=0$.

23. A general equation of a circle is $x^2+y^2+ax+by+c=0$. We substitute each of the three given points into this to obtain the system:

$$\begin{aligned} 2a+3b+c&=-13\\ 3a+2b+c&=-13\\ -4a+3b+c&=-25, \end{aligned}$$

which has solution $a=2$, $b=2$, $c=-23$. The equation is

$$x^2+y^2+2x+2y-23=0.$$

25. A general equation of a circle is $x^2+y^2+ax+by+c=0$. We substitute each of the three given points into this to obtain the system:

$$\begin{aligned} 7a+b+c&=-50\\ -a+7b+c&=-50\\ c&=0, \end{aligned}$$

which has solution $a=-6$, $b=-8$, $c=0$. The equation is $x^2+y^2-6x-8y=0$ or $(x-3)^2+(y-4)^2=25$, with $C(3,4)$, $r=5$.

27. $\sqrt{(x-6)^2+y^2}=2\sqrt{x^2+(y-3)^2}\Rightarrow x^2+y^2+4x-8y=0$ or $(x+2)^2+(y-4)^2=20$. $C(-2,4)$, $r=\sqrt{20}=2\sqrt{5}$.

29. $A = 2(x-1)y$

$= 2(x-1)\sqrt{36-x^2}$

$$A' = \frac{-2(2x-9)(x+4)}{\sqrt{36-x^2}} = 0$$

$\Leftrightarrow (2x-9)(x+4) = 0 \Leftrightarrow x - -4,\ \frac{9}{2}$

$y' > 0$ if $x > \frac{9}{2}$, $y' < 0$ if $x < \frac{9}{2} \Rightarrow$ maximum value at $x = \frac{9}{2}$

The dimensions are $\frac{7}{2}$ by $\frac{3\sqrt{7}}{2}$.

31. Let $P(x_1, y_1)$ be the exterior point, and the center of the circle be $O(h, k)$. Let Q be either of the points of tangency from P to the circle and s be the length of PQ. $\triangle PQO$ is a right triangle with hypotenuse OP. Then $OP^2 = PQ^2 + OQ^2$, or $(x_1 - h)^2 + (y_1 - k)^2 = s^2 + a^2$.

33. From a theorem in geometry, PT is the mean proportion between PM and PN. ($\Delta PTN \approx \Delta PTM$). Therefore, $\frac{PN}{PT} = \frac{PT}{PM}$, or $(PT)^2 = (PN)(PM)$.

8.3 PARABOLAS

1. $x^2 = 4(2)y$

$x^2 = 8y$

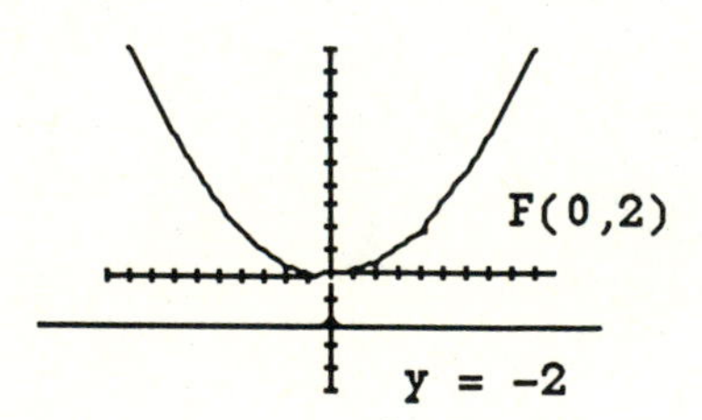

3. $x^2 = -4(4)y$

$x^2 = -16y$

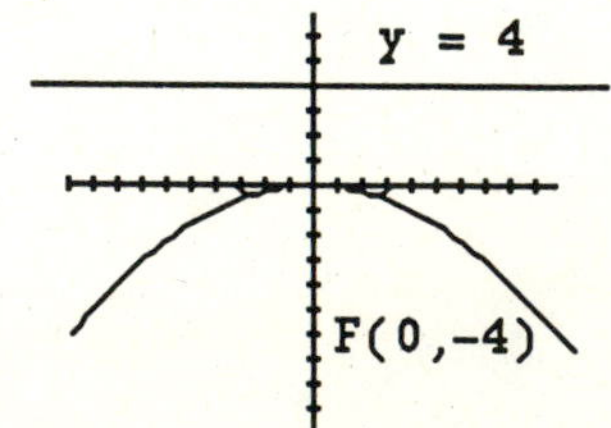

5. $(x + 2)^2 = 4(y - 3)$

$x^2 + 4x - 4y + 16 = 0$

$F(-2,4)$, $y = 2$

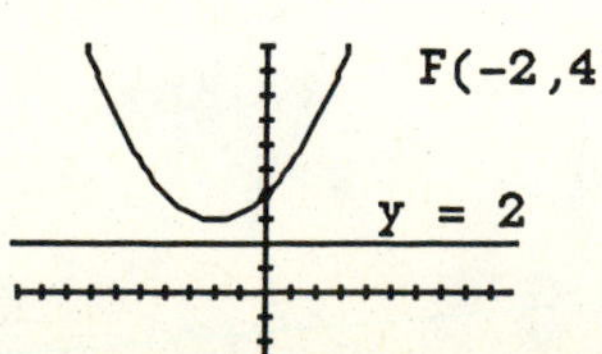

7. $(y - 1)^2 = 12(x + 3)$

$y^2 - 2y - 12x - 35 = 0.$

$F(0,1)$, $x = -6$

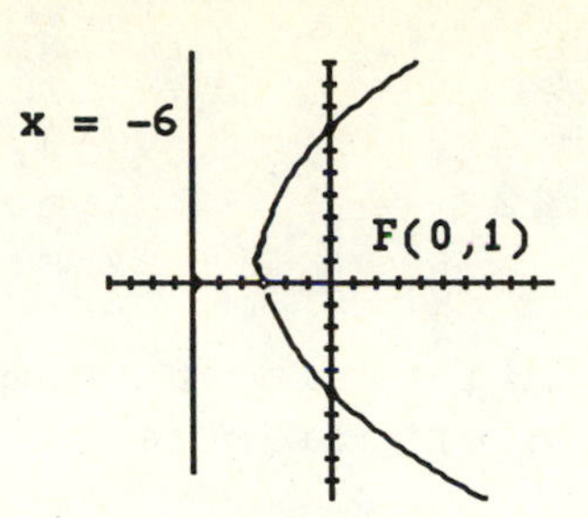

9. $y^2 = 8(x - 2)$

$y^2 - 8x + 16 = 0$

$F(4,0)$, $x = 0$

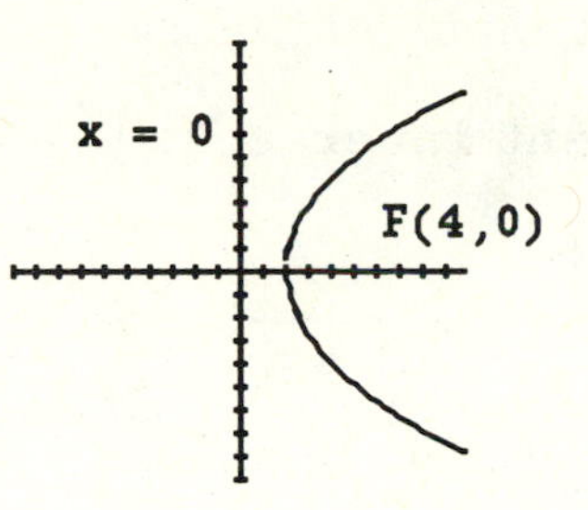

11. $(y - 1)^2 = -16(x + 3)$

$y^2 - 2y + 16x + 49 = 0$

$F(-7,1)$, $x = 1$

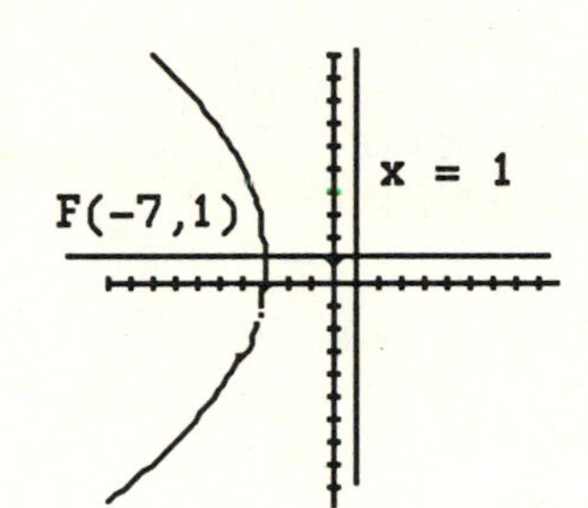

13. $(y - 1)^2 = 4x$

$y^2 - 2y - 4x + 1 = 0$

$F(1,1)$, $x = -1$

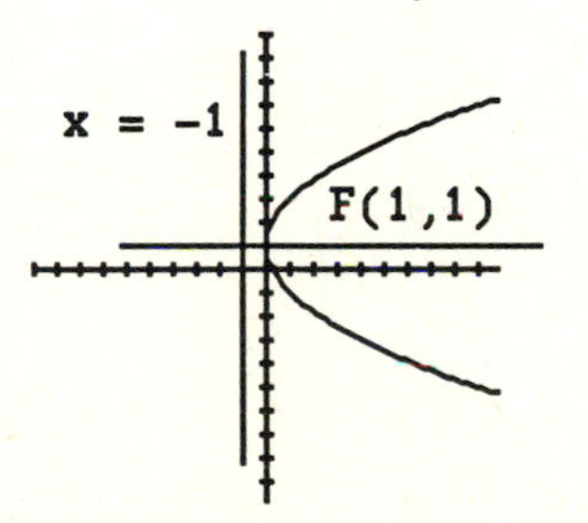

15. $y^2 = 8x$. $4p = 8 \Rightarrow p = 2$. Vertex is $V(0,0)$, focus is $F(2,0)$, directrix is $x = -2$, axis is $y = 0$.

17. $x^2 = 100y$. $4p = 100 \Rightarrow p = 25$. $V(0,0)$, $F(0,25)$, directrix: $y = -25$, axis: $x = 0$.

19. $x^2 - 2x + 8y - 7 = 0$
$x^2 - 2x + 1 = -8y + 7 + 1$
$(x - 1)^2 = -8(y - 1)$

$V(1,1)$ $F(1,-1)$
directrix: $y = 3$
axis: $x = 1$

21. $y^2 + 4x - 8 = 0$
$y^2 = -4x + 8$
$y^2 = -4(x - 2)$

$V(2,0)$ $F(1,0)$
directrix: $x = 3$
axis: $y = 0$

23. $x^2 + 2x - 4y - 3 = 0$
$x^2 + 2x + 1 = 4y + 3 + 1$
$(x + 1)^2 = 4(y + 1)$

$V(-1,-1)$ $F(-1,0)$
directrix: $y = -2$
axis: $x = -1$

25. $y^2 - 4y - 8x - 12 = 0$
$y^2 - 4y + 4 = 8x + 12 + 4$
$(y - 2)^2 = 8(x + 2)$

V(-2,2) F(0,2)
directrix: $x = -4$
axis: $y = 2$

27. $y^2 < x$ is the region interior to the parabola $y^2 = x$, and $y^2 > x$ is the region exterior to the parabola.

29. Let $P_1(-p, y_1)$ be any point on the line $x = -p$, and let $P(x, y)$be the point where the tangent intersects $y^2 = 4x$. Since $\frac{dy}{dx} = \frac{2p}{y}$,

the slope of this tangent is: $\frac{y - y_1}{x + p} = \frac{2p}{y}$. We substitute

for x and solve for y: $y(y - y_1) = 2p\left(\frac{y^2}{4p} + p\right) \Rightarrow y^2 - (2y_1)y - 4p^2 = 0 \Rightarrow$

$y = y_1 \pm \sqrt{y_1^2 + 4p^2}$. The slopes of the two tangents from P_1 to the curve are $m_1 = \frac{2p}{y_1 + \sqrt{y_1^2 + 4p^2}}$ and $m_2 = \frac{2p}{y_1 - \sqrt{y_1^2 + 4p^2}}$. Notice that $m_1 m_2 = -1$, so that the tangents are perpendicular.

31. $V = \pi \int_0^h x^2\, dy = \pi \int_0^h \frac{b^2}{4h} y\, dy = \frac{\pi b^2}{4h}\left[\frac{y^2}{2}\right]_0^h = \frac{\pi b^2 h}{8}$

The volume of the cone $= \frac{1}{3}\pi r^2 h = \frac{1}{3}\pi\left(\frac{b}{2}\right)^2 h = \frac{\pi b^2 h}{12}$.

Notice that $\frac{3}{2}\left(\frac{\pi b^2 h}{12}\right) = \frac{\pi b^2 h}{8}$

33. If $y^2 = 4px$, then $\frac{dy}{dx} = \frac{2p}{y}$. The slope of the tangent to the graph at $P(x_1, y_1)$ is $m = \frac{2p}{y_1}$, and the equation of this tangent is $y - y_1 = \frac{2p}{y_1}(x - x_1)$. Set $y = 0$ to find the x-intercept of this line:

$-y_1 = \frac{2p}{y_1}(x - x_1)$ or $x = -\frac{y_1^2}{2p} + x_1 = -\frac{4px_1}{2p} + x_1 = -x_1$.

8.4 ELLIPSES

1. $c^2 = a^2 - b^2$

   $4 = 16 - b^2 \Rightarrow b^2 = 12$

   $\frac{y^2}{16} + \frac{x^2}{12} = 1$

   $e = \frac{c}{a} = \frac{1}{2}$

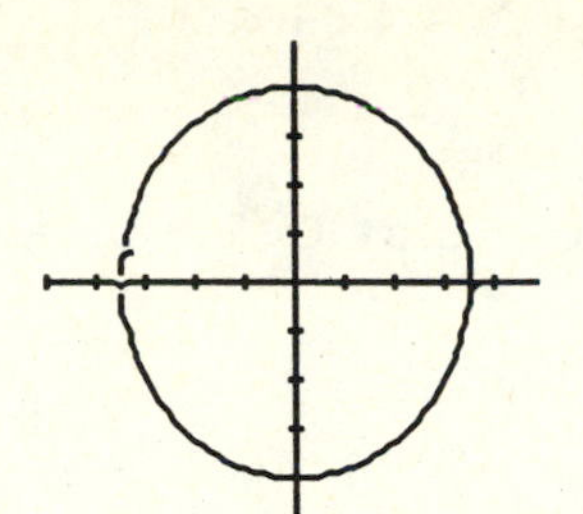

3. $4 = 9 - b^2 \Rightarrow b^2 = 5$

   $\frac{x^2}{5} + \frac{(y-2)^2}{9} = 1$

   $e = \frac{2}{3}$

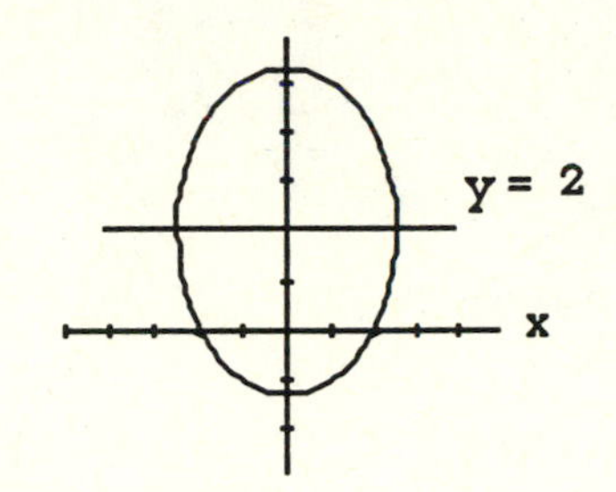

5. $\frac{(x-7)^2}{4} + \frac{(y-5)^2}{25} = 1$. Center is at $C(7,5)$ and the major axis is vertical. Vertices are $V(7, 5 \pm 5) = V_1(7,10)$ and $V_2(7,0)$. $c^2 = 25 - 4 = 21 \Rightarrow c = \sqrt{21}$. Foci are $F(7, 5 \pm \sqrt{21})$.

7. $\frac{(x+1)^2}{9} + \frac{(y+4)^2}{25} = 1$. Center is at $C(-1,-4)$ and the major axis is vertical. Vertices are $V(-1,-4 \pm 5) = V_1(-1,-9)$ and $V_2(-1,1)$. $c^2 = 25 - 9 = 16 \Rightarrow c = 4$. Foci are $F(-1,-4 \pm 4) = F_1(-1,0)$ and $F_2(-1,-8)$.

9. $25(x-3)^2 + 4(y-1)^2 = 100$

   $\frac{(x-3)^2}{4} + \frac{(y-1)^2}{25} = 1$. Center is at $C(3,1)$ and the major axis is vertical. Vertices are $V(3, 1 \pm 5) = V_1(3,6)$ and $V_2(3,-4)$. $c^2 = 25 - 4 = 21 \Rightarrow c = \sqrt{21}$. Foci are $F(3, 1 \pm \sqrt{21})$.

11. $x^2 + 10x + 25y^2 = 0 \Rightarrow x^2 + 10x + 100 + 25y^2 = 25$ or

    $\frac{(x+5)^2}{25} + \frac{y^2}{1} = 1$. Center is at $C(-5,0)$ and the major axis is horizontal. Vertices are $V(5 \pm 5, 0) = V_1(10,0)$ and $V_2(0,0)$. $c^2 = 25 - 1 = 24 \Rightarrow c = 2\sqrt{6}$. Foci are $F(-5 \pm 2\sqrt{6}, 0)$.

13. $x^2 + 9y^2 - 4x + 18y + 4 = 0 \Rightarrow (x^2 - 4x + 4) + 9(y^2 + 2y + 1) = -4 + 4 + 9$ or

$\dfrac{(x-2)^2}{9} + \dfrac{(y+1)^2}{1} = 1$. Center is at $C(2,-1)$ and the major axis is horizontal. Vertices are $V(2 \pm 3, -1) = V_1(5,-1)$ and $V_2(-1,-1)$.

$c^2 = 9 - 1 = 8 \Rightarrow c = 2\sqrt{2}$. Foci are $F(2 \pm 2\sqrt{2}, -1)$.

15. $4x^2 + y^2 - 16x + 4y + 16 = 0 \Rightarrow 4(x^2 - 4x + 4) + (y^2 + 4y + 4) = -16 + 16 + 4$ or

$\dfrac{(x-2)^2}{1} + \dfrac{(y+2)^2}{4} = 1$. Center is at $C(2,-2)$ and the major axis is vertical. Vertices are $V(2, -2 \pm 2) = V_1(2,0)$ and $V_2(2,-4)$.

$c^2 = 4 - 1 = 3 \Rightarrow c = \sqrt{3}$. Foci are $F(2, -2 \pm \sqrt{3})$.

17. $9x^2 + 16y^2 - 18x - 96y + 9 = 0 \Rightarrow 9(x^2 - 2x + 1) + 16(y^2 - 6y + 9) = -9 + 144 + 9$ or

$\dfrac{(x+1)^2}{16} + \dfrac{(y-3)^2}{9} = 1$. Center is at $C(-1,3)$ and the major axis is horizontal. Vertices are $V(-1 \pm 4, 3) = V_1(-5,3)$ and $V_2(3,3)$.

$c^2 = 16 - 9 = 7 \Rightarrow c = \sqrt{7}$. Foci are $F(-1 \pm \sqrt{7}, 3)$.

19.

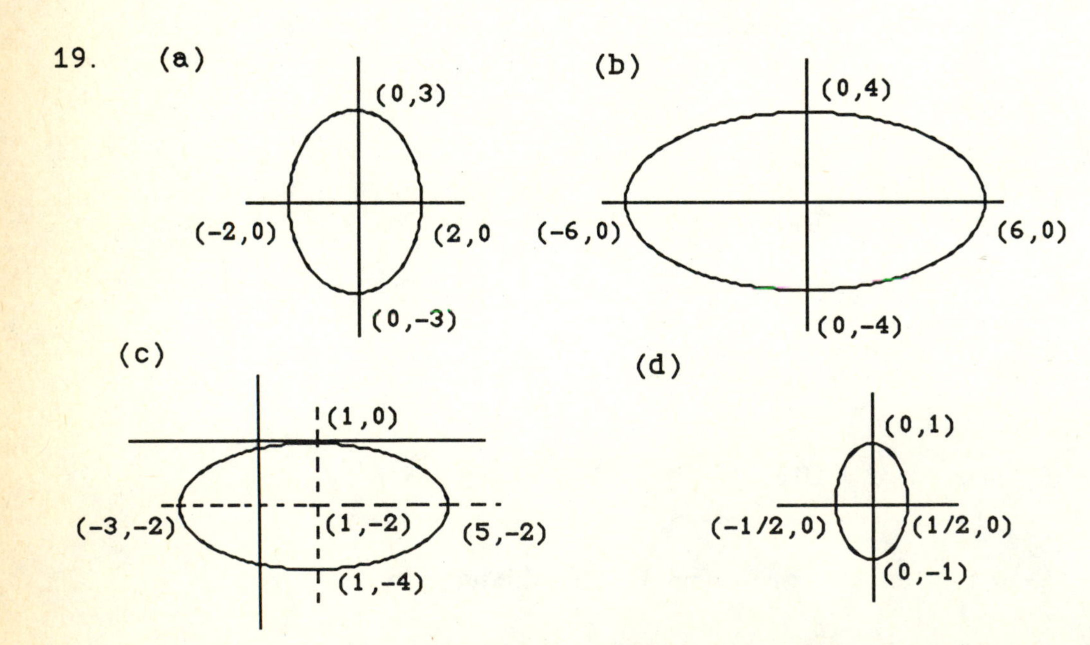

(e)

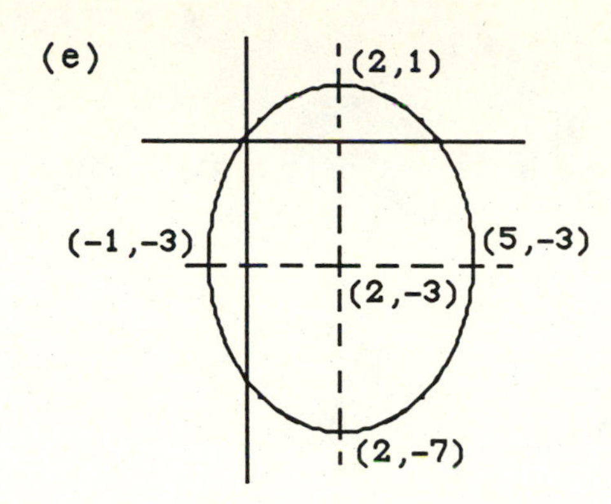

21. One axis is from A(1,1) to B(1,7) and is 6 units long. The other axis is from C(3,4) to D(-1,4) and is 4 units long. Therefore. $a = 3$, $b = 2$ and its major axis is vertical. The center is the point C(1,4). The ellipse is

$$\frac{(x-1)^2}{4} + \frac{(y-4)^2}{9} = 1.$$

The foci would be $F(1, 4 \pm \sqrt{5})$

23. $9x^2 + 16y^2 \le 144$

$$\frac{x^2}{16} + \frac{y^2}{9} \le 1$$

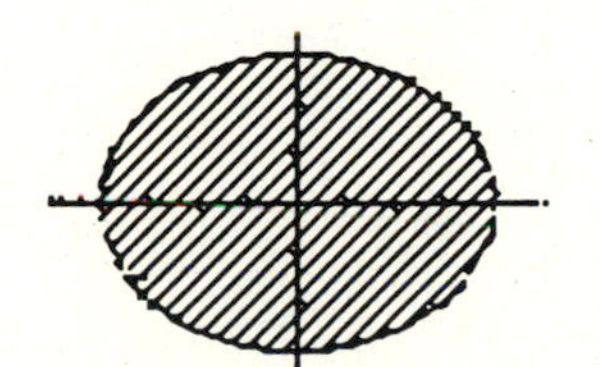

25. The intersection of the graphs of

$$\frac{x^2}{4} + \frac{y^2}{1} = 1 \text{ and } \frac{x^2}{9} + \frac{y^2}{4} = 1$$

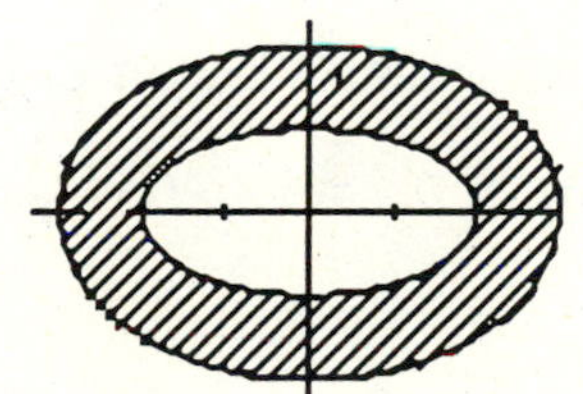

27. $\frac{c}{a} = \frac{4}{5} \Rightarrow c^2 = \frac{16}{25}a^2$

Then $b^2 = a^2 - \frac{16}{25}a^2 = \frac{9}{25}a^2$

or $\frac{b^2}{a^2} = \frac{9}{25}$.

We will draw the ellipse

$$\frac{x^2}{25} + \frac{y^2}{9} = 1.$$

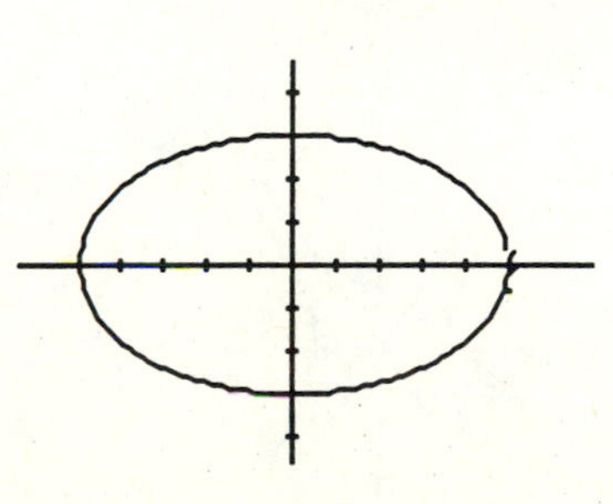

29. (a) Using figures in the text, $a = \frac{36.18}{2} = 18.09$ AU; $b = \frac{9.12}{2} = 4.56$ AU;

$c = \sqrt{(18.09)^2 + (4.56)^2} = 17.50$. If the sun is at the origin, the center is $(c, 0)$ and the equation is

$$\frac{(x - 17.5)^2}{(18.09)^2} + \frac{y^2}{(4.56)^2} = 1.$$

(b) Minimum distance $= a - c = 0.5842$ AU $\approx 5.41 \times 10^7$ miles.

(c) Maximum distance $= a + c = 35.59$ AU $\approx 3.30 \times 10^9$ miles.

31. $$A = 2\int_0^3 4\sqrt{1 - \frac{x^2}{9}}\, dx = \frac{8}{3}\int_0^3 \sqrt{9 - x^2}\, dx$$

$$= \frac{8}{3}\int_0^{\pi/2} \sqrt{9 - 9\sin^2 u}\; 3\cos u\, du = 24\int_0^{\pi/2} \cos^2 u\, du$$

$$= 24\left[\frac{1}{2} + \frac{1}{4}\sin 2u\right]_0^{\frac{\pi}{2}} = 6\pi.$$

$$M_x = \int_{-3}^{3} \frac{1}{2}\left(\frac{4\sqrt{9 - x^2}}{3}\right)^2 dx = \frac{8}{9}\int_{-3}^{3} (9 - x^2)\, dx = \frac{8}{9}\left(9x - \frac{1}{3}x^3\right)\Big]_{-3}^{3} = 32.$$

$\bar{y} = \frac{32}{6\pi} = \frac{16}{3\pi}$. $\bar{x} = 0$ by symmetry.

33. The ellips must pass through $(0, 0) \Rightarrow c = 0$. It also contains the point $(-1, 2) \Rightarrow -a + 2b = -8$. To be tangent to the x–axis means that the center must be on the y–axis, so $a = 0$. Then $b = -4$ and the equation is $4x^2 + y^2 - 4y = 0$.

8.5 HYPERBOLAS

1. $\frac{x^2}{9} - \frac{y^2}{16} = 1$

$y = \pm \frac{4}{3}x$

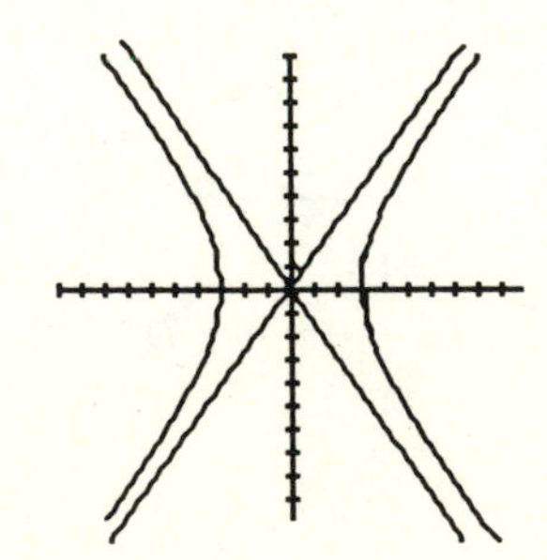

3. $\frac{y^2}{9} - \frac{x^2}{16} = 1$

$y = \pm \frac{3}{4}x$

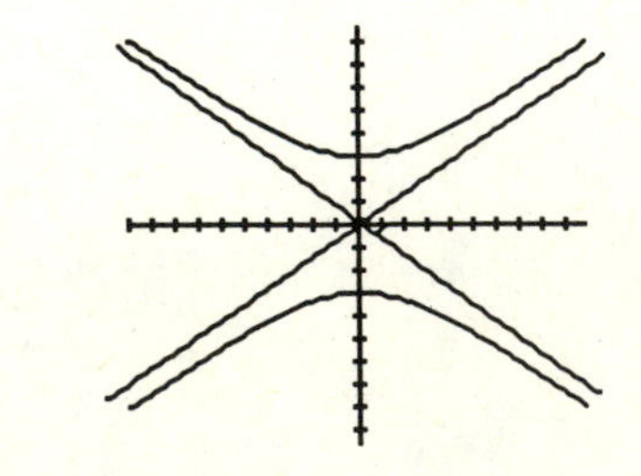

5. $9(x-2)^2 - 4(y+3)^2 = 36$

$$\frac{(x-2)^2}{4} - \frac{(y+3)^2}{9} = 1$$

Center: $C(2,-3)$

Vertices: $(4,-3)$ and $(0,-3)$

Foci: $(2 \pm \sqrt{13}, -3)$; $e = \frac{\sqrt{13}}{2}$

$y + 3 = \pm \frac{4}{3}(x - 4)$

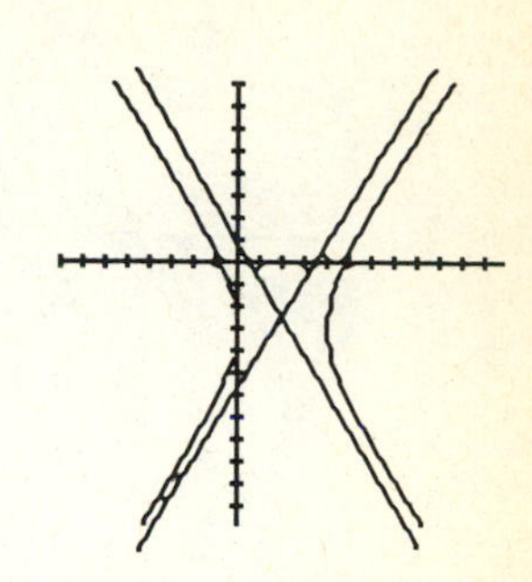

7. $4(x-2)^2 - 9(y+3)^2 = 36$

$$\frac{(x-2)^2}{9} - \frac{(y+3)^2}{4} = 1$$

Center: $C(2,-3)$

Vertices: $(5,-3)$ and $(1,-3)$

Foci: $(2 \pm \sqrt{13}, -3)$; $e = \frac{\sqrt{13}}{3}$

$y + 3 = \pm\frac{2}{3}(x-2)$

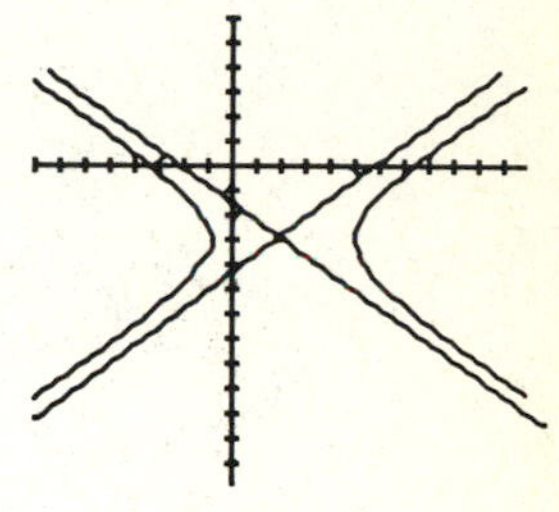

9. $16(y-3)^2 - 9(x-4)^2 = 144$

$$\frac{(y-3)^2}{9} - \frac{(x-4)^2}{16} = 1$$

Center: $(4,3)$

Vertices: $(4,6)$ and $(4,0)$

Foci: $(4,8)$ and $(4,-2)$ $\quad y-3 = \pm\frac{3}{4}(x-4)$

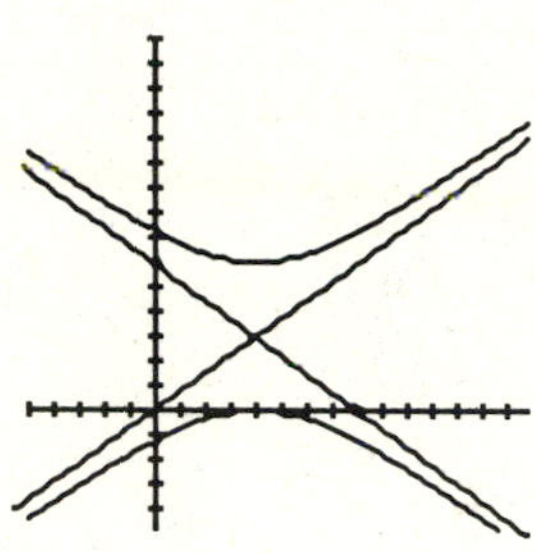

11. $5x^2 - 4y^2 + 20x + 8y = 4$

$5(x^2 + 4x + 4) - 4(y^2 - 2y + 1) = 4 + 20 - 4$

$$\frac{(x+2)^2}{4} - \frac{(y-1)^2}{5} = 1$$

Center: $C(-2,1)$

Vertices: $(0,1)$ and $(-4,1)$

Foci: $(-5,1)$ and $(1,1)$ $\quad y-1 = \pm\frac{\sqrt{5}}{2}(x+2)$; $e = \frac{3}{2}$

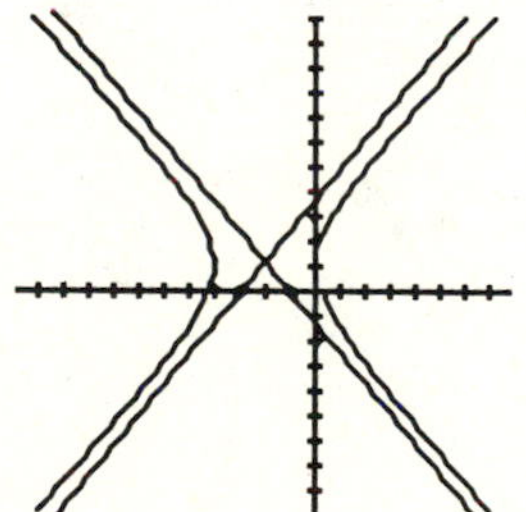

13. $4y^2 = x^2 - 4x$

$4y^2 - (x^2 - 4x + 4) = -4$

$$\frac{(x-2)^2}{4} - \frac{y^2}{1} = 1$$

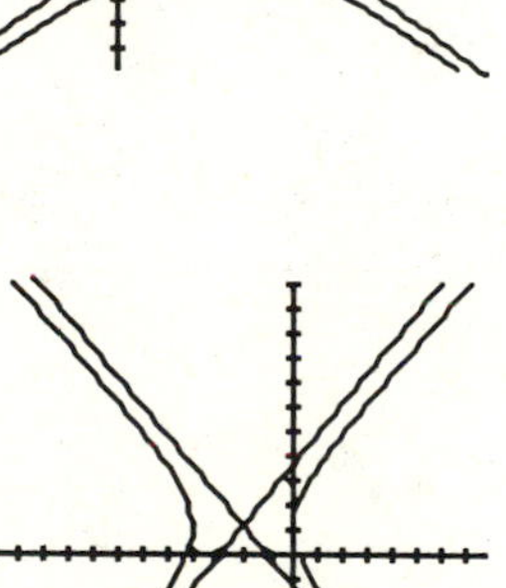

Center: $C(2,0)$

Vertices: $(0,0)$ and $(4,0)$ Foci: $(2 \pm \sqrt{5}, 0)$; $e = \frac{\sqrt{5}}{2}$; $2y = \pm(x-2)$.

15. Foci at (0, 0) and (0, 4) $\Rightarrow$ Center is (0, 2). Equation is:

$\frac{(y-2)^2}{a^2} - \frac{x^2}{b^2} = 1$. The point (12, 9) belongs to the hyperbola, giving

$\frac{49}{a^2} - \frac{144}{b^2} = 1$. Now c = 2, so $a^2 + b^2 = 4$, or $a^2 = 4 - b^2$. Substituting,

$$\frac{49}{4-b^2} - \frac{144}{b^2} = 1 \Rightarrow b^4 + 189b^2 - 576 = 0 \Rightarrow (b^2 - 3)(b^2 + 192) = 0.$$

Therefore, $b^2 = 3$, $a^2 = 1$ and $\frac{(y-2)^2}{1} - \frac{x^2}{3} = 1$ is required equation.

17. (a) If the boat is at point P, then

PA - PB| = 1400 ms x 980 ft/s = 1.372×10^6 ft. = 259.8 miles.
The boat is on a hyperbola with foci A and B, and with
2a = 259.8 miles, or a = 129.9 miles.

19. (a) $V = \pi \int_3^5 (x^2 - 9)\,dx = \pi\left[\frac{1}{3}x^3 - 9x\right]_3^5 = \frac{44\pi}{3}$

(b) $V = \pi \int_{-4}^{4} (25 - x^2)\,dy = \pi \int_{-4}^{4} (25 - (16 - y^2))\,dy$

$$= 2\pi \int_0^4 (16 - y^2)\,dy = 2\pi\left[16y - \frac{1}{3}y^2\right]_0^4 = \frac{256\pi}{3}$$

21. $3x^2 - 4y^2 = 12$ and $x = 5 \Rightarrow y = \pm\frac{3\sqrt{7}}{2}$.

$$V = 2\left(\pi \int_0^{\frac{3\sqrt{7}}{2}} (25 - x^2)\,dy\right) = 2\pi \int_0^{\frac{3\sqrt{7}}{2}} \left(25 - \left(4 + \frac{4}{3}y^2\right)\right) dy$$

$$= 2\pi\left[21y - \frac{4}{9}y^3\right]_0^{\frac{3\sqrt{7}}{2}} = 42\pi\sqrt{7}$$

23. The circles radii increase at the same rate, so their difference is constant.

8.6 THE GRAPHS OF QUADRATIC EQUATIONS

1. $\cot 2\alpha = 0 \Rightarrow 2\alpha = \frac{\pi}{2} \Rightarrow \alpha = \frac{\pi}{4}$.

$$x = \frac{\sqrt{2}}{2}x' - \frac{\sqrt{2}}{2}y';\ y = \frac{\sqrt{2}}{2}x' + \frac{\sqrt{2}}{2}y'$$

$$\left(\frac{\sqrt{2}}{2}x' - \frac{\sqrt{2}}{2}y'\right)\left(\frac{\sqrt{2}}{2}x' + \frac{\sqrt{2}}{2}y'\right) = 2$$

$$\frac{1}{2}x'^2 - \frac{1}{2}y'^2 = 2 \text{ or } x'^2 - y'^2 = 4$$

The graph is a hyperbola.

3. $\cot 2\alpha = \frac{1}{\sqrt{3}} \Rightarrow 2\alpha = \frac{\pi}{3} \Rightarrow \alpha = \frac{\pi}{6}$.

$$x = \frac{\sqrt{3}}{2}x' - \frac{1}{2}y';\qquad y = \frac{1}{2}x' + \frac{\sqrt{3}}{2}y'$$

$$3\left(\frac{\sqrt{3}}{2}x' - \frac{1}{2}y'\right)^2 + 2\sqrt{3}\left(\frac{\sqrt{3}}{2}x' - \frac{1}{2}y'\right)\left(\frac{1}{2}x' + \frac{\sqrt{3}}{2}y'\right) + \left(\frac{1}{2}x' + \frac{\sqrt{3}}{2}y'\right)$$

$$-8\left(\frac{\sqrt{3}}{2}x' - \frac{1}{2}y'\right) + 8\sqrt{3}\left(\frac{1}{2}x' + \frac{\sqrt{3}}{2}y'\right) = 0.$$

$$4x'^2 - 8\left(\frac{\sqrt{3}}{2}x' - \frac{1}{2}y'\right)\ 8\sqrt{3}\left(\frac{1}{2}x' + \frac{\sqrt{3}}{2}y'\right) = 0$$

$4x'^2 + 16y' = 0$ or $x'^2 = -4y'$; a parabola

5. $\cot 2\alpha = 0 \Rightarrow \alpha = \frac{\pi}{4}$.

$$A' = A\cos^2\alpha + B\sin\alpha\cos\alpha + C\sin^2\alpha = \left(\frac{1}{\sqrt{2}}\right)^2 - 2\left(\frac{1}{\sqrt{2}}\right)\left(\frac{1}{\sqrt{2}}\right) + \left(\frac{1}{\sqrt{2}}\right)^2 = 0$$

$$C' = A\cos^2\alpha - B\sin\alpha\cos\alpha + C\sin^2\alpha = \left(\frac{1}{\sqrt{2}}\right)^2 + 2\left(\frac{1}{\sqrt{2}}\right)\left(\frac{1}{\sqrt{2}}\right) + \left(\frac{1}{\sqrt{2}}\right)^2 = 2$$

$F' = F = 2.\ \therefore\ 2y'^2 = 2$ or $y = \pm 1$, two parallel lines

7. $\cot 2\alpha = 0 \Rightarrow \alpha = \frac{\pi}{4}$.

$$A' = A\cos^2\alpha + B\sin\alpha\cos\alpha + C\sin^2\alpha = \left(\frac{1}{\sqrt{2}}\right)^2 + 2\left(\frac{1}{\sqrt{2}}\right)\left(\frac{1}{\sqrt{2}}\right) + \left(\frac{1}{\sqrt{2}}\right)^2 = 2$$

$$C' = A\cos^2\alpha - B\sin\alpha\cos\alpha + C\sin^2\alpha = 0$$

$$D' = D\cos\alpha + E\sin\alpha = -4\sqrt{2}\left(\frac{1}{\sqrt{2}}\right) + 4\sqrt{2}\left(\frac{1}{\sqrt{2}}\right) = 0$$

$E' = -D\cos\alpha + E\sin\alpha = 8;\ F' = F = 4$

$\therefore\ 2x'^2 + 8y' + 4 = 0$ or $x'^2 = -4\left(y' + \frac{1}{2}\right)$, a parabola

9. $\cot 2\alpha = 0 \Rightarrow \alpha = \frac{\pi}{4}$.

$$A' = A\cos^2\alpha + B\sin\alpha\cos\alpha + C\sin^2\alpha = 3\left(\frac{1}{\sqrt{2}}\right)^2 + 2\left(\frac{1}{\sqrt{2}}\right)\left(\frac{1}{\sqrt{2}}\right) + 3\left(\frac{1}{\sqrt{2}}\right)^2 = 4$$

$$C' = A\cos^2\alpha - B\sin\alpha\cos\alpha + C\sin^2\alpha = 3\left(\frac{1}{\sqrt{2}}\right)^2 - 2\left(\frac{1}{\sqrt{2}}\right)\left(\frac{1}{\sqrt{2}}\right) + 3\left(\frac{1}{\sqrt{2}}\right)^2 = 2$$

$F' = F = 19$

$\therefore\ 4x'^2 + 2y'^2 = 19$, an ellipse

11. $\cot 2\alpha = \frac{14-2}{16} = \frac{3}{4}$. $\therefore\ \cos 2\alpha = \frac{3}{5}$.

$$\sin\alpha = \sqrt{\frac{1-\cos\alpha}{2}} = \sqrt{\frac{1-\frac{3}{5}}{2}} = \frac{1}{\sqrt{5}}$$

$$\cos\alpha = \sqrt{\frac{1+\cos\alpha}{2}} = \sqrt{\frac{1+\frac{3}{5}}{2}} = \frac{2}{\sqrt{5}}$$

13. $a^2 = x^2 + y^2 = (x'\cos\alpha - y'\sin\alpha)^2 + (x'\sin\alpha + y'\cos\alpha)^2 =$

$x'^2\cos^2\alpha - 2x'y'\sin\alpha\cos\alpha + y'^2\sin^2\alpha + x'^2\sin^2\alpha$

$+ 2x'y'\sin\alpha\cos\alpha + y'^2\cos^2\alpha =$

$x'^2(\cos^2\alpha + \sin^2\alpha) + y'^2(\cos^2\alpha + \sin^2\alpha) = x'^2 + y'^2$

15. $\cot 2\alpha = 0 \Rightarrow \alpha = \frac{\pi}{4}$.

$$A' = A\cos^2\alpha + B\sin\alpha\cos\alpha + C\sin^2\alpha = = \left(\frac{1}{\sqrt{2}}\right)^2 + 2\left(\frac{1}{\sqrt{2}}\right)\left(\frac{1}{\sqrt{2}}\right) + \left(\frac{1}{\sqrt{2}}\right)^2 = 2$$

$C' = A\cos^2\alpha - B\sin\alpha\cos\alpha + C\sin^2\alpha = 0$

$F' = F = 1$

$\therefore\ 2x'^2 = 1$ or $x' = \pm\frac{1}{\sqrt{2}}$, two parallel line

8.7 PARABOLA, ELLIPSE, OR HYPERBOLA?

1. $A = 1$, $B = 0$, $C = -1$; $B^2 - 4AC = 4$; hyperbola

3. $A = 0$, $B = 0$, $C = 1$; $B^2 - 4AC = 0$; parabola

5. $A = 1$, $B = 0$, $C = 4$; $B^2 - 4AC = -16$; ellipse

7. $A = 2$, $B = 0$, $C = -1$; $B^2 - 4AC = 8$; hyperbola

9. $A = 1$, $B = 0$, $C = 1$; $B^2 - 4AC = -4$; ellipse

11. $A = 3$, $B = 6$, $C = 3$; $B^2 - 4AC = 0$; parabola

13. $A = 2$, $B = 0$, $C = 3$; $B^2 - 4AC = -24$; ellipse

15. $A = 25$, $B = 0$, $C = -4$; $B^2 - 4AC = 400$; hyperbola

17. $A = 3$, $B = 12$, $C = 12$; $B^2 - 4AC = 0$; parabola

19. $D'^2 + E'^2 = (D\cos\alpha + E\sin\alpha)^2 + (-D\sin\alpha + E\cos\alpha)^2$
$= D^2\cos^2\alpha + 2DE\cos\alpha\sin\alpha + E^2\sin^2\alpha + D^2\sin^2\alpha$
$-2DE\cos\alpha\sin\alpha + E^2\cos\alpha\sin\alpha$
$= D^2(\cos^2\alpha + \sin^2\alpha) + E^2(\cos^2\alpha + \sin^2\alpha) = D^2 + E^2.$

21. $A' = A\cos^2\alpha + B\cos\alpha\sin\alpha + C\sin^2\alpha$
$C' = A\sin^2\alpha - B\cos\alpha\sin\alpha + C\cos^2\alpha$
$A' + C' = A(\sin^2\alpha + B\cos^2\alpha) + C(\sin^2\alpha + B\cos^2\alpha) = A + C$
$A' - C' = (A - C)\cos^2\alpha + (A - C)\sin^2\alpha + 2B\cos\alpha\sin\alpha$
$= (A - C)\cos 2\alpha + B\sin 2\alpha$
$(A' - C')^2 = (A - C)^2\cos^2 2\alpha + 2B(A - C)\sin 2\alpha\cos 2\alpha + B^2\sin^2 2\alpha$
$B'^2 = B^2\cos^2 2\alpha + 2B(C - A)\cos 2\alpha\sin 2\alpha + (C - A)^2\sin^2 2\alpha$
$(A' - C')^2 + B'^2 = (A - C)^2 + B^2(\cos^2 2\alpha + \sin^2 2\alpha) = (A - C)^2 + B^2$
$(A' + C')^2 = (A + C)^2$. Subtracting the last two, and simplifying, we have that $B'^2 - 4A'B' = B^2 - 4AC$.

8.8 SECTIONS OF A CONE

1. $PQ = PA\cos\beta$ and $PQ = PD\cos\alpha \Rightarrow PA\cos\beta = PD\cos\alpha$ or

$\dfrac{PA}{PD} = \dfrac{\cos\alpha}{\cos\beta}$. Since $\alpha = \beta$, $PA = PD$. Since $PA = PF$, $e = 1$ and we have a parabola.

3. The plane of the circle and the plane perpendicular to the axis of the cone are the same, so there is no line of intersection on which to locate the point D.

1. $x = \cos t$ and $y = \sin t \Rightarrow x^2 = \cos^2 t$ and $y^2 = \sin^2 t$ so
$x^2 + y^2 = \sin^2 t + \cos^2 t = 1$.

3. $x = 4\cos t$ and $y = 2\sin t \Rightarrow \frac{x}{4} = \cos t$ and $\frac{y}{2} = \sin t \Rightarrow$
$\left(\frac{x}{4}\right)^2 = \cos^2 t$ and $\left(\frac{y}{2}\right)^2 = \sin^2 t$ so $\left(\frac{x}{4}\right)^2 + \left(\frac{y}{2}\right)^2 = \sin^2 t + \cos^2 t = 1$.

5. $x = \cos 2t = 1 - 2\sin^2 t$ and $y = \sin t$, so $x = 1 - 2y^2$, $|x| \le 1$, $|y| \le 1$

7. $x = -\sec t$ and $y = \tan t \Rightarrow x^2 = \sec^2 t$ and $y = \tan^2 t$ so
$x^2 - y^2 = \sec^2 t - \tan^2 t = 1$. Since $-\frac{\pi}{2} < t < \frac{\pi}{2}$, the curve
includes only the left branch of the hyperbola.

9. $y = 1 - \cos t \Rightarrow \cos t = 1 - y \Rightarrow t = \cos^{-1}(1 - y)$. Therefore
$x = t - \sin t = \cos^{-1}(1 - y) - \sin(\cos^{-1}(1 - y)) = \cos^{-1}(1 - y) - \sqrt{2y - y^2}$.

11. $x = t^3 \Rightarrow t = x^{\frac{1}{3}}$ so $y = t^2 = x^{\frac{2}{3}}$.

13. $x = \sec^2 t - 1 = \tan^2 t = y^2$. Therefore $x = y^2$.

15. $x = t + 1 \Rightarrow t = x - 1$. Therefore, $y = t^2 + 4 = (x - 1)^2 + 4$.
$t \ge 0 \Rightarrow x \ge 1$ so the curve consists of only the right half of the parabola.

17. If $x^2 + y^2 = a^2$, then $2x + 2y\frac{dy}{dx} = 0$ or $\frac{dy}{dx} = -\frac{x}{y}$. Let $t = \frac{dy}{dx}$.
$-\frac{x}{y} = t \Rightarrow x = -yt$. $y^2t^2 + y^2 = a^2 \Rightarrow y = \frac{a}{\sqrt{1+t^2}}$ and $x = \frac{-at}{\sqrt{1+t^2}}$.

19. $PT = \text{arc}(AT) = s$. $\angle PTB = \frac{1}{2}s = t$.
$x = OB + BC = OB + DP = a\cos t + at\sin t = a(\cos t + t\sin t)$
$y = PC = TB - TD = a\sin t - at\cos t = a(\sin t - t\cos t)$

21. (a) The hypocycloid can be obtained from the equation of the epicycloid in problem 20 by replacing b with $-b$ to obtain

$$x = (a - b)\cos\theta + b\cos\left(\frac{a - b}{b}\right)\theta$$

$$y = (a - b)\sin\theta - b\sin\left(\frac{a - b}{b}\right)\theta$$

(b) If $b = \frac{a}{4}$, then $x = \frac{3a}{4}\cos\theta + \frac{a}{4}\cos 3\theta$ and $y = \frac{3a}{4}\sin\theta - \frac{a}{4}\sin 3\theta$.

$$x = \frac{a}{4}[3\cos\theta + \cos(2\theta + \theta)]$$
$$= \frac{a}{4}(3\cos\theta + \cos 2\theta\cos\theta - \sin 2\theta\sin\theta)$$
$$= \frac{a}{4}[(3\cos\theta + (\cos^2\theta - \sin^2\theta)\cos\theta - 2\sin^2\theta\cos\theta]$$
$$= \frac{a}{4}[3\cos\theta + \cos^3\theta - 3\sin^2\theta\cos\theta]$$
$$= \frac{a}{4}[3\cos\theta + \cos^3\theta - 3(1 - \cos^2\theta)\cos\theta]$$
$$= \frac{a}{4}[3\cos\theta + \cos^3\theta - 3\cos\theta + 3\cos^3\theta]$$
$$= a\cos^3\theta$$

$$y = \frac{a}{4}[3\sin\theta - \sin(2\theta + \theta)]$$
$$= \frac{a}{4}(3\sin\theta - \sin 2\theta\cos\theta - \cos 2\theta\sin\theta)$$
$$= \frac{a}{4}[(3\sin\theta - (\cos^2\theta - \sin^2\theta)\sin\theta - 2\sin\theta\cos^2\theta]$$
$$= \frac{a}{4}[3\sin\theta + \sin^3\theta - 3\cos^2\theta\sin\theta]$$
$$= \frac{a}{4}[3\sin\theta + \sin^3\theta - 3(1 - \sin^2\theta)\sin\theta]$$
$$= \frac{a}{4}[3\sin\theta + \sin^3\theta - 3\sin\theta + 3\sin^3\theta]$$
$$= a\sin^3\theta$$

23. (a) $x = x_0 + (x_1 - x_0)t \Rightarrow t = \frac{x - x_0}{x_1 - x_0}$. $y = y_0 + (y_1 - y_0)t \Rightarrow t = \frac{y - y_0}{y_1 - y_0}$.

$\frac{x - x_0}{x_1 - x_0} = \frac{y - y_0}{y_1 - y_0} \Rightarrow \frac{y - y_0}{x - x_0} = \frac{y_1 - y_0}{x_1 - x_0} = m$. Therefore, $y - y_0 = m(x - x_0)$.

(b) $x = tx_0$; $y = ty_0$;

(c) $x = -1 + t$; $y = t$ or $x = t$, $y = 1 + t$

25. $x = t$ and $y = t^2 \Rightarrow y = x^2$. $S = (x-2)^2 + \left(y - \frac{1}{2}\right)^2 = (x-2)^2 + \left(x^2 - \frac{1}{2}\right)^2$.

$S'(x) = 4x^3 - 4 = 0 \Leftrightarrow x = 1$. $S''(x) = 12x^2 \Rightarrow S(1)$ is a minimum.

The point closest to $\left(2, \frac{1}{2}\right)$ is $(1,1)$.

27. $x = a(t - \sin t) \Rightarrow \dfrac{dx}{dt} = a(1 - \cos t) \Rightarrow \left(\dfrac{dx}{dt}\right)^2 = a^2(1 - 2\cos t + \cos^2 t)$

$y = a(1 - \cos t) \Rightarrow \dfrac{dy}{dx} = a \sin t \Rightarrow \left(\dfrac{dy}{dx}\right)^2 = a^2 \sin^2 t.$

$$s = \int_0^{2\pi} \sqrt{\left(\frac{dx}{dt}\right)^2 + \left(\frac{dy}{dt}\right)^2}\, dt = \int_0^{2\pi} \sqrt{2a^2(1 - \cos t)}\, dt$$

$$= a\sqrt{2}\int_0^{2\pi} \sqrt{2}\sqrt{\frac{1 - \cos t}{2}}\, dt = 2a \int_0^{2\pi} \left|\sin \frac{t}{2}\right| dt$$

$$= -4a\left[\cos \frac{t}{2}\right]_0^{2\pi} = 8a \left(\text{Note: } \sin \frac{t}{2} \geq 0 \text{ for } 0 \leq t \leq 2\pi\right)$$

29. $V = \pi \displaystyle\int_0^{2\pi} y^2\, dx = \pi\int_0^{2\pi} y^2 \left(\frac{dx}{dt}\right) dt = \pi \int_0^{2\pi} (1 - \cos t)^2 (1 - \cos t)\, dt$

$$= \pi \int_0^{2\pi} (1 - 3\cos t + 3\cos^2 t - \cos^3 t)\, dt$$

$$I_1 = \pi\int_0^{2\pi} dt = 2\pi^2$$

$$I_2 = \pi\int_0^{2\pi} (-3\cos t)\, dt = -3\pi \sin t\Big]_0^{2\pi} = 0$$

$$I_3 = \pi\int_0^{2\pi} 3\cos^2 t\, dt = 3\pi\left[\frac{1}{2}t + \frac{1}{4}\sin 2t\right]_0^{2\pi} = 3\pi^2$$

$$I_4 = \pi\int_0^{2\pi} \cos^3 t\, dt = -\pi\int_0^{2\pi} (1 - \sin^2 t)\cos t\, dt = 0$$

$$V = I_1 + I_2 + I_3 + I_4 = 5\pi^2$$

31. From problem 27, $s = 8a$ for one arch. Therefore, the total length is $S = 8\left(\dfrac{5280}{2\pi}\right) = 6{,}723$.

8.M MISCELLANEOUS PROBLEMS

1. Let $u = kx$ and $v = ky$, so that $x = \frac{u}{k}$ and $y = \frac{v}{k}$. Then

$$x^2 + xy + y^2 = 3 \Rightarrow \left(\frac{u}{k}\right)^2 + \left(\frac{u}{k}\right)\left(\frac{v}{k}\right) + \left(\frac{v}{k}\right)^2 = 3 \Rightarrow u^2 + uv + v^2 = 3k^2.$$

3. $\sqrt{(x - x_1)^2 + (y - y_1)^2} = k\sqrt{(x - x_0)^2 + (y - y_0)^2} \Rightarrow$

$x^2 - 2xx_1 + x_1^2 + y^2 - 2yy_1 + y_1^2 = k^2[x^2 - 2xx_0 + x_0^2 + y^2 - 2yy_0 + y_0^2]$.

If $k = 1$, this reduces to $Ax + By = C$, where $A = 2x_0 - 2x_1$, $B = 2y_0 - 2y_1$, and $C = x_0^2 + y_0^2 - x_1^2 - y_1^2$. If $k \neq 1$, this reduces to $ax^2 + ay^2 + bx + cy = d$, where $a = 1 - k^2$, $b = 2k^2x_0 - 2x_1$, $c = 2k^2y_0 - 2y_1$, and $d = kx_0^2 + ky_0^2 - x_1^2 - y_1^2$.

5. Let $(x - h)^2 + (y - k)^2 = r^2$ be the equation of the circle. Then $\frac{dy}{dx} = -\frac{x - h}{y - k}$, and for $y = x^2$, $\frac{dy}{dx} = 2x$. To be tangent at the point $(2, 4) \Rightarrow -\frac{2 - h}{4 - k} = 4 \Rightarrow h = 18 - 4k$. If the circle passes through $(0, 1)$, then $r^2 = h^2 + (k - 1)^2$; similarly, $r^2 = (h - 2)^2 + (4 - k)^2$. Thus $h^2 + (k - 1)^2 = (h - 2)^2 + (4 - k)^2 \Rightarrow 4h + 6k = 19$. Solving the two linear equations together $\Rightarrow k = \frac{53}{10}$ and $h = -\frac{16}{5}$.

7. The vertex is the point $\left(\frac{7}{2}, 0\right)$.

$\therefore\ y^2 = 2\left(x - \frac{7}{2}\right)$.

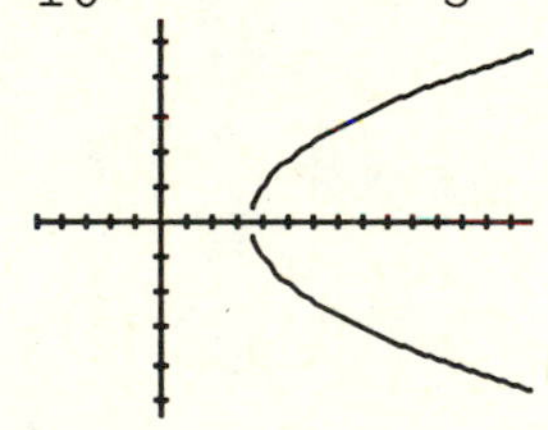

9. Let $P(x_1, y_1)$ be any point not the origin on the curve $y^2 = kx$. $\frac{dy}{dx} = \frac{k}{2y}$ so the slope of the line through P and Q is $m = \frac{k}{2y_1}$.

Therefore, $\frac{y - y_1}{x - x_1} = \frac{k}{2y_1} \Rightarrow y - y_1 = \frac{k}{2y_1}(x - x_1)$. At point A, $x = 0$

so $y = y_1 - \frac{kx_1}{2y_1} = y_1 - \left(\frac{k}{2y_1}\right)\left(\frac{y_1^2}{k}\right) = y_1 - \frac{1}{2}y_1 = \frac{1}{2}y_1$.

Since $\triangle AQO \approx \triangle PQR$, $AQ = \frac{1}{2}PQ$ and A bisects PQ.

11. Let $A(x_1, y_1)$ and $B(x_2, y_2)$ be any two points on the parabola $x^2 = 4py$.

Then $m = \dfrac{y_1 - y_2}{x_1 - x_2} = \dfrac{\dfrac{x_1^2}{4p} - \dfrac{x_2^2}{4p}}{x_1 - x_2} = \dfrac{(x_1 + x_2)(x_1 - x_2)}{4p(x_1 - x_2)} = \dfrac{1}{4p}(x_1 + x_2)$

$= \dfrac{1}{2p}\left(\dfrac{x_1 + x_2}{2}\right) = \dfrac{1}{2p}x_m$ where x_m is the midpoint of the segment AB.

The equation is: $x = 2pm$

13. $S = x^2 + (y-4)^2 = y^3 + (y-4)^2$.

$\dfrac{dS}{dy} = 3y^2 + 2(y-4) = 3y^2 + 2y - 8 = (3y-4)(y+2) = 0 \Leftrightarrow y = -2$ or $\dfrac{4}{3}$.

$\dfrac{d^2S}{dy^2} = 6y + 2 > 0$ if $y = \dfrac{4}{3} \Rightarrow$ a minimum value.

There are two points, $\left(\pm\dfrac{8\sqrt{3}}{9}, \dfrac{4}{3}\right)$.

15. Let the coordinates of Q be $(x_1, y_1) = \left(x_1, \sqrt{4px_1}\right)$.

Then the coordinates of the vertices are:

$D\left(-p, \sqrt{4px_1}\right)$, $A(-x_1, 0)$, $F(p, 0)$, and $Q\left(x_1, \sqrt{4px_1}\right)$.

The midpoint of DF is $\left(\dfrac{p-p}{2}, \dfrac{\sqrt{4px_1}}{2}\right) = \left(0, \dfrac{\sqrt{4px_1}}{2}\right)$.

The midpoint of AQ is $\left(\dfrac{x_1 - x_1}{2}, \dfrac{\sqrt{4px_1}}{2}\right) = \left(0, \dfrac{\sqrt{4px_1}}{2}\right)$.

Therefore DF and AQ bisect each other.

The slope of DF is $m_1 = -\dfrac{\sqrt{4x_1}}{2p}$. The slope of AQ is $m_2 = \dfrac{\sqrt{4px_1}}{2x_1}$.

DF is perpendicular to AQ since $\left(\dfrac{\sqrt{4px_1}}{2x_1}\right)\left(-\dfrac{\sqrt{4x_1}}{2p}\right) = -1$.

Therefore $\square$QDAF is a rhombus.

17. Let $y^2 = 4px$ be the parabola. Using the results in problem 9, $QM = MP$. Then $\triangle QMV \cong \triangle PTM \Rightarrow PT = QV$.

19. (a) $F_1P = \sqrt{4^2 + 3^2} = 5$. $F_2P = \sqrt{1^2 + 2^2} = \sqrt{5}$

(b) $OF_1 = 3$ and $Of_2 = 5$. $OF_1 + OF_2 = 8 > 5 + \sqrt{5} = 2a$. The point lies outside.

21. $x^2 + 9y^2 - 6x - 36y = 99 = 0 \Rightarrow (x^2 - 6x + 9) + 9(y^2 - 4y + 4) = 99 + 9 + 36.$

$\frac{(x-3)^2}{144} + \frac{(y-2)^2}{16} = 1.$ Center is $C(3, 2)$, vertices are $(15, 2)$ and $(-9, 2)$,

the foci are $(3 \pm 8\sqrt{2}, 2)$, and the eccentricity is $\frac{2\sqrt{2}}{3}$.

23. If one vertex is $(3, 1)$ and focus is $(1, 1)$ then $a - c = 2$. Solving this equation with $e = \frac{2}{3} = \frac{c}{a}$ gives $\frac{c}{2+c} = \frac{2}{3}$ or $c = 4$. Then $a = 6$, $b^2 = 36 - 16 = 20$, the center is $(-3, 1)$ and the equation is

$$\frac{(x+3)^2}{36} + \frac{(y-1)^2}{20} = 1.$$

25. $x^2 + 4y^2 - 4x - 8y + 4 = 0 \Rightarrow x^2 - 4x + 4 + 4(y^2 - 2y + 1) = 4$ or $(x-2)^2 + 4(y-1)^2 = 4$. The new center is $(2, 1)$.

27. (a) $K_{circle} = \int_0^a \sqrt{a^2 - x^2}\, dx$ (b) $K_{ellipse} = \int_0^a \frac{b}{a}\sqrt{a^2 - x^2}\, dx.$

Therefore $K_{ellipse} = \frac{b}{a} K_{circle} = \frac{b}{a}(\pi a^2) = \pi ab.$

29. $\frac{d_{11}}{c} + \frac{d_{12}}{c} = \frac{30}{c}$ and $\frac{d_{21}}{c} + \frac{d_{22}}{c} = \frac{30}{c} \Rightarrow$ that the locations P and Q lie on an ellipse with $2a = 30$ and $c = 10$. $b^2 = 225 - 100 = 125$ and the equation

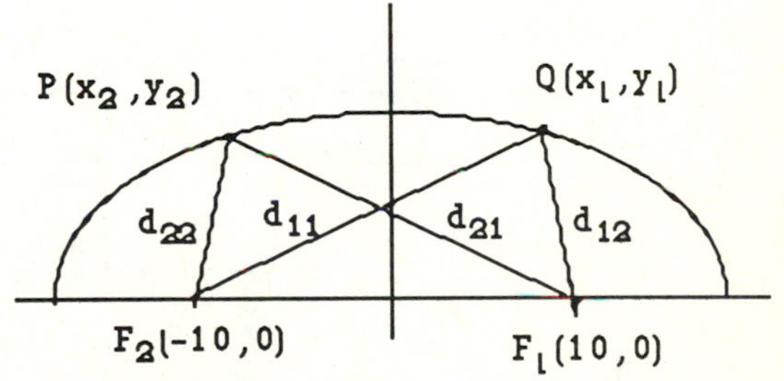

is $\frac{x^2}{225} + \frac{y^2}{125} = 1$. Moreover, $x_2 = x_1 + v_0 t = x_1 + 10$, and $x_2 = -x_1$, so $x_2 = -5$. $y_2 = \sqrt{125\left(1 - \frac{1}{9}\right)} = \pm\frac{10\sqrt{10}}{3}$. Assuming $y_2 > 0$, the plane is at the position $\left(-5, \frac{10\sqrt{10}}{3}\right)$.

31. $9x^2 - 4y^2 - 18x - 16y + 29 = 0 \Rightarrow 9(x^2 - 2x + 1) - 4(y^2 - 4y + 4) = -29 + 9 - 16$

$9(x-1)^2 - 4(y+2)^2 = -4$ or $\frac{(y+2)^2}{9} - \frac{(x-1)^2}{4} = 1$. The center is $C(1, -2)$, the vertices are $(1, -5)$ and $(1, 1)$, the foci are $(1, -2 \pm \sqrt{13})$, amd the asymptotes are $y + 2 = \pm\frac{3}{2}(x - 1)$.

33. Let $\frac{x^2}{t^2} + \frac{y^2}{t^2 - c^2} = 1$ and $\frac{x^2}{u^2} - \frac{y^2}{c^2 - u^2} = 1$ be any members of the family of ellipses and hyperbolas. Then $\frac{dy_e}{dx} = -\frac{x}{y}\left(\frac{t^2 - c^2}{t^2}\right)$ and $\frac{dy_h}{dx} = \frac{x}{y}\left(\frac{c^2 - u^2}{u^2}\right)$. At any point (x, y), $\left(\frac{dy_e}{dx}\right)\left(\frac{dy_h}{dx}\right) = -\frac{x^2}{y^2}\left(\frac{t^2 - c^2}{t^2}\right)\left(\frac{c^2 - u^2}{u^2}\right)$.

The curves intersect where $\frac{x^2}{t^2} + \frac{y^2}{t^2 - c^2} = \frac{x^2}{u^2} - \frac{y^2}{c^2 - u^2} \Rightarrow$

$x^2\left(\frac{1}{u^2} - \frac{1}{t^2}\right) = y^2\left(\frac{1}{t^2 - c^2} + \frac{1}{c^2 - u^2}\right) \Rightarrow \frac{x^2}{y^2} = \frac{u^2 t^2}{(t^2 - c^2)(c^2 - u^2)}$. Thus

$$\left(\frac{dy_e}{dx}\right)\left(\frac{dy_h}{dx}\right) = -\left(\frac{u^2 t^2}{(t^2 - c^2)(c^2 - u^2)}\right)\left(\frac{t^2 - c^2}{t^2}\right)\left(\frac{c^2 - u^2}{u^2}\right) = -1.$$

35. $\angle A = 2\angle B$. $\frac{y}{x + c} = \tan A = \tan 2B$. Using the double-angle tangent formula,

$$\frac{y}{c - x} = \frac{2\left(\frac{y}{x + c}\right)}{1 - \left(\frac{y}{x + c}\right)^2} = \frac{2y(x + c)}{(x + c)^2 - y^2}.$$ Simplifying, we get

$3x^2 - y^2 + 2cx - c^2 = 0$. The path is the right branch of this hyperbola.

37. The listener is on one brance of a hyperbola with foci the locations of the rifle and the target. Since the time for the bullet to hit the target remains constant, the difference between the distances from the listener to the rifle and to the target remains constant.

39. $\alpha = \frac{\pi}{4}$ so $x = \frac{x'}{\sqrt{2}} - \frac{y'}{\sqrt{2}}$ and $y = \frac{x'}{\sqrt{2}} + \frac{y'}{\sqrt{2}}$. Therefore

$1 = xy = \left(\frac{x'}{\sqrt{2}} - \frac{y'}{\sqrt{2}}\right)\left(\frac{x'}{\sqrt{2}} + \frac{y'}{\sqrt{2}}\right) = \frac{x'^2}{2} - \frac{y'^2}{2}$. $a = \sqrt{2}$, $c = 2$ and $e = \sqrt{2}$.

41. $xy = x + y \Rightarrow x\dfrac{dy}{dx} + y = 1 + \dfrac{dy}{dx}$ or $\dfrac{dy}{dx} = \dfrac{1-y}{x-1}$.

$\left.\dfrac{dy}{dx}\right|_{\left(-2,\frac{2}{3}\right)} = -\dfrac{1}{9}$. The slope of the normal is then $m = 9$,

and one equation is $y - \dfrac{2}{3} = 9(x+2)$ or $3y - 2yx - 56 = 0$. We need to

find the other point where the slope to the curve is $-\dfrac{1}{9}$:

$\dfrac{1-y}{x-1} = -\dfrac{1}{9} \Rightarrow x = 1 + 9y - 9$. Substituting into the curve,

$y(1 + 9y - 9) = 1 + 9y - 9 + y \Rightarrow y = \dfrac{2}{3}$ or $\dfrac{4}{3}$. If $y = \dfrac{4}{3}$, then $x = 4$ and

the other normal is $y - \dfrac{4}{3} = 9(x-4)$ or $3y - 27x + 104 = 0$.

43. (a) symmetric to the origin $\Rightarrow$

$Ax^2 + Bxy + Cy^2 + Dx + Ey + F = A(-x)^2 + B(-x)(-y) + C(-y)^2 + D(-x) + E(-y) +$

$-Dx - Ey = Dx + Ey \Rightarrow D = E = 0$.

(b) If $Ax^2 + Bxy + Cy^2 + F = 0$ and $(1,0)$ belongs to the curve, then

$A + F = 0$ or $F = -A$. The point $(2,1)$ belongs to the curve, so

$4A - 2B + C + F = 0$ so $B = 4A$.

(c) $2Ax + B\left(x\dfrac{dy}{dx} + y\right) + 2Cy\dfrac{dy}{dx} = 0$ and we know that the

slope of the tangent at $(-2,1)$ is 0, so substituting we get $C = 5A$.

Let $A = 1$. Then $x^2 + 4xy + 5y^2 - 1 = 0$.

45. If we do not rotate the axes, we can reason as follows: $\sqrt{2}y - 2xy = 2 \Rightarrow$

$y = \dfrac{2}{\sqrt{2} - 2x}$. Since $\lim\limits_{x\to\frac{1}{\sqrt{2}}} \dfrac{2}{\sqrt{2}-2x} = \infty$, $x = \dfrac{1}{\sqrt{2}}$ is vertical asymptote.

$\lim\limits_{x\to\infty} \dfrac{2}{\sqrt{2}-2x} = 0 \Rightarrow y = 0$ is horizontal asymptote. Hence the

center is $\left(\dfrac{1}{\sqrt{2}}, 0\right)$. One axis passes through the center with slope $= -1$,

and has equation $2y + 2x = \sqrt{2}$. This intersects the curve at the

vertices $\left(1 + \dfrac{1}{\sqrt{2}}, -1\right)$ and $\left(-1 + \dfrac{1}{\sqrt{2}}, 1\right)$. The other axis has slope $= 1$

and equation $2x - 2y = \sqrt{2}$. The distance from the center to a vertex

$= \sqrt{2} = a$. The hyperbolar is rectangular, so $e = \sqrt{2}$, hence $c = 2$

The foci are then $\left(\sqrt{2} + \dfrac{1}{\sqrt{2}}, -\sqrt{2}\right)$ and $\left(-\sqrt{2} + \dfrac{1}{\sqrt{2}}, \sqrt{2}\right)$.

Alternatively, we rotate the axes. Then $\cot 2\alpha = 0 \Rightarrow \alpha = \frac{\pi}{4}$.

$A' = A\cos^2\alpha + B\sin\alpha\cos\alpha + C\sin^2\alpha = -2\left(\frac{1}{\sqrt{2}}\right)\left(\frac{1}{\sqrt{2}}\right) = -1$

$C' = A\cos^2\alpha - B\sin\alpha\cos\alpha + C\sin^2\alpha = 1$

$D' = D\cos\alpha + E\sin\alpha = \sqrt{2}\left(\frac{1}{\sqrt{2}}\right) = 1$

$E' = -D\cos\alpha + E\sin\alpha = 1$; $F' = F = -2$

$\therefore -x'^2 + y'^2 \quad x' + y' = 2$, or $\dfrac{\left(y' + \frac{1}{2}\right)^2}{2} - \dfrac{\left(x' - \frac{1}{2}\right)^2}{2} = 1$.

With reference to these axes, we have $C\left(\frac{1}{2}, -\frac{1}{2}\right)$, $V\left(\frac{1}{2}, -\frac{1}{2} \pm \sqrt{2}\right)$,

$F\left(\frac{1}{2}, -\frac{1}{2} \pm 2\right)$, $e = \sqrt{2}$, $y + \frac{1}{2} = \pm\left(x - \frac{1}{2}\right)$.

47. $\left(\frac{x'}{\sqrt{2}} - \frac{y'}{\sqrt{2}}\right)^{1/2} + \left(\frac{x'}{\sqrt{2}} + \frac{y'}{\sqrt{2}}\right)^{1/2} = a^{1/2}$

$$\left[\left(\frac{x'}{\sqrt{2}} - \frac{y'}{\sqrt{2}}\right)^{1/2}\right]^2 = \left[a^{1/2} - \left(\frac{x'}{\sqrt{2}} + \frac{y'}{\sqrt{2}}\right)^{1/2}\right]^2$$

$$\frac{x'}{\sqrt{2}} - \frac{y'}{\sqrt{2}} = a - 2a^{1/2}\left(\frac{x'}{\sqrt{2}} + \frac{y'}{\sqrt{2}}\right)^{1/2} + \frac{x'}{\sqrt{2}} + \frac{y'}{\sqrt{2}}$$

$$\left[-\frac{2y'}{\sqrt{2}} - a\right]^2 = \left[-2a^{1/2}\left(\frac{x'}{\sqrt{2}} + \frac{y'}{\sqrt{2}}\right)^{1/2}\right]^2$$

$2y'^2 + 2a\sqrt{2}y' + a^2 = 2a\sqrt{2}x' + 2a\sqrt{2}y'$ or $2y'^2 = 2a\sqrt{2}x' - a^2$.

This is a part of a parabola in the first quadrant ($x > 0$, $y > 0$).

49. The solution set is the union of points which belong to the circle, line or parabola.

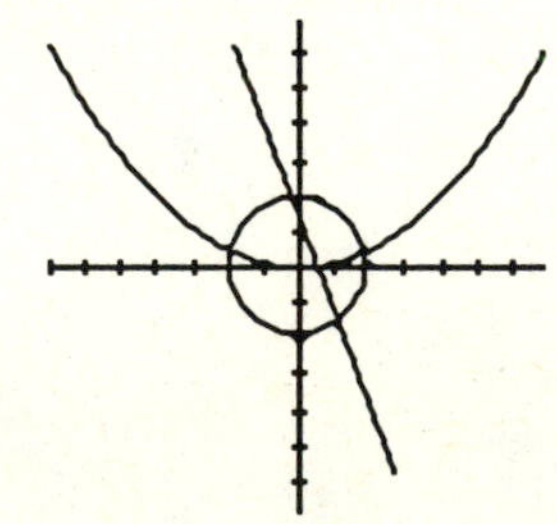

51. The solution set is the union of points which belong to the circle or line.

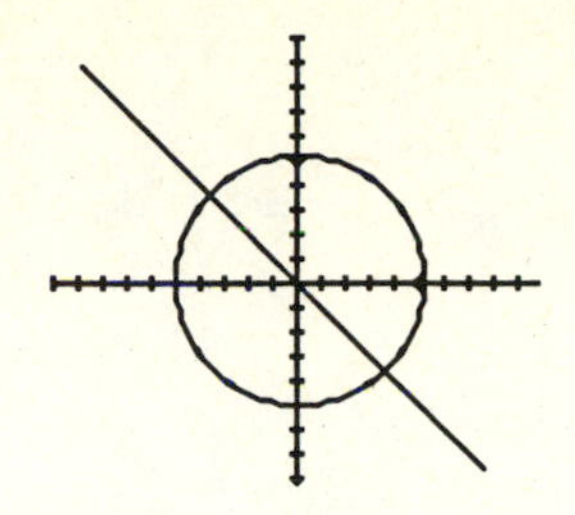

53. The solution set is the ellipse and the points interior to it.

55.

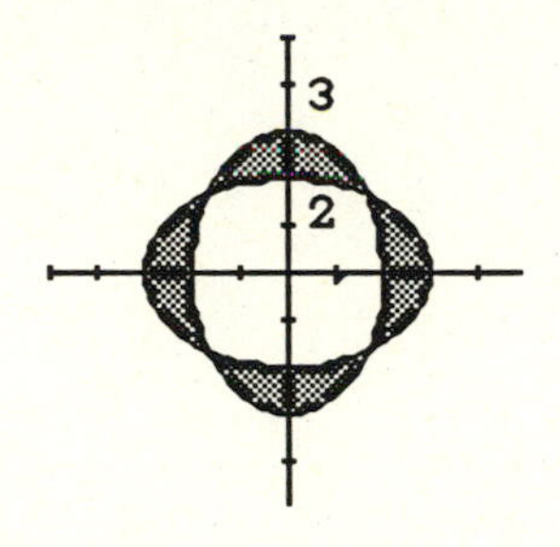

57. $x^4 - (y^2 - 9)^2 = 0$

$(x^2 + y^2 - 9)(x^2 - y^2 + 9) = 0$

The points on the circle and the hyperbola.

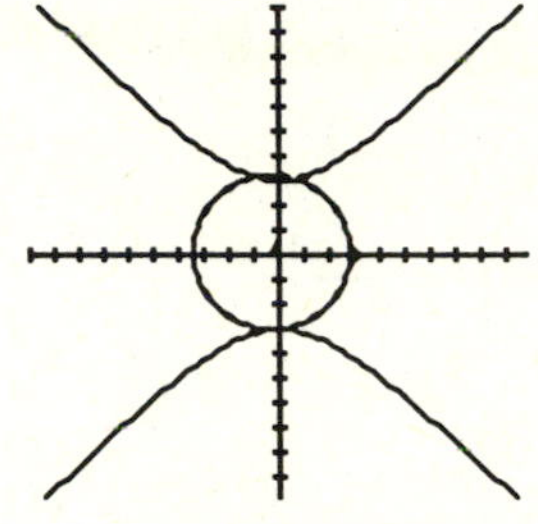

59. $|OA| = a$. $\dfrac{|AN|}{a} = \tan t$ and $\dfrac{|AM|}{a} = \sin t$. $MN = OP$, so

$|OP|^2 = |AN|^2 - |AM|^2 = a^2\tan^2 t - a^2 \sin^2 t \Rightarrow$

$|OP| = \sqrt{a^2\tan^2 t - a^2\sin^2 t} = \dfrac{a\sin^2 t}{\cos t}$. $\dfrac{x}{|OP|} = \cos\left(\dfrac{\pi}{2} - t\right)$ and

$\dfrac{y}{|OP|} = \sin\left(\dfrac{\pi}{2} - t\right) = \cos t$. $\therefore x = \dfrac{a\sin^3 t}{\cos t}$ and $y = a\sin^2 t$.

61. Use the equations for a trochoid derived in Section 8.9.

$x = a\phi - b\sin\phi$ and $y = a - b\cos\phi$, where $a = 4$, $b = 2$ and $\omega = 2$ rad/s $\Rightarrow$

$\dfrac{d\phi}{dt} = 2 \Rightarrow \phi = 2t$. Therefore, $x = 8t - 2\sin 2t$ and $y = 4 - 2\cos 2t$.

65. (a) $s = 4\int_0^{\frac{\pi}{2}} \sqrt{a^2 \sin^2 t + b^2 \cos^2 t}\, dt = 4\int_0^{\frac{\pi}{2}} \sqrt{a^2 (1 - \cos^2 t) + b^2 \cos^2 t}\, dt$

$= 4\int_0^{\frac{\pi}{2}} \sqrt{a^2 - (a^2 - b^2)\cos^2 t}\, dt = 4\int_0^{\frac{\pi}{2}} \sqrt{a^2 - c^2 \cos^2 t}\, dt$

$= 4\int_0^{\frac{\pi}{2}} \sqrt{a^2 \left(1 - \frac{c^2}{a^2}\cos^2 t\right)}\, dt = 4a\int_0^{\frac{\pi}{2}} \sqrt{1 - e^2 \cos^2 t}\, dt$

67. $xy = 2 \Rightarrow x\frac{dy}{dx} + y = 0 \Rightarrow \frac{dy}{dx} = -\frac{y}{x}.$

$x^2 - y^2 = 3 \Rightarrow 2x - 2y\frac{dx}{dy} = 0 \Rightarrow \frac{dy}{dx} = \frac{x}{y}.$

If (x_0, y_0) is a point of intersection,

the product of the slopes $\left(\frac{x_0}{y_0}\right)\left(-\frac{y_0}{x_0}\right) = -1$ so

the curves are orthogonal.

69. $2x^2 + 3y^2 = a^2 \Rightarrow \frac{dy}{dx} = -\frac{2x}{3y}.$

$Ky^2 = x^3 \Rightarrow \frac{dy}{dx} = \frac{3x^2}{2Ky} = \frac{3x^2 y}{2Ky^2} = \frac{3x^2 y}{2x^3} = \frac{3y}{2x}.$

$\left(-\frac{2x}{3y}\right)\left(\frac{3y}{2x}\right) = -1$ independently of a and K.

CHAPTER 9

HYPERBOLIC FUNCTIONS

9.1 DEFINITIONS AND IDENTITIES

1. $\sinh u = -\frac{3}{4}$

(a) $\cosh^2 u - \sinh^2 u = 1 \Rightarrow \cosh^2 u = 1 + \left(-\frac{3}{4}\right)^2 \Rightarrow \cosh u = \frac{5}{4}$

(b) $\tanh u = \frac{\sinh u}{\cosh u} = -\frac{3}{5}$ (c) $\coth u = \frac{1}{\tanh u} = -\frac{5}{3}$

(d) $\operatorname{sech} u = \frac{1}{\cosh u} = \frac{4}{5}$ (e) $\operatorname{csch} u = \frac{1}{\sinh u} = -\frac{4}{5}$

3. $\tanh u = -\frac{7}{25}$

(a) $1 - \tanh^2 u = \operatorname{sech}^2 u \Rightarrow 1 - \left(-\frac{7}{25}\right)^2 = \operatorname{sech}^2 u \Rightarrow \operatorname{sech} u = \frac{24}{25}$

(b) $\cosh u = \frac{1}{\operatorname{sech} u} = \frac{25}{24}$

(c) $\coth u = \frac{1}{\tanh u} = -\frac{25}{7}$ (d) $\frac{\sinh u}{\frac{25}{24}} = -\frac{7}{25} \Rightarrow \sinh u = -\frac{7}{24}$

(e) $\operatorname{csch} u = \frac{1}{\sinh u} = -\frac{24}{7}$

5. $\operatorname{sech} u = \frac{3}{5}$ (a) $\cosh u = \frac{5}{3}$

(b) $\cosh^2 u - \sinh^2 u = 1 \Rightarrow \frac{25}{9} - \sinh^2 u = 1 \Rightarrow \sinh u = \pm\frac{4}{3}$

(c) $\operatorname{csch} u = \frac{1}{\sinh u = \pm\frac{3}{4}}$ (d) $\tanh u = \frac{\sinh u}{\cosh u} = \frac{\pm\frac{4}{3}}{\pm\frac{5}{3}} = \pm\frac{4}{5}$

(e) $\coth u = \frac{1}{\tanh u} = \pm\frac{5}{4}$

7. $2\cosh(\ln x) = 2\left[\frac{e^{\ln x} + e^{-\ln x}}{2}\right] = x + \frac{1}{x}$

9. $\tanh(\ln x) = \frac{e^{\ln x} - e^{-\ln x}}{e^{\ln x} + e^{-\ln x}} = \frac{x - \frac{1}{x}}{x + \frac{1}{x}} = \frac{x^2 - 1}{x^2 + 1}$

11. $\cosh 5x + \sinh 5x = e^{5x}$

13. $\cosh 3x = \sinh 3x = e^{-3x}$

15. $\tanh(-x) = \dfrac{\sinh(-x)}{\cosh(-x)} = \dfrac{-\sinh x}{\cosh x} = -\tanh x \Rightarrow$ odd

17. $\operatorname{sech} -(x) = \dfrac{1}{\cosh(-x)} = \dfrac{1}{\cosh x} = \operatorname{sech} x \Rightarrow$ even

19. $\cosh(-3x) = \cosh 3x \Rightarrow$ even

21. $\sinh(-x)\cosh(-x) = (-\sinh x)(\cosh x) = -\sinh x \cosh x \Rightarrow$ odd

23. $\operatorname{sech}(-x) + \cosh(-x) = \operatorname{sech} x + \cosh x \Rightarrow$ even

25. $\cosh x = \sinh x + \frac{1}{2} \Rightarrow e^{-x} = \frac{1}{2} \Rightarrow x = -\ln\frac{1}{2}$

27. $\sinh u \cosh v + \cosh u \sinh v = \left(\dfrac{e^u - e^{-u}}{2}\right)\left(\dfrac{e^v + e^{-v}}{2}\right) + \left(\dfrac{e^u + e^{-u}}{2}\right)\left(\dfrac{e^v - e^{-v}}{2}\right)$

$= \frac{1}{4}(e^{u+v} - e^{-u+v} + e^{u-v} - e^{-u-v} + e^{u+v} + e^{-u+v} - e^{u-v} - e^{-u-v})$

$= \frac{1}{4}(2e^{u+v} - 2e^{-u-v}) = \dfrac{e^{u+v} - e^{-(u+v)}}{2} = \sinh(u+v)$

29. $\sinh(u - v) = \sinh(u + (-v) = \sinh u \cosh(-v) + \cosh u \sinh(-v)$
$= \sinh u \cosh v - \cosh u \sinh v$

31. $\sinh(u + v) + \sinh(u - v)$
$= (\sinh u \cosh v + \cosh u \sinh v) + (\sinh u \cosh v - \cosh u \sinh v)$
$= 2 \sinh u \cosh v$
$\therefore \sinh u \cosh v = \frac{1}{2} \sinh(u + v) + \frac{1}{2} \sinh(u - v)$

33. $\cosh(u + v) + \cosh(u - v)$
$= (\cosh u \cosh v + \sinh u \sinh v) + (\cosh u \cosh v - \sinh u \sinh v)$
$= 2 \cosh u \cosh v$
$\therefore \cosh u \cosh v = \frac{1}{2} \cosh(u + v) + \frac{1}{2} \cosh(u - v)$

35. $\sinh 3u = \sinh(2u + u) = \sinh 2u \cosh u + \sinh u \cosh 2u$

$= (2\sinh u \cosh u) \cosh u + \sinh u (\cosh^2 u + \sinh^2 u)$

$= 2 \sinh u \cosh^2 u + \sinh u \cosh^2 u + \sinh^3 u$

$= \sinh^3 u + 3 \sinh u \cosh^2 u$

37. $\cosh^2 u - \cosh^2 v = (1 + \sinh^2 u) - (1 + \sinh^2 v) = \sinh^2 u - \sinh^2 v$

39. $x = -\cosh u,\ y = \sinh u,\ -\infty < u < \infty$

$\left.\begin{array}{l} x^2 = \cosh^2 u \\ y^2 = \sinh^2 u \end{array}\right\} \Rightarrow \quad x^2 - y^2 = \cosh^2 u - \sinh^2 u = 1$

$\cosh u > 0 \Rightarrow -\cosh u < 0 \Rightarrow x < 0 \Rightarrow$ left-hand branch

41. $r = \sqrt{\cosh^2 u + \sinh^2 u} = \sqrt{\left(\frac{e^u + e^{-u}}{2}\right)^2 + \left(\frac{e^u - e^{-u}}{2}\right)^2}$

$$= \sqrt{\frac{e^{2u} + 2 + e^{-2u}}{4} + \frac{e^{2u} - 2 + e^{-2u}}{4}} = \sqrt{\frac{e^{2u} + e^{-2u}}{2}} = \sqrt{\cosh 2u}$$

43. $-\frac{\pi}{2} < \theta < \frac{\pi}{2} \Rightarrow \cos\theta > 0.$

Then $\sec\theta = \sqrt{1 + \tan^2\theta} = \sqrt{1 + \sinh^2 x} = \sqrt{\cosh^2 x} = |\cosh x| = \cosh x$

$\cos\theta = \frac{1}{\sec\theta} = \frac{1}{\cosh x} = \operatorname{sech} x$

$\frac{\sin\theta}{\cos\theta} = \tan\theta \Rightarrow \sin\theta = \tan\theta\cos\theta = \sinh x \operatorname{sech} x = \frac{\sinh x}{\cosh x} = \tanh x$

$\cot\theta = \frac{1}{\tan\theta} = \frac{1}{\sinh x} = \operatorname{csch} x$

$\csc\theta = \frac{1}{\sin\theta} = \frac{1}{\tanh x} = \coth x$

9.2 DERIVATIVES AND INTEGRALS

1. $y = \sinh 3x \Rightarrow \frac{dy}{dx} = 3\cosh 3x$

3. $y = \cosh^2 5x - \sinh^2 5x = 1 \Rightarrow \frac{dy}{dx} = 0$

5. $y = \coth(\tan x) \Rightarrow \frac{dy}{dx} = -\operatorname{csch}^2(\tan x)\sec^2 x$

7. $y = 4\operatorname{csch}\left(\frac{x}{4}\right)$

$\frac{dy}{dx} = -4\operatorname{csch}\left(\frac{x}{4}\right)\coth\left(\frac{x}{4}\right)\left(\frac{1}{4}\right) = -\operatorname{csch}\left(\frac{x}{4}\right)\coth\left(\frac{x}{4}\right)$

9. $y = \operatorname{sech}^2 x + \tanh^2 x = 1 \Rightarrow \frac{dy}{dx} = 0$

11. $y = \sin^{-1}(\tanh x)$

$\frac{dy}{dx} = \sqrt{\frac{1}{1 - \tanh^2 x}} \cdot \frac{d}{dx}(\tanh x) = \sqrt{\frac{1}{1 - \tanh^2 x}} \cdot \operatorname{sech}^2 x$

$= \frac{1.}{\operatorname{sech} x} \cdot \operatorname{sech}^2 x = \operatorname{sech} x$

13. $y = \ln|\tanh\left(\frac{x}{2}\right)|$

$$\frac{dy}{dx} = \frac{1}{|\tanh\left(\frac{x}{2}\right)|}\cdot\operatorname{sech}^2\left(\frac{x}{2}\right)\cdot\left(\frac{1}{2}\right) = \frac{1}{2}\cdot\frac{\cosh\left(\frac{x}{2}\right)}{\sinh\left(\frac{x}{2}\right)}\cdot\frac{1}{\cosh^2\left(\frac{x}{2}\right)}$$

$$= \frac{1}{2\sinh\left(\frac{x}{2}\right)\left(\cosh\left(\frac{x}{2}\right)\right)} = \frac{1}{\sinh x} = \operatorname{csch} x$$

15. $y = x\sinh 2x - \left(\frac{1}{2}\right)\cosh 2x$

$$\frac{dy}{dx} = \sinh 2x + 2x\cosh 2x - 2\left(\frac{1}{2}\right)\sinh 2x = 2x\cosh 2x$$

17. $\int \operatorname{sech} x\, dx = \sin^{-1}(\tanh x) + C$ because

$$\frac{d}{dx}(\sin^{-1}(\tanh x)) = \frac{1}{\sqrt{1-\tanh^2 x}}\cdot\operatorname{sech}^2 x = \operatorname{sech} x$$

19. $\int_{-1}^{1} \cosh 5x\, dx = \frac{1}{5}\sinh 5x\Big]_{-1}^{1} = \frac{1}{5}(\sinh 5 - \sinh(-5)) = \frac{2}{5}\sinh 5$

21. $\int_{-3}^{3} \sinh x\, dx = 0$ since $\sinh x$ is an odd function.

23. $\int_0^1 x\cosh x\, dx = x\sinh x\Big]_0^1 - \int_0^1 \sinh x\, dx = \sinh 1 - \cosh 1 + 1 = 1 - e^{-1}$

Let $u = x \Rightarrow du = dx$, $dv = \cosh x\, dx \Rightarrow v = \sinh x$

25. $$\int_0^{\frac{1}{2}} \frac{\sinh x}{e^x}\, dx = \int_0^{\frac{1}{2}} e^{-x}\left(\frac{e^x - e^{-x}}{2}\right)dx = \int_0^{\frac{1}{2}}\left(\frac{1}{2} - \frac{1}{2}e^{-2x}\right)dx$$

$$= \frac{1}{2}x + \frac{1}{4}e^{-2x}\Big]_0^{\frac{1}{2}} = \frac{1}{4} + \frac{1}{4e} - \frac{1}{4} = \frac{1}{4e}$$

27. $\int_0^1 \sinh\sqrt{x}\, dx = \int_0^1 2y\sinh y\, dy = 2y\cosh y - 2\int_0^1 \cosh y\, dy$

Let $y = \sqrt{x} \Rightarrow dy = \frac{1}{2\sqrt{x}}dx \Rightarrow dx = 2\sqrt{x}dy = 2y\, dy$

Limits: $x = 0 \Rightarrow y = \sqrt{0} = 0$; $x = 1 \Rightarrow y = \sqrt{1} = 1$

Let $u = 2y$ $dv = \sinh y\, dy$

$du = 2\, dy$ $v = \cosh y$

$= 2y\cosh y - 2\sinh y\Big]_0^1 = 2\cosh 1 - 2\sinh 1 = 2e^{-1}$

29. $\int \cosh(2x+1)\, dx = \frac{1}{2}\sinh(2x+1) + C$

31. $\displaystyle\int \frac{\sinh x\,dx}{\cosh^4 x} = \int \cosh^{-4} x \sinh x\,dx = -\frac{1}{3}\cosh^{-3} x + C = -\frac{1}{3}\operatorname{sech}^3 x + C$

33. $\displaystyle\int \frac{e^x - e^{-x}}{e^x + e^{-x}}\,dx = \int \frac{\sinh x}{\cosh x}\,dx = \ln|\cosh x| + C = \ln(\cosh x) + C$

35. $\displaystyle\int \frac{\sinh\sqrt{x}}{\sqrt{x}}\,dx = 2\cosh\sqrt{x} + C$

$\quad$ Let $u = \sqrt{x} \Rightarrow du = \dfrac{1}{2\sqrt{x}}\,dx \Rightarrow 2\,du = \dfrac{1}{\sqrt{x}}\,dx$

$\quad \displaystyle\int 2\sinh u\,du = 2\cosh u + C$

37. $\displaystyle\int \sqrt{\cosh x - 1}\,dx = \int \sqrt{\left(2\cosh^2 \frac{x}{2} - 1\right) - 1}\,dx$

$$= \sqrt{2}\int \sqrt{\cosh^2 \frac{x}{2} - 1}\,dx = \sqrt{2}\int \sinh\frac{x}{2}\,dx = 2\sqrt{2}\cosh\frac{x}{2} + C$$

39. There are three possibilities:

(1) $\displaystyle\int 2\cosh x \sinh x\,dx = \cosh^2 x + C$

(2) $\displaystyle\int 2\cosh x \sinh x\,dx = \sinh^2 x + C$

(3) $\displaystyle\int 2\cosh x \sinh x\,dx = \int \sinh 2x\,dx = \frac{1}{2}\cosh 2x + C$

41. $\displaystyle\int \frac{\sinh x}{1 + \cosh x}\,dx = \ln(1 + \cosh x) + C$

43. $\displaystyle\int x^2 \cosh x\,dx = x^2 \sinh x - 2\int x \sinh x\,dx$

$\quad$ Let $u = x^2 \Rightarrow du = 2x\,dx$, $dv = \cosh x\,dx \Rightarrow v = \sinh x$

$$= x^2 \sinh x - 2\left(x\cosh x - \int \cosh x\,dx\right)$$

$\quad$ Let $u = x \Rightarrow du = dx$, $dv = \sinh x\,dx \Rightarrow v = \cosh x$

$$= x^2 \sinh x - 2x\cosh x + 2\sinh x + C$$

45. $\displaystyle\int \operatorname{sech}^3 5x \tanh 5x\,dx = \int \operatorname{sech}^2 5x\,(\operatorname{sech} 5x \tanh 5x)\,dx = -\frac{1}{15}\operatorname{sech}^3 x + C$

47. $\displaystyle\int \sinh^4 3x\,dx = \frac{1}{12}\sinh^3 3x \cosh 3x - \frac{3}{4}\int \sinh^2 3x\,dx$

$$= \frac{1}{12}\sinh^3 3x \cosh 3x - \frac{1}{16}\sinh 6x + \frac{3}{8}x + C$$

49. $\displaystyle\int e^{3x}\cosh 2x\,dx = \frac{e^{3x}}{2}\left[\frac{e^{2x}}{5} + \frac{e^{-2x}}{1}\right] + C = \frac{e^{5x}}{10} + \frac{e^{x}}{2} + C$

51. (a) $\int \operatorname{csch} x\,dx = \int \operatorname{csch} x\left(\frac{\operatorname{csch} x + \coth x}{\operatorname{csch} x + \coth x}\right)dx$

$$= \int \frac{\operatorname{csch}^2 x + \operatorname{csch} x \coth x}{\operatorname{csch} x + \coth x}\,dx = -\ln|\operatorname{csch} x + \coth x| + C$$

(b) $-\ln|\operatorname{csch} x + \coth x| + C = \ln\left|\tanh\frac{x}{2}\right| + C$

$$= \ln\left|\frac{\sinh\frac{x}{2}}{\cosh\frac{x}{2}}\right| + C = \ln\left|\sinh\frac{x}{2}\right| - \ln\left|\cosh\frac{x}{2}\right| + C$$

53. $V = \pi\int_0^3 (\cosh^2 x - \sinh^2 x)\,dx$

$$= \pi\int_0^3 dx = \pi x\Big]_0^3 = 3\pi$$

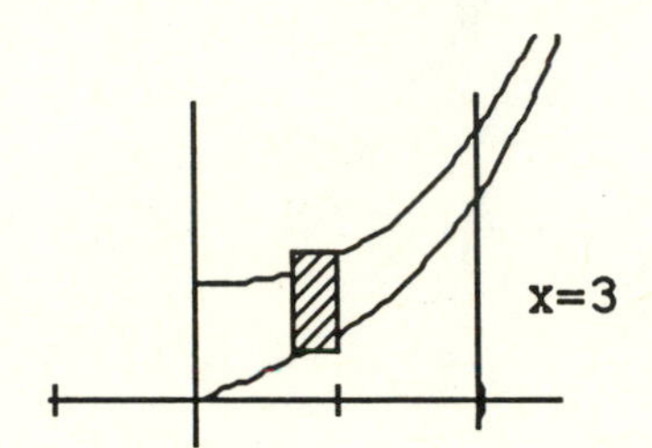

55. $S = 2\pi\int_0^{\ln 2} \cosh x\sqrt{1+\sinh^2 x}\,dx$

$$= 2\pi\int_0^{\ln 2} \cosh^2 x\,dx = 2\pi\int_0^{\ln 2} \frac{\cosh 2x + 1}{2}\,dx$$

$$= \pi\left[\frac{\sinh 2x}{2} + x\right]_0^{\ln 2} = \pi\left[\frac{e^{2\ln 2} - e^{-2\ln 2}}{4} + \ln 2\right]$$

$$= \frac{15}{16}\pi + \pi\ln 2$$

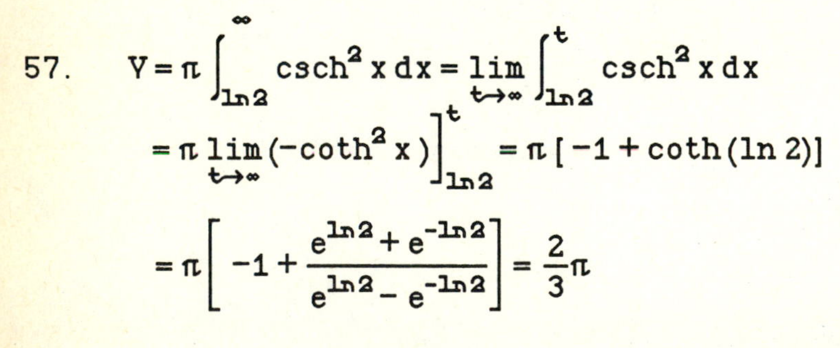

57. $V = \pi\int_{\ln 2}^{\infty} \operatorname{csch}^2 x\,dx = \lim_{t\to\infty}\int_{\ln 2}^{t} \operatorname{csch}^2 x\,dx$

$$= \pi\lim_{t\to\infty}(-\coth^2 x)\Big]_{\ln 2}^{t} = \pi[-1 + \coth(\ln 2)]$$

$$= \pi\left[-1 + \frac{e^{\ln 2} + e^{-\ln 2}}{e^{\ln 2} - e^{-\ln 2}}\right] = \frac{2}{3}\pi$$

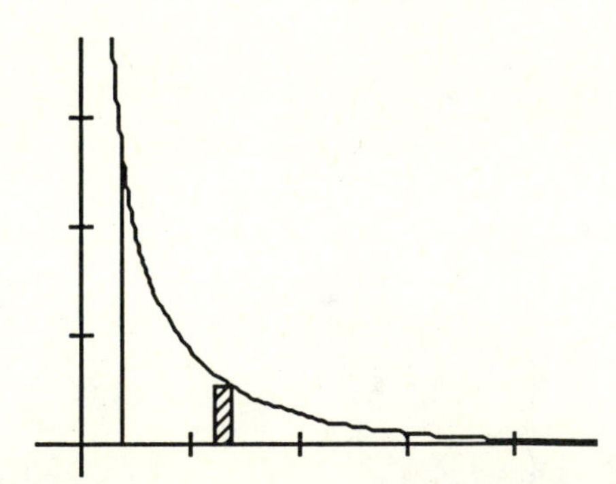

59. $\bar{x} = 0$ by symmetry.

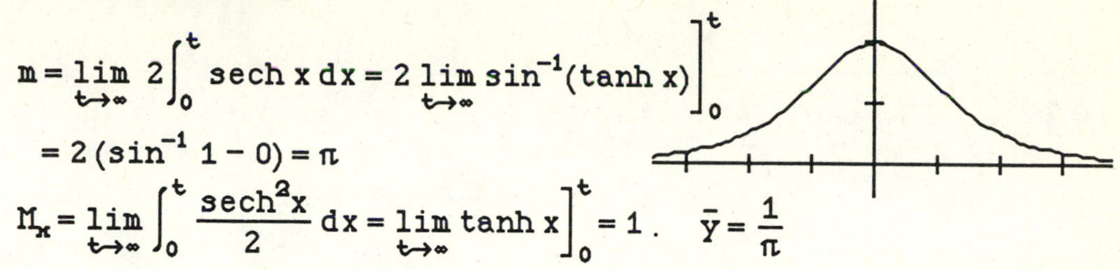

$$m = \lim_{t\to\infty} 2\int_0^t \operatorname{sech} x\,dx = 2\lim_{t\to\infty} \sin^{-1}(\tanh x)\Big]_0^t$$

$$= 2(\sin^{-1} 1 - 0) = \pi$$

$$M_x = \lim_{t\to\infty} \int_0^t \frac{\operatorname{sech}^2 x}{2}\,dx = \lim_{t\to\infty} \tanh x\Big]_0^t = 1. \quad \bar{y} = \frac{1}{\pi}$$

61. $$V = \pi\int_0^\infty (1 - \tanh^2 x)\,dx$$

$$= \pi\lim_{t\to\infty}\int_0^t \operatorname{sech}^2 x\,dx = \pi\lim_{t\to\infty}[\tanh x\Big]_0^t$$

$$= \pi(1 - 0) = \pi$$

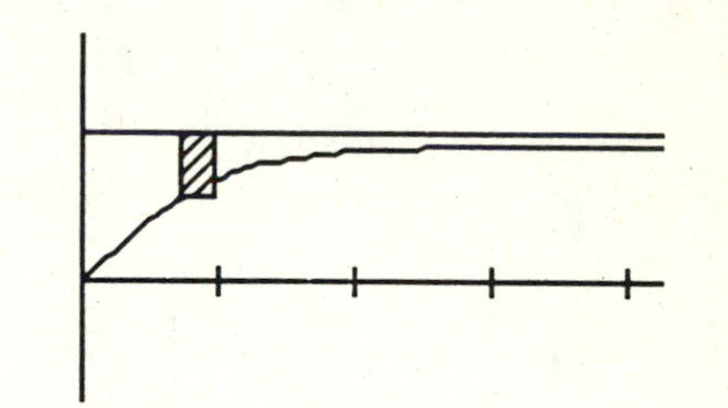

63. (a) $y = A\cosh x + B\sinh x + C\cos x + D\sin x$

$y' = A\sinh x + B\cosh x - C\sin x + D\cos x$

$y'' = A\cosh x + B\sinh x - C\cos x - D\sin x$

$y^{(3)}(x) = A\sinh x + B\cosh x + C\sin x - D\cos x$

$y^{(4)}(x) = A\cosh x + B\sinh x + C\cos x + D\sin x$

$y^{(4)} - y = 0$

(b) $y^{(4)} = y \Rightarrow y^{(4)} - y = 0$

This yields the system

$$A + C = 0$$
$$B + D = 0$$
$$A - C = 2$$
$$B - D = 2,$$

which has solution $A = 1$, $B = 1$, $C = -1$, $D = -1$

65. $x = -\frac{2}{\sqrt{3}}\sinh\frac{t}{\sqrt{3}} \qquad \frac{dx}{dt} = -\frac{2}{\sqrt{3}}\cdot\frac{1}{\sqrt{3}}\cosh\frac{t}{\sqrt{3}}$

$y = \frac{1}{\sqrt{3}}\sinh\frac{t}{\sqrt{3}} + \cosh\frac{t}{\sqrt{3}} \qquad \frac{dy}{dt} = \frac{1}{\sqrt{3}}\cdot\frac{1}{\sqrt{3}}\cosh\frac{t}{\sqrt{3}} + \frac{1}{\sqrt{3}}\sinh t$

$$\frac{dx}{dt} + 2\frac{dy}{dt} + x = -\frac{2}{3}\cosh\frac{t}{\sqrt{3}} + 2\left(\frac{1}{3}\cosh\frac{t}{\sqrt{3}} + \frac{1}{\sqrt{3}}\sinh\frac{1}{\sqrt{3}}\right)$$
$$- \frac{2}{\sqrt{3}}\sinh\frac{t}{\sqrt{3}} = 0$$

$$\frac{dx}{dt} - \frac{dy}{dt} + y = -\frac{2}{3}\cosh\frac{t}{\sqrt{3}} - \frac{1}{3}\cosh\frac{t}{\sqrt{3}} - \frac{1}{\sqrt{3}}\sinh\frac{1}{\sqrt{3}}$$
$$+ \frac{1}{\sqrt{3}}\sinh\frac{t}{\sqrt{3}} + \cosh\frac{t}{\sqrt{3}} = 0$$

$x(0) = -\frac{2}{\sqrt{3}}\sinh 0 = 0 \quad y(0) = \frac{1}{\sqrt{3}}\sinh 0 + \cosh 0 = 1$

9.3 HANGING CABLES

1. $s = \int_0^{x_1} \sqrt{1 + \sinh^2\left(\frac{x}{a}\right)}\, dx = \int_0^{x_1} \cosh \frac{x}{a}\, dx = a \sinh \frac{x}{a}\Big]_0^{x_1} = a \sinh \frac{x_1}{a}$

3. By problem 1, the arc length is $a \sinh \frac{x_1}{a}$, so the area of the rectangle is $a \cdot a \sinh \frac{x_1}{a} = a^2 \sinh \frac{x_1}{a}$.

5. $\bar{x} = 0$ by symmetry.

$$s = \int_{-x_1}^{x_1} \sqrt{1 + \sinh^2 \frac{x}{a}}\, dx = \cosh \frac{x}{a}\Big]_{-x_1}^{x_1} = 2 \sinh \frac{x_1}{a}$$

$$M_x = \int_{-x_1}^{x_1} a \cosh^2 \frac{x}{a}\, dx = \frac{a}{2} \int_{-x_1}^{x_1} \left(\cosh \frac{2x}{a} + 1\right) dx = \frac{1}{2}\left[x + \frac{a}{2} \sinh \frac{2x}{a}\right]_{-x_1}^{x_1}$$

$$= x_1 + \frac{a}{2} \sinh \frac{2x_1}{a}. \quad \bar{y} = \frac{M_x}{s} = \frac{x_1}{2} \operatorname{csch} \frac{x_1}{a} + \frac{a}{2} \cosh \frac{x_1}{a} = \frac{x_1}{2} \operatorname{csch} \frac{x_1}{a} + \frac{y_1}{2}$$

7. (a) If $s = a \sinh \frac{x}{a}$ then $\sinh \frac{x}{a} = \frac{x}{a}$ and $x = a \sinh^{-1}\left(\frac{x}{a}\right)$.

$$y = a \cosh \frac{x}{a} = \sqrt{a^2 + a^2 \sinh^2 \frac{x}{a}} = \sqrt{s^2 + a^2}$$

(b) $\frac{dx}{ds} = \frac{a}{\sqrt{1 + \frac{s^2}{a^2}}} \cdot \frac{1}{a} = \frac{a}{\sqrt{a^2 + s^2}} \qquad \frac{dy}{ds} = \frac{s}{\sqrt{s^2 + a^2}}$

$$\left(\frac{dx}{ds}\right)^2 + \left(\frac{dy}{ds}\right)^2 = \frac{a^2}{a^2 + s^2} + \frac{s^2}{a^2 + s^2} = 1$$

9. (a) We will use Newton's Method on the function $f(x) = \cosh x - \frac{x}{2} - 1$.

$x_{n+1} = x_n - \frac{f(x_n)}{f'(x_n)} = x_n - \frac{2 \cosh x_n - x_n + 2}{2 \sinh x_n - 1}$. Take $x_1 = 1$.

$x_2 = 0.9362$, $x_3 = 0.9309$ and $F(0.9309) = 0.00002$. The curves cross at approximately $(0.9309, 1.4654)$

(b) From part (a), $\frac{50}{a} \approx 0.930$ or $a \approx 53.8$.

$$s = \int_{-50}^{50} \sqrt{1+\sinh^2 \frac{x}{a}}\, dx = \int_{-50}^{50} \cosh \frac{x}{a}\, dx = a \sinh \frac{x}{a} \Big]_{-50}^{50} = 115 \text{ ft}$$

(c) $T = wy = wa \cosh \frac{x}{a} \approx (0.3)(53.8) \approx 16.1$ lbs.

9.4 INVERSE HYPERBOLIC FUNCTIONS

1. $\sinh^{-1}(0) = \ln(0 + \sqrt{0+1}) = \ln 1 = 0$

3. $\sinh^{-1}\left(-\frac{4}{3}\right) = \ln\left(-\frac{4}{3} + \sqrt{\frac{16}{9}+1}\right) = \ln\left(-\frac{4}{3} + \frac{5}{3}\right) = \ln \frac{1}{3}$

5. $\cosh^{-1}\left(\frac{5}{4}\right) = \ln\left(\frac{5}{4} + \sqrt{\frac{25}{16}-1}\right) = \ln\left(\frac{5}{4} + \frac{3}{4}\right) = \ln 2$

7. $\cosh^{-1}\left(\frac{2}{\sqrt{3}}\right) = \ln\left(\frac{2}{\sqrt{3}} + \sqrt{\frac{4}{3}-1}\right) = \ln\left(\frac{2}{\sqrt{3}} + \frac{1}{\sqrt{3}}\right) = \frac{1}{2}\ln 3$

9. $\tanh^{-1}\left(\frac{1}{2}\right) = \frac{1}{2}\ln \frac{1+\frac{1}{2}}{1-\frac{1}{2}} = \frac{1}{2}\ln 3$

11. $\coth^{-1}(-2) = \frac{1}{2}\ln \frac{-2+1}{-2-1} = \frac{1}{2}\ln \frac{1}{3} = -\frac{1}{2}\ln 3$

15. $\operatorname{csch}^{-1}\left(\frac{5}{12}\right) = \ln\left(\frac{12}{5} + \sqrt{1 + \frac{25}{144}}\cdot\frac{12}{5}\right) = \ln 5$

17. $y = \sinh^{-1}(2x)$ $\quad \frac{dy}{dx} = \frac{2}{\sqrt{1+4x^2}}$

19. $y = \cosh^{-1}(\sec x)$ $\quad \frac{dy}{dx} = \frac{\sec x \tan x}{\sqrt{\sec^2 x - 1}} = \frac{\sec x \tan x}{\tan x} = \sec x$

21. $y = \operatorname{sech}^{-1}(\sin 2x)$ $\quad \frac{dy}{dx} = \frac{-2\cos 2x}{\sin 2x\sqrt{1-\sin^2 2x}} = -2\csc 2x$

23. $y = \sinh^{-1}\sqrt{x-1}$ $\quad \frac{dy}{dx} = \frac{1}{\sqrt{1+x-1}}\cdot\frac{1}{2\sqrt{x-1}} = \frac{1}{2\sqrt{x(x-1)}}$

25. $y = \sinh^{-1}\frac{1}{x}$ $\quad \frac{dy}{dx} = -\frac{1}{x^2}\frac{1}{\sqrt{1+\frac{1}{x^2}}} = -\frac{1}{x^2}\cdot\frac{|x|}{\sqrt{x^2+1}} = -\frac{1}{|x|\sqrt{x^2+1}}$

27. $y = \sinh^{-1}(\tan x)$ $\quad \frac{dy}{dx} = \frac{\sec^2 x}{\sqrt{1+\tan^2 x}} = |\sec x|$

29. $y = \sqrt{1+x^2} - \sinh^{-1}\frac{1}{x} = \sqrt{1+x^2} + \operatorname{csch}^{-1} x$

$$\frac{dy}{dx} = \frac{x}{\sqrt{1+x^2}} + \frac{1}{|x|\sqrt{x^2+1}}$$

31. $y = 2\cosh^{-1}\left(\frac{x}{2}\right) + \frac{x}{2}\sqrt{x^2-4}$

$$\frac{dy}{dx} = \frac{2\cdot\frac{1}{2}}{\sqrt{\frac{x^2}{4}-1}} + \frac{1}{2}\sqrt{x^2-4} + \left(\frac{x}{2}\right)\left(\frac{1}{2}\right)\left(\frac{2x}{\sqrt{x^2-4}}\right)$$

$$= \frac{2}{\sqrt{x^2-4}} + \frac{\sqrt{x^2-4}}{2} + \frac{x^2}{2\sqrt{x^2-4}} = \frac{x^2}{\sqrt{x^2-4}}$$

33. $\displaystyle\int_0^{\frac{4}{3}} \frac{dx}{\sqrt{1+4x^2}} = \frac{1}{2}\sinh^{-1} 2x\Big]_0^{\frac{4}{3}} = \frac{1}{2}\left[\sinh^{-1}\frac{8}{3} - \sinh^{-1} 0\right]$

$$= \frac{1}{2}\left[\ln\left(\frac{8}{3} + \sqrt{\frac{64}{9}+1}\right) - \ln(0+1)\right] = \frac{1}{2}\ln\left(\frac{8}{3} + \frac{\sqrt{73}}{3}\right)$$

35. $\displaystyle\int_0^{.5} \frac{dx}{1-x^2} = \tanh^{-1} x\Big]_0^{.5} = \frac{1}{2}\ln\left|\frac{\frac{3}{2}}{\frac{1}{2}}\right| - \frac{1}{2}\ln 1 = \frac{1}{2}\ln 3$

37. $\displaystyle\int_1^2 \frac{dx}{x\sqrt{4+x^2}} = \int_1^2 \frac{dx}{2x\sqrt{1+\left(\frac{x}{2}\right)^2}} = -\frac{1}{2}\operatorname{csch}^{-1}\frac{x}{2}\Big]_1^2$

$$= -\frac{1}{2}\left[\operatorname{csch}^{-1} 1 - \operatorname{csch}^{-1}\frac{1}{2}\right] = -\frac{1}{2}\left[\ln(1+\sqrt{2}) - \ln\left(2 + \frac{\sqrt{\frac{5}{4}}}{\frac{1}{2}}\right)\right]$$

$$= -\frac{1}{2}[\ln(1+\sqrt{2}) - \ln(2+\sqrt{5})]$$

39. $\displaystyle\int_{-1}^1 \sinh^{-1} x\,dx = x\sinh^{-1} x\Big]_{-1}^1 - \int_{-1}^1 \frac{x}{\sqrt{1+x^2}}\,dx$

$$= x\sinh^{-1} x - \sqrt{1+x^2}\Big]_{-1}^1 = \sinh^{-1} 1 + \sinh^{-1}(-1) - \sqrt{2} + \sqrt{2} = 0$$

41. $\int \sqrt{x^2+1}\, dx = \int \frac{x^2+1}{\sqrt{x^2+1}}\, dx = \int \frac{x^2}{\sqrt{x^2+1}}\, dx + \int \frac{1}{\sqrt{x^2+1}}\, dx$

$$\int \sqrt{x^2}+1 = x\sqrt{x^2+1} - \int \sqrt{x^2+1}\, dx + \sinh^{-1} x + C$$

$$2\int \sqrt{x^2+1}\, dx = x\sqrt{x^2+1} + \sinh^{-1} x + C$$

$$\int \sqrt{x^2+1}\, dx = \frac{x}{2}\sqrt{x^2+1} + \frac{1}{2}\sinh^{-1} x + C$$

43. $\int_2^5 4x \coth^{-1} x\, dx = 2x^2 \coth^{-1} x \Big]_2^5 - \int_2^5 \frac{2x^2}{1-x^2}\, dx$

$$= 2x^2 \coth^{-1} x \Big]_2^5 + \int_2^5 \left(2 - \frac{2}{1-x^2}\right) dx$$

$$= (2x^2-2)\coth^{-1} x + 2x \Big]_2^5 = 24 \ln \frac{3}{2} - 3 \ln 3 + 6$$

45. $\int \frac{dx}{\sqrt{x^2-4x+3}} = \int \frac{dx}{\sqrt{(x-2)^2-1}} = \cosh^{-1}(x-2) + C$, if $x > 3$

47. $\int \frac{du}{\sqrt{u^2-1}} = \int \frac{\sec\theta \tan\theta\, d\theta}{\tan\theta} = \int \sec\theta\, d\theta = \ln|\sec\theta + \tan\theta|$

$$= \ln|u^2 + \sqrt{u^2-1}| + C$$

49. $\int \frac{du}{u\sqrt{u^2-1}} = \int \frac{\sec^2\theta\, d\theta}{\tan\theta \sec\theta} = \int \csc\theta\, d\theta = \ln|\csc\theta - \cot\theta| + C$

$$= \ln\left|\frac{\sqrt{u^2+1}}{u} - \frac{1}{u}\right| + C$$

51. Use Formula #42 in the textbook. Then

$$\int \frac{\sqrt{x^2-25}}{x}\, dx = \sqrt{x^2-25} - 5\sec^{-1}\left|\frac{x}{5}\right| + C$$

53. Use Formula # 44 in the textbook. Then

$$\int \frac{x^2}{\sqrt{x^2-25}}\, dx = \frac{25}{2}\cosh^{-1}\frac{x}{5} + \frac{x}{2}\sqrt{x^2-25} + C$$

55. Use Formula #37 in the textbook. Then

$$\int \sqrt{x^2-4}\, dx = \frac{x}{2}\sqrt{x^2-4} - 2\cosh^{-1}\frac{x}{2} + C$$

57. $s = \int_0^1 \sqrt{1 + 4x^2}\, dx = \int_0^1 2\sqrt{\frac{1}{4} + x^2}\, dx = 2\left(\frac{x}{2}\sqrt{\frac{1}{4} + x^2} + \frac{\frac{1}{4}}{2}\sinh^{-1}\frac{x}{\frac{1}{\frac{1}{2}}}\right)\Bigg|_0^1$

$= \frac{\sqrt{5}}{2} + \frac{1}{4}\ln(2 + \sqrt{5})$

59. $\int_{\frac{3}{4}}^1 \sqrt{1 + \left(\frac{1}{x}\right)^2}\, dx = \int_{\frac{3}{4}}^1 \frac{\sqrt{x^2+1}}{x}\, dx = \sqrt{x^2+1} - \sinh^{-1}\left|\frac{1}{x}\right|\Bigg]_{\frac{3}{4}}^1$

$= \sqrt{2} - \ln(1 + \sqrt{2}) - \frac{5}{4} + \ln 3$

61. Solve $x\frac{d^2y}{dx^2} = \sqrt{1 + \left(\frac{dy}{dx}\right)^2}$, $y = 0$, $\frac{dy}{dx} = 0$ when $x = 1$.

Let $\frac{dy}{dx} = p$, $\frac{dp}{dx} = \frac{d^2y}{dx^2}$. Then $x\frac{dp}{dx} = \sqrt{1 + p^2} \Rightarrow \frac{dp}{\sqrt{1+p^2}} = \frac{dx}{x}$ or

$\sinh^{-1}p = \ln|x| + C$. $p = o$ when $x = 1$, so $C = 0$ and thus $p = \sinh(\ln x)$.

$y = \int \sinh(\ln x)\, dx = \int \frac{e^{\ln x} - e^{-\ln x}}{2}\, dx = \frac{1}{2}\int\left(x - \frac{1}{x}\right) dx.$

$y = \frac{x^2}{4} - \frac{1}{2}\ln|x| + C$. Since $0 = \frac{1}{4} - 0$, $C = -\frac{1}{4}$. Therefore, $y = \frac{x^2}{4} - \frac{1}{2}\ln|x| - \frac{1}{4}$

63. Solve $x = \sinh y = \frac{1}{2}(e^y - e^{-y})$ for y. Multiply both sides of $2x = e^y - e^{-y}$ by e^y to form an equation quadratic in e^y.

Then, $e^{2y} - 2xe^y - 1 = 0 \Leftrightarrow y = \frac{-(-2x) \pm \sqrt{4x^2 + 4}}{2}$

$= x + \sqrt{x^2 + 1}$ (since we need the positive root). Therefore,
$y = \ln(x + \sqrt{x^2 + 1})$ or $\sinh^{-1} x = x + \sqrt{x^2 + 1}$.

65. Prove: $\frac{d}{dx}(\sinh^{-1}u) = \frac{1}{\sqrt{1+u^2}}\frac{du}{dx}$.

If $\sinh y = x$ then $\frac{d}{dx}\sinh y = 1$ or $\cosh y\frac{dy}{dx} = 1$. Therefore,

$\frac{dy}{dx} = \frac{1}{\cosh y} = \frac{1}{\sqrt{1 + \sinh^2 y}} = \frac{1}{\sqrt{1+x^2}}$. Applying the Chain Rule,

$\frac{dy}{dx} = \frac{dy}{du}\cdot\frac{du}{dx}$, to this equation gives the desired result.

9.M MISCELLANEOUS

1. Prove: $\cosh x\, 2x = \cosh^2 x + \sinh^2 x$

$\cosh^2 x + \sinh^2 x = \frac{1}{2}(e^x + e^{-x})^2 + \frac{1}{2}(e^x + e^{-x})^2$

$= \frac{1}{2}(e^{2x} + 2 + e^{-2x}) + \frac{1}{2}(e^{2x} - 2 + e^{-2x}) = \frac{1}{2}(e^{2x} + e^{-2x}) = \cosh 2x.$

3. $\lim_{x\to\infty}(\cosh x - \sinh x) = \lim_{x\to\infty}\left(\frac{1}{2}(e^x + e^{-x}) - \frac{1}{2}(e^x + e^{-x})\right) = \lim_{x\to\infty} e^{-x} = 0.$

5. If $\operatorname{csch} x = -\frac{9}{40}$, then $\sinh x = -\frac{40}{9}$ and $\cosh^2 x = \sqrt{1 + \frac{1600}{81}} = \frac{41}{9}$.

$\tanh x = \frac{\sinh x}{\cosh x} = -\frac{40}{41}.$

7. Since the radius = 1, $RP^2 + PQ^2 = RQ^2 = 1$. RP is equal to tanh x, so $\tanh^2 x + PQ^2 = 1$, $PQ = \sqrt{1 - \tanh^2 x} = \operatorname{sech} x$.

9. $y = \tanh\left(\frac{1}{2}\ln|x|\right) = \tanh(\ln\sqrt{x}) = \frac{e^{\ln\sqrt{x}} - e^{-\ln\sqrt{x}}}{e^{\ln\sqrt{x}} + e^{\ln\sqrt{x}}} = \frac{\sqrt{x} - \frac{1}{\sqrt{x}}}{\sqrt{x} + \frac{1}{\sqrt{x}}} = \frac{x-1}{x+1}.$

As the $\lim_{x\to\infty}\frac{x-1}{x+1} = 1$, the horizontal asymptote is $y = 1$.

11.

$y = \cosh 2x$	$y = \sinh 2x$	$y = \cos 2x$
$\frac{dy}{dx} = 2\sinh 2x$	$\frac{dy}{dx} = 2\cosh 2x$	$\frac{dy}{dx} = -2\sin 2x$
$\frac{d^2y}{dx^2} = 4\cosh 2x$	$\frac{d^2y}{dx^2} = 4\sinh 2x$	$\frac{d^2y}{dx^2} = -4\cos 2x$
$\frac{dy^3}{dx^3} = 8\sinh 2x$	$\frac{dy^3}{dx^3} = 8\cosh 2x$	$\frac{dy^3}{dx^3} = 8\sin 2x$
$\frac{dy^4}{dx^4} = 16\cosh 2x$	$\frac{dy^4}{dx^4} = 16\sinh 2x$	$\frac{dy^4}{dx^4} = 16\cos 2x$
$\frac{dy^4}{dx^4} = y$	$\frac{dy^4}{dx^4} = y$	$\frac{dy^4}{dx^4} = y$

13. If $y = \sinh^2 3x$, then $\frac{dy}{dx} = 6\sinh 3x\cosh 3x$

15. $\sin^{-1} x = \operatorname{sech} y$

$$\frac{1}{\sqrt{1-x^2}} = -\operatorname{sech} y \tanh y \frac{dy}{dx} \text{ or } \frac{dy}{dx} = \frac{-1}{\operatorname{sech} y \tanh y\sqrt{1-x^2}}$$

17. $\tan^{-1} y = \tanh^{-1} x$

$$\frac{1}{1+y^2}\frac{dy}{dx} = \frac{1}{1-x^2} \text{ or } \frac{dy}{dx} = \frac{1+y^2}{1-x^2} = \frac{1+\tan^2(\tanh^{-1} x)}{1-x^2} = \frac{\sec^2(\tanh^{-1} x)}{1-x^2}$$

19. $x = \cosh(\ln y) = \dfrac{e^{\ln y} + e^{-\ln y}}{2} = \dfrac{y + \frac{1}{y}}{2} = \dfrac{y^2+1}{2y}$

$$1 = \frac{(2y)^2\frac{dy}{dx} - 2(y^2+1)\frac{dy}{dx}}{4y^2} \text{ or } \frac{dy}{dx} = \frac{2y^2}{y^2-1}.$$

21. If $y = \sinh^{-1}(\tan x)$ then $\dfrac{dy}{dx} = \dfrac{\sec^2 x}{\sqrt{1+\tan^2 x}} = |\sec x|.$

23. $\displaystyle\int_{-\ln 2}^{0} \frac{d\theta}{\sinh\theta + \cosh\theta} = \int_{-\ln 2}^{0} \frac{d\theta}{\sinh\theta + \cosh\theta} \cdot \frac{\cosh\theta - \sinh\theta}{\cosh\theta - \sinh\theta}$

$$= \int_{-\ln 2}^{0} (\cosh\theta - \sinh\theta)\, d\theta = \sinh\theta - \cosh\theta\Big]_{-\ln 2}^{0}$$

$$= -1 - \left[\frac{e^{\ln 2} + e^{\ln 2}}{2}\right] + \left[\frac{e^{\ln 2} - e^{\ln 2}}{2}\right] = 1$$

25. $\displaystyle\int_0^{\frac{\ln 3}{2}} \sinh^3 x\, dx = \int_0^{\frac{\ln 3}{2}} (\cosh^2 x - 1)\sinh x\, dx = \frac{1}{3}\cosh^3 x - \cosh x\Big]_0^{\frac{\ln 3}{2}}$

$$= \frac{1}{3}\left(\frac{e^{\frac{\ln 3}{2}} + e^{-\frac{\ln 3}{2}}}{2}\right)^3 - \frac{e^{\frac{\ln 3}{2}} + e^{-\frac{\ln 3}{2}}}{2} + \frac{2}{3}$$

$$= \frac{1}{24}\left(\frac{4}{3}\sqrt{3}\right)^3 - \frac{2}{3}\sqrt{3} + \frac{2}{3} = \frac{-10\sqrt{3}+18}{27}.$$

27. $\displaystyle\int_1^{\sqrt{2}} \frac{e^{2x}-1}{e^{2x}+1}\, dx = \int_1^{\sqrt{2}} \frac{e^{2x}-1}{e^{2x}+1} \cdot \frac{e^{-x}}{e^{-x}}\, dx = \int_1^{\sqrt{2}} \frac{\sinh x}{\cosh x}\, dx = \int_1^{\sqrt{2}} \tanh x\, dx$

$$= \ln|\cosh x|\Big]_1^{\sqrt{2}} = \ln|\cosh\sqrt{2} - \cosh 1|.$$

29. $\displaystyle\int_0^1 \frac{dx}{\sqrt{x} - x\sqrt{x}} = \lim_{t\to 0^+} \int_t^k \frac{dx}{\sqrt{x} - x\sqrt{x}} + \lim_{t\to 1^-} \int_k^t \frac{dx}{\sqrt{x} - x\sqrt{x}}$

$$\lim_{t\to 1^-} \int_k^t \frac{dx}{\sqrt{x} - x\sqrt{x}} = \lim_{t\to 1^-} \int_k^t \frac{2du}{1-u^2} = \lim_{t\to 1^-} (\tanh^{-1} u)\Big]_k^t = \infty.$$

Since both parts must converge for the integral to converge,

$\displaystyle\int_0^1 \frac{dx}{\sqrt{x} - x\sqrt{x}}$ diverges.

31. $\displaystyle\int_{\pi/3}^{\pi/2} \frac{\sin x\,dx}{1-\cos^2 x} = -\tanh^{-1}(\cos x)\Big]_{\pi/3}^{\pi/2} = -\tanh^{-1}0 + \tanh^{-1}\left(\frac{1}{2}\right) = \ln\sqrt{3}.$

33. $\displaystyle\lim_{x\to\infty}(\cosh^{-1}x - \ln x) = \lim_{x\to\infty}\left[\left(x+\sqrt{x^2-1}\right) - \ln x\right]$

$$= \lim_{x\to\infty} \ln\left|\frac{x+\sqrt{x^2-1}}{x}\right| = \lim_{x\to\infty}\left|1+\frac{\sqrt{x^2-1}}{x}\right| = \ln 2$$

CHAPTER 10

POLAR COORDINATES

10.1 THE POLAR COORDINATE SYSTEM

1. The following groups represent the same points:
   a,c; b,d; e,j,k; f,i,l; m,o; n,p

3. (a) $(2, \frac{\pi}{2} + 2k\pi)$ or $(-2, -\frac{\pi}{2} + 2k\pi)$, k an integer.
   (b) $(2, 2k\pi)$ or $(-2, (2k+1)\pi)$, k an integer.
   (c) $(-2, \frac{\pi}{2} + 2k\pi)$ or $(2, \frac{3\pi}{2} + 2k\pi)$, k an integer.
   (d) $(2, (2k+1)\pi)$ or $(-2, 2k\pi)$, k an integer.

5. (a) $x = r\cos\theta = \sqrt{2}\cos\frac{\pi}{4} = 1$ $\quad y = r\sin\theta = \sqrt{2}\sin\frac{\pi}{4} = 1$
   (b) $x = 1\cos 0 = 1$ $\quad y = 1\sin 0 = 0$
   (c) $r = 0 \Rightarrow x = 0$ and $y = 0$
   (d) $x = -\sqrt{2}\cos\frac{\pi}{4} = -1$ $\quad y = -\sqrt{2}\sin\frac{\pi}{4} = -1$
   (e) $x = -3\cos\frac{5\pi}{6} = \frac{3\sqrt{3}}{2}$ $\quad y = -3\sin\frac{5\pi}{6} = -\frac{3}{2}$
   (f) $x = 5\cos(\tan^{-1}\frac{4}{3}) = 5(\frac{3}{5}) = 3$ $\quad y = 5\sin(\tan^{-1}\frac{4}{3}) = 5(\frac{4}{5})\ 4$
   (g) $x = -1\cos 7\pi = 1$ $\quad y = -1\sin 7\pi = 0$
   (h) $x = 2\sqrt{3}\cos\frac{2\pi}{3} = -\sqrt{3}$ $\quad y = 2\sqrt{3}\sin\frac{2\pi}{3} = 3$

7. $r = 2$

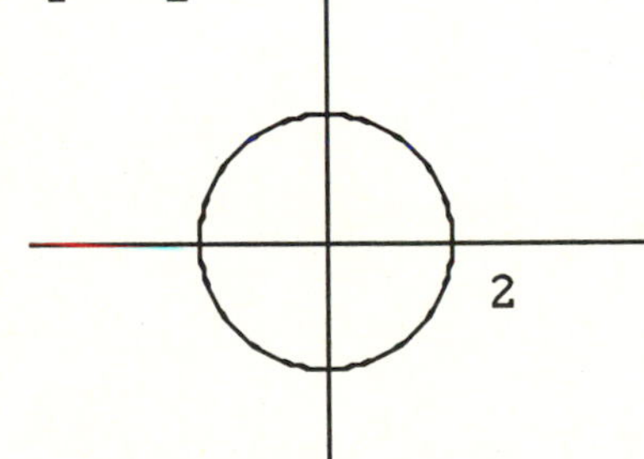

9. $r \geq 1$

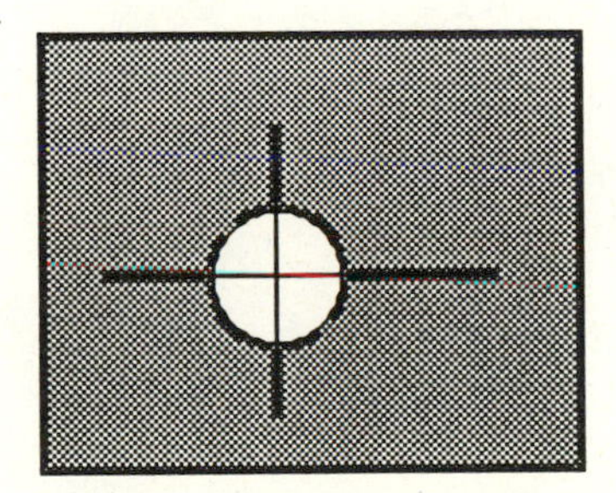

11. $0 \leq \theta \leq \frac{\pi}{6}$, $r \geq 0$

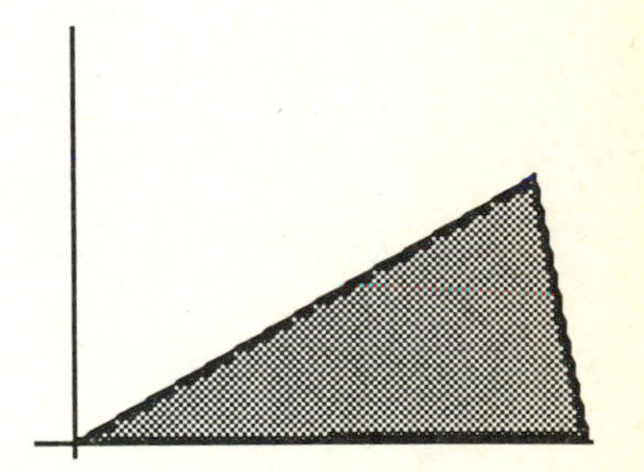

13. $\theta = \frac{\pi}{3}$, $-1 \leq r \leq 3$

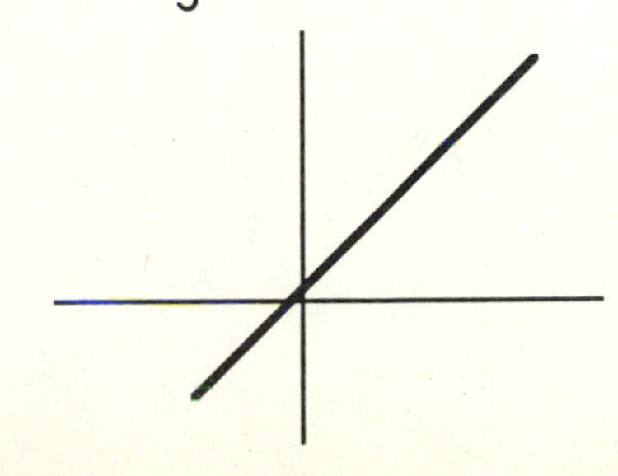

15. $\theta = \frac{\pi}{2}$, $r \geq 0$

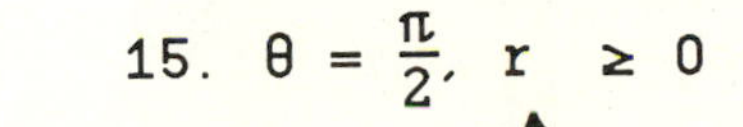

17. $0 \leq \theta \leq \pi$, $r = 1$

1

19. $\frac{\pi}{4} \le \theta \le \frac{3\pi}{4}$, $0 \le r \le 1$

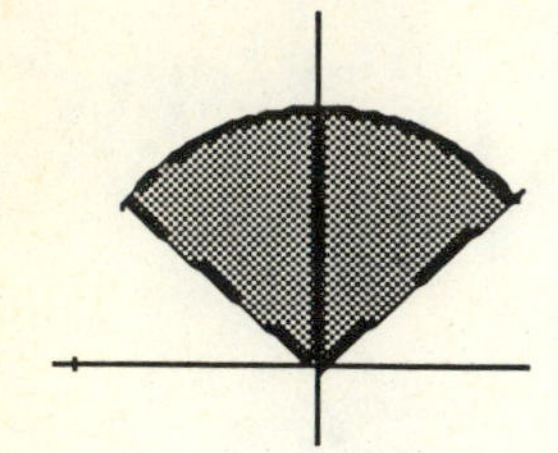

21. $-\frac{\pi}{2} \le \theta \le \frac{\pi}{2}$, $1 \le r \le 2$

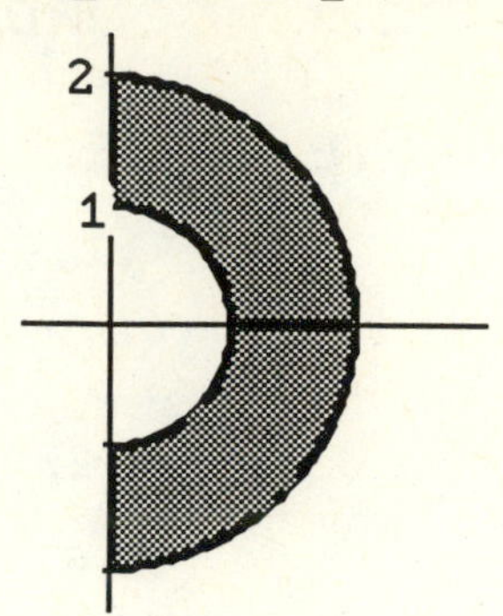

23. $x = r\cos\theta$ so the vertical line $x = 2$
25. $y = r\sin\theta$ so the horizontal line $y = 4$
27. $y = 0$ or the x-axis
29. $r\cos\theta + r\sin\theta = 1$ is the line $x + y = 1$
31. $r^2 = 1$ is the circle $x^2 + y^2 = 1$
33. $r\sin\theta = e^{r\cos\theta}$ is the function $y = e^x$
35. $r = \dfrac{5}{\sin\theta - 2\cos\theta} \Leftrightarrow r\sin\theta - 2r\cos\theta = 5$ or $y - 2x = 5$
37. $r = 4\sin\theta \Leftrightarrow \sqrt{x^2 + y^2} = \dfrac{4y}{\sqrt{x^2 + y^2}}$

    $x^2 + y^2 - 4y = 0$ or $x^2 + (y - 2)^2 = 4$
39. $r = 4\tan\theta\sec\theta = \dfrac{4\sin\theta}{\cos\theta} \cdot \dfrac{1}{\cos\theta} \cdot \dfrac{r}{r}$

    $r\cos\theta = \dfrac{4r\sin\theta}{r\cos\theta}$ or $x = \dfrac{4y}{x}$ or $x^2 = 4y$
41. $r\cos\theta = 7$
43. $r\cos\theta = r\sin\theta \Rightarrow \tan\theta = 1$ or $\theta = \frac{\pi}{4}$.
45. A circle centered at origin with radius 2: $r = 2$
47. $\dfrac{x^2}{9} + \dfrac{y^2}{4} = 1 \Leftrightarrow \dfrac{r^2\cos^2\theta}{9} + \dfrac{r^2\sin^2\theta}{4} = 1$
49. $y^2 = 4x \Leftrightarrow r^2\sin^2\theta = 4r\cos\theta \Leftrightarrow r = \dfrac{4\cos\theta}{\sin^2\theta} = 4\cot\theta\csc\theta$
51. $3 = r\cos(\theta - \frac{\pi}{3}) = r(\cos\theta\cos\frac{\pi}{3} + \sin\theta\sin\frac{\pi}{3})$

    $= r\cos\theta \cdot \frac{1}{2} + r\sin\theta \cdot \frac{\sqrt{3}}{2}$ $\therefore\ x + \sqrt{3}\,y = 6$
53. $\sqrt{2} = r\sin(\frac{\pi}{4} - \theta) = r(\sin\frac{\pi}{4}\cos\theta - \cos\frac{\pi}{4}\sin\theta)$

    $= r\cos\theta \cdot \frac{1}{\sqrt{2}} - r\sin\theta \cdot \frac{1}{\sqrt{2}}$ $\therefore\ x - y = 2$

10.2 GRAPHING IN POLAR COORDINATES

1. $\cos\theta = \cos(-\theta) \Rightarrow$ symmetry to x=axis.
   $|\cos\theta| \le 1 \Rightarrow 0 \le r \le 2a$

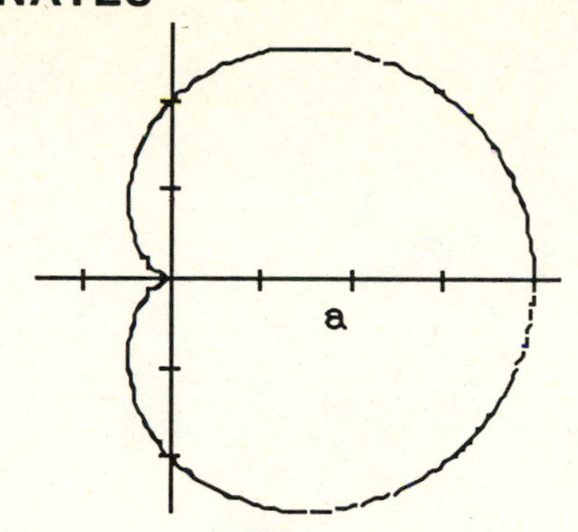

3. Replacing (r,θ) by $(-r,-\theta)$:
   $-r = \sin 2(-\theta) = -\sin 2\theta$ or
   $r = \sin 2\theta$. $\therefore$ The graph has symmetry to y-axis.
   $\theta = \frac{\pi}{4} \Rightarrow r = a$

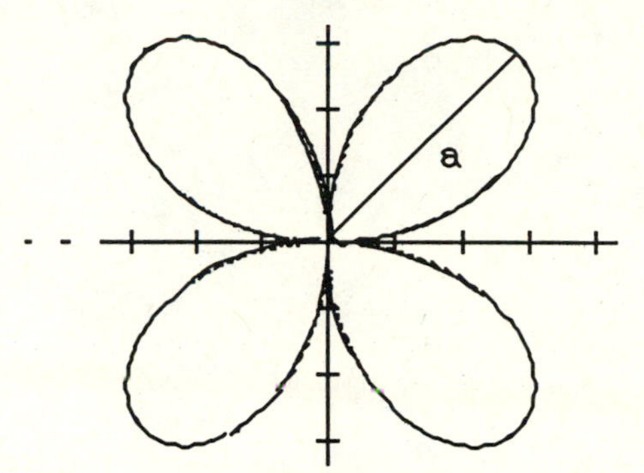

5. Replacing θ by $\pi-\theta$:
   $r = a(2 + \sin(\pi-\theta))$
   $= a(2 + \sin\theta) \Rightarrow$
   symmetry to y-axis

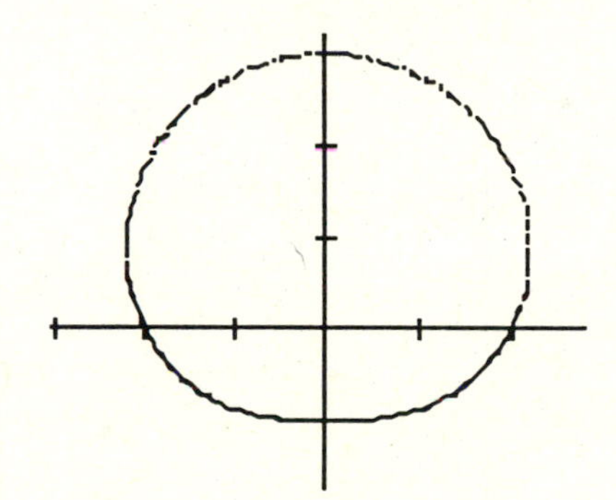

7. Spiral of Archimedes
   $(r,\theta) = (-r,-\theta) \Rightarrow$ symmetry to y-axis.

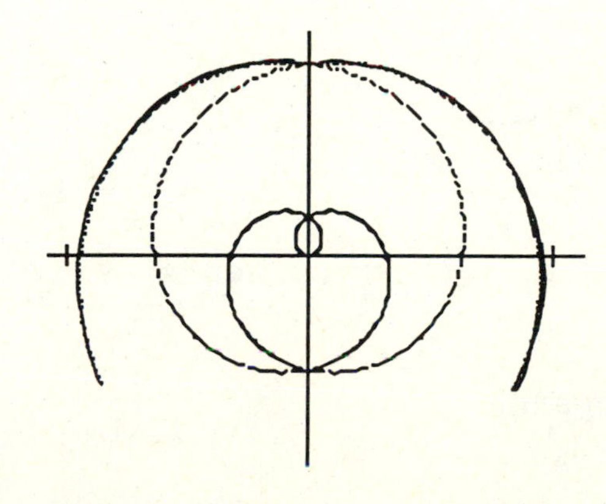

9. (a) $r = \frac{1}{2} + \cos\theta$

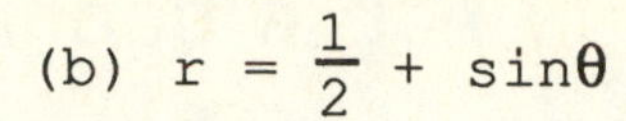

(b) $r = \frac{1}{2} + \sin\theta$

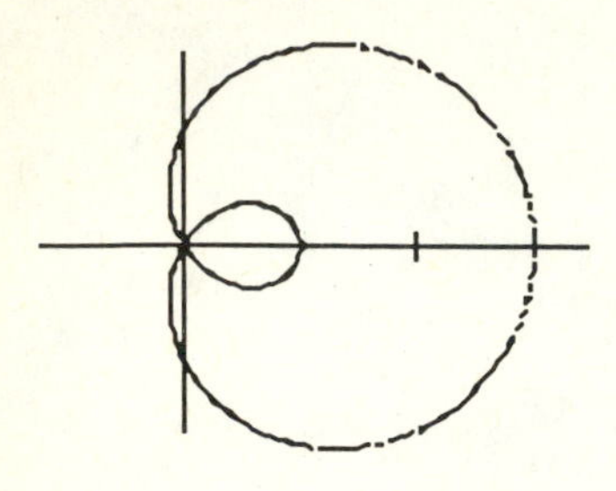

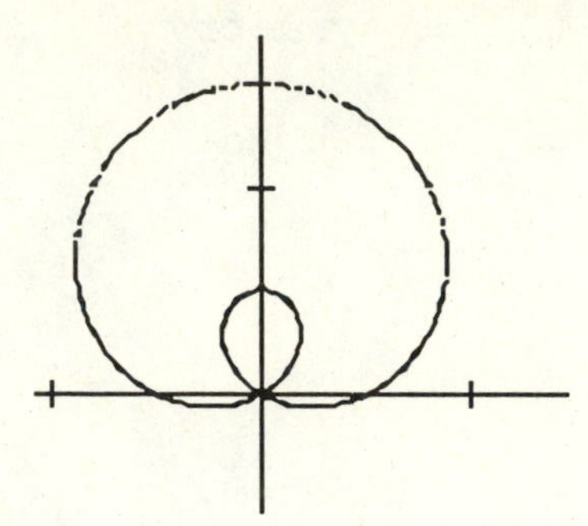

11. (a) $r = \frac{3}{2} + \cos\theta$ (b) $r = \frac{3}{2} - \sin\theta$

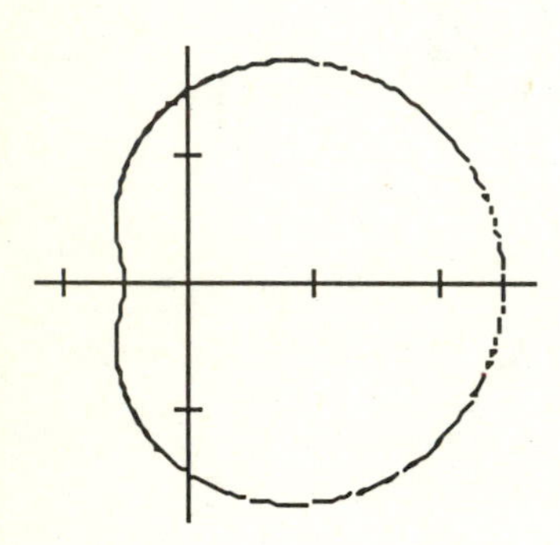

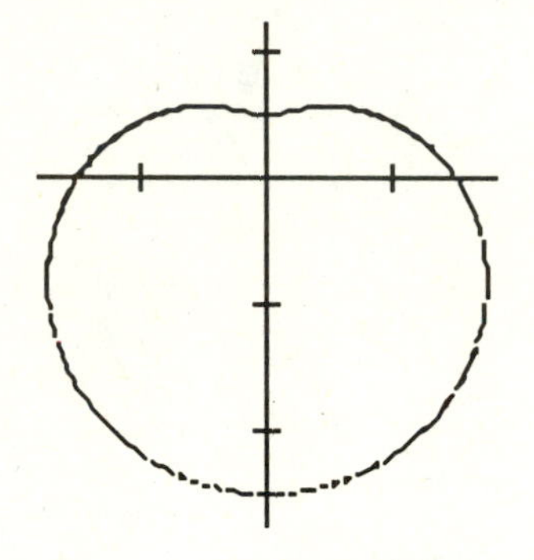

13. Region: $0 \leq r \leq 2 - 2\cos$

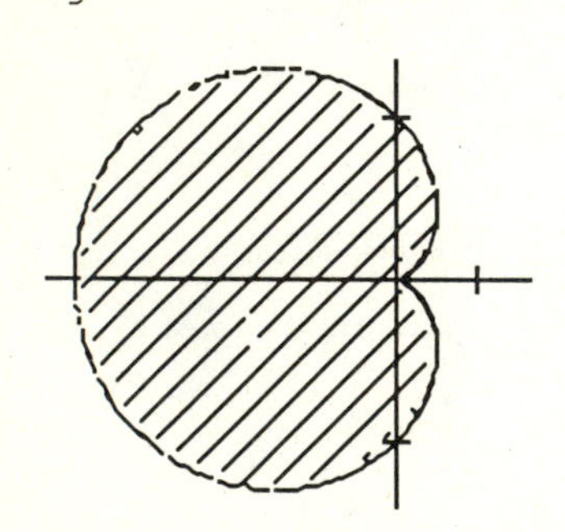

15. When named by $(2, \frac{3\pi}{4})$, the point does not satisfy the equation $r = 2\sin 2\theta$, since $2\sin 2(\frac{3\pi}{4}) = 2\sin\frac{3\pi}{2} = -2$. If the point is renamed as $(-2, -\frac{\pi}{4})$, then it will satisfy the equation.

17.

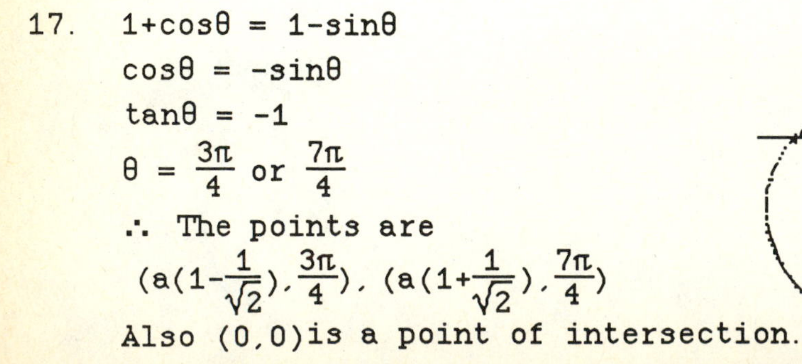

$1+\cos\theta = 1-\sin\theta$

$\cos\theta = -\sin\theta$

$\tan\theta = -1$

$\theta = \frac{3\pi}{4}$ or $\frac{7\pi}{4}$

$\therefore$ The points are

$(a(1-\frac{1}{\sqrt{2}}), \frac{3\pi}{4}), (a(1+\frac{1}{\sqrt{2}}), \frac{7\pi}{4})$

Also (0,0) is a point of intersection.

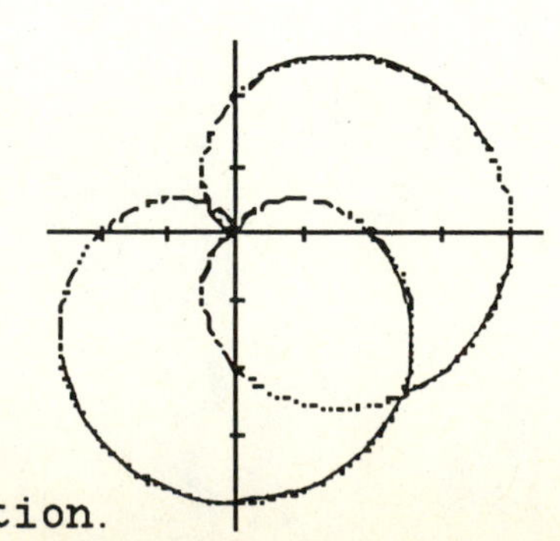

19. $r = 1 - \cos\theta$ and $r^2 = \cos\theta \Rightarrow (1 - \cos\theta)^2 = \cos\theta$

$1 - 2\cos\theta + \cos^2\theta = \cos\theta \Rightarrow \cos^2\theta - 3\cos\theta + 1 = 0$

$\cos\theta = \dfrac{3 - \sqrt{5}}{2}$. $r = 1 - \dfrac{3 - \sqrt{5}}{2} = \dfrac{-1 + \sqrt{5}}{2}$. The points of intersection are

$(0, 0)$, $\left(\dfrac{-1 + \sqrt{5}}{2}, \cos^{-1}\left(\dfrac{3 - \sqrt{5}}{2}\right)\right)$ and $\left(\dfrac{-1 + \sqrt{5}}{2}, 2\pi - \cos^{-1}\left(\dfrac{3 - \sqrt{5}}{2}\right)\right)$

21. $2\sin 2\theta = 1 \Leftrightarrow \sin 2\theta = \dfrac{1}{2} \Rightarrow 2\theta = \dfrac{\pi}{6} \Rightarrow \theta = \dfrac{\pi}{12}$, etc.

There are 8 points of intersection between the circle and the four-leafed rose. These are

$\left(1, \pm\dfrac{\pi}{12}\right), \left(1, \pm\dfrac{5\pi}{12}\right), \left(1, \pm\dfrac{13\pi}{12}\right)$ and $\left(1, \pm\dfrac{17\pi}{12}\right)$.

23. 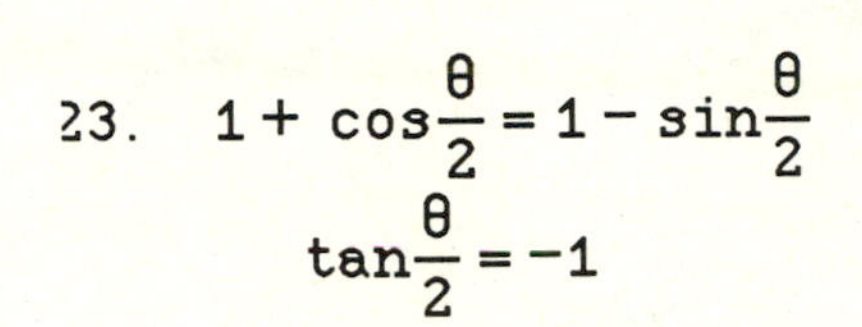

$1 + \cos\dfrac{\theta}{2} = 1 - \sin\dfrac{\theta}{2}$

$\tan\dfrac{\theta}{2} = -1$

$\dfrac{\theta}{2} = \dfrac{3\pi}{4}$ or $\dfrac{7\pi}{2} \Rightarrow \theta = \dfrac{3\pi}{2}$ or $\dfrac{\pi}{2}$

$r = 1 \pm \dfrac{\sqrt{2}}{2}$

There are 5 points of intersection in all.

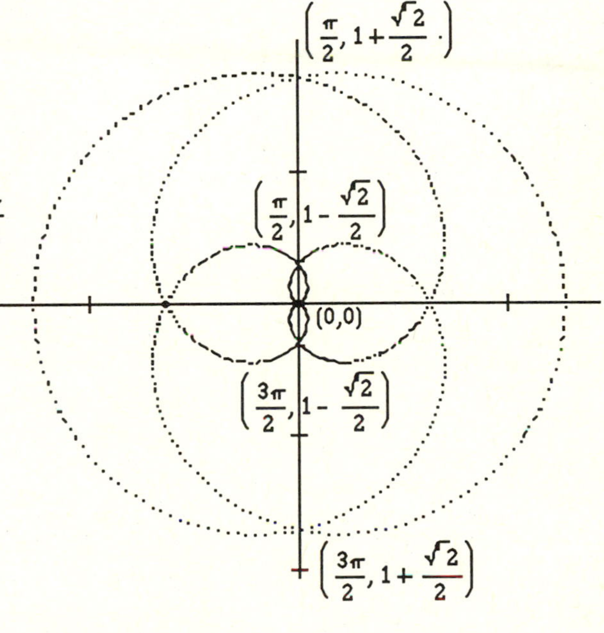

25. We shall show that whenever (r, θ) satisfies one equation, then $(-r, \cos(\pi + \theta))$ satisfies the other. Since these are different names for the same point, the equations must represent the same curve. If (r_1, θ_1) satisfies $r = 1 + \cos\theta$, then $r = -1 + \cos(\pi + \theta_1) =$ $-1 - \cos\theta_1 = -(1 + \cos\theta_1) = -r_1$ satisfies $r = -1 + \cos\theta$. Conversely, if (r_1, θ_1) satisfies $r = -1 + \cos\theta$, then $r = 1 + \cos(\pi + \theta_1) =$ $1 - \cos\theta_1 = -(-1 + \cos\theta_1) = -r_1$ satisfies $r = 1 + \cos\theta$.

27. arc OT = arc DP $\Rightarrow \angle OBC = \angle PAD$. Then $\triangle ABC \cong \triangle PAD \Rightarrow OT = DP$. $\therefore OP \parallel AB$

$\Rightarrow \theta = \angle OBC$. $TB = \frac{1}{2}(2a - r)$. Thus $\dfrac{a - \frac{1}{2}r}{a} = \cos\theta \Rightarrow r = 2a(1 - \cos\theta)$.

10.3 POLAR EQUATIONS OF CONIC SECTIONS AND OTHER CURVES

1. $r = 4\cos\theta \Leftrightarrow \sqrt{x^2 + y^2} = \dfrac{4x}{\sqrt{x^2 + y^2}}$

   $x^2 + y^2 = 4x$ or $(x - 2)^2 + y^2 = 4$

3. $r = -2\cos\theta \Leftrightarrow \sqrt{x^2 + y^2} = \dfrac{-2x}{\sqrt{x^2 + y^2}}$

   $x^2 + y^2 = -2x$ or $(x + 1)^2 + y^2 = 1$

5. $r = \sin 2\theta = 2\sin\theta\cos\theta \Leftrightarrow \sqrt{x^2 + y^2} = \dfrac{2xy}{x^2 + y^2} \Rightarrow (x^2 + y^2)^{\frac{3}{2}} = 2xy$

   $\therefore (x^2 + y^2)^3 = 4x^2y^2$

7. $r^2 = 8\cos 2\theta = 8(2\cos^2\theta - 1) \Leftrightarrow x^2 + y^2 = \dfrac{16x^2}{x^2 + y^2} - 8$

   $(x^2 + y^2)^2 = 8(x^2 - y^2)$

9. $r = 8(1 - 2\cos\theta) \Rightarrow \sqrt{x^2 + y^2} = 8\left(1 - \dfrac{2x}{\sqrt{x^2 + y^2}}\right)$

   $x^2 + y^2 = 8\left(\sqrt{x^2 + y^2} - 2x\right)$ or $(x^2 + y^2 + 16x)^2 = 64(x^2 + y^2)$.

11. $r = 4(1 - \cos\theta) \Rightarrow \sqrt{x^2 + y^2} = 4 - \dfrac{4x}{\sqrt{x^2 + y^2}} \Rightarrow (x^2 + y^2 + 4x)^2 = 16(x^2 + y^2)$.

13. $r(3 - 6\cos\theta) = 12 \Rightarrow \sqrt{x^2 + y^2}\left(1 - \dfrac{2x}{\sqrt{x^2 + y^2}}\right) = 4 \Rightarrow$

    $\sqrt{x^2 + y^2} = 2x + 4 \Rightarrow x^2 + y^2 = 4x^2 + 16x + 16 \Rightarrow 3x^2 - y^2 + 16x + 16 = 0$.

15. $r(2 + \cos\theta) = 4 \Rightarrow 2\sqrt{x^2 + y^2} + x = 4 \Rightarrow 4(x^2 + y^2) = 16 - 8x + x^2 \Rightarrow$

    $3x^2 + 4y^2 + 8x = 16$.

17. $r(3 + 3\cos\theta) = 2 \Rightarrow 3\sqrt{x^2 + y^2} = 2 - 3x \Rightarrow 9(x^2 + y^2) = 4 - 12x + 9x^2 \Rightarrow$

    $9y^2 = 4 - 12x$.

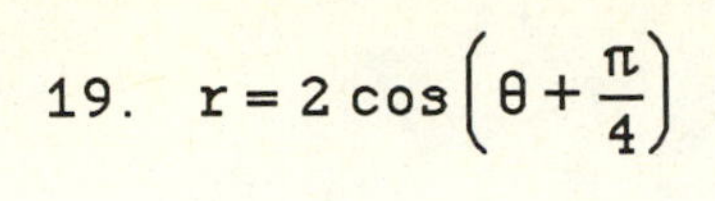

19. $r = 2\cos\left(\theta + \frac{\pi}{4}\right)$

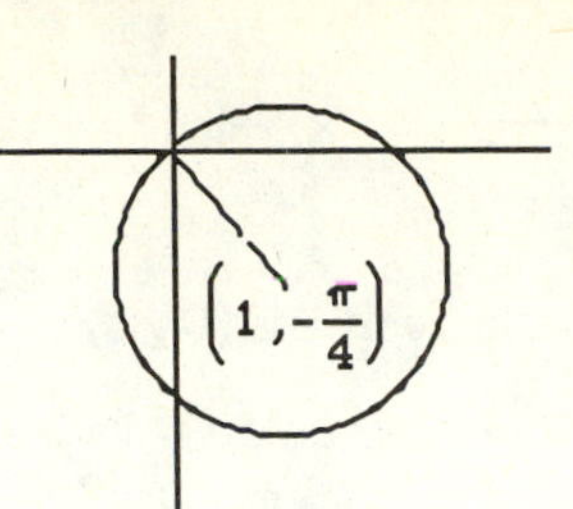

21. $r = 5\sec\left(\frac{\pi}{3} - \theta\right)$ or

$r\cos\left(\theta - \frac{\pi}{3}\right) = 5$

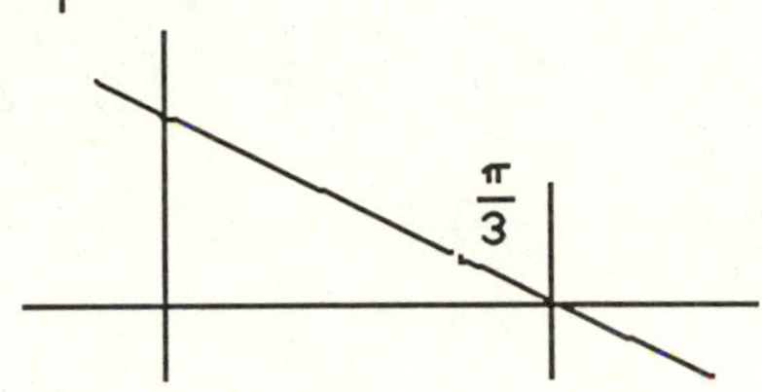

23. A cardioid with its axis of symmetry rotated counterclockwise through and angle of $\frac{\pi}{6}$ radians.

25. (d) 27. (l) 29. (k) 31. (i)

33. Consider any ray $\theta = k$. This ray will intersect the spiral $r = a\theta$ in two successive points $(a(k + 2n\pi), k + 2n\pi)$ and $(a(k + 2(n + 1)\pi), k + 2(n + 1)\pi)$. Then $a(k + 2(n + 1)\pi) - a(k + 2n\pi) = ak + 2an\pi + 2a\pi - ak - 2an\pi = 2a\pi$, a constant width.

35. Mercury: $r = \dfrac{0.3871(1 - (0.2056)^2)}{1 - (0.2056)\cos\theta} = \dfrac{0.3707}{1 - (0.2056)\cos\theta}$

Venus: $r = \dfrac{0.7233(1 - (0.0068)^2)}{1 - (0.0068)\cos\theta} = \dfrac{0.7233}{1 - (0.0068)\cos\theta}$

Earth: $r = \dfrac{1.0000(1 - (0.0167)^2)}{1 - (0.0167)\cos\theta} = \dfrac{0.9997}{1 - (0.0167)\cos\theta}$

Mars: $r = \dfrac{1.524(1 - (0.0934)^2)}{1 - (0.0934)\cos\theta} = \dfrac{1.5107}{1 - (0.0934)\cos\theta}$

Jupiter: $r = \dfrac{5.203(1 - (0.0484)^2)}{1 - (0.0484)\cos\theta} = \dfrac{5.1908}{1 - (0.0484)\cos\theta}$

Saturn: $r = \dfrac{9.539(1 - (0.0543)^2)}{1 - (0.0543)\cos\theta} = \dfrac{9.5109}{1 - (0.0543)\cos\theta}$

Uranus: $r = \dfrac{19.18(1 - (0.0460)^2)}{1 - (0.0460)\cos\theta} = \dfrac{19.139}{1 - (0.0460)\cos\theta}$

Neptune: $r = \dfrac{30.06(1 - (0.0082)^2)}{1 - (0.0082)\cos\theta} = \dfrac{30.0580}{1 - (0.0082)\cos\theta}$

37. $r = 2\sin\theta \Rightarrow \sqrt{x^2+y^2} = \dfrac{2y}{\sqrt{x^2+y^2}} \Rightarrow x^2+y^2 = 2y$

or $x^2+(y-1)^2 = 1$ is a circle with center $(0,1)$ and radius $= 1$.

$r = \sec\theta \Rightarrow r\cos\theta = 1 \Rightarrow x = 1$ is a vertical line tangent to the circle at the point $(1,1)$ or $\left(\sqrt{2}, \frac{\pi}{4}\right)$.

39. In the equation $r = \dfrac{ke}{1-e\cos\theta}$, $k = 4$ and $e = 1$. Therefore, $r = \dfrac{4}{1-\cos\theta}$

41. $k = -9$, so $r = \dfrac{-9\left(\frac{5}{4}\right)}{1-\frac{5}{4}\cos\theta} = \dfrac{-45}{4-5\cos\theta}$. The vertices are at $(5,0)$ and $(-45,\pi) = (45,0)$. The other focus is at $(50,0)$

10.4 INTEGRALS IN POLAR COORDINATEs

1. $$A = \frac{1}{2}\int r^2 d\theta = \frac{1}{2}\int_0^{\frac{\pi}{4}} \cos^2\theta d\theta = \frac{1}{2}\int_0^{\frac{\pi}{4}} \frac{1}{2}(1 + \cos 2\theta)\, d\theta$$

$$= \frac{1}{4} + \frac{1}{8}\sin 2\theta\Big]_0^{\frac{\pi}{4}} = \frac{\pi+2}{16}$$

3. $$A = \int_0^{2\pi} \frac{1}{2}(4 + 2\cos\theta)^2 d\theta = 2\int_0^{\pi}\frac{1}{2}(16 + 16\cos\theta + 4\cos^2\theta)\, d\theta$$

$$= 4\int_0^{\pi}[4 + 4\cos\theta + \frac{1}{2}(1 + \cos 2\theta)]\, d\theta$$

$$= 16\theta + 16\sin\theta + 2\theta + \sin 2\theta\,\Big]_0^{\pi} = 18\pi$$

5. $$A = \int_0^{\pi}\frac{1}{2}(2a\sin\theta)^2 d\theta = 2a^2\int_0^{\pi}\left(\frac{1}{2} - \frac{1}{2}\cos 2\theta\right) d\theta$$

$$= 2a^2\left[\frac{1}{2} - \frac{1}{4}\sin 2\theta\right]_0^{\pi} = 2a^2\left(\frac{\pi}{2}\right) = \pi a^2$$

7. $2a^2 \cos 2\theta = a^2$

$\cos 2\theta = \frac{1}{2}$

$2\theta = \frac{\pi}{3}$

$\theta = \frac{\pi}{6}$

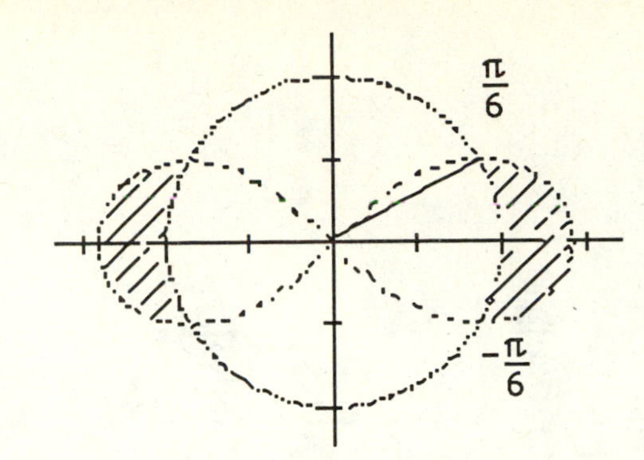

$$A = 4\int_0^{\frac{\pi}{6}} \frac{1}{2}[(2a^2\cos 2\theta) - a^2]d\theta$$

$$= 2a^2 \int_0^{\frac{\pi}{6}} (2\cos 2\theta - 1)d\theta = 2a^2(\sin 2\theta - \theta)\Big]_0^{\frac{\pi}{6}} = \frac{a^2(3\sqrt{3} - \pi)}{3}$$

9. $2a\cos\theta = 2a\sin\theta$

$\frac{\sin\theta}{\cos\theta} = \frac{2a}{2a}$

$\tan\theta = 1$

$\theta = \frac{\pi}{4}$

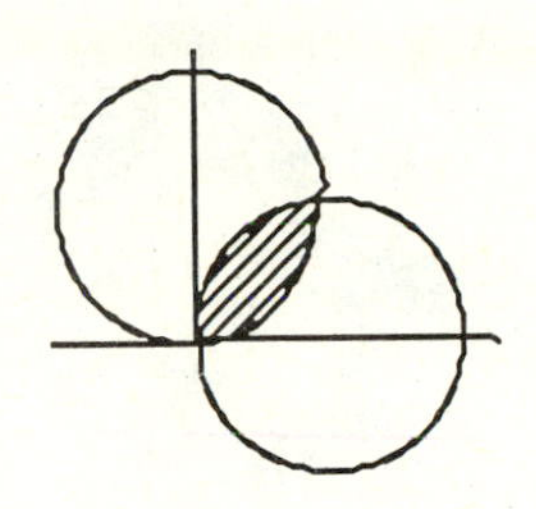

$$A = 2\cdot\frac{1}{2}\int_0^{\frac{\pi}{4}}(2a\sin\theta)^2 d\theta = 4a^2\int_0^{\frac{\pi}{4}}\left(\frac{1}{2} - \frac{1}{2}\cos 2\theta\right)d\theta$$

$$= 2a^2\left[\theta - \frac{1}{2}\sin 2\theta\right]_0^{\frac{\pi}{4}} = \frac{a^2(\pi - 2)}{2}$$

11. $-2\cos\theta = 1$

$\cos\theta = -\frac{1}{2}$

$\theta = \frac{2\pi}{3}, \frac{4\pi}{3}$

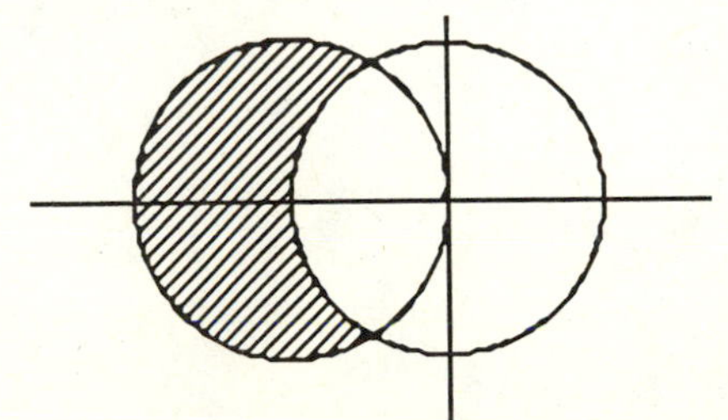

$$A = \int_{\frac{2\pi}{3}}^{\frac{4\pi}{3}} \frac{1}{2}[(-2\cos\theta)^2 - 1]d\theta = \frac{1}{2}\int_{\frac{2\pi}{3}}^{\frac{4\pi}{3}}(4\cos^2\theta - 1)d\theta$$

$$= \frac{1}{2}\int_{\frac{2\pi}{3}}^{\frac{4\pi}{3}}[2(1 + \cos 2\theta) - 1]\,d\theta \quad = \frac{1}{2}\left[\theta + \sin 2\theta\right]_{\frac{2\pi}{3}}^{\frac{4\pi}{3}}$$

$$= \frac{1}{2}\left(\frac{4\pi}{3} - \frac{2\pi}{3} + \frac{\sqrt{3}}{2} - \left(-\frac{\sqrt{3}}{2}\right)\right)$$

$$= \frac{1}{2}\left(\frac{2\pi}{3} + \sqrt{3}\right) = \frac{2\pi + 3\sqrt{3}}{6}$$

13. $a(1 + \cos\theta) = a(1 - \cos\theta)$

$\cos\theta = 0$

$\theta = \frac{\pi}{2}, \frac{3\pi}{2}$

$$A = 4\cdot\frac{1}{2}\int_0^{\frac{\pi}{2}}[a(1 - \cos\theta)]^2 d\theta$$

$$= 2a^2\int_0^{\frac{\pi}{2}}(1 - 2\cos\theta + \cos^2\theta)\,d\theta =$$

$$2a^2\int_0^{\frac{\pi}{2}}\left[1 - 2\cos\theta + \frac{1}{2}(1 + \cos 2\theta)\right]d\theta$$

$$= 2a^2\left[\frac{3}{2}\theta - 2\sin\theta + \frac{1}{4}\sin 2\theta\right]_0^{\frac{\pi}{2}} = \frac{(3\pi - 8)a^2}{2}$$

15. $A = 2\cdot\frac{1}{2}\int_0^{\frac{\pi}{4}}4\sin 2\theta\,d\theta = 4\left[-\frac{1}{2}\cos 2\theta\right]_0^{\frac{\pi}{4}} = 2$

17. In one loop,

$$A = \frac{1}{2}\int_0^{\frac{\pi}{3}}(2a^2\sin 3\theta)\,d\theta$$

$$= a^2\left[-\frac{1}{3}\cos 3\theta\right]_0^{\frac{\pi}{3}} = \frac{2a^2}{3}$$

In six loops, $6\left(\frac{2a^2}{3}\right) = 4a^2$

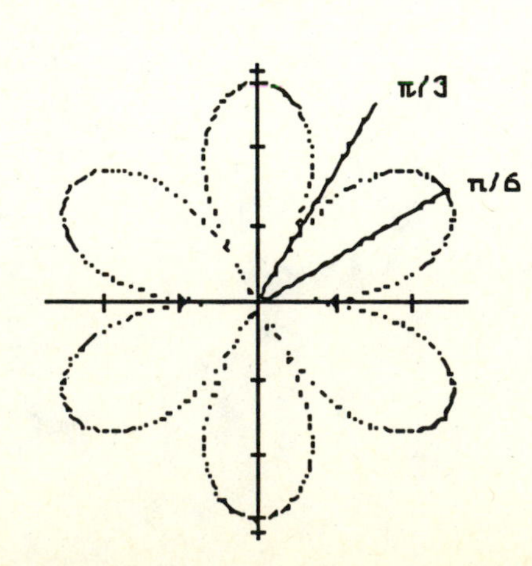

19. $A = \frac{1}{2}\int_0^{\pi}[\sqrt{\theta}e^{\theta}]^2 d\theta = \frac{1}{2}\int_0^{\pi}\theta e^{2\theta}d\theta$

$= \frac{1}{2}\left[\frac{\theta}{2}e^{2\theta} - \frac{1}{4}e^{2\theta}\right]_0^{\pi} = \frac{\pi e^{2\pi}}{4} - \frac{e^{2\pi}}{8} + \frac{1}{8}$

21. (a) $r = a \Rightarrow dr = 0. \therefore ds^2 = r^2d\theta^2 + dr^2 = a^2d\theta^2$

$$L = \int ds = \int_0^{2\pi} a\,d\theta = [a\theta]_0^{2\pi} = 2\pi a$$

(b) $r = a\cos\theta \Rightarrow dr = -a\sin\theta d\theta$

$ds^2 = r^2d\theta^2 + dr^2$

$= a^2\cos^2\theta d\theta^2 + a^2\sin^2\theta d\theta^2 = a^2d\theta^2$

$$L = \int_{-\frac{\pi}{2}}^{\frac{\pi}{2}} a\,d\theta = [a\theta]_{-\frac{\pi}{2}}^{\frac{\pi}{2}} = \pi a$$

(c) $r = a\sin\theta \Rightarrow dr = a\cos\theta d\theta$

$ds^2 = r^2d\theta^2 + dr^2$

$= a^2\cos^2\theta d\theta^2 + a^2\sin^2\theta d\theta^2 = a^2d\theta^2$

$$L = \int_0^{\pi} a\,d\theta = [a\theta]_0^{\pi} = \pi a$$

23. $r = a\sin^2\frac{\theta}{2}$, $0 \le \theta \le \pi$, $\Rightarrow dr = a\sin\frac{\theta}{2}\cos\frac{\theta}{2}\,d\theta$

$ds^2 = r^2d\theta^2 + dr^2 = a^2\sin^4\frac{\theta}{2}d\theta^2 + a^2\sin^2\frac{\theta}{2}\cos^2\frac{\theta}{2}d\theta^2$

$= a^2\sin^2\frac{\theta}{2}d\theta^2(\sin^2\frac{\theta}{2} + \cos^2\frac{\theta}{2}) = a^2\sin^2\frac{\theta}{2}d\theta^2$

NOTE: $ds = a|\sin\frac{\theta}{2}|d\theta$ but that $\sin\frac{\theta}{2} \ge 0$ for $0 \le \theta \le \pi$.

$$L = a\int_0^{\pi}\sin\frac{\theta}{2}d\theta = \left[-2a\cos\frac{\theta}{2}\right]_0^{\pi} = 2a$$

25. $r = a\sin^3\frac{\theta}{3},\ 0 \le \theta \le 3\pi, \quad \Rightarrow dr = 3a\sin^2\frac{\theta}{3}\cos\frac{\theta}{3}(\frac{1}{3})\,d\theta$

$$ds^2 = r^2d\theta^2 + dr^2 = a^2\sin^6\frac{\theta}{3}\,d\theta^2 + a^2\sin^4\frac{\theta}{3}\cos^2\frac{\theta}{3}\,d\theta^2$$

$$= a^2\sin^4\frac{\theta}{3}(\sin^2\frac{\theta}{3} + \cos^2\frac{\theta}{3})\,d\theta^2 = a^2\sin^4\frac{\theta}{3}\,d\theta^2$$

$$L = \int_0^{3\pi} a\sin^2\frac{\theta}{3}d\theta = a\int_0^{3\pi}\frac{1}{2}(1 - \cos\frac{2\theta}{3})\,d\theta$$

$$= a\left[\frac{1}{2}\theta - \frac{3}{4}\sin\frac{2\theta}{3}\right]_0^{3\pi} = \frac{3a\pi}{2}$$

27. $r^2 = 2a^2\cos 2\theta \quad \Rightarrow \quad 2rdr = -2(2a^2\sin 2\theta)d\theta \quad \Rightarrow r^2dr^2 = 4a^4\sin^2 2\theta d\theta^2$

Then $yds^2 = r\sin\theta ds^2 = r\sin\theta(r^2d\theta^2 + dr^2)$

$$S = \int 2\pi y ds = 2\pi\int_0^{\frac{\pi}{4}} r\sin\theta\sqrt{r^2d\theta^2 + dr^2} = 2\pi\int_0^{\frac{\pi}{4}}\sin\theta\sqrt{r^4d\theta^2 + r^2dr^2}$$

$$= 2\pi\int_0^{\frac{\pi}{4}}\sin\theta\sqrt{4a^4\cos^2 2\theta d\theta^2 + 4a^4\sin^2 2\theta d\theta^2}$$

$$= 2\pi\int_0^{\frac{\pi}{4}} 2a^2\sin\theta d\theta = 4\pi a^2\left[-\cos\theta\right]_0^{\frac{\pi}{4}}$$

$$= 2\pi a^2(2 - \sqrt{2}).$$

The total surface area is $4\pi a^2(2 - \sqrt{2})$

29. $r = 1 + \cos\theta \Rightarrow dr = -\sin\theta d\theta \quad \Rightarrow dr^2 = \sin^2\theta d\theta^2$

$\therefore\ ds^2 = r^2d\theta^2 + dr^2 = (1 + \cos\theta)^2d\theta^2 + \sin^2\theta d\theta^2$

$\qquad = 1 + 2\cos\theta + (\sin^2\theta + \cos^2\theta)\,d\theta^2 = (2 + 2\cos\theta)d\theta^2$

$$S = 2\pi\int_0^{\frac{\pi}{2}} yds = 2\pi\int_0^{\frac{\pi}{2}} r\sin\theta\sqrt{2 + 2\cos\theta d\theta}$$

$$= 2\pi\int_0^{\frac{\pi}{2}} r\sin\theta\sqrt{2(2\cos^2\frac{\theta}{2})}\,d\theta = 4\pi\int_0^{\frac{\pi}{2}} r\left|\cos\frac{\theta}{2}\right|\sin\theta d\theta$$

$$= 4\pi\int_0^{\frac{\pi}{2}}(1 + \cos\theta)\cos\frac{\theta}{2}\sin\theta d\theta = 4\pi\int_0^{\frac{\pi}{2}}2\cos^2\frac{\theta}{2}\cos\frac{\theta}{2}2\sin\frac{\theta}{2}\cos\frac{\theta}{2}d\theta$$

$$= 16\pi\int_0^{\frac{\pi}{2}}2\cos^4\frac{\theta}{2}\sin\frac{\theta}{2}d\theta = 16\pi\left[-\frac{2}{5}\cos^5\frac{\theta}{2}\right]_0^{\frac{\pi}{2}} = \frac{4\pi(8 - \sqrt{2})}{5}$$

10.M MISCELLANEOUS

1. $r = a\theta$ is the Spiral of Archimedes. See Problem 7, Article 10.2.

3. (a) $r = a\sec\theta = a(\frac{1}{\cos\theta}) \Rightarrow r\cos\theta = a$. $\therefore$ The graph is the vertical line $x = a$.

   (b) $r = a\csc\theta = a(\frac{1}{\sin\theta}) \Rightarrow r\sin\theta = a$. $\therefore$ The graph is the horizontal line $y = a$.

   (c) $r = a\sec\theta + a\csc\theta =$

   $$= a(\frac{1}{\cos\theta} + \frac{1}{\sin\theta})$$

   $$1 = \frac{a}{r}(\frac{1}{\cos\theta} + \frac{1}{\sin\theta})$$

   $$= a(\frac{1}{x} + \frac{1}{y})$$

   $$xy = a(y + x) \Rightarrow y = \frac{ax}{x - a}$$

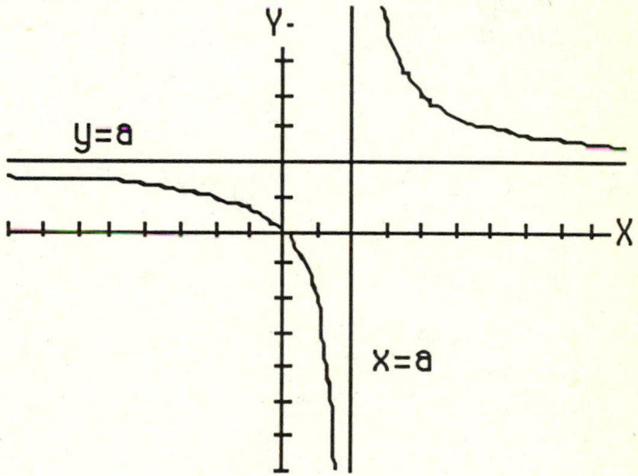

5. $r^2 + 2r(\cos\theta + \sin\theta) = 7$

   $x^2 + y^2 + 2(x + y) = 7$ Circle with center(-1,-1, radius = 3

   $(x + 1)^2 + (y + 1)^2 = 9$

7. $r\cos\frac{\theta}{2} = a$

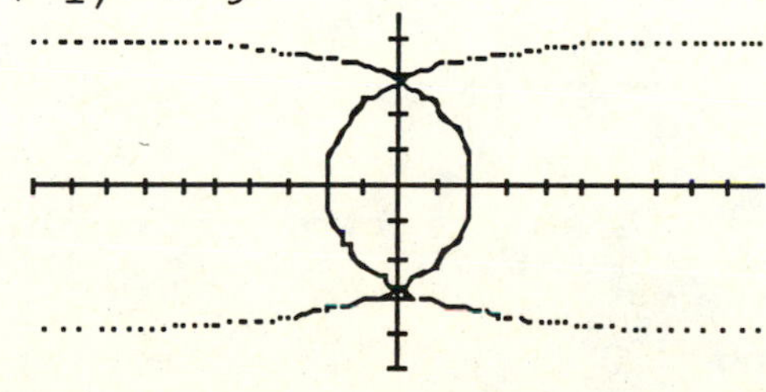

9. $r^2 = 2a^2 \sin 2\theta$
A lemniscate symmetrical to pole (r by -r)

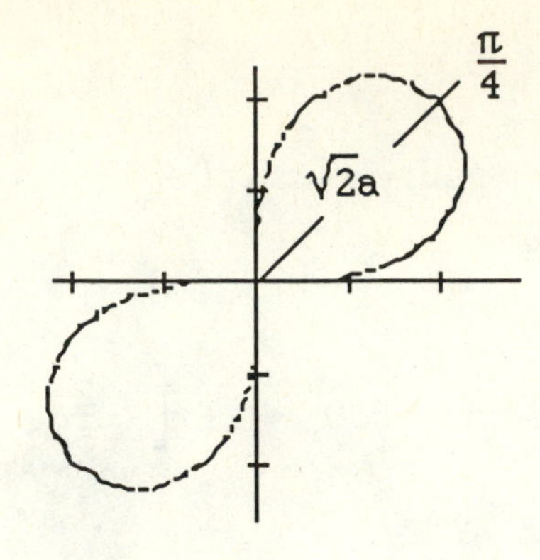

11. (a) $r = \cos 2\theta$ (b) $r^2 = \cos 2\theta$

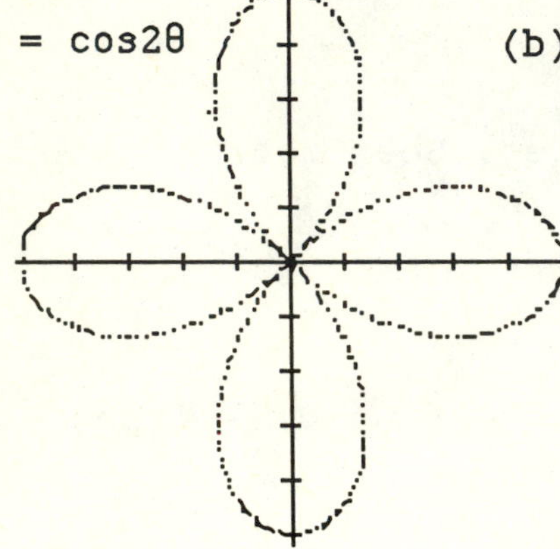

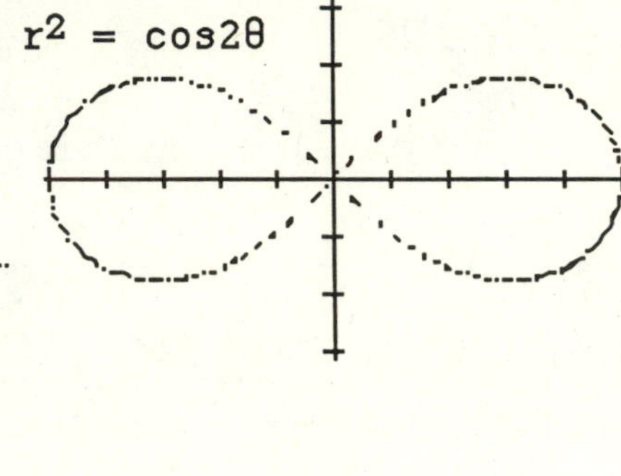

13. (a) $r = \dfrac{2}{1 - \cos\theta}$ (b) $r = \dfrac{2}{1 + \sin\theta}$

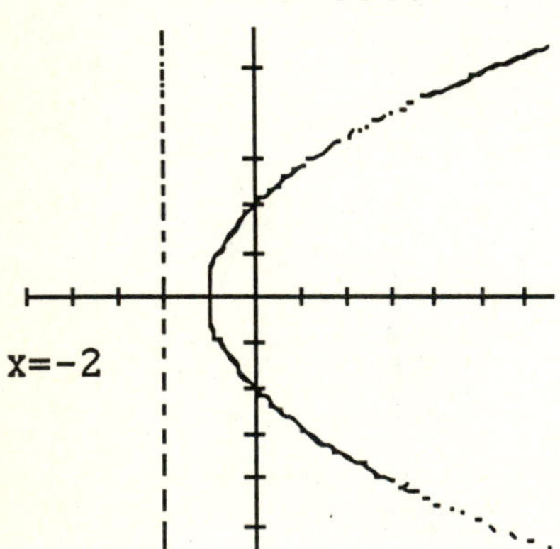

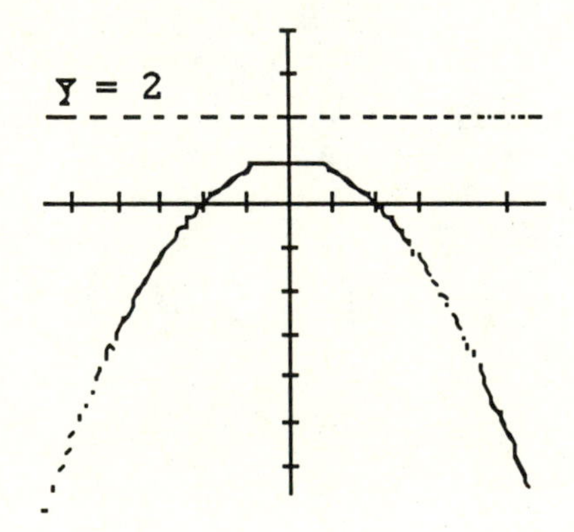

15. $a\cos\theta = a\sin\theta \Leftrightarrow \dfrac{a}{a} = \dfrac{\sin\theta}{\cos\theta} \Leftrightarrow \tan\theta = 1$ or $\theta = \dfrac{\pi}{4}$.

The point is $(\frac{a\sqrt{2}}{2}, \frac{\pi}{4})$. Also the two circles intersect at the origin.

17. $1 = 1 - \sin\theta \Leftrightarrow \sin\theta = 0$ or $\theta = 0, \pi$. The points are $(a,\pi), (a,0)$

19. There are no points of simultaneous intersection, since $a\cos\theta = a(1 + \cos\theta) \Rightarrow 0 = a$. However the graphs do intersect at the origin.

21. There are no simultaneous solutions. The curves do intersect at the origin and the four points

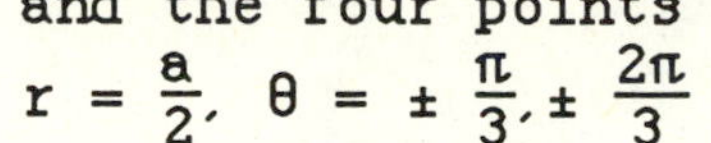

$r = \frac{a}{2}$, $\theta = \pm \frac{\pi}{3}, \pm \frac{2\pi}{3}$

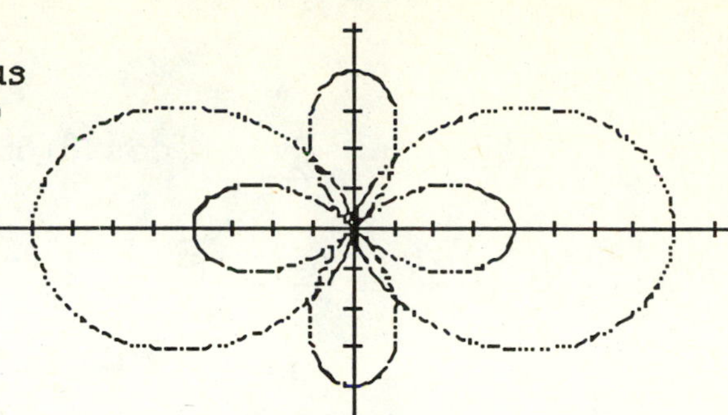

23. (a) $r = \dfrac{1}{1 - \cos\theta}$

$r - r\cos\theta = 1$

$\sqrt{x^2 + y^2} = 1 + x$

$x^2 + y^2 = 1 + 2x + x^2$

$y^2 - 2x = 1$

(b) $r = \dfrac{1}{1 + \cos\theta}$

$r + r\cos\theta = 1$

$\sqrt{x^2 + y^2} = 1 - x$

$x^2 + y^2 = 1 - 2x + x^2$

$y^2 + 2x = 1$

25. (a) $r = \dfrac{1}{1 - 2\cos\theta}$

$r - 2r\cos\theta = 1$

$\sqrt{x^2 + y^2} = 1 + 2x$

$x^2 + y^2 = 1 + 4x + 4x^2$

$3x^2 - y^2 + 4x + 1 = 0$

(b) $r = \dfrac{1}{1 + 2\cos\theta}$

$r + 2r\cos\theta = 1$

$\sqrt{x^2 + y^2} = 1 - 2x$

$x^2 + y^2 = 1 - 4x + 4x^2$

$3x^2 - y^2 - 4x + 1 = 0$

27. The directrix is the line $x = 2$. $\therefore\ r = \dfrac{2}{1 + \cos\theta}$

29. The diameter is $2a$. $\therefore\ r = -2a\cos\theta$

31. The eccentricity is $\frac{1}{3}$, since $a = 3$ and $c = 1$. Using equation 16 in Section 10.3, we have

$$r = \frac{3(1 - \frac{1}{9})}{1 - \frac{1}{3}\cos\theta} = \frac{8}{3 - \cos\theta}$$

33. $A = 4\cdot\frac{1}{2}\displaystyle\int_0^{\frac{\pi}{4}} a^2\cos 2\theta\, d\theta = 2a^2\left[\frac{1}{2}\sin 2\theta\right]_0^{\frac{\pi}{4}} = a^2$

35. $A = 4\cdot\frac{1}{2}\displaystyle\int_0^{\frac{\pi}{2}} a^2(1 + \cos 2\theta)^2 d\theta = 2\int_0^{\frac{\pi}{2}} a^2(1 + 2\cos 2\theta + \cos^2 2\theta)\, d\theta$

$$= 2a^2\int_0^{\frac{\pi}{2}} 1 + 2\cos2\theta + \frac{1}{2}(1 + \cos4\theta)\,d\theta$$

$$= 2a^2\left[\frac{3\theta}{2} + \frac{\sin2\theta}{4} + \frac{\sin4\theta}{8}\right]_0^{\frac{\pi}{2}} = \frac{3a^2\pi}{2}$$

37. $$A = 6\cdot\frac{1}{2}\int_0^{\frac{\pi}{3}} 2a^2\sin3\theta\,d\theta = 6\left[-\frac{a^2}{3}\cos3\theta\right]_0^{\frac{\pi}{3}} = 4a^2$$

39. $A = R_1 + R_2$, where the regions are above and below the x-axis.

$$R_1 = \frac{a^2}{2}\int_0^{\pi}(1 + \sin\theta)^2 - (\sin\theta)^2 d\theta$$

$$= \frac{a^2}{2}\int_0^{\pi}(1 + 2\sin\theta)d\theta$$

$$= \frac{a^2}{2}\left[\theta - 2\cos\theta\right]_0^{\pi} = \frac{a^2\pi}{2}$$

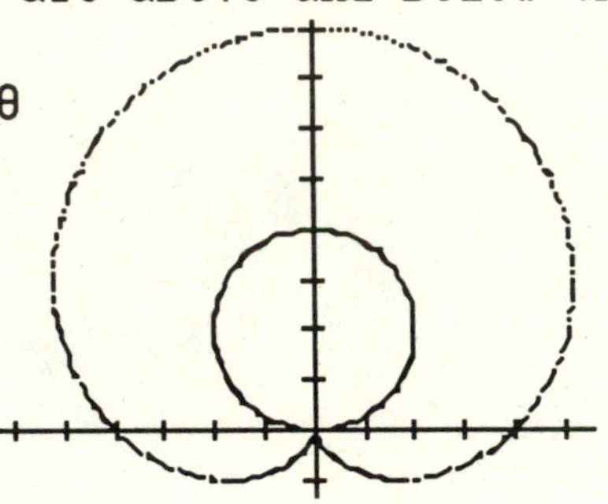

$$R_2 = 2\cdot\frac{1}{2}\int_{-\frac{\pi}{2}}^{0} a^2(1 + \sin\theta)^2 d\theta = \int_{-\frac{\pi}{2}}^{0} a^2(1 + 2\sin\theta + \frac{1}{2}(1 - \cos2\theta))\,d\theta$$

$$= a^2\left[\frac{3\theta}{2} - 2\cos\theta - \frac{\sin2\theta}{4}\right]_0^{\frac{\pi}{2}} = \frac{3a^2\pi}{4}.$$ The total area is then $\frac{5a^2\pi}{4}$

41. $$A = 4\cdot\frac{1}{2}\int_0^{\frac{\pi}{2}}[2a\sin^2\frac{\theta}{2}]^2 d\theta = 8a^2\int_0^{\frac{\pi}{2}}\left(\frac{1 - \cos\theta}{2}\right)^2 d\theta$$

$$= 2a^2\int_0^{\frac{\pi}{2}}(1 - 2\cos\theta + \frac{1}{2}(1 + \cos2\theta))d\theta$$

$$= 2a^2\left[\frac{3\theta}{2} - 2\sin\theta + \frac{\sin2\theta}{4}\right]_0^{\frac{\pi}{2}} = \frac{a^2(3\pi - 8)}{2}$$

43. $r = a\cos^3\frac{\theta}{3} \Rightarrow dr = 3a\cos^2\frac{\theta}{3}(-\sin\frac{\theta}{3})\frac{1}{3}d\theta$

$$ds^2 = r^2d\theta^2 + dr^2 = a^2(\cos^3\frac{\theta}{3})^2d\theta^2 + a^2(\cos^2\frac{\theta}{3}(-\sin\frac{\theta}{3}))^2d\theta^2$$

$$= a^2\cos^4\frac{\theta}{3}d\theta^2(\cos^2\frac{\theta}{3} + \sin^2\frac{\theta}{3})$$

$$\therefore \quad ds = a(\cos^2\frac{\theta}{3})\,d\theta$$

$$\rho = 2\int_0^{\frac{3\pi}{2}} a\cos^2\frac{\theta}{3}d\theta = 2a\int_0^{\frac{3\pi}{2}}\left(\frac{1}{2}+\frac{1}{2}\cos\frac{2\theta}{3}\right)d\theta$$

$$= 2a\left[\frac{1}{2}\theta + \frac{3}{4}\sin\frac{2\theta}{3}\right]_0^{\frac{3\pi}{2}} = \frac{3a\pi}{2}$$

45. $\beta = \psi_2 - \psi_1$

$\tan\beta = \tan(\psi_2 - \psi_1)$

$$= \frac{\tan\psi_2 - \tan\psi_1}{1 + \tan\psi_2\tan\psi_1}$$

The curves will be orthogonal when $\tan\beta$ is undefined, or when

$$\tan\psi_2 = \frac{-1}{\tan\psi_1}$$

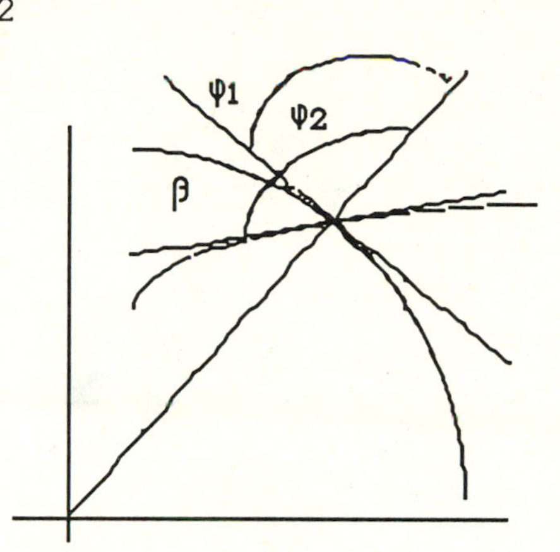

47. $$\tan\varphi = \frac{r}{\frac{dr}{d\theta}} = \frac{2a\sin3\theta}{6a\cos3\theta} = \frac{1}{3}\tan3\theta$$

When $\theta = \frac{\pi}{3}$, $\tan\varphi = \frac{1}{3}\tan\pi = 0$ so $\varphi = 0$.

49. $$\tan\varphi_1 = \frac{\sqrt{3}\cos\theta}{-\sqrt{3}\sin\theta} = -\cot\theta = -\frac{1}{\sqrt{3}} \text{ at } \theta = \frac{\pi}{3}$$

$$\tan\varphi_2 = \frac{\sin\theta}{\cos\theta} = \tan\theta = \sqrt{3} \text{ at } \theta = \frac{\pi}{3}.$$ Since these values are negative reciprocals, the tangents are perpendicular.

51. $$r_1 = \frac{1}{1 - \cos\theta} \Rightarrow \frac{dr_1}{d\theta} = -\frac{\sin\theta}{(1 - \cos\theta)^2}$$

$$r_2 = \frac{3}{1 + \cos\theta} \Rightarrow \frac{dr_2}{d\theta} = \frac{3\sin\theta}{(1 + \cos\theta)^2}$$

$$\frac{1}{1 - \cos\theta} = \frac{3}{1 + \cos\theta} \Leftrightarrow 1 + \cos\theta = 3 - 3\cos\theta \Leftrightarrow 4\cos\theta = 2$$

$\Leftrightarrow \cos\theta = \frac{1}{2}$ or $\theta = \pm\frac{\pi}{3}$. Intersection is at points $(2, \pm\frac{\pi}{3})$

$$\tan\varphi_1 = \frac{\frac{1}{1 - \cos\theta}}{\frac{-\sin\theta}{(1 - \cos\theta)^2}} = -\frac{1 - \cos\theta}{\sin\theta} = -\frac{1}{\sqrt{3}} \text{ at } \theta = \frac{\pi}{3}$$

$$\tan\varphi_2 = \frac{\dfrac{3}{1 + \cos\theta}}{\dfrac{3\sin\theta}{(1 + \cos\theta)^2}} = \frac{1 + \cos\theta}{\sin\theta} = \sqrt{3} \text{ at } \theta = \frac{\pi}{3}$$

$\tan\beta$ is undefined at $\theta = \frac{\pi}{3}$ since $1 + \left(\frac{-1}{\sqrt{3}}\right)(\sqrt{3}) = 1 - 1 = 0$

53. $$r_1 = \frac{a}{1 + \cos\theta} \Rightarrow \frac{dr_1}{d\theta} = \frac{a\sin\theta}{(1 + \cos\theta)^2}$$

$$r_2 = \frac{b}{1 - \cos\theta} \Rightarrow \frac{dr_2}{d\theta} = -\frac{b\sin\theta}{(1 - \cos\theta)^2}$$

$$\tan\varphi_1 = \frac{\dfrac{a}{1 + \cos\theta}}{\dfrac{a\sin\theta}{(1 + \cos\theta)^2}} = \frac{1 + \cos\theta}{\sin\theta}$$

$$\tan\varphi_2 = \frac{\dfrac{b}{1 - \cos\theta}}{\dfrac{-b\sin\theta}{(1 - \cos\theta)^2}} = \frac{1 - \cos\theta}{-\sin\theta}$$

$$1 + \tan\varphi_1\tan\varphi_2 = 1 + \left(\frac{1 + \cos\theta}{\sin\theta}\right)\left(\frac{1 - \cos\theta}{-\sin\theta}\right) = 1 - \frac{1 - \cos^2\theta}{\sin^2\theta} = 0$$

55. $$r = 3\sec\theta \Rightarrow r = \frac{3}{\cos\theta}$$

$$\frac{3}{\cos\theta} = 4 + 4\cos\theta \Leftrightarrow 3 = 4\cos\theta + 4\cos^2\theta \Leftrightarrow (2\cos\theta + 3)(2\cos\theta - 1) = 0$$

$\cos\theta = \frac{1}{2}$ or $\theta = \frac{\pi}{3}, \frac{5\pi}{3}$.

$$\tan\varphi_2 = \frac{4(1 + \cos\theta)}{-4\sin\theta} = -\frac{1 + \cos\theta}{\sin\theta} = -\sqrt{3} \text{ at } \frac{\pi}{3}$$

$$\tan\varphi_1 = \frac{3\sec\theta}{3\sec\theta\tan\theta} = \cot\theta = \frac{1}{\sqrt{3}} \text{ at } \frac{\pi}{3}.$$

$\tan\beta$ is undefined and $\beta = \frac{\pi}{2}$

57. $$\frac{1}{1 - \cos\theta} = \frac{1}{1 - \sin\theta} \Leftrightarrow 1 - \cos\theta = 1 - \sin\theta$$

$\Leftrightarrow \cos\theta = \sin\theta$ or $\theta = \frac{\pi}{4}$

$$\tan\varphi_1 = \frac{\dfrac{1}{1 - \cos\theta}}{\dfrac{-\sin\theta}{(1 - \cos\theta)^2}} = \frac{1 - \cos\theta}{-\sin\theta}, \quad \tan\varphi_2 = \frac{\dfrac{1}{1 - \sin\theta}}{\dfrac{\cos\theta}{(1 - \sin\theta)^2}} = \frac{1 - \sin\theta}{\cos\theta}$$

At $\theta = \frac{\pi}{4}$, $\tan\varphi_1 = 1 - \sqrt{2}$ and $\tan\varphi_2 = \sqrt{2} - 1$

$$\tan\beta = \frac{(\sqrt{2} - 1) - (1 - \sqrt{2})}{1 + (\sqrt{2} - 1)(1 - \sqrt{2})} = \frac{2\sqrt{2} - 2}{2\sqrt{2} - 2} = 1 \quad \therefore \ \beta = \frac{\pi}{4}$$

59. (a) $\tan a = \dfrac{r}{\frac{dr}{d\theta}} \Rightarrow \dfrac{dr}{r} = \dfrac{d\theta}{\tan a}$

$$\ln r = \frac{1}{\tan a}\theta + C \quad \text{or } r = Ae^{\frac{\theta}{\tan a}}$$

$$A = \frac{1}{2}\int_{\theta_1}^{\theta_2} A^2 e^{\frac{2\theta}{\tan a}} d\theta =$$

$$= \left[\frac{1}{4}A^2 \tan a\, e^{\frac{2\theta}{\tan a}}\right]_{\theta_2}^{\theta_1} = \frac{\tan a}{4}(r_2^2 - r_1^2) \text{ since}$$

$$r_2^2 = A^2 e^{\frac{2\theta_2}{\tan a}} \text{ and } r_1^2 = A^2 e^{\frac{2\theta_1}{\tan a}}. \quad K = \frac{\tan a}{4}$$

(b) $\tan a = \dfrac{r}{\frac{dr}{d\theta}} \Rightarrow dr = \dfrac{r\,d\theta}{\tan a} \Rightarrow dr^2 = \dfrac{r^2 d\theta^2}{\tan^2 a}$

$$ds^2 = r^2\,d\theta^2 + dr^2 = r^2\,d\theta^2 + \frac{r^2 d\theta^2}{\tan^2 a} = r^2\,d\theta^2\left(\frac{\sec^2 a}{\tan^2 a}\right)$$

$$s = \int_{\theta_1}^{\theta_2} Ae^{\frac{\theta}{\tan a}} \frac{\sec a}{\tan a} d\theta = A \sec a\, e^{\frac{\theta}{\tan a}}\Big]_{\theta_1}^{\theta_2}$$

$= K(r_2 - r_1)$, constant $K = \sec a$.

61. By Prob. 1, Art. 10.4, the area is $\dfrac{3\pi a^2}{2}$. $\bar{y} = 0$ by symmetry.

$$M_y = \frac{1}{3}\int_0^{2\pi} [a(1 + \cos\theta)]^3 \cos\theta d\theta$$

$$= \frac{a^3}{3}\int_0^{2\pi} [(\cos\theta + 3\cos^2\theta + 3\cos^3\theta + \cos^4\theta)d\theta$$

$$I_1 = \int_0^{2\pi} \cos\theta d\theta = [\sin\theta]_0^{2\pi} = 0$$

$$I_2 = \int_0^{2\pi} 3\cos^2\theta d\theta = \frac{3}{2}\int_0^{2\pi}(1 + \cos 2\theta)d\theta = \frac{3}{2}(\theta + \frac{\sin 2\theta}{2})\Big|_0^{2\pi} = 3\pi$$

$$I_3 = \int_0^{2\pi} 3\cos^3\theta d\theta = 3\int_0^{2\pi}(1 - \sin^2\theta)\cos\theta d\theta = 3\sin\theta - \sin^3\theta\Big]_0^{2\pi} = 0$$

$$I_4 = \int_0^{2\pi}\cos^4\theta d\theta = \frac{1}{4}\int_0^{2\pi}(1 + 2\cos 2\theta + \cos^2 2\theta)\,d\theta$$

$$= \frac{1}{4}\left[\theta + \sin 2\theta + \frac{\theta}{2} + \frac{\sin 4\theta}{8}\right]_0^{2\pi} = \frac{3\pi}{4}$$

$$My = \frac{a^3}{3}(3\pi + \frac{3\pi}{4}) = \frac{5\pi a^3}{4}\ ; \ \bar{x} = \frac{\frac{5\pi a^3}{4}}{\frac{3\pi a^2}{2}} = \frac{5a}{6}$$

63. $r = a(1 + \cos\theta) \Rightarrow dr = -a\sin\theta d\theta \Rightarrow dr^2 = a^2\sin^2\theta d\theta^2$

$\therefore ds^2 = r^2 d\theta^2 + dr^2 = a^2(1 + \cos\theta)^2 d\theta^2 + a^2\sin^2\theta d\theta^2$

$= a^2(1 + 2\cos\theta + (\sin^2\theta + \cos^2\theta))\,d\theta^2 = 2a^2(1 + \cos\theta)d\theta^2$

$$L = \int_0^{2\pi}\sqrt{2a^2(2\cos^2\frac{\theta}{2})}\,d\theta = 2a\int_0^{2\pi}|\cos\frac{\theta}{2}|d\theta = 4a\int_0^{\pi}\cos\frac{\theta}{2}d\theta = 8a\sin\frac{\theta}{2}\Big]_0^{\pi} = 8a$$

$\bar{y} = 0$ by symmetry. $\int_0^{2\pi} r\cos\theta ds = \int_0^{2\pi} a(1 + \cos\theta)(\cos\theta)(2a|\cos\frac{\theta}{2}|)\,d\theta$

$$= 2\int_0^{\pi} a(2\cos^2\frac{\theta}{2})(\cos\theta)(2a|\cos\frac{\theta}{2}|)\,d\theta = 8a^2\int_0^{\pi}(\cos^3\frac{\theta}{2})(\cos\theta)d\theta$$

$$= 8a^2\int_0^{\pi}(\cos^3\frac{\theta}{2})(2\cos^2\frac{\theta}{2} - 1)\,d\theta = 8a^2\int_0^{\pi}(2\cos^4\frac{\theta}{2} - \cos^2\frac{\theta}{2})\cos\frac{\theta}{2}d\theta$$

$$= 8a^2\int_0^{\pi}[2(1-\sin^2\frac{\theta}{2})^2 - (1 - \sin^2\frac{\theta}{2})]\cos\frac{\theta}{2}d\theta$$

$$= 8a^2\int_0^{\pi}(1-3\sin^2\frac{\theta}{2} + 2\sin^4\frac{\theta}{2})\cos\frac{\theta}{2}d\theta$$

$$= 8a^2\left[2\sin\frac{\theta}{2} - 2\sin^3\frac{\theta}{2} + \frac{4}{5}\sin^5\frac{\theta}{2}\right]_0^{\pi} = \frac{32a^2}{5};\ \bar{x} = \frac{32a^2}{5}\cdot\frac{1}{8a} = \frac{4a}{5}.$$

CHAPTER 11

INFINITE SEQUENCES AND INFINITE SERIES

11.1 SEQUENCES OF NUMBERS

1. $a_1 = \frac{2}{3}$; $a_2 = \frac{3}{5}$; $a_3 = \frac{4}{7}$; $a_4 = \frac{5}{9}$; $\lim\limits_{x \to \infty} \frac{n+1}{2n+1} = \frac{1}{2}$

3. $a_1 = -\frac{1}{3}$; $a_2 = -\frac{3}{5}$; $a_3 = -\frac{5}{7}$; $a_4 = -\frac{7}{9}$; $\lim\limits_{x \to \infty} \frac{1-2n}{1+2n} = -1$

5. $a_1 = \frac{1}{2}$; $a_2 = \frac{1}{2}$; $a_3 = \frac{1}{2}$; $a_4 = \frac{1}{2}$; $\lim\limits_{x \to \infty} \frac{2^n}{2^{n+1}} = \frac{1}{2}$

7. $x_1 = 1$; $x_{n+1} = x_n + \left(\frac{1}{2}\right)^n$; $x_2 = 1 + \frac{1}{2} = \frac{3}{2}$; $x_3 = \frac{3}{2} + \left(\frac{1}{2}\right)^2 = \frac{7}{4}$;

   $x_4 = \frac{7}{4} + \left(\frac{1}{2}\right)^3 = \frac{15}{8}$; $x^5 = \frac{15}{8} + \left(\frac{1}{2}\right)^4 = \frac{31}{16}$; $x_6 = \frac{31}{16} + \left(\frac{1}{2}\right)^5 = \frac{63}{32}$

9. $x_1 = 2$; $x_{n+1} = \frac{x_n}{2}$; $x_2 = \frac{2}{2} = 1$; $x_3 = \frac{1}{2}$; $x_4 = \frac{\frac{1}{2}}{2} = \frac{1}{4}$;

   $x_5 = \frac{\frac{1}{4}}{2} = \frac{1}{8}$; $x_6 = \frac{\frac{1}{8}}{2} = \frac{1}{16}$

11. $x_1 = x_2 = 1$; $x_{n+2} = x_{n+1} + x_n$; $x_3 = 1+1 = 2$; $x_4 = 2+1 = 3$;

    $x_5 = 3+2 = 5$; $x_6 = 5+3 = 8$

13. (a) $(1)^2 - 2(1)+2 = -1$; $(3)^2 - 2(2)^2 = 1$; If $a^2 - 2b^2 = 1$, then

    $(a + 2b)^2 - 2(a + b)^2 = a^2 - 4ab + 4b^2 - 2a^2 - 4ab - 2b^2 =$

    $2b^2 - a^2 = -1$. If $a^2 - 2b^2 = -1$, then $(a + 2b)^2 - 2(a + b)^2$

    $2b^2 - a^2 = -(a^2 - 2b^2) = -(-1) = 1$

    (b) $\left(\frac{a+2b}{a+b}\right)^2 - 2 = \frac{a^2+4ab+4b^2}{a^2+2ab+b^2} - 2 = \frac{2b^2-a^2}{(a+b)^2} = \frac{\pm 1}{(y_n)^2}$

    $$= \lim_{n\to\infty} r_n = \lim_{n\to\infty} \sqrt{2 \pm \left(\frac{1}{y_n}\right)^2} = \sqrt{2}$$

15. Let $f(x) = \cos x - x$. $f'(x) = -\sin x - 1$ and using the formula $x_{n+1} = x_n - \frac{f(x_n)}{f'(x_n)}$, we have $x_1 = 0.755222417$, $x_2 = 0.739141666$, $x_3 = 0.739085134$, $x_4 = 0.739085133$.

17. 0.876726216; $x^2 - \sin x = 0$

19. (a) Using $y = \tan x - 2x$, the root is $x = 1.65561185$
(b) Yes, $\frac{\pi}{3} \approx 1.05 < 1.66$
(c) If $x < \tan x < 2x$ for all $0 < x < b$,
then $\int x\,dx < \int \tan x\,dx < \int 2x\,dx$ or $\frac{x^2}{2} < \ln|\sec x| < x^2$

11.2 LIMIT THEOREMS

1. $0, -\frac{1}{4}, -\frac{2}{9}, -\frac{3}{16}$; converges to 0

3. $\frac{1}{3}, \frac{1}{9}, \frac{1}{27}, \frac{1}{81}$; converges to 0

5. $1, -\frac{1}{3}, \frac{1}{5}, -\frac{1}{7}$; converges to 0

7. 0,-1,0,1,0; diverges

9. $1, -\frac{1}{\sqrt{2}}, \frac{1}{\sqrt{3}}, -\frac{1}{2}$; converges to 0

11. converges to 0

13. converges to 1

15. diverges because the terms oscillate between ±1.

17. converges to $-\frac{2}{3}$

19. $\sqrt{\frac{2n}{n+1}} = \sqrt{\frac{2}{1+\frac{1}{n}}}$ converges to $\sqrt{2}$

21. converges to 0

23. diverges, terms oscillate between 1 and -1

25. $\frac{n^2}{(n+1)^2} = \left(\frac{n}{n+1}\right)^2$ converges to 1

27 $\lim \frac{1-5n^4}{n^4+8n^3} = \frac{\frac{1}{n^4} - 5}{1 + \frac{8}{n}} = -5$

29. $\lim\limits_{n\to\infty} \tanh n = \lim\limits_{n\to\infty} \frac{e^n - e^{-n}}{e^n + e^{-n}} = \lim\limits_{n\to\infty} \frac{1 - \frac{1}{e^{2n}}}{1 + \frac{1}{e^{2n}}} = 1$

31. $\lim_{n\to\infty} \frac{2(n+1)+1}{2n+1} = \lim_{n\to\infty} \frac{2n+3}{2n+1} = \lim_{n\to\infty} \frac{2+\frac{3}{n}}{2+\frac{1}{n}} = 1$

33. $\lim_{n\to\infty} 5 = 5$

35. $\lim_{n\to\infty} (.5)^n = \lim_{n\to\infty} \left(\frac{1}{2}\right)^n = 0$

37. $\lim_{n\to\infty} \frac{n^n}{(n+1)^{n+1}} = \lim_{n\to\infty} \left(\frac{n}{n+1}\right)^n \left(\frac{1}{n+1}\right) = (1)(0) = 0$

39. $\lim_{n\to\infty} \sqrt{2-\frac{1}{n}} = \sqrt{2}$

41. $\lim_{n\to\infty} \frac{3^n}{n^3} = \lim_{n\to\infty} \frac{(\ln 3)3^n}{3n^2} = \lim_{n\to\infty} \frac{(\ln 3)^2 3^n}{6n} = \lim_{n\to\infty} \frac{(\ln 3)^3 3^n}{6} = \infty$

43. $\lim_{n\to\infty} (\ln n - \ln (n+1)) = \lim_{n\to\infty} \ln \left(\frac{n}{n+1}\right) = \ln \lim_{n\to\infty} \left(\frac{n}{n+1}\right) = \ln 1 = 0$

45. $\lim_{n\to\infty} \frac{n^2-2n+1}{n-1} = \lim_{n\to\infty} \frac{1-\frac{2}{n}+\frac{1}{n^2}}{\frac{1}{n}-\frac{1}{n^2}} = \frac{1}{0} = \infty$

47. $\lim_{n\to\infty} \left(-\frac{1}{2}\right)^n = 0$

49. $\lim_{n\to\infty} \tan^{-1} n = \frac{\pi}{2}$

51. $\lim_{n\to\infty} n\sin\frac{1}{n} = \lim_{n\to\infty} \frac{\sin\frac{1}{n}}{\frac{1}{n}} = \lim_{n\to\infty} \frac{\cos\frac{1}{n}\left(-\frac{1}{n^2}\right)}{-\frac{1}{n^2}} = \cos 0 = 1$

53. $\lim_{n\to\infty} \frac{n^2}{2n-1}\sin\frac{1}{n} = \lim_{n\to\infty} \left(\frac{n}{2n-1}\right)\left(n\sin\frac{1}{n}\right) = \left(\frac{1}{2}\right)(1) = \frac{1}{2}$

55. Note that $\frac{n!}{n^n} = \frac{1\cdot 2\cdot 3\cdot \ldots \cdot n}{n\cdot n\cdot n \ldots \cdot n} < \frac{1}{n}$ since $\frac{2\cdot 3\cdot \ldots \cdot n}{n\cdot n\cdot \ldots \cdot n} < 1$

$\therefore\ 0 < \lim_{n\to\infty} \frac{n!}{n^n} < \lim_{n\to\infty} \frac{1}{n} = 0$

57. $\lim_{n\to\infty} (x)^{1/n} = 1$ for all $x > 0$. Fix $x > 0$ and consider $a_n = x^{1/n}$.

$\lim_{n\to\infty} (\ln a_n) = \lim_{n\to\infty} \frac{1}{n}\ln x = \ln x \lim_{n\to\infty} \frac{1}{n} = 0$. Therefore,

$\lim_{n\to\infty} (x)^{1/n} = \lim_{n\to\infty} e^{\frac{1}{n}\ln x} = e^0 = 1$

59. $\lim_{n\to\infty} nf\left(\frac{1}{n}\right) = \lim_{n\to\infty} \frac{f\left(\frac{1}{n}\right)}{\frac{1}{n}} = \lim_{n\to\infty} \frac{f'\left(\frac{1}{n}\right)\left(-\frac{1}{n^2}\right)}{-\frac{1}{n^2}} = \lim_{n\to\infty} f'\left(\frac{1}{n}\right) = f'(0)$

61. Let $f(\frac{1}{n}) = e^{\frac{1}{n}} - 1$. Then $f(x) = e^x - 1$ and $f'(x) = e^x$.

$\lim_{n\to\infty} (e^{\frac{1}{n}} - 1) = 1$ since $f'(0) = 1$

63. Suppose that $\{a_n\}$ converges to some finite value L. Let $\varepsilon = 1$ be given. There exists $N_1 > 0$ for which

(1) $|a_n - L| < 1$ or $a_n < L + 1$ for all $n > N_1$.

There also exists $N_2 > 0$ for which

(2) $f(n) = a_n > L + 1$ for all $n > N_2$.

Let $N > \max\{N_1, N_2\}$. For all $n > N$, both equations (1) and (2) are true, which cannot be. Thus the sequence $\{a_n\}$ cannot converge, and hence must diverge.

65. The conclusion is that $|f(a_n) - f(L)| < \varepsilon$, so that the sequence $\{f(a_n)\}$ converges to $f(L)$.

11.3 LIMITS THAT ARISE FREQUENTLY

1. $\lim_{n\to\infty} \frac{1 + \ln n}{n} = \lim_{n\to\infty} \left(\frac{1}{n} + \frac{\ln n}{n}\right) = 0 + 0 = 0$

3. $\lim_{n\to\infty} \frac{(-4)^n}{n!} = 0$ by Formula 6 with $x = -4$.

5. $\lim_{n\to\infty} (.5)^n = 0$ by Formula 4 for $x = 0.5$.

7. $\lim_{n\to\infty} \left(1 + \frac{7}{n}\right)^n = e^7$, by Formula 5, for $x = 7$

9. $\lim_{n\to\infty} \frac{\ln(n+1)}{n} = \lim_{n\to\infty} \frac{\frac{1}{n+1}}{1} = 0$

11. $\lim_{n\to\infty} \frac{n!}{10^{6n}} = \lim_{n\to\infty} \frac{1}{\frac{10^{6n}}{n!}} = \infty$.

13. $\lim_{n\to\infty} \sqrt[2n]{n} = \lim_{n\to\infty} \left(n^{\frac{1}{n}}\right)^{\frac{1}{2}} = 1$

15. $\lim_{n\to\infty} \frac{1}{3^{2n-1}} = 0$

17. $\lim_{n\to\infty} \left(\frac{n}{n+1}\right)^n = \lim_{n\to\infty} \left(\frac{n+1}{n}\right)^{-n} = \lim_{n\to\infty} \left[\left(1 + \frac{1}{n}\right)^n\right]^{-1} = e^{-1}$

19. $\lim_{n\to\infty} \frac{\ln(2n+1)}{n} = \lim_{n\to\infty} \frac{\frac{2}{2n+1}}{1} = 0$

21. $\lim_{n\to\infty} \sqrt[n]{\frac{x^n}{2n+1}}$, $x > 0$, $= x\left(\lim_{n\to\infty} \sqrt[n]{\frac{1}{2n+1}}\right) = x$

23. Consider $a_n = \ln(n^2 + n)^{1/n}$. $\lim_{n\to\infty} \ln(n^2+n)^{1/n} = \lim_{n\to\infty} \frac{1}{n}\ln(n^2+n)^{1/n} =$

$\lim_{n\to\infty} \frac{2n+1}{n^2+n} = 0$. $\therefore \lim_{n\to\infty} e^{\frac{1}{n}\ln(n^2+n)} = e^0 = 1$.

25. $\lim_{n\to\infty}\left(\frac{3}{n}\right)^{\frac{1}{n}} = \lim_{n\to\infty}\frac{3^{1/n}}{n^{1/n}} = 1$

27. $\lim_{n\to\infty}\left(1-\frac{1}{n}\right)^n = e^{-1}$

29. $\lim_{n\to\infty}\frac{\ln(n^2)}{n} = \lim_{n\to\infty}\frac{2\ln n}{n} = 2\lim_{n\to\infty}\frac{\ln n}{n} = 0$

31. Diverges, since $\lim_{n\to\infty} \ln n = \infty$ and $\lim_{n\to\infty} n^{1/n} = 1$.

33. $\int_1^n \frac{1}{x^p}\,dx = \left.\frac{x^{-p+1}}{-p+1}\right]_1^n = \frac{1}{-p+1}[n^{-p+1} - 1]$. For $p > 1$,

$\frac{1}{-p+1}\lim_{n\to\infty}[n^{-p+1} - 1] = (-1)\frac{1}{-p+1} = \frac{1}{p-1}$.

35. $N \geq 9124$

37. $|\sqrt[c]{n} - 1| < \varepsilon \Leftrightarrow \frac{1}{n^c} < \varepsilon \Leftrightarrow n^c > \frac{1}{\varepsilon}$ or $n > \sqrt[c]{\frac{1}{\varepsilon}}$

$\therefore$ Let $N = \varepsilon^{-\frac{1}{c}}$. For $n > N$, $0 < \frac{1}{n^c} < \varepsilon$ so $\frac{1}{n^c}$ converges to 0.

About the hint: $\frac{1}{n^{0.04}} < .001 \Leftrightarrow n^{0.04} > 10^3 \Leftrightarrow n > (10^3)^{25} = 10^{75}$

11.4 INFINITE SERIES

1. $S_n = \frac{2\left(1-\left(\frac{1}{3}\right)^n\right)}{1-\frac{1}{3}} = 3\left(1-\left(\frac{1}{3}\right)^n\right)$. $r = \frac{1}{3} < 1$, so $S = \frac{2}{1-\frac{1}{3}} = 3$

3. $S_n = \frac{1(1-e^{-n})}{1-e^{-1}}$. $r = \frac{1}{e} < 1$ so $S = \frac{1}{1-\frac{1}{e}} = \frac{e}{e-1}$

5. $S_n = \dfrac{1(1-(-2)^n)}{1-(-2)} = \dfrac{1-(-2)^n}{3}$. $r = 2 > 1$ so series diverges.

7. $\ln\frac{1}{2} + \ln\frac{2}{3} + \ln\frac{3}{4} + \ldots + \ln\frac{n}{n+1} =$

$\ln 1 - \ln 2 + \ln 2 - \ln 3 + \ln 3 - \ln 4 + \ldots + \ln n - \ln(n+1)$

$\therefore \lim_{n\to\infty} S_n = \lim_{n\to\infty}(-\ln(n+1)) = -\infty$. so series diverges

9. (a) $\displaystyle\sum_{n=-2}^{\infty} \frac{1}{(n+4)(n+5)}$ (b) $\displaystyle\sum_{n=0}^{\infty} \frac{1}{(n+2)(n+3)}$ (c) $\displaystyle\sum_{n=5}^{\infty} \frac{1}{(n-3)(n-2)}$

11. $S = \dfrac{1}{1-\frac{1}{4}} = \dfrac{4}{3}$

13. $S = \dfrac{\frac{7}{4}}{1-\frac{1}{4}} = \dfrac{7}{3}$

15. $S = \dfrac{5}{1-\frac{1}{2}} + \dfrac{1}{1-\frac{1}{3}} = 10 + \dfrac{3}{2} = \dfrac{23}{2}$

17. $S = \dfrac{1}{1-\frac{2}{5}} = \dfrac{5}{3}$

19. $\displaystyle\sum_{n=1}^{\infty} \frac{4}{(4n-3)(4n+1)} = \sum_{n=1}^{\infty}\left(\frac{1}{4n-3} - \frac{1}{4n+1}\right) =$

$\left(1-\frac{1}{5}\right) + \left(\frac{1}{5}-\frac{1}{9}\right) + \left(\frac{1}{9}-\frac{1}{13}\right) + \ldots + \left(\frac{1}{4n-3} - \frac{1}{4n+1}\right) + \ldots$

$\therefore \lim_{n\to\infty} S_n = \lim_{n\to\infty}\left(1 - \frac{1}{4n+1}\right) = 1$

21. $\displaystyle\sum_{n=3}^{\infty} \frac{4}{(4n-3)(4n+1)} = \sum_{n=3}^{\infty}\left(\frac{1}{4n-3} - \frac{1}{4n+1}\right) =$

$\left(\frac{1}{9}-\frac{1}{13}\right) + \left(\frac{1}{13}-\frac{1}{17}\right) + \ldots + \left(\frac{1}{4n-3} - \frac{1}{4n+1}\right) + \ldots$

$\therefore \lim_{n\to\infty} S_n = \lim_{n\to\infty}\left(\frac{1}{9} - \frac{1}{4n+1}\right) = \frac{1}{9}$

23. (a) $0.234234234\ldots = \dfrac{234}{10^3} + \dfrac{234}{10^6} + \ldots$

$$S = \frac{\frac{234}{10^3}}{1 - \frac{1}{10^3}} = \frac{234}{999} = \frac{26}{111}$$

(b) Yes, every repeating decimal is a geometric series, with ratio $= 10^{-n}$, where n is the number of repeating digits.

27. $\sum_{n=1}^{\infty} (-1)^{n+1} \frac{3}{2^n} = \frac{\frac{3}{2}}{1-\left(-\frac{1}{2}\right)} = 1$

29. $\sum_{n=0}^{\infty} \cos n\pi = 1 - 1 + 1 - 1 + 1 \ldots$ diverges because $\lim_{n\to\infty} (-1)^n \neq 0$

31. $\sum_{n=0}^{\infty} e^{-2n} = \frac{1}{1-e^{-2}} = \frac{e^2}{e^2-1}$

33. $\sum_{n=1}^{\infty} (-1)^{n+1} n$ diverges because $\lim_{n\to\infty} (-1)^{n+1} n \neq 0$

35. $\sum_{n=0}^{\infty} \frac{2^n - 1}{3^n} = \sum_{n=0}^{\infty} \left(\frac{2}{3}\right)^n - \sum_{n=0}^{\infty} \left(\frac{1}{3}\right)^n = \frac{1}{1-\frac{2}{3}} - \frac{1}{1-\frac{1}{3}} = \frac{3}{2}$

37. $\sum_{n=0}^{\infty} \frac{n!}{1000^n}$ diverges because $\lim_{n\to\infty} \frac{n!}{1000^n} = \infty$

39. $a = 1$ and $r = -x$, since $\frac{1}{1+x} = \frac{1}{1-(-x)} = \sum_{n=0}^{\infty} (-1)^n x^n$, $|x| < 1$.

41. The area of each square is one-half that of the preceding square. Thererore, the sum of the areas is
$S = 4 + 2 + \frac{1}{2} + \ldots = \frac{4}{1 - \frac{1}{2}} = 8$.

43. Let $\sum_{n=1}^{\infty} n$ and $\sum_{n=1}^{\infty} (-n)$ be the two divergent series. But the sum $\sum_{n=1}^{\infty} (n + (-n)) = \sum_{n=0}^{\infty} 0$ converges.

45. Let $\sum_{n=1}^{\infty} \frac{1}{2^n} = 1$ and $\sum_{n=1}^{\infty} \frac{1}{3^n} = \frac{1}{2}$ be the two convergent series.

Then $\sum_{n=1}^{\infty} \frac{\frac{1}{2^n}}{\frac{1}{3^n}} = \sum_{n=1}^{\infty} \left(\frac{3}{2}\right)^n$ diverges.

47. If $\sum_{n=1}^{\infty} a_n$ converges, then $\lim_{n\to\infty} a_n = 0$. Therefore, $\lim_{n\to\infty}\left(\frac{1}{a_n}\right) \neq 0$ and $\sum_{n=1}^{\infty}\left(\frac{1}{a_n}\right)$ diverges.

11.5 TESTS FOR CONVERGENCE OF SERIES WITH NONNEGATIVE TERMS

Note: The reasons given for convergence may not be the only ones that apply.

1. $\sum_{n=1}^{\infty} \frac{1}{10^n}$ converges. Geometric series with $r = \frac{1}{10} < 1$.

3. $\sum_{n=1}^{\infty} \frac{\sin^2 n}{2^n}$ converges, since $\sin^2 n \le 1 \Rightarrow \frac{\sin^2 n}{2^n} \le \frac{1}{2^n}$ and $\sum_{n=1}^{\infty} \frac{1}{2^n}$ converges.

5. $\sum_{n=1}^{\infty} \frac{1+\cos n}{n^2}$ converges by comparison with $\sum_{n=1}^{\infty} \frac{2}{n^2} =$

   $2\sum_{n=1}^{\infty} \frac{1}{n^2}$ which is a multiple of a p-series with $p = 2 > 1$.

7. $\sum_{n=1}^{\infty} \frac{\ln n}{n}$ diverges since $\ln n > 1$ for $n > e \Rightarrow \frac{\ln n}{n} > \frac{1}{n}$ and $\sum_{n=1}^{\infty} \frac{1}{n}$ diverges.

9. $\sum_{n=1}^{\infty} \frac{2^n}{3^n}$ converges, geometric series with $r = \frac{2}{3} < 1$.

11. $\sum_{n=1}^{\infty} \frac{1}{1+\ln n}$ diverges by comparison with $\sum_{n=1}^{\infty} \frac{1}{1+n}$, since

    $\ln n < n$ for $n > 1 \Rightarrow 1 + \ln n < 1 + n \Rightarrow \frac{1}{1+\ln n} > \frac{1}{1+n}$.

13. $\sum_{n=1}^{\infty} \frac{2^n}{n+1}$ diverges, since $\lim_{n\to\infty} \frac{2^n}{n+1} \neq 0$.

15. $\sum_{n=1}^{\infty} \frac{1}{\sqrt{n^3+1}}$ converges, since $\sum_{n=1}^{\infty} \frac{1}{n^{3/2}}$ converges and

$$\lim_{n\to\infty} \frac{\frac{1}{\sqrt{n^3+1}}}{\frac{1}{\sqrt{n^3}}} = \lim_{n\to\infty} \sqrt{\frac{n^3}{n^3+1}} = 1.$$

17. $\sum_{n=1}^{\infty} \frac{n}{n^2+1}$ diverges by comparison with $\sum_{n=1}^{\infty} \frac{1}{n}$, since

$$\lim_{n\to\infty} \frac{\frac{n}{n^2+1}}{\frac{1}{n}} = \lim_{n\to\infty} \frac{n^2}{n^2+1} = 1.$$

19. $\sum_{n=1}^{\infty} \left(1+\frac{1}{n}\right)^n$ diverges, since $\lim_{n\to\infty}\left(1+\frac{1}{n}\right)^n = e \neq 0$.

21. $\sum_{n=1}^{\infty} \frac{1-n}{n\cdot 2^n}$ converges, because it is the difference of

$\sum_{n=1}^{\infty} \frac{1}{n\cdot 2^n}$ which converges because $\frac{1}{n\cdot 2^n} < \frac{1}{2^n}$ and $\sum_{n=1}^{\infty} \frac{1}{2^n}$.

23. $\sum_{n=1}^{\infty} \frac{1}{3^{n-1}+1}$ converges by comparison to $\sum_{n=1}^{\infty} \frac{1}{3^{n-1}} = \frac{3}{2}$.

25. $\sum_{n=1}^{\infty} \frac{1}{2n-1}$ diverges since $\int_1^{\infty} \frac{dx}{2x-1} = \lim_{t\to\infty} \int_1^t \frac{dx}{2x-1} =$

$\lim_{t\to\infty} \frac{1}{2} \ln|2x-1| \Big]_1^t = \lim_{t\to\infty} \frac{1}{2}[2t-1] = \infty.$

27. $S_n \le 1 + \ln(365\cdot 24\cdot 60\cdot 60\cdot 13\cdot 10^9) \approx 41.55$

29. For $n \ge 1$, $\frac{1}{n} \le 1$ so $\frac{a_n}{n} \le a_n$. $\therefore \sum_{n=1}^{\infty} \frac{a_n}{n}$ converges if $\sum_{n=1}^{\infty} a_n$ converges.

31. If $\{S_n\}$ is nonincreasing with lower bound M, then $\{-S_n\}$ is a nondecreasing sequence with upper bound -M. By Theorem 1, $\{-S_n\}$ converges, and hence $\{S_n\}$ converges. If $\{S_n\}$ has no lower bound, then $\{-S_n\}$ has no upper bound and diverges. Hence $\{S_n\}$ also diverges.

33. (a) $\sum_{n=2}^{\infty} \frac{1}{n \ln n}$ diverges, since $\sum_{n=2}^{\infty} 2^n \cdot \frac{1}{2^n \ln 2^n} = \sum_{n=2}^{\infty} \frac{1}{n \ln 2}$

$= \frac{1}{\ln 2} \sum_{n=2}^{\infty} \frac{1}{n}$ diverges.

(b) $\sum_{n=1}^{\infty} 2^n \cdot \frac{1}{(2^n)^p} = \sum_{n=1}^{\infty} 2^{n(1-p)}$ converges if $p > 1$ or $1 - p < 0$ and diverges if $p \leq 1$ or $1 - p \geq 0$.

35. Since $\sum_{n=2}^{\infty} \frac{1}{n \ln n}$ diverges, the limit comparison test and the fact that $\lim_{n\to\infty} \frac{n \ln n}{p_n} = 1$ states that $\sum_{n=1}^{\infty} \frac{1}{p_n}$ diverges.

11.6 SERIES WITH NONNEGATIVE TERMS: RATIO AND ROOT TESTS

Note: The reasons given for convergence may not be the only ones that apply.

1. $\sum_{n=1}^{\infty} \frac{n^2}{2^n}$ converges, since $\lim_{n\to\infty} \left| \frac{(n+1)^2}{2^{n+1}} \cdot \frac{2^n}{n^2} \right| = \lim_{n\to\infty} \frac{1}{2} \cdot \left(\frac{n+1}{n}\right)^2 = \frac{1}{2} < 1$.

3. $\sum_{n=1}^{\infty} \frac{n^{10}}{10^n}$ converges, since $\lim_{n\to\infty} \left(\frac{n^{10}}{10^n}\right)^{\frac{1}{n}} = \frac{1}{10}(1)^{10} = \frac{1}{10} < 1$.

5. $\sum_{n=1}^{\infty} \left(\frac{n-2}{n}\right)^n$ diverges since $\lim_{n\to\infty} \left(\frac{n-2}{n}\right)^n = e^{-2} \neq 0$.

7. $\sum_{n=1}^{\infty} n!e^{-n}$ diverges since $\lim_{n\to\infty} \left| \frac{(n+1)!}{e^{n+1}} \cdot \frac{e^n}{n!} \right| = \lim_{n\to\infty} \frac{n+1}{e} = \infty$.

9. $\sum_{n=1}^{\infty} \left(\frac{n-3}{n}\right)^n$ diverges since $\lim_{n\to\infty} \left(1 - \frac{3}{n}\right)^n = e^{-3} \neq 0$.

11. $\sum_{n=1}^{\infty} \sin\frac{1}{n}$ diverges, since $\lim_{n\to\infty} \frac{\sin\frac{1}{n}}{\frac{1}{n}} = \lim_{n\to\infty} \frac{\left(-\frac{1}{n^2}\right)\cos\frac{1}{n}}{-\frac{1}{n^2}} = \cos 0 = 1$.

13. $\sum_{n=1}^{\infty} \left(1 - \cos\frac{1}{n}\right)$ converges, since $\lim_{n\to\infty} \frac{1 - \cos\frac{1}{n}}{\frac{1}{n^2}} = \lim_{n\to\infty} \frac{\left(-\frac{1}{n^2}\right)\sin\frac{1}{n}}{-\frac{2}{n^3}}$

$= \lim_{n\to\infty} \frac{n \sin\frac{1}{n}}{2} = \frac{1}{2}\lim_{n\to\infty} \frac{\sin\frac{1}{n}}{\frac{1}{n}} = \frac{1}{2}$.

15. $\sum_{n=1}^{\infty} \tan\left(\frac{\ln n}{n}\right)$ diverges, since $\lim_{n\to\infty} \frac{\tan\left(\frac{\ln n}{n}\right)}{\frac{\ln n}{n}}$

$= \lim_{n\to\infty} \frac{\sec^2\left(\frac{\ln n}{n}\right) D\left(\frac{\ln n}{n}\right)}{D\left(\frac{\ln n}{n}\right)} = \sec^2 0 = 1$ and $\sum_{n=1}^{\infty} \frac{\ln n}{n}$ diverges.

17. $\sum_{n=1}^{\infty} \ln \frac{n+2}{n+1} = \sum_{n=1}^{\infty} [\ln(n+2) - \ln(n+1)]$ diverges since

$\ln 3 - \ln 2 + \ln 4 - \ln 3 + \ln 5 - \ln 4 + ... + \ln(n+1) - \ln n + \ln(n+2) - \ln(n+1)$

$= -\ln 2 + \ln(n+2) = \infty$

19. $\sum_{n=1}^{\infty} n \sin\left(\frac{1}{n^2}\right)$ diverges, because $\lim_{n\to\infty} \frac{n \sin\left(\frac{1}{n^2}\right)}{\frac{1}{n}}$

$= \lim_{n\to\infty} \frac{\sin\left(\frac{1}{n^2}\right)}{\frac{1}{n^2}} = \lim_{n\to\infty} \frac{\cos\left(\frac{1}{n^2}\right) D_n(n^{-2})}{D_n(n^{-2})} = \cos 0 = 1$

21. $\sum_{n=1}^{\infty} \frac{(n+1)(n+2)}{n!}$ converges because

$\lim_{n\to\infty} \frac{(n+2)(n+3)}{(n+1)!} \cdot \frac{n!}{(n+1)(n+2)} = \lim_{n\to\infty} \frac{n+3}{(n+1)^2} = 0 < 1$

23. $\sum_{n=1}^{\infty} \frac{(n+3)!}{3!n!3^n}$ converges because

$\lim_{n\to\infty} \frac{(n+4)!}{3!(n+1)!3^{n+1}} \cdot \frac{3!n!3^n}{(n+3)!} = \lim_{n\to\infty} \frac{n+4}{3(n+1)} = \frac{1}{3} < 1$

25. $\sum_{n=1}^{\infty} \frac{1}{(2n+1)!}$ converges because $\lim_{n\to\infty} \frac{(2n+1)!}{(2n+3)!} =$

$\lim_{n\to\infty} \frac{1}{(2n+2)(2n+3)} = 0 < 1$

27. $\sum_{n=1}^{\infty} \frac{|nx^n|}{2^n}$ converges for $-2 < x < 2$, because:

$\lim_{n\to\infty} \left| \frac{(n+1)x^{n+1}}{2^{n+1}} \cdot \frac{2^n}{nx^n} \right| = \lim_{n\to\infty} \frac{n+1}{n} \cdot \frac{1}{2} \cdot |x| < 1 \Leftrightarrow |x| < 2$

If $x = \pm 2$, $\sum_{n=1}^{\infty} |n|$ diverges because $\lim_{n\to\infty} |n| = \infty$.

29. $\sum_{n=1}^{\infty}\left(\frac{x^2+1}{3}\right)^n$ converges for $-\sqrt{2} < x < \sqrt{2}$, because a geometric series converges for $|r| < 1$. $\frac{x^2+1}{3} < 1 \Leftrightarrow x^2+1 < 3 \Leftrightarrow x^2 < 2$ or $|x| < \sqrt{2}$

31. $\sum_{n=1}^{\infty}\frac{x^{2n+1}}{n^2}$ converges for $-1 \le x \le 1$ because:

$$\lim_{n\to\infty}\frac{x^{2n+3}}{(n+1)^2}\cdot\frac{n^2}{x^{2n+1}} = \lim_{n\to\infty}\left(\frac{n}{n+1}\right)^2\cdot x^2 < 1 \Leftrightarrow |x| < 1$$

If $x = 1$, $\sum_{n=1}^{\infty}\frac{1}{n^2}$ converges, and if $x = -1$, $\sum_{n=1}^{\infty}\frac{(-1)^{2n+1}}{n^2}$ converges.

33. Converges, because $\lim_{n\to\infty}\left|\frac{a_{n+1}}{a_n}\right| = \lim_{n\to\infty}\left|\frac{\frac{(1+\sin n)a_n}{n}}{a_n}\right| = \lim_{n\to\infty}\frac{1+\sin n}{n} = 0 < 1$

35. Diverges, because $a_1 = 3$, $a_2 = \frac{3}{2}$, $a_3 = 1$, $a_4 = \frac{3}{4}$, ...is the series $\sum_{n=1}^{\infty}\frac{3}{n}$.

37. Converges, because $\lim_{n\to\infty}\left|\frac{a_{n+1}}{a_n}\right| = \lim_{n\to\infty}\left|\frac{(1+\ln n)a_n}{na_n}\right| = \lim_{n\to\infty}\frac{1+\ln n}{n} = \lim_{n\to\infty}\frac{\frac{1}{n}}{1} = 0 < 1$

39. $\sum_{n=1}^{\infty}\frac{2^n n!n!}{(2n)!}$ converges, because $\lim_{n\to\infty}\frac{2^{n+1}(n+1)!(n+1)!}{(2n+2)!}\cdot\frac{(2n)!}{2^n n!n!}$

$$= \lim_{n\to\infty}\frac{2(n+1)^2}{(2n+1)(2n+2)} = \frac{1}{2} < 1.$$

41. $a_1 = 1$; $a_2 = \frac{1\cdot 2}{3\cdot 4}$; $a_3 = \frac{2\cdot 3}{4\cdot 5}\cdot\frac{1\cdot 2}{3\cdot 4}$; $a_4 = \frac{3\cdot 4}{5\cdot 6}\cdot\frac{2\cdot 3}{4\cdot 5}\cdot\frac{1\cdot 2}{3\cdot 4}$;

$a_5 = \frac{4\cdot 5}{6\cdot 7}\cdot\frac{3\cdot 4}{5\cdot 6}\cdot\frac{2\cdot 3}{4\cdot 5}\cdot\frac{1\cdot 2}{3\cdot 4}$; $a_n = \frac{12}{(n+1)(n+3)(n+2)^2}$

$\sum_{n=1}^{\infty}\frac{12}{(n+1)(n+3)(n+2)^2}$ converges by comparison with $\sum_{n=1}^{\infty}\frac{12}{n^4}$

11.7 ABSOLUTE CONVERGENCE

1. $\sum_{n=1}^{\infty}\frac{1}{n^2}$ converges absolutely, p-series with $p = 2 > 1$.

3. $\sum_{n=1}^{\infty}\left|\frac{1-n}{n^2}\right|$ diverges, since it can be expressed as

$\sum_{n=1}^{\infty}\frac{1}{n^2}$ which converges, and $\left(-\sum_{n=1}^{\infty}\frac{1}{n}\right)$ which diverges.

5. $\sum_{n=1}^{\infty}\frac{-1}{n^2+2n+1}$ converges absolutely, since $\sum_{n=1}^{\infty}\frac{1}{(n+1)^2}$

converges by comparison to $\sum_{n=1}^{\infty}\frac{1}{n^2}$.

7. $\sum_{n=1}^{\infty}\frac{|\cos n\pi|}{n\sqrt{n}} = \sum_{n=1}^{\infty}\frac{1}{n^{3/2}}$ is a p-series for $p = \frac{3}{2} > 1$ and hence

converges. Therefore $\sum_{n=1}^{\infty}\frac{\cos n\pi}{n\sqrt{n}}$ converges absolutely.

9. $\sum_{n=1}^{\infty}\frac{(-1)^n}{(2n)!}$ converges absolutely, since $\lim_{n\to\infty}\left|\frac{(-1)^{n+1}}{(2n+2)!}\cdot\frac{(2n)!}{(-1)^n}\right| =$

$\lim_{n\to\infty}\frac{1}{(2n+2)(2n+1)} = 0 < 1$.

11. $\sum_{n=1}^{\infty}\left|(-1)^n\frac{n}{n+1}\right|$ is divergent, since $\lim_{n\to\infty}\frac{n}{n+1} \neq 0$.

13. $\sum_{n=1}^{\infty}(5)^{-n}$ converges, geometric series with $r = \frac{1}{5} < 1$

15. $\sum_{n=1}^{\infty}\frac{(-100)^n}{n!}$ converges absolutely, since $\lim_{n\to\infty}\left|\frac{(-100)^{n+1}}{(n+1)!}\cdot\frac{n!}{(-100)^n}\right|$

$= \lim_{n\to\infty} 100\cdot\frac{1}{n+1} = 0 < 1$

17. $\sum_{n=1}^{\infty}\frac{2-n}{n^3} = 1 + \sum_{n=2}^{\infty}\frac{n-2}{n^3}$ converges absolutely, because $\sum_{n=2}^{\infty}\frac{1}{n^2}$ is a

p-series, $p = 2 > 1$, and $-2\sum_{n=2}^{\infty}\frac{1}{n^3}$ is a multiple of a p-series, $p=3 > 1$.

19. Since $|a_n| \geq a_n$, if $\sum_1^{\infty} a_n$ diverges then $\sum_1^{\infty}|a_n|$ must diverge.

21. Let $\sum_1^\infty a_n$ and $\sum_1^\infty b_n$ converge absolutely. Then

(a) $|a_n + b_n| \le |a_n| + |b_n| \Rightarrow \sum_1^\infty |a_n + b_n| \le \sum_1^\infty |a_n| + \sum_1^\infty |b_n|$

and hence $\sum_1^\infty (a_n + b_n)$ converges absolutely.

(b) $|a_n - b_n| = |a_n + (-b_n)| \le |a_n| + |-b_n| = |a_n| + |b_n|$.

Hence $\sum_1^\infty |a_n - b_n|$ converges and $\sum_1^\infty (a_n - b_n)$ converges absolutely.

(c) $|ka_n| = |k||a_n| \Rightarrow \sum_1^\infty |ka_n| = |k|\sum_1^\infty |a_n|$ converges. Hence

$\sum_1^\infty ka_n$ converges absolutely.

11.8 ALTERNATING SERIES AND CONDITIONAL CONVERGENCE

1. $\sum_{n=1}^\infty (-1)^{n+1}\frac{1}{n^2}$ converges because it converges absolutely, since

$\sum_{n=1}^\infty \frac{1}{n^2}$ converges (p-series for $p = 2 > 1$).

3. $\sum_{n=1}^\infty (-1)^{n-1} = \sum_{n=1}^\infty (-1)^{n-1}(1)$ diverges, since $\lim_{n\to\infty}(1) = 1 \ne 0$.

5. $\sum_{n=1}^\infty (-1)^{n+1}\frac{\sqrt{n}+1}{n+1}$ converges since $\lim_{n\to\infty}\frac{\sqrt{n}+1}{n+1} = \lim_{n\to\infty}\frac{\frac{1}{\sqrt{n}} + \frac{1}{n}}{1 + \frac{1}{\sqrt{n}}} = 0$

and $\left\{\frac{\sqrt{n}+1}{n+1}\right\}$ is a decreasing sequence.

7. $\sum_{n=1}^\infty (-1)^{n+1}\frac{1}{n^{3/2}}$ converges since it converges absolutely.

9. $\sum_{n=1}^{\infty}(-1)^n \ln\left(1+\frac{1}{n}\right)$ converges because $\lim_{n\to\infty} \ln\left(1+\frac{1}{n}\right) = \ln 1 = 0$.

If $f(x) = \ln\left(1+\frac{1}{x}\right)$ then $f'(x) = \frac{x}{x+1}\cdot\frac{x(1)-(x+1)(1)}{x^2} =$

$\frac{-1}{x(x+1)} < 0$ for $x > 1$. Hence $\left\{\ln\left(1+\frac{1}{n}\right)\right\}$ is decreasing.

11. $\sum_{n=1}^{\infty}(-1)^{n-1}(0.1)^n$ converges absolutely because it is a geometric series with $r = .1 < 1$.

13. $\sum_{n=1}^{\infty}(-1)^{n+1}\frac{n}{n^3+1}$ converges absolutely by comparison with $\sum_{n=1}^{\infty}\frac{1}{n^2}$,

since $n^3 \le n^3 + 1 \Rightarrow \frac{1}{n^3+1} \le \frac{1}{n^3} \Rightarrow \frac{n}{n^3+1} \le \frac{n}{n^3} = \frac{1}{n^2}$

15. $\sum_{n=1}^{\infty}(-1)^{n+1}\frac{1}{n+3}$ converges conditionally, since $\left\{\frac{1}{n+3}\right\}$

is a decreasing sequence with limit 0, but $\sum_{n=1}^{\infty}\frac{1}{n+3}$ diverges.

17. $\sum_{n=1}^{\infty}(-1)^{n+1}\frac{3+n}{5+n}$ diverges since $\lim_{n\to\infty}\frac{3+n}{5+n} = 1 \neq 0$.

19. $\sum_{n=1}^{\infty}(-1)^{n+1}\frac{1+n}{n^2}$ converges conditionally, since $\left\{\frac{1+n}{n^2}\right\}$ is a decreasing sequence converging to 0. But the series

$\sum_{n=1}^{\infty}\frac{1+n}{n^2}$ diverges by comparison to $\sum_{n=1}^{\infty}\frac{1}{n}$ since $\lim_{n\to\infty}\frac{\frac{1+n}{n^2}}{\frac{1}{n}} = \lim_{n\to\infty}\frac{1+n}{n} = 1$.

21. $\sum_{n=1}^{\infty}n^2\left(\frac{2}{3}\right)^n$ converges absolutely since $\lim_{n\to\infty}\frac{(n+1)^2\left(\frac{2}{3}\right)^{n+1}}{n^2\left(\frac{2}{3}\right)^2}$

$= \lim_{n\to\infty}\frac{2}{3}\cdot\left(\frac{n+1}{n}\right)^2 = 1$.

23. $\sum_{n=1}^{\infty}(-1)^n \frac{\tan^{-1} n}{n^2+1}$ converges absolutely, since $|\tan^{-1} n| < \frac{\pi}{2}$

$$\Rightarrow \sum_{n=1}^{\infty}\left|\frac{\tan^{-1} n}{n^2+1}\right| \le \frac{\pi}{2}\sum_{n=1}^{\infty}\frac{1}{n^2+1} \le \frac{\pi}{2}\sum_{n=1}^{\infty}\frac{1}{n^2}.$$

25. $\sum_{n=1}^{\infty}\left(\frac{1}{n} - \frac{1}{2n}\right) = \frac{1}{2}\sum_{n=1}^{\infty}\frac{1}{n}$ diverges.

27. $\sum_{n=1}^{\infty}(-1)^{n+1}(\sqrt{n+1} - \sqrt{n})$ converges conditionally since:

$\frac{\sqrt{n+1} - \sqrt{n}}{1} \cdot \frac{\sqrt{n+1} + \sqrt{n}}{\sqrt{n+1} + \sqrt{n}} = \frac{1}{\sqrt{n+1} + \sqrt{n}}$ and $\left\{\frac{1}{\sqrt{n+1} + \sqrt{n}}\right\}$ is a decreasing sequence which converges to 0. But $\frac{1}{\sqrt{n+1} + \sqrt{n}} \ge \frac{1}{3\sqrt{n}}$

so $\sum_{n=1}^{\infty}\frac{1}{\sqrt{n+1} + \sqrt{n}}$ diverges.

29. $a_5 = \frac{1}{5}$; error ≤ 0.2

31. $a_5 = \frac{(10^{-10})}{5}$; error $\le 2 \times 10^{-11}$

33. If $n = 5$, $\frac{1}{9!} = .0000028 = 2.8 \times 10^{-6}$. Therefore the sum of the first 4 terms is sufficiently accurate.
$1 - \frac{1}{2!} + \frac{1}{4!} - \frac{1}{4!} + \frac{1}{8!} = 0.540302579.$

35. (a) The condition $a_n \ge a_{n+1}$ is not met.

(b) $S = \frac{\frac{1}{3}}{1 - \frac{1}{3}} - \frac{\frac{1}{2}}{1 - \frac{1}{2}} = \frac{1}{2} - 1 = -\frac{1}{2}$

37. $\sum_{j=n+1}^{\infty}(-1)^{j+1}a_j = (-1)^{n+1}(a_{n+1} - a_{n+2}) + (-1)^{n+3}(a_{n+3} - a_{n+4}) + \ldots$

$= (-1)^{n+1}(a_{n+1} - a_{n+2}) + (a_{n+3} - a_{n+4}) + \ldots$

Each grouped term is positive, so the remainder has the same sign as $(-1)^{n+1}$, which is the sign of the first unused term.

39. Let $a_n = b_n = (-1)^n\frac{1}{\sqrt{n}}$. Then $\sum_{1}^{\infty}(-1)^n\frac{1}{\sqrt{n}}$ converges but

$\sum_{1}^{\infty}a_n b_n = \sum_{1}^{\infty}\frac{1}{n}$ diverges.

41. The rearranged sequences of partial sums would be:

$-\frac{1}{2}$ with $S_1 = -\frac{1}{2}$

$-\frac{1}{2} + 1$ with $S_2 = \frac{1}{2}$

$-\frac{1}{2} + 1 - \frac{1}{4} - \frac{1}{6} - \frac{1}{8}$ with $S_5 = -\frac{1}{24}$

$-\frac{1}{2} + 1 - \frac{1}{4} - \frac{1}{6} - \frac{1}{8} + \frac{1}{3} + \frac{1}{5} + \frac{1}{7}$ with $S_8 = \frac{498}{840}$

11.9 RECAPITULATION

1. $\sum_{n=1}^{\infty}(-1)^{n-1}\frac{1}{\ln(n+1)}$ converges since $\lim_{n\to\infty}\frac{1}{\ln(n+1)} = 0$ and $\frac{1}{\ln(n+2)} \le \frac{1}{\ln(n+1)}$ for all n.

3. This is the series $\sum_{n+1}^{\infty}(-1)^{n+1}\frac{1}{n}$ which converges.

5. Diverges, because $\lim_{n\to\infty}\frac{(n+1)^{n+1}}{(n+1)!}\cdot\frac{n!}{n^n} = \lim_{n\to\infty}\frac{(n+1)^n(n+1)}{(n+1)n!}\cdot\frac{n!}{n^n}$ $= \lim_{n\to\infty}\left(\frac{n+1}{n}\right)^n = e > 1$

7. Converges, because $\lim_{n\to\infty}\frac{(n+1)^2}{2^{n+1}}\cdot\frac{2^n}{n^2} = \frac{1}{2} < 1$.

9. Converges; it is the geometric series $\sum_{n=1}^{\infty}\left(\frac{1}{2}\right)^n$.

11. Diverges; it is the series $2\sum_{n=1}^{\infty}\frac{1}{n}$.

13. Diverges; $a_n \le 1 \Rightarrow 1 + a_n \le 2 \Rightarrow \frac{1}{1+a_n} \ge \frac{1}{2}$. Thus $\lim_{n\to\infty} a_{n+1} \ne 0$.

15. Diverges; $\lim_{n\to\infty} na_n = \infty \ne 0$.

17. Diverges; $\lim_{n\to\infty} a_n \ne 0$ since a_n = either +1 or −1.

19. Converges by comparison to $\sum_{n=1}^{\infty}\frac{1}{n^2}$ since $|\sin n| \le 1$.

21. (a) Let $t_1 = c_1$, $t_2 = c_1 + c_2 \Rightarrow c_2 = t_2 - t_1$, $t_3 = c_1 + c_2 + c_3 \Rightarrow$

$c_3 = t_3 - (c_1 + c_2) = t_3 - t_2$, and in general, $t_n - t_{n-1} = \sum_{k=1}^{n} c_k$.

$$s_{2n+1} = \sum_{k=1}^{2n+1} \frac{c_k}{k} = c_1(1) + c_2\left(\frac{1}{2}\right) + \ldots + c_{2n+1}\left(\frac{1}{2n+1}\right)$$

$$= t_1(1) + \frac{1}{2}(t_2 - t_1) + \ldots + t_{2n}\left(\frac{1}{2n} - \frac{1}{2n+1}\right) + t_{2n+1}\left(\frac{1}{2n+1}\right)$$

$$= \sum_{k=1}^{2n} \frac{t_k}{k(k+1)} + \frac{t_{2n+1}}{2n+1}$$

$\sum_{k=1}^{2n} \left|\frac{t_k}{k(k+1)}\right| \le M \sum_{k=1}^{2n} \frac{1}{k(k+1)}$ converges absolutely, and $\lim_{n\to\infty} \frac{t_{2n+1}}{2n+1} = 0$,

so $\lim_{n\to\infty} s_{2n+1}$ exists. If $s_{2n} = \sum_{k=1}^{2n} \frac{c_k}{k}$, then $\lim_{n\to\infty}(s_{2n+1} - s_{2n})$

$= \lim_{n\to\infty} \frac{c_{2n+1}}{2n+1} = 0$, so $\sum_{k=1}^{\infty} \frac{c_k}{k}$ converges to $\sum_{k=1}^{\infty} \frac{t_k}{k(k+1)}$.

$$= t_1(1) + \frac{1}{2}(t_2 - t_1) + \ldots + t_{2n}\left(\frac{1}{2n} - \frac{1}{2n+1}\right) + t_{2n+1}\left(\frac{1}{2n+1}\right)$$

$$= \sum_{k=1}^{2n} \frac{t_k}{k(k+1)} + \frac{t_{2n+1}}{2n+1}$$

(b) In $\sum_{n=1}^{\infty} (-1)^{n+1} \frac{1}{n}$, take $\{c_k\} = \{1, -1, 1, -1, ..\}$ so that

$\sum_{k=1}^{n} c_k \le 1$, so that the alternating harmonic series converges.

(c) If $\{c_k\} = \{1, -1, -1, 1, 1, -1, -1, 1, 1, \ldots\}$ then $\sum_{k=1}^{n} c_k \le 1$, so

the series converges.

11.10 ESTIMATING THE SUM OF A SERIES

3. The ratios $\frac{a_{n+1}}{a_n} = \frac{n+1}{3n}$ form a decreasing sequence which converges to $\frac{1}{3}$. Take $r_1 = \frac{1}{3}$ and $r_2 = \frac{N+1}{3N}$.

(a) $\frac{10}{3^{10}}\cdot\frac{\frac{1}{3}}{1-\frac{1}{3}} \le R_{10} \le \frac{10}{3^{10}}\cdot\frac{\frac{11}{30}}{1-\frac{11}{30}}$; $8.47\times 10^{-5} \le R_{10} \le 1.07\times 10^{-5}$

(b) $\frac{100}{3^{100}}\cdot\frac{\frac{1}{3}}{1-\frac{1}{3}} \le R_{100} \le \frac{100}{3^{100}}\cdot\frac{\frac{101}{300}}{1-\frac{101}{300}}$; $9.70\times 10^{-47} \le R_{100} \le 9.85\times 10^{-47}$

11.M MISCELLANEOUS

1. $\sum_{k=2}^{n} \ln\left(1-\frac{1}{k^2}\right) = \sum_{k=2}^{n}\left[\ln\left(1+\frac{1}{k}\right)+\ln\left(1-\frac{1}{k}\right)\right] =$

$\sum_{k=2}^{n}[\ln(k+1)-\ln k+\ln(k-1)-\ln k] = \ln(n+1)-\ln n-\ln 2$

$= \ln\frac{n+1}{2n}$. $\therefore \lim_{n\to\infty}\ln\frac{n+1}{2n} = \ln\frac{1}{2}$.

3. $\sum_{n=1}^{\infty}(x_{n+1}-x_n) = \lim_{n\to\infty}\sum_{j=1}^{n}(x_{j+1}-x_j) = \lim_{n\to\infty}(x_{n+1}-x_1) =$

$\left(\lim_{n\to\infty} x_{n+1}\right) - x_1$. Therefore, the series and the sequence either both converge or both diverge.

5. $\frac{1}{1-x} = -\frac{\frac{1}{x}}{1-\frac{1}{x}}$. Using $a = r = \frac{1}{x}$, this equals

$-\left(\frac{1}{x}+\frac{1}{x^2}+\frac{1}{x^3}+\ldots+\frac{1}{x^n}+\ldots\right)$ for $\frac{1}{|x|} < 1$ or $|x| > 1$.

7. $\sum_{n=1}^{\infty}(-1)^n \tanh n$ diverges because:

$\lim_{n\to\infty}\tanh n = \lim_{n\to\infty}\frac{\sinh n}{\cosh n} = \lim_{n\to\infty}\frac{e^x-e^{-x}}{e^x+e^{-x}} = \lim_{n\to\infty}\frac{e^{2x}-1}{e^{2x}+1} = \lim_{n\to\infty}\frac{2e^{2x}}{2e^{2x}} = 1 \ne 0$

9. $\sum_{n=1}^{\infty}\frac{n}{2(n+1)(n+2)}$ diverges because $\sum_{n=1}^{\infty}\frac{1}{n}$ diverges and

$\lim_{n\to\infty}\frac{\frac{n}{2(n+1)(n+2)}}{\frac{1}{n}} = \lim_{n\to\infty}\frac{n^2}{2(n^2+3n+2)} = \frac{1}{2}$

11. $\sum_{n=2}^{\infty} \frac{1}{n(\ln n)^2}$ converges because $\int_2^{\infty} \frac{dx}{x(\ln x)^2} = \lim_{t\to\infty} \int_2^t \frac{dx}{x(\ln x)^2}$

$= \lim_{t\to\infty}\left[-\frac{1}{\ln x}\right]_2^t = \lim_{t\to\infty}\left[-\frac{1}{\ln t} + \frac{1}{\ln 2}\right] = \frac{1}{\ln 2}$

13. $\sum_{n=1}^{\infty} \frac{n}{1000n^2+1}$ diverges by comparison to $\sum_{n=1}^{\infty}\frac{1}{n}$, since

$\lim_{n\to\infty} \frac{n}{1000n^2+1}\cdot\frac{n}{1} = \frac{1}{1000}$

15. $\sum_{n=1}^{\infty} \frac{1}{n\sqrt{n^2+1}}$ converges by comparison to $\sum_{n=1}^{\infty}\frac{1}{n^2}$ because $\lim_{n\to\infty} \frac{1}{n\sqrt{n^2+1}}\cdot\frac{n^2}{1} = 1$

17. Diverges, because $\frac{1\cdot3\cdot5\cdot\ldots\cdot(2n-1)}{2\cdot4\cdot6\cdot\ldots\cdot(2n)} \geq \frac{1\cdot2\cdot4\cdot\ldots\cdot(2n-2)}{2\cdot4\cdot6\cdot\ldots\cdot(2n)} = \frac{1}{2n}$

and $\frac{1}{2}\sum_{n=1}^{\infty}\frac{1}{n}$ diverges.

19. $\sum_{n=1}^{\infty} \frac{n+1}{n!}$ converges because $\lim_{n\to\infty} \frac{n+2}{(n+1)!}\cdot\frac{n!}{n+1} = 0 < 1$

21. (a) $\sum_{n=1}^{\infty}\frac{a_n}{n} = a_1 + \frac{1}{2}a_2 + \frac{1}{3}a_3 + \frac{1}{4}a_4 + \frac{1}{5}a_5 + \frac{1}{6}a_6 + \ldots + \frac{1}{n}a_n + \ldots$

$\geq a_1 + \frac{1}{2}a_2 + \frac{1}{3}a_4 + \frac{1}{4}a_4 + \frac{1}{5}a_8 + \frac{1}{6}a_8 + \frac{1}{7}a_8 + \frac{1}{8}a_8 + \ldots + \frac{1}{n}a_n + \ldots$

$= a_1 + \frac{1}{2}a_2 + \left(\frac{1}{3}+\frac{1}{4}\right)a_4 + \left(\frac{1}{5}+\frac{1}{6}+\frac{1}{7}+\frac{1}{8}\right)a_8 + \left(\frac{1}{9}+\frac{1}{10}+ \ldots + \frac{1}{16}\right)a_{16} + \ldots$

$\geq \frac{1}{2}(a_2 + a_4 + a_8 + a_{16} + \ldots)$ which diverges.

(b) Let $a_n = \frac{1}{\ln n}$, $n \geq 2$. Then

(i) $a_2 \geq a_3 \geq a_4 \geq \ldots$ and

(ii) $\frac{1}{\ln 2} + \frac{1}{\ln 4} + \frac{1}{\ln 8} + \frac{1}{\ln 16} + \ldots = \frac{1}{\ln 2} + \frac{1}{2\ln 2} + \frac{1}{3\ln 2} + \frac{1}{4\ln 4} + \ldots$

$= \frac{1}{\ln 2}\left(1 + \frac{1}{2} + \frac{1}{3} + \frac{1}{4} + \ldots\right)$ which diverges.

By part (a), $1 + \sum_{n=2}^{\infty}\frac{1}{n\ln n}$ diverges.

23. $\sum_{n=3}^{\infty} \frac{1}{n\ln n\,(\ln(\ln n))^p}$ converges since $\int_3^{\infty} \frac{dx}{x\ln x\,(\ln(\ln x))^p} =$

$\frac{1}{-p+1}(\ln(\ln x))^{-p+1}\Big]_3^{\infty}$ converges only for $p < 1$.

CHAPTER 12

POWER SERIES

12.2 TAYLOR POLYNOMIALS

1.

n	$f^n(x)$	$f^n(0)$
0	e^{-x}	1
1	$-e^{-x}$	-1
2	e^{-x}	1
3	$-e^{-x}$	-1
4	e^{-x}	1

$$P_3(x) = 1 - x + \frac{x^2}{2} - \frac{x^3}{6}$$

$$P_4(x) = 1 - x + \frac{x^2}{2} - \frac{x^3}{6} + \frac{x^4}{24}$$

3.

n	$f^n(x)$	$f^n(0)$
0	$\cos x$	1
1	$-\sin x$	0
2	$-\cos x$	-1
3	$\sin x$	0
4	$\cos x$	1

$$P_3(x) = 1 - \frac{x^2}{2}$$

$$P_4(x) = 1 - \frac{x^2}{2} + \frac{x^4}{24}$$

5.

n	$f^n(x)$	$f^n(0)$
0	$\sinh x$	0
1	$\cosh x$	1
2	$\sinh x$	0
3	$\cosh x$	1
4	$\sinh x$	0

$$P_3(x) = x + \frac{x^3}{6}$$

$$P_4(x) = x + \frac{x^3}{6}$$

7.

n	$f^n(x)$	$f^n(0)$
0	$x^4 - 2x + 1$	1
1	$4x^3 - 2$	-2
2	$12x^2$	0
3	$24x$	0
4	24	24

$$P_3(x) = 1 - 2x$$

$$P_4(x) = 1 - 2x + x^4$$

9.

n	$f^n(x)$	$f^n(0)$
0	$x^2 - 2x + 1$	1
1	$2x - 2$	-2
2	2	2
3	0	0
4	0	0

$$P_3(x) = 1 - 2x + x^2$$

$$P_4(x) = P_3(x)$$

11. $f(x) = x^2$; $f'(x) = 2x$; $f''(x) = 2$; $f^n(x) = 0$ for $n > 2$
$f(0) = 0$; $f'(0) = 0$; $f''(0) = 2$

$$x^2 = 0 + 0x + \frac{2}{2}x^2 + 0 + \ldots = x^2$$

13. $f(x) = (1+x)^{\frac{3}{2}}$ $\qquad f(0) = 1$

$f'(x) = \frac{3}{2}(1+x)^{\frac{1}{2}}$ $\qquad f'(0) = \frac{3}{2}$

$f''(x) = \frac{1}{2}\cdot\frac{3}{2}(1+x)^{-\frac{1}{2}}$ $\qquad f''(0) = \frac{1\cdot 3}{2^2}$

$f'''(x) = -\frac{1}{2}\cdot\frac{1}{2}\cdot\frac{3}{2}(1+x)^{-\frac{3}{2}}$ $\qquad f'''(0) = \frac{(-1)\cdot 1\cdot 3}{2^3}$

$f^4(x) = \frac{3\cdot 1\cdot(-1)(-3)}{2^4}(1+x)^{-\frac{5}{2}}$ $\qquad f^4(x) = \frac{3\cdot 1\cdot(-1)(-3)}{2^4}$

$$f(x) = 1 + \sum_{n=1}^{\infty} \frac{(5-2n)}{n!2^n}x^n$$

15. $f(x) = e^{10} + e^{10}(x-10) + \frac{e^{10}}{2!}(x-10)^2 + \ldots$

$$= \sum_{n=0}^{\infty} \frac{e^{10}}{n!}(x-10)^n$$

17.

n	$f^n(x)$	$f^n(1)$
0	ln x	0
1	x^{-1}	1
2	$-x^{-2}$	-1
3	$2x^{-3}$	2
4	$-3\cdot 2x^{-4}$	-6

$\ln x = \ln 1 + \frac{1}{1}(x-1) - \frac{1}{1^2\cdot 2}(x-1)^2 + \frac{2}{1^3\cdot 3!}(x-1)^3 - \ldots$

$$= \sum_{n=1}^{\infty}(-1)^{n-1}\frac{(x-1)^n}{n}$$

19.

n	$f^n(x)$	$f^n(-1)$
0	x^{-1}	-1
1	$-x^{-2}$	-1
2	$2x^{-3}$	-2
3	$-3\cdot 2x^{-4}$	-6
4	$4\cdot 3\cdot 2x^{-5}$	-24

$$f(x) = -\sum_{n=0}^{\infty}(x+1)^n$$

21. $f(x) = \tan x$ $\qquad f(\frac{\pi}{4}) = 1$

$f'(x) = \sec^2 x$ $\qquad f'(\frac{\pi}{4}) = 2$

$f''(x) = 2\sec^2 x \tan x$ $\qquad f''(\frac{\pi}{4}) = 4$

$\tan x = 1 + 2(x - \frac{\pi}{4}) + 2(x - \frac{\pi}{4})^2 + \ldots$

12.3 TAYLOR'S THEOREM WITH REMAINDER: SINES, COSINES, AND e^x

1. $e^{\frac{x}{2}} = \sum_{n=0}^{\infty}\left(\frac{x}{2}\right)^n \cdot \frac{1}{n!} = \sum_{n=0}^{\infty} \frac{x^n}{2^n n!}$

3. $5\cos\frac{x}{\pi} = 5\sum_{n=0}^{\infty}(-1)^n\left(\frac{x}{\pi}\right)^{2n}\frac{1}{(2n)!} = 5\sum_{n=0}^{\infty}(-1)^n\frac{x^{2n}}{\pi^{2n}(2n)!}$

5. $\frac{x^2}{2} - 1 + \cos x = \frac{x^2}{2} - 1 + 1 - \frac{x^2}{2} + \frac{x^4}{4!} - \ldots = \sum_{n=2}^{\infty}(-1)^n\frac{x^{2n}}{(2n)!}$

7. (a) $\cos(-x) = \sum_{n=0}^{\infty}(-1)^n\frac{(-x)^{2n}}{(2n)!} = \sum_{n=0}^{\infty}(-1)^n\frac{x^{2n}}{(2n)!} = \cos x$

(b) $\sin(-x) = \sum_{n=0}^{\infty}(-1)^n\frac{(-x)^{2n+1}}{(2n+1)!} = -\sum_{n=0}^{\infty}(-1)^n\frac{x^{2n+1}}{(2n+1)!} = -\sin x$

9. Using the formula for a geometric series with $a = 1$ and $r = -x$:

$\frac{1}{1 + x} = 1 - x + x^2 + R_n(x)$

11. $f(x) = (1 + x)^{\frac{1}{2}}$; $f(0) = 1$; $f'(x) = \frac{1}{2}(1 + x)^{-\frac{1}{2}}$; $f'(0) = \frac{1}{2}$

$f''(x) = -\frac{1}{4}(1 + x)^{-\frac{3}{2}}$; $f''(0) = -\frac{1}{4}$

$(1 + x)^{\frac{1}{2}} \approx 1 + \frac{1}{2}x - \frac{1}{8}x^2 + R_n(x)$

13. $\left|\frac{x^5}{5!}\right| < 5\times10^{-4} \Leftrightarrow |x^5| < 120\cdot5\times10^{-4} = .06 \Leftrightarrow |x| < 0.5697$

15. (a) $R_2(x) = \frac{-\cos c}{3!}x^3$. $\therefore$ $|R_2(x)| = \frac{\cos c}{6}|x|^3$. If $|x| < 10^{-3}$,

then $|R_2(x)| \le \frac{1}{6}(10^{-3})^3 < 1.67 \times 10^{-10}$.

(b) $-10^{-3} < x < 10^{-3} \Rightarrow \cos c > 0 \Rightarrow -\cos c < 0$.

If $-10^{-3} < x < 0$, $\frac{-\cos c}{3!}x^3 < 0 \Rightarrow x > \sin x$

If $10^{-3} > x > 0$, $\frac{-\cos c}{3!}x^3 > 0 \Rightarrow \sin x > x$

17. If $|x| < 0.1$, then $e^c < e^{.1}$. Therefore,

$|R_2(x)| \le \frac{e^{.1}(.1)^3}{3!} < 0.000184$.

19. $|R_4(x)| \le \frac{|\sinh c||x^5|}{5!} = \frac{.521(0.5)^5}{5!} < 0.0003$

21. $(f + g)(x) = \sum_{n=0}^{\infty}\frac{(f + g)^n(0)}{n!}x^n = \sum_{n=0}^{\infty}\frac{f^n(0) + g^n(0)}{n!}x^n$.

23. (a) $e^{i\pi} = \cos\pi + i\sin\pi = -1 + 0i$

(b) $e^{\frac{i\pi}{4}} = \cos\frac{\pi}{4} + i\sin\frac{\pi}{4} = \frac{\sqrt{2}}{2} + \frac{\sqrt{2}}{2}i$

(c) $e^{-\frac{i\pi}{2}} = \cos\left(-\frac{\pi}{2}\right) + i\sin\left(-\frac{\pi}{2}\right) = 0 - i$

(d) $e^{i\pi} \cdot e^{-\frac{i\pi}{2}} = e^{\frac{i\pi}{2}} = \cos\frac{\pi}{2} + i\sin\frac{\pi}{2} = 0 + i$

25. $\cos^3\theta = \left(\frac{e^{i\theta} + e^{-i\theta}}{2}\right)^3 = \frac{1}{8}(e^{3i\theta} + 3e^{2i\theta}e^{-i\theta} + 3e^{i\theta}e^{-2i\theta} + e^{-3i\theta})$

$= \frac{1}{8}(\cos 3\theta + i\sin 3\theta + 3[\cos\theta + i\sin\theta] + 3[\cos(-\theta) + i\sin(-\theta)] + \cos(-3\theta) + i\sin(-3\theta))$

$= \frac{1}{8}(2\cos 3\theta + 6\cos\theta) = \frac{1}{4}\cos 3\theta + \frac{3}{4}\cos\theta$

$\sin^3\theta = \left(\frac{e^{i\theta} - e^{-i\theta}}{2i}\right)^3 = \frac{1}{8i^3}(e^{3i\theta} - 3e^{2i\theta}e^{-i\theta} + 3e^{i\theta}e^{-2i\theta} - e^{-3i\theta})$

$= -\frac{1}{8i}(\cos 3\theta + i\sin 3\theta - 3[\cos\theta + i\sin\theta] + 3[\cos(-\theta) + i\sin(-\theta)] - \cos(-3\theta) - i\sin(-3\theta))$

$= -\frac{1}{8i}(2i\sin 3\theta - 6i\sin\theta) = -\frac{1}{4}\sin 3\theta + \frac{3}{4}\sin\theta$

27. $\int e^{(a+ib)x}dx = \frac{1}{a+ib}e^{(a+ib)x} + C = \frac{a-ib}{a^2+b^2}e^{(a+ib)x} + C$

$= \frac{a}{a^2+b^2}e^{(a+ib)x} + \frac{-ib}{a^2+b^2}e^{(a+ib)x}$

$= \frac{ae^{ax}}{a^2+b^2}\cos bx - \frac{ib\,e^{ax}}{a^2+b^2}\cos bx + \frac{iae^{ax}}{a^2+b^2}\sin bx + \frac{be^{ax}}{a^2+b^2}\sin bx$

$= \frac{1}{a^2+b^2}e^{ax}(a\cos bx + b\sin bx) + \frac{i}{a^2+b^2}e^{ax}(a\sin bx - b\cos bx)$

$\int e^{(a+ib)x}dx = \int e^{ax}(\cos bx + i\sin bx)dx = \int e^{ax}\cos bx\,dx + i\int e^{ax}\sin bx\,dx$

Equating real part to real part and imaginary part to imaginary part in the two expressions for the integral, we have:

$$\int e^{ax}\cos bx\,dx = \frac{1}{a^2+b^2}e^{ax}(a\cos bx + b\sin bx)$$

$$\int e^{ax}\sin bx\,dx = \frac{1}{a^2+b^2}e^{ax}(a\sin bx - b\cos bx)$$

12.4 EXPANSION POINTS, THE BINOMIAL THEOREM, ARCTANGENTS & Π

1. $31^\circ = \frac{31\pi}{180} \approx 0.5411$ radians.

   $\cos 31^\circ = 1 - \frac{(.5411)^2}{2} + \frac{(.5411)4}{24} \approx 0.857$ with error

   $|R_5(x)| \leq \frac{(.5411)^6}{6!} < 0.000035$

3. Using $\sin(2\pi + x) = \sin x$, $\sin 6.3 = \sin(2\pi + 0.0168) = \sin 0.0168$

   We need only the first term $\sin 0.0168 = 0.0168$ with error

   $|R_2(x)| \leq \frac{(0.0168)^3}{6} < 7.9 \times 10^{-7}$

5. $\ln 1.25 = \ln(1 + .25) = .25 - \frac{(.25)^2}{2} + \frac{(.25)^3}{3} - \frac{(.25)^4}{4} = .223$

   with error $|R_5(x)| \leq \frac{(.25)^5}{5} < 0.0002$

7. $\ln(1+2x) = 2x - \frac{(2x)^2}{2} + \frac{(2x)^3}{3} - \ldots = \sum_{n=1}^{\infty}(-1)^{n-1}\frac{2^n x^n}{n}$ converges

   for all $-1 < 2x \leq 1$ or $-\frac{1}{2} < x \leq \frac{1}{2}$.

9. $$\int_0^{0.1} \frac{\sin x}{x}dx = \int_0^{0.1} \frac{1}{x}\left(x - \frac{x^3}{3!} + \frac{x^5}{5!} - \ldots\right)dx$$

   $$= \int_0^{0.1}\left(1 - \frac{x^2}{3!} + \frac{x^4}{5!} - \frac{x^6}{7!} + \ldots\right)dx$$

   $$= x - \frac{x^3}{3\cdot 3!} + \frac{x^5}{5\cdot 5!} - \Big]_0^{0.1} = 0.1$$

   with error $|R_2(x)| \leq \frac{(0.1)^3}{18} = 0.00006$

11. The parabola $x^2 = 2a(y - a)$ solved for y is $y = a + \frac{x^2}{2a}$

    These are the first two terms of the Maclaurin's Series for $y = \cosh x$. The remainder is:

    $$|R_3(x)| \leq \frac{|a||\cosh c|}{4!3^4} = 0.0005|a|$$

13. This is the series for $\ln(1 + x)$ evaluated at $x = \frac{1}{2}$, which converges to $\ln(\frac{3}{2})$

15. $\pi = 48\tan^{-1}\frac{1}{18} + 32\tan^{-1}\frac{1}{57} - 20\tan^{-1}\frac{1}{239} =$

    $$= 48\left[.0556 - \frac{(.0556)^3}{3}\right] + 32\left[.0175 - \frac{(.0175)^3}{3}\right] - 20\left[.0042 - \frac{(.0042)^3}{3}\right]$$

    $= 2.66604 + 0.55994 - 0.0839 = 3.142$

17. $a_0 = 1$ $\qquad b_0 = \frac{1}{\sqrt{2}} \approx 0.707107$

$a_1 = \frac{1 + \frac{1}{\sqrt{2}}}{2} = 0.853553$ $\qquad b_1 = \sqrt{\frac{1}{\sqrt{2}}} = 0.840896$

$a_2 = \frac{a_1 + b_1}{2} = 0.847225$ $\qquad b_2 = \sqrt{a_1 b_1} = 0.847201$

$a_3 = \frac{a_2 + b_2}{2} = 0.847213$ $\qquad b_3 = \sqrt{a_2 b_2} = 0.847213$

$$c_3 = \frac{4a_3b_3}{1 - 4(a_1^2 - b_1^2) - 8(a_2^2 - b_2^2) - 16(a_3^2 - b_3^2)}$$

$$= \frac{2.871079}{1 - .085788 - .000325 - 0} = 3.1416127$$

19. By dividing $1 - t^2$ into 1, and expressing the answer as quotient plus remainder over divisor, we obtain the identity

$$\frac{1}{1 - t^2} = 1 + t^2 + t^4 + \ldots + t^{2n} + \frac{t^{2n+2}}{1 - t^2}$$

Another way of obtaining this identity is to use the formula for the nth partial sum of a geometric series

$$1 + t^2 + t^4 + \ldots + t^{2n} = \frac{1 - (t^2)^{n+1}}{1 - t^2}$$

Then

$$\int_0^x \frac{dt}{1 - t^2} = \int_0^x \left[1 + t^2 + t^4 + \ldots + t^{2n} + \frac{t^{2n+2}}{1 - t^2}\right] dt$$

$$= t + \frac{t^3}{3} + \frac{t^5}{5} + \ldots + \frac{t^{2n+1}}{2n+1}\Bigg]_0^x + \int_0^x \frac{t^{2n+2}}{1 - t^2} dt$$

Thus $\tanh^{-1} x = x + \frac{x^3}{3} + \frac{x^5}{5} + \ldots + \frac{x^{2n+1}}{2n + 1} + R$, with

$$R = \int_0^x \frac{t^{2n+2}}{1 - t^2} dt$$

21. (a) $\frac{d}{dx}(1 - x)^{-1} = \frac{d}{dx}\left(1 + x + x^2 + \ldots + x^n + \frac{x^{n+1}}{1 - x}\right)$

$(1 - x)^{-2} = 1 + 2x + 3x^2 + \ldots + nx^{n-1} + R$, where

$$R = \frac{(1 - x)(n + 1)x^n + x^{n+1}}{(1 - x)^2}$$

(b) $\lim_{n\to\infty} x^{n+1} = 0$ for $|x| < 1$. We consider $\lim_{n\to\infty} (n + 1)x^n =$

$$\lim_{n\to\infty} \frac{n + 1}{x^{-n}} = \lim_{n\to\infty} \frac{1}{-nx^{-n-1}} = \lim_{n\to\infty} \frac{x^{n+1}}{-n} = 0$$

(c) $\sum_{n=1}^{\infty} n\left(\frac{5}{6}\right)^{n-1}\left(\frac{1}{6}\right) = \frac{1}{6}\left(\frac{1}{1 - \left(\frac{5}{6}\right)^2}\right) = 6$

(d) $\sum_{n=1}^{\infty} np^{n-1}q = \frac{q}{(1 - p)^2} = \frac{q}{q^2} = \frac{1}{q}$

12.5 CONVERGENCE OF POWER SERIES, INTEGRATION, DIFFERENTIATION, MULTIPLICATION AND DIVISION

1. $\sum_{n=0}^{\infty} x^n$ converges for $-1 < x < 1$ because $\lim_{n\to\infty} \frac{|x^{n+1}|}{|x^n|} = \lim_{n\to\infty} |x| < 1$

for all x such that $|x| < 1$. If $x = 1$, then $\sum_{n=0}^{\infty} (1)^n$ diverges, and

if $x = -1$, $\sum_{n=0}^{\infty} (-1)^n$ diverges.

3. $\sum_{n=0}^{\infty} \frac{nx^n}{2^n}$ converges for $-2 < x < 2$ because $\lim_{n\to\infty} \left| \frac{(n+1)x^{n+1}}{2^{n+1}} \cdot \frac{2^n}{nx^n} \right| =$

$\lim_{n\to\infty} \left| \frac{n+1}{n} \cdot \frac{1}{2} x \right| = \frac{1}{2}|x| < 1$ for all x such that $|x| < 2$. If $x = \pm 2$, then

$\sum_{n=0}^{\infty} \frac{n2^n}{2^n} = \sum_{n=0}^{\infty} n$ diverges, and $\sum_{n=0}^{\infty} \frac{n(-2)^n}{2^n} = \sum_{n=0}^{\infty} (-1)^n n$ diverges.

5. $\sum_{n=0}^{\infty} \frac{(-1)^n x^{2n+1}}{(2n+1)!}$ converges for all x because $\lim_{n\to\infty} \left| \frac{(n+1)x^{n+1}}{2^{n+1}} \cdot \frac{2^n}{nx^n} \right| =$

$\lim_{n\to\infty} \left| \frac{x^{2n+3}}{(2n+3)!} \cdot \frac{(2n+1)!}{x^{2n+1}} \right| = \lim_{n\to\infty} \frac{1}{(2n+3)(2n+1)} x^2 = 0 < 1$ for all x.

7. $\sum_{n=0}^{\infty} \frac{n^2 (x+2)^n}{2^n}$ converges for $-4 < x < 0$ because $\lim_{n\to\infty} \left| \frac{(n+1)^2 (x+2)^{n+1}}{2^{n+1}} \cdot \frac{2^n}{n^2 (x+2} \right.$

$\lim_{n\to\infty} \left| \left(\frac{n+1}{n}\right)^2 \cdot \frac{1}{2}(x+2) \right| < 1$ if $|x+2| < 2$ or $-4 < x < 0$.

$\sum_{n=0}^{\infty} \frac{n^2}{2^n} \cdot 2^n = \sum_{n=0}^{\infty} n^2$ diverges and $\sum_{n=0}^{\infty} \frac{n^2}{2^n}(-2)^n = \sum_{n=0}^{\infty} (-1)n^2$ diverges

9. $\sum_{n=0}^{\infty} \frac{(-1)^n x^{2n+1}}{2n+1}$ converges for $-1 < x < 1$ because

$\lim_{n\to\infty} \left| \frac{x^{2n+3}}{2n+3} \cdot \frac{2n+1}{x^{2n+1}} \right| = \lim_{n\to\infty} \left| \left(\frac{2n+3}{2n+1}\right) x^2 \right| < 1$ if $|x| < 1$.

$\sum_{n=0}^{\infty} (-1)^n \frac{1}{2n+1}$ converges by alternating series test.

$\sum_{n=0}^{\infty} (-1)^n \frac{(-1)^{2n+1}}{2n+1} = \sum_{n=0}^{\infty} (-1)^{n+1} \frac{1}{2n+1}$ also converges.

11. $\sum_{n=0}^{\infty} \frac{\cos nx}{2^n}$ converges absolutely for all x since

$\frac{|\cos nx|}{2^n} \le \frac{1}{2^n}$ and $\sum_{n=0}^{\infty} \frac{1}{2^n}$ converges.

13. $\sum_{n=0}^{\infty} \frac{x^n e^n}{n+1}$ converges for $-\frac{1}{e} \le x < \frac{1}{e}$ because $\lim_{n\to\infty} \left| \frac{x^{n+1}e^{n+1}}{n+2} \cdot \frac{n+1}{x^n e^n} \right| =$

$\lim_{n\to\infty} \left(\frac{n+1}{n+2}\right) |x|\, e < 1$ if $|x| < \frac{1}{e}$. If $x = \frac{1}{e}$, $\sum_{n=0}^{\infty} \frac{1}{n+1}$ diverges,

and if $x = -\frac{1}{e}$, $\sum_{n=0}^{\infty} (-1)^n \frac{1}{n+1}$ converges.

15. $\sum_{n=0}^{\infty} n^n x^n$ converges for $x = 0$ only because

$\lim_{n\to\infty} \sqrt[n]{n^n x^n} = \lim_{n\to\infty} |nx| = \infty$ for all x except $x = 0$.

17. $\sum_{n=0}^{\infty} (-2)^n (n+1)(x-1)^n$ converges for $\frac{1}{2} < x < \frac{3}{2}$ because

$\lim_{n\to\infty} \left| \frac{(-2)^{n+1}(n+2)(x-1)^{n+1}}{(-2)^n (n+1)(x-1)^n} \right| = \lim_{n\to\infty} |-2| \left(\frac{n+2}{n+1}\right) |x-1| < 1$ if

$|x-1| < \frac{1}{2}$. At $x = \frac{3}{2}$ and $\frac{1}{2}$, $\sum_{n=0}^{\infty} (-1)^n (n+1)$ and $\sum_{n=0}^{\infty} (n+1)$ both diverge.

19. $\sum_{n=0}^{\infty} \left(\frac{x^2-1}{2}\right)^n$ converges for $-\sqrt{3} < x < \sqrt{3}$ because $\lim_{n\to\infty} \left| \frac{(x^2-1)^{n+1}}{2^{n+1}} \cdot \frac{2^n}{(x^2-1)^n} \right| =$

$\lim_{n\to\infty} \frac{1}{2} |x^2-1| < 1$ if $|x^2-1| < 3$ or if $|x| < \sqrt{3}$.

At $x = \pm\sqrt{3}$, $\sum_{n=0}^{\infty} (1)^n$ diverges.

21. The series converges to $e^{3x} + 6$

23 (a) $\cos x = 1 - \frac{x^2}{2!} + \frac{x^4}{4!} + \ldots (-1)^n \frac{x^{2n}}{(2n)!}$

$\frac{d}{dx}(\cos x) = -\frac{2x}{2!} + \frac{4x^3}{4!} + \ldots (-1)^n (2n \frac{x^{2n-1}}{(2n-1)!}$

$= -1 + \frac{x^3}{3!} - \frac{x^5}{5!} + \ldots (-1) \frac{x^{2n-1}}{(2n-1)!} = -\sin x$

(b) $\int_0^x \cos t\,dt = \int_0^x (1 - \frac{t^2}{2!} + \frac{t^4}{4!} - \ldots + (1)^n \frac{t^{2n}}{(2n)!})\,dt$

$$= t - \frac{1}{3}\cdot\frac{t^3}{2!} + \frac{1}{5}\cdot\frac{t^5}{4!} - \ldots + (-1)^n \frac{1}{2n+1}\cdot\frac{t^{2n+1}}{(2n)!}\Bigg]_0^x$$

$$= x - \frac{x^3}{3!} + \frac{x^5}{5!} - \ldots + (-1)^n \frac{x^{2n+1}}{(2n+1)!} = \sin x$$

(c) $y = e^x = 1 + x + \frac{x^2}{2!} + \frac{x^3}{3!} + \ldots + \frac{x^n}{n!} + \ldots$

$$y' = 1 + \frac{2x}{2!} + \frac{3x^2}{3!} + \ldots + \frac{nx^{n-1}}{n!} + \ldots$$

$$= 1 + x + \frac{x^2}{2!} + \frac{x^3}{3!} + \ldots + \frac{x^n}{n!} + \ldots = e^x$$

25. $\frac{1}{1-x^2} = 1 + x^2 + x^4 + x^6 + \ldots + (x^{2n}) + \ldots = \sum_{n=0}^{\infty} x^{2n}$, $|x| < 1$.

Note that $\frac{d}{dx}(1-x^2)^{-1} = \frac{2x}{(1-x^2)^2}$. So, for $|x| < 1$

$$\frac{2x}{(1-x^2)^2} = 2x + 4x^3 + 6x^5 + \ldots + 2nx^{2n-1} = \sum_{n=1}^{\infty} (2n)x^{2n-1}$$

27. $\int_0^{.2} \sin x^2\,dx = \int_0^{.2} \left(x^2 - \frac{x^6}{3!} + \frac{x^{10}}{5!} - \ldots\right)dx = \frac{1}{3}x^3 - \frac{1}{7\cdot 3!}x^7 - \ldots\Bigg]_0^{.2}$

$= 0.0027$ with error $|E| \le \frac{1}{7\cdot 3!}(.2)^7 = 3 \times 10^{-7}$

29. $\int_0^{.1} x^2 e^{-x^2}\,dx = \int_0^{.1} x^2\left(1 - x^2 + \frac{x^4}{2!} - \frac{x^6}{3!} + \ldots\right)dx = \int_0^{.1}\left(x^2 - x^4 + \frac{x^6}{2!} - \ldots\right)dx$

$$= \frac{1}{3}x^3 - \frac{1}{5}x^5 + \frac{1}{7\cdot 2!}x^7 - \ldots\Bigg]_0^{.1} = \frac{1}{3}(.1)^3 = 0.00033$$

with error $|E| \le \frac{1}{5}(.1)^5 = 2 \times 10^{-6}$

31. $\int_0^{.4} \frac{1-e^{-x}}{x}\,dx = \int_0^{.4} \frac{1}{x}\left[1 - \left(1 - x + \frac{x^2}{2} - \frac{x^3}{3!} + \ldots\right)\right]dx$

$$= \int_0^{.4}\left(1 - \frac{x}{2!} + \frac{x^2}{3!} - \frac{x^3}{4!} + \ldots\right)dx = 1 - \frac{1}{2\cdot 2!}x^2 + \frac{1}{3\cdot 3!}x^3 - \frac{1}{4\cdot 4!}x^4 + \ldots\Bigg]_0^{.4}$$

$= 0.3636$ with error $|E| \le \frac{(.4)^4}{4\cdot 4!} < 0.0003$

33. $\displaystyle\int_0^{.1} \frac{1}{\sqrt{1+x^4}}dx = \int_0^{.1}\left(1 - \frac{1}{2}x^4 + \frac{\left(-\frac{1}{2}\right)\left(-\frac{3}{2}\right)}{2}x^8 \ldots\right)dx$

$\displaystyle = \int_0^{.1}\left(1 - \frac{1}{2}x^4 + \frac{3}{8}x^8 - \ldots\right)dx = x - \frac{1}{10}x^5 + \frac{3}{72}x^9 - \ldots\Big]_0^{.1}$

$= 0.1$ with error $|E| \le 1 \times 10^{-6}$

35. (a) $\displaystyle \sinh^{-1}x = \int_0^x \frac{dt}{\sqrt{1+t^2}} = \int_0^x \left(1 - \frac{1}{2}t^2 + \frac{3}{8}t^4 - \frac{5}{16}t^6 + \ldots\right)dt$

$\displaystyle = t - \frac{1}{6}t^3 + \frac{3}{40}t^5 - \frac{5}{112}t^7 + \ldots\Big]_0^x = x - \frac{1}{6}x^3 + \frac{3}{40}x^5 - \frac{5}{112}x^7 + \ldots$

(b) $\displaystyle \sinh^{-1}.25 = .25 - \frac{(.25)^3}{6} = 0.247$

with error $\displaystyle |E| \le \frac{3(.25)^5}{40} = 0.00007$

37. Problem 15 of this section is an example of a series which converges only for $x = 0$.

39. We are given that $\sum a_n$ converges for $-r < x < r$. Let a be any point such that $-r < a < r$. There exists r_1 such that $-r < -r_1 < a < r_1 < r$. The $\sum a_n$ converges for r_1 and hence absolutely for a, by Theorem 1. But a was any point between $-r$ and r, and hence the series converges absolutely for all x such that $-r < x < r$.

41. $\displaystyle e^x \sin x = \left(1 + x + \frac{x^2}{2!} + \frac{x^3}{3!} + .+\frac{x^n}{n!}\right)\left(x - \frac{x^3}{3!} + \frac{x^5}{5!} - .+(-1)^n\frac{x^{2n+1}}{(2n+1)!}\right)$

$\displaystyle = 1\left(x - \frac{x^3}{3!} + \frac{x^5}{5!} - \ldots\right) + x\left(x - \frac{x^3}{3!} + \frac{x^5}{5!} - \ldots\right) +$

$\displaystyle \frac{x^2}{2}\left(x - \frac{x^3}{3!} + \frac{x^5}{5!} - \ldots\right) + \frac{x^3}{3!}\left(x - \frac{x^3}{3!} + ..\right) + \frac{x^4}{4!}\left(x - \frac{x^3}{3!} + ..\right)$

$\displaystyle = \left(x - \frac{x^3}{3!} + \frac{x^5}{5!} - ..\right) + \left(x^2 - \frac{x^4}{3!} + ..\right) + \left(\frac{x^3}{2} - \frac{x^5}{2\cdot 3!} + \ldots\right)$

$\displaystyle + \left(\frac{x^4}{3!} + ..\right) + \left(\frac{x^5}{4!} - ..\right)$

$\displaystyle = x + x^2 + \frac{1}{3}x^3 - \frac{1}{30}x^5 \ \ldots$

Check: $\displaystyle (e^x)(e^{ix}) = e^{(1+i)x} = 1 + (1+i)x + \frac{(1+i)^2x^2}{2} + \frac{(1+i)^3x^3}{3!}$

$\displaystyle + \frac{(1+i)^4x^4}{4!} + \frac{(1+i)^5x^5}{5!} + \ldots.$

$\displaystyle = 1 + (1+i)x + \frac{2ix^2}{2} + \frac{(2i-2)x^3}{3!} - \frac{4x^4}{4!} + \frac{(-4-4i)x^5}{5!} + ..$

The imaginary part of this series is the same as the one obtained in the first part of the problem.

43. $\int_0^x \tan t\,dt = \int_0^x \left(t + \frac{t^3}{3} + \frac{2t^5}{15} + \ldots\right)dt = \frac{1}{2}t^2 + \frac{1}{12}t^4 + \frac{2}{90}t^6 + \ldots\Big]_0^x$

$\ln|\sec x| = \frac{1}{2}x^2 + \frac{1}{2}x^4 + \frac{1}{45}x^6 + \ldots$

45. (a) $\frac{r_2}{r_1} = \sec\frac{\pi}{3}$

$\frac{r_3}{r_2} = \sec\frac{\pi}{4}$

$\frac{r_4}{r_3} = \sec\frac{\pi}{5}$

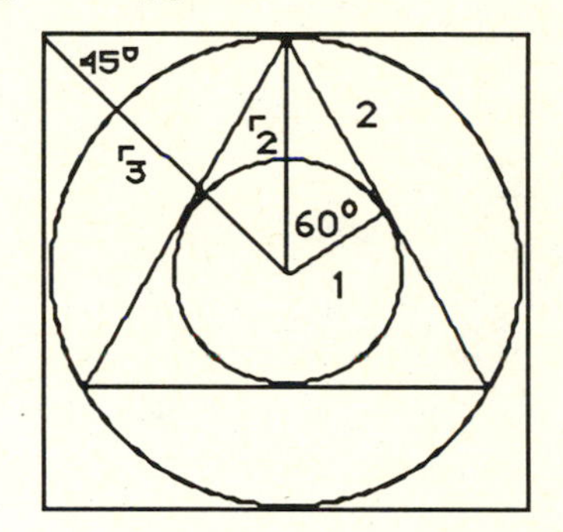

In general, if a (n+1)-sided polygon is inscribed in a circle, the central angle is $\frac{2\pi}{n+1}$ and the ratio is $\frac{r_n}{r_{n-1}} = \sec\frac{1}{2}\left(\frac{2\pi}{n+1}\right)$

(b) $r_n = r_{n-1}\sec\frac{\pi}{n+1} = \sec\frac{\pi}{n+1}\left(r_{n-2}\sec\frac{\pi}{n}\right) = \sec\frac{\pi}{n+1}\sec\frac{\pi}{n}\left(r_{n-3}\sec\frac{\pi}{n-1}\right)$

$= \sec\frac{\pi}{n+1}\sec\frac{\pi}{n}\sec\frac{\pi}{n-1}\ldots\sec\frac{\pi}{4}\left(r_1\sec\frac{\pi}{3}\right)$

$\ln r_n = \ln r_1 + \ln\sec\frac{\pi}{3} + \ln\sec\frac{\pi}{4} + \ldots + \ln\sec\frac{\pi}{n+1}$

(c) $\lim_{n\to\infty}\frac{\ln\sec\frac{\pi}{n}}{\frac{1}{n^2}} = \lim_{n\to\infty} n^2\left(\frac{\pi^2}{2n^2} + \frac{\pi^4}{12n^4} + \ldots\right) = \frac{\pi^2}{2}.$

$\sum_{n=3}^{\infty}\ln\sec\frac{\pi}{n}$ converges by comparison to $\sum_{n=1}^{\infty}\frac{1}{n^2}.$

12.6 INDETERMINATE FORMS

1. $\lim_{h\to 0}\frac{\sinh}{h} = \lim_{h\to 0}\frac{h - \frac{h^3}{3!} + \frac{h^5}{5!} - \ldots}{h} = \lim_{h\to 0}\left(1 - \frac{h^2}{3!} - \frac{h^4}{5!} - \ldots\right) = 1$

3. $\lim_{t\to 0}\frac{1 - \cos t - \frac{t^2}{2}}{t^4} = \lim_{t\to 0}\frac{1}{t^4}\left[1 - \frac{t^2}{2} - \left(1 - \frac{t^2}{2} + \frac{t^4}{4!} - \frac{t^6}{6!} + \ldots\right)\right]$

$= \lim_{t\to 0}\left[-\frac{1}{4!} + \frac{t^2}{6!} - \ldots\right] = -\frac{1}{24}$

5. $\lim_{x\to 0}\frac{x^2}{1 - \cosh x} = \lim_{x\to 0}\frac{x^2}{1 - \left(1 + \frac{x^2}{2!} + \frac{x^4}{4!} + \ldots\right)} = \lim_{x\to 0}\frac{1}{-\frac{1}{2} - \frac{x^2}{4!} - \ldots} = -2$

7. $$\lim_{x\to 0}\frac{1-\cos x}{\sin x} - \lim_{x\to 0}\frac{1-\left(1-\frac{x^2}{2}+\frac{x^4}{4!}-\ldots\right)}{x-\frac{x^3}{3!}+\frac{x^5}{5!}-\ldots} = \lim_{x\to 0}\frac{x^2\left(\frac{1}{2}-\frac{x^2}{4!}+\ldots\right)}{x\left(1-\frac{x^2}{3!}+\ldots\right)} = 0$$

9. $$\lim_{z\to 0}\frac{\sin(z^2)-\sinh(z^2)}{z^6} = \lim_{z\to 0}\frac{\left(z^2-\frac{z^6}{3!}+\frac{z^{10}}{5!}-\ldots\right)-\left(z^2+\frac{z^6}{3!}+\frac{z^{10}}{5!}+\ldots\right)}{z^6}$$

$$= \lim_{z\to 0}\frac{-\frac{2z^6}{3!}-\frac{2z^{10}}{5!}-\ldots}{z^6} = -\frac{1}{3}$$

11. $$\lim_{x\to 0}\frac{\sin x - x + \frac{x^3}{6}}{x^5} = \lim_{x\to 0}\frac{-x+\frac{x^3}{6}+\left(x-\frac{x^3}{3!}+\frac{x^5}{5!}-\frac{x^7}{7!}\ldots\right)}{x^5} = \frac{1}{120}$$

13. $$\lim_{x\to 0}\frac{x-\tan^{-1}x}{x^3} = \lim_{x\to 0}\frac{x-\left(x-\frac{x^3}{3}+\frac{x^5}{5}-\ldots\right)}{x^3} = \frac{1}{3}$$

15. $$\lim_{x\to\infty} x^2\left(e^{-1/x^2}-1\right) = \lim_{x\to\infty} x^2\left[\left(1-\frac{1}{x^2}+\frac{1}{2x^4}-\frac{1}{6x^6}+\ldots\right)-1\right] = -1$$

17. $$\lim_{x\to 0}\frac{\tan 3x}{x} = \lim_{x 0}\frac{3x+\frac{(3x)^3}{3}+\frac{2(3x)^5}{15}+\ldots}{x} = 3$$

19. $$\lim_{x\to\infty}\frac{x^{100}}{e^x} = \lim_{x\to\infty}\frac{x^{100}}{1+x+\frac{x^2}{2!}+\ldots+\frac{x^{101}}{101!}+\ldots} = 0$$

21. (a) For $x \geq 0$, $e^{x^2} \geq 1$. Therefore, $\int_0^x e^{t^2}dt \geq \int_0^x dt = x$

and $\int_0^x e^{t^2}dt$ diverges.

(b) $$\lim_{x\to\infty} x\int_0^x e^{t^2-x^2}dt = \lim_{x\to\infty} x\int_0^x e^{t^2}e^{-x^2}dt = \lim_{x\to\infty} xe^{-x^2}\int_0^x e^{t^2}dt$$

$$= \lim_{x\to\infty}\frac{x\int_0^x e^{t^2}dt}{e^{x^2}} = \lim_{x\to\infty}\frac{\int_0^x e^{t^2}dt + xe^{x^2}}{2xe^{x^2}} = \lim_{x\to\infty}\frac{e^{x^2}+e^{x^2}+2x^2e^{x^2}}{4x^2e^{x^2}+2e^{x^2}}$$

$$= \lim_{x\to\infty}\frac{2+2x^2}{2+4x^2} = \frac{1}{2}$$

23.

x	$\sin x$	$\frac{6x}{6+x^2}$
± 1.0	± 0.84147	± 0.85714
± 0.1	± 0.09983	± 0.09983
± 0.01	± 0.00999	± 0.00999

$\sin x \approx \frac{6x}{6+x^2}$ is better

12.M MISCELLANEOUS

1. (a) $\frac{x^2}{1+x^2} = x^2\left(\frac{1}{1-(-x)}\right) = x^2(1 - x + x^2 - x^3 + .. + (-1)^n x^n) = \sum_{n=0}^{\infty} (-1)^n x^{n+2}$.

(b) No, the radius of convergence for the series is $-1 < x < 1$.

3. $$e^{\sin x} = 1 + \left(x - \frac{x^3}{3!} + \frac{x^5}{5!} - \ldots\right) + \frac{1}{2}\left(x - \frac{x^3}{3!} + \frac{x^5}{5!} - \ldots\right)^2 + \frac{1}{6}\left(x - \frac{x^3}{3!} + \frac{x^5}{5!} - \ldots\right)^3 + \frac{1}{24}\left(x - \frac{x^3}{3!} + \frac{x^5}{5!} - \ldots\right)^4 + ..$$

$$= 1 + \left(x - \frac{x^3}{3!} + \frac{x^5}{5!} - \ldots\right) + \frac{1}{2}\left(x^2 - \frac{x^4}{3} + \frac{x^6}{36} - ..\right) + \frac{1}{6}\left(x^3 + 3x^2 \cdot \frac{x^3}{3!} + ..\right) + \frac{1}{24}(x^4 + \ldots) + \ldots$$

$$= 1 + x + \frac{1}{2}x^2 - \frac{1}{8}x^4 + \ldots$$

5 (a) $\ln(\cos x) = \ln[1 - (1 - \cos x)]$

$$= -\left(\frac{x^2}{2} - \frac{x^4}{24} + \frac{x^6}{720} - ..\right) - \frac{1}{2}\left(\frac{x^2}{2} - \frac{x^4}{24} + ..\right)^2 - \frac{1}{3}\left(\frac{x^2}{2} - \frac{x^4}{24} + ..\right)^3 - \ldots$$

$$= -\frac{x^2}{2} + \frac{x^4}{24} - \frac{x^6}{720} + \ldots - \frac{x^4}{8} + \frac{x^6}{48} - \ldots - \frac{x^6}{24} - \ldots$$

$$= -\frac{x^2}{2} - \frac{x^4}{12} - \frac{x^6}{45} - \ldots$$

(b) $$\int_0^{.1} \ln(\cos x)\,dx = \int_0^{.1}\left(-\frac{x^2}{2} - \frac{x^4}{12} - \frac{x^6}{45} - \ldots\right)dx$$

$$= -\frac{1}{6}x^3 - \frac{1}{60}x^5 - \frac{1}{315}x^7\Big]_0^{.1} = -0.00017$$

7. $$\int_0^1 e^{-(x^2)}\,dx = \int_0^1\left(1 - x^2 + \frac{x^4}{2} - \frac{x^6}{6} + \frac{x^8}{24} - \ldots\right)dx$$

$$= x - \frac{1}{3}x^3 + \frac{1}{10}x^5 - \frac{1}{42}x^7 + \frac{1}{216}x^9 - \ldots\Big]_0^1 = 0.747$$

9. $f(x) = \frac{1}{1-x}$ $\quad f(2) = -1$

$f'(x) = \frac{1}{(1-x)^2}$ $\quad f'(2) = 1$

$f''(x) = 2(1-x)^3$ $\quad f''(2) = -2$

$f^{(n)}(x) = \frac{n!}{(1-x)^n}$ $\quad f^{(n)}(2) = (-1)^{n+1} n!$

$f(x) = \sum_{n=0}^{\infty} (-1)^{n+1}(x-2)^n$ converges for $1 < x < 3$ since

$$\lim_{n\to\infty} \left| \frac{(x-2)^{n+1}}{(x-2)^n} \right| < 1 \Leftrightarrow |x-2| < 1$$

11. $f(x) = \cos x$ $\quad f\left(\frac{\pi}{3}\right) = \frac{1}{2}$

$f'(x) = -\sin x$ $\quad f\left(\frac{\pi}{3}\right) = -\frac{\sqrt{3}}{2}$

$f''(x) = -\cos x$ $\quad f''\left(\frac{\pi}{3}\right) = -\frac{1}{2}$

$f'''(x) = \sin x$ $\quad f'''\left(\frac{\pi}{3}\right) = \frac{\sqrt{3}}{2}$

$$\cos x = \frac{1}{2} - \frac{\sqrt{3}}{2}\left(x - \frac{\pi}{3}\right) - \frac{1}{2}\cdot\frac{1}{2}\left(x - \frac{\pi}{3}\right)^2 + \frac{1}{3!}\cdot\frac{\sqrt{3}}{2}\left(x - \frac{\pi}{3}\right)^3 + \ldots$$

$$= \frac{1}{2}\left(1 - \frac{1}{2}\left(x - \frac{\pi}{3}\right)^2 - \ldots\right) + \frac{\sqrt{3}}{2}\left(-\left(x - \frac{\pi}{3}\right) + \left(x - \frac{\pi}{3}\right)^3 - \ldots\right)$$

$$= \frac{1}{2}\sum_{n=0}^{\infty} \frac{(-1)^n}{(2n)!}\left(x - \frac{\pi}{3}\right)^n + \frac{\sqrt{3}}{2}\sum_{n=0}^{\infty} \frac{(-1)^{n+1}}{(2n+1)!}\left(x - \frac{\pi}{3}\right)^{2n+1}$$

13. $f'(x) = g(x) \Rightarrow f''(x) = g'(x) = f(x)$. Then $f'''(x) = f'(x) = g(x)$. In general, $f^{(2n+1)}(x) = g(x)$ and $f^{(2n)}(x) = f(x)$. Therefore,

$$f(x) = f(0) + f'(0)x + \frac{f''(0)}{2!}x^2 + \frac{f'''(0)}{3!}x^3 + \ldots.$$

$$= f(0) + g(0)x + \frac{f(0)}{2!}x^2 + \frac{g(0)}{3!}x^3 = 1 + \frac{1}{2}x^2 + \frac{1}{4!}x^4 + \frac{1}{6!}x^6 +$$

$f(1) = 1.543$

15. $f(x) = e^{(e^x)}$ $\qquad f(0) = e$

$f'(x) = e^x e^{(e^x)} = e^{(x+e^x)}$ $\qquad f'(0) = e$

$f''(x) = (1 + e^x)\, e^{(x+e^x)}$ $\qquad f''(0) = 2e$

$f'''(x) = e^x e^{(x+e^x)} + (1 + e^x)^2 e^{(x+e^x)}$ $\qquad f'''(0) = 5e$

$f(x) = e + ex + ex^2 + \frac{5e}{6} x^3 + \ldots$

17. $(1 + x)^{\frac{1}{3}} = 1 + \frac{1}{3}x + \frac{\left(\frac{1}{3}\right)\left(-\frac{2}{3}\right)}{2}x^2 - \ldots$ begins alternating after the first term, and the error $E \le \left| -\frac{1}{9}\left(\frac{1}{10}\right)^2 \right| = 0.0011$

19. $\lim_{x\to 0} \left(\frac{\sin x}{x}\right)^{\frac{1}{x^2}} = e^{-1/6}$. To see this, consider $\lim_{x\to 0} \left[\ln \left(\frac{\sin x}{x}\right)^{\frac{1}{x^2}} \right]$

$$= \lim_{x\to 0} \left[\frac{1}{x^2} \ln \left(\frac{\sin x}{x}\right) \right] = \lim_{x\to 0} \left[\frac{\ln \left(\frac{\sin x}{x}\right)}{x^2} \right] = \lim_{x\to 0} \frac{\left(\frac{x}{\sin x}\right)\left(\frac{x \cos x - \sin x}{x^2}\right)}{2x}$$

$$= \lim_{x\to 0} \frac{x \cos x - \sin x}{2x^2 \sin x} = \lim_{x\to 0} \frac{\cos x - x \sin x - \cos x}{4x \sin x + 2x^2 \cos x}$$

$$= \lim_{x\to 0} \frac{-\sin x}{4 \sin x + 2x \cos x} = \lim_{x\to 0} \frac{-\cos x}{4 \cos x - 2x \sin x + 2 \cos x} = -\frac{1}{6}$$

21. $\lim_{n\to\infty} \left| \frac{(x + 2)^{n+1}}{3^{n+1}(n + 1)} \cdot \frac{3^n n}{(x + 2)^n} \right| = \lim_{n\to\infty} \left(\frac{n}{n + 1}\right)\left(\frac{1}{3}\right) | x + 2 | < 1$

if $\frac{1}{3} | x + 2 | < 1$ or $-5 < x < 1$. At $x = 1$, $\sum_{n=1}^{\infty} \frac{1}{n}$ diverges, and

at $x = -5$, $\sum_{n=1}^{\infty} (-1)^n \frac{1}{n}$ converges. The convergenge is for $-5 \le x < 1$.

23. $\sum_{n=1}^{\infty} \frac{x^n}{n^n}$ converges for all x, since $\lim_{n\to\infty} \sqrt[n]{\left| \frac{x^n}{n^n} \right|} = 0$

25. $\sum_{n=0}^{\infty} \frac{n+1}{2n+1} \frac{(x-3)^n}{2^n}$ converges for $1 < x < 5$ because

$$\lim_{n\to\infty} \left| \frac{(n+2)(x-3)^{n+1}}{(2n+3)\,2^{n+1}} \cdot \frac{(2n+1)\,2^n}{(n+1)(x-3)^n} \right| = \lim_{n\to\infty} \left(\frac{n+2}{n+1}\right)\left(\frac{2n+1}{2n+3}\right)\left(\frac{1}{2}\right) |x-3| < 1$$

if $|x-3| < 1$ or is $1 < x < 5$. Since $\lim_{n\to\infty} \frac{n+1}{2n+1} = \frac{1}{2} \neq 0$, the series diverges at both endpoints.

27. $\sum_{n=1}^{\infty} \frac{(-1)^{n-1}(x-1)^n}{n^2}$ converges for $0 \le x \le 2$.

$$\lim_{n\to\infty} \left| \frac{(x-1)^{n+1}}{(n+1)^2} \cdot \frac{n^2}{(x-1)^n} \right| = \lim_{n\to\infty} \left(\frac{n+1}{n}\right)^2 |x-1| < 1 \text{ if } 0 < x < 2.$$

At $x = 0$, $\sum_{n=1}^{\infty} \frac{(-1)^{n-1}(-1)^n}{n^2} = \sum_{n=1}^{\infty} \frac{1}{n^2}$ converges (p-series, $p = 2$).

At $x = 2$, $\sum_{n=1}^{\infty} \frac{(-1)^{n-1}}{n^2}$ converges (alternating series test).

29. $\sum_{n=1}^{\infty} \frac{(x-2)^{3n}}{n!}$ converges for all x.

$$\lim_{n\to\infty} \left| \frac{(x-2)^{3n+3}}{(n+1)!} \cdot \frac{n!}{(x-2)^{3n}} \right| = \lim_{n\to\infty} \frac{1}{n+1} |x-2|^3 = 0 \text{ for all } x.$$

31. $\sum_{n=1}^{\infty} \frac{1}{n}\left(\frac{x-1}{x}\right)^n$ converges for all $x \ge \frac{1}{2}$.

$$\lim_{n\to\infty} \left| \frac{\left(\frac{x-1}{x}\right)^{n+1} \frac{1}{n+1}}{\left(\frac{x-1}{x}\right)^{n} \frac{1}{n}} \right| \lim_{n\to\infty} \left(\frac{n}{n+1}\right) \left| \frac{x-1}{x} \right| < 1 \Leftrightarrow -1 < \frac{x-1}{x} < 1.$$

Case I: $\frac{x-1}{x} < 1 \Leftrightarrow \frac{x-1}{x} - 1 < 0 \Leftrightarrow -\frac{1}{x} < 0 \Leftrightarrow x > 0$

Case II: $\frac{x-1}{x} > -1 \Leftrightarrow \frac{x-1}{x} + 1 > 0 \Leftrightarrow \frac{2x-1}{x} > 0 \Leftrightarrow x < 0 \text{ or } x > \frac{1}{2}$.

The intersection of these solutions sets is $x \geqslant \frac{1}{2}$.

At $x = \frac{1}{2}$, $\sum_{n=1}^{\infty} (-1)^n \frac{1}{n}$ converges.

33. If $\sum_{n=1}^{\infty} a_n$ converges and $a_n > 0$, show that $\sum_{n=1}^{\infty} \frac{a_n}{1+a_n}$ converges.

$\lim_{n\to\infty} \frac{\frac{a_n}{1+a_n}}{a_n} = \lim_{n\to\infty} \frac{1}{1+a_n} = 1$ since $\lim_{n\to\infty} a_n = 0$. By the Limit Comparison

Test, $\sum_{n=1}^{\infty} \frac{a_n}{1+a_n}$ since $\sum_{n=1}^{\infty} a_n$ does.

35. If $\sum_{n=1}^{\infty} |a_n|$ converges, and $a_n > -1$, prove $\prod_{n=1}^{\infty}(1+a_n)$ converges.

The convergence of $\sum_{n=1}^{\infty} |a_n|$ means that $\lim_{n\to\infty} |a_n| = 0$. Let $N > 0$ be such

that $|a_n| < \frac{1}{2}$ for all $n > N$ and consider $\sum_{n=N}^{\infty} \ln(1+a_n)$.

$$|\ln(1+a_n)| = |a_n - \frac{a_n^2}{2} + \frac{a_n^3}{3} - \frac{a_n^4}{4} + \ldots| \le |a_n| + |\frac{a_n^2}{2}| + |\frac{a_n^3}{3}| + |\frac{a_n^4}{4}| + .$$

$$< |a_n| + |a_n|^2 + |a_n|^3 + |a_n|^4 + \ldots$$

$$= \frac{|a_n|}{1-|a_n|} < 2|a_n| \text{ since, for } n > N,\ 1-|a_n| \ge \frac{1}{2}.$$

Therefore, $\sum_{n=N}^{\infty} \ln(1+a_n) \le \left|\sum_{n=N}^{\infty} \ln(a_n)\right| \le \sum_{n=N}^{\infty} 2|a_n|$

$\le 2\sum_{n=1} |a_n|$ which is convergent. Hence $\prod_{n=1}^{\infty}(1+a_n)$ converges.

37. $\frac{\tan^{-1}x}{1-x} = \frac{1}{1-x}\cdot\tan^{-1}x = (1+x+x^2+x^3+x^4+x^5+\ldots)\left(x-\frac{x^3}{3}+\frac{x^5}{5}-\ldots\right)$

$$= x + x^2 + x^3 + x^4 + x^5 - \frac{x^3}{3} - \frac{x^4}{3} - \frac{x^5}{3} + \frac{x^5}{5} + \ldots$$

$$= x + x^2 + \frac{2}{3}x^3 + \frac{2}{3}x^4 + \frac{13}{15}x^5 + \ldots$$

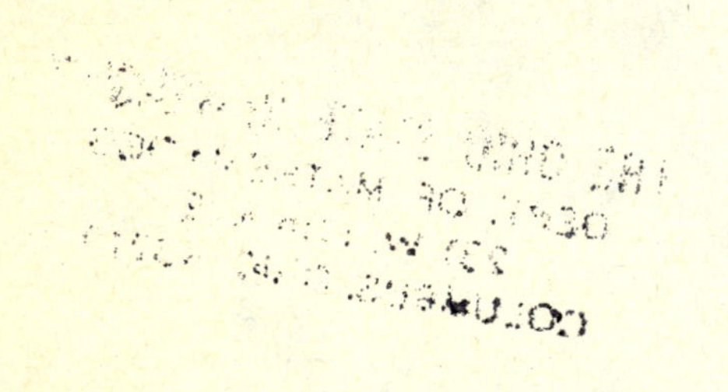